Flavour in food

Related titles:

Texture in food, volume 1: Semi-solid foods
(ISBN-13: 978-1-85573-673-3; ISBN-10: 1-85573-673-X)
Understanding and controlling the texture of semi-solid foods such as yoghurt and ice cream is a complex process. With a distinguished international team of contributors, this important collection summarises some of the most significant research in this area. The first part of the book looks at the behaviour of gels and emulsions, how they can be measured and their textural properties improved. The second part of the collection discusses the control of texture in particular foods such as yoghurt, ice cream, spreads and sauces.

Texture in food, volume 2: Solid foods
(ISBN-13: 978-1-85573-724-2; ISBN-10: 1-85573-724-8)
This authoritative book summarises the wealth of recent research on what influences texture in solid foods and how it can be controlled to maximise product quality. The first part of the book reviews research on understanding how consumers experience texture when they eat, and how they perceive and describe key textural qualities such as crispness. Part II considers the instrumental techniques used for analysing texture. It includes chapters on force/deformation and sound input techniques, near infrared spectroscopy (NIR), nuclear magnetic resonance (NMR) and magnetic resonance imaging (MRI). The final part examines how the texture of particular foods may be better understood and improved. A number of chapters review ways of controlling the texture of fruits and vegetables, whilst others cover cereal foods such as bread, rice, pasta and fried food.

Colour in food, Improving quality
(ISBN-13: 978-1-85573-590-3; ISBN-10: 1-85573-590-3)
The colour of a food is central to consumer perceptions of quality. This important collection reviews key issues in controlling colour quality in food, from the chemistry of colour in food to measurement issues, improving natural colour and the use of colourings to improve colour quality.

Details of these books and a complete list of Woodhead titles can be obtained by:

- visiting our web site at www.woodheadpublishing.com
- contacting Customer Services (email: sales@woodhead-publishing.com; fax: +44 (0) 1223 893694; tel.: +44 (0) 1223 891358 ext. 30; address: Woodhead Publishing Limited, Abington Hall, Abington, Cambridge CB1 6AH, England)

Flavour in food

Edited by
Andrée Voilley and Patrick Etiévant

CRC Press
Boca Raton Boston New York Washington, DC

WOODHEAD PUBLISHING LIMITED
Cambridge England

Published by Woodhead Publishing Limited
Abington Hall, Abington
Cambridge CB1 6AH
England
www.woodheadpublishing.com

Published in North America by CRC Press LLC
6000 Broken Sound Parkway, NW
Suite 300
Boca Raton, FL 33487
USA

First published 2006, Woodhead Publishing Limited and CRC Press LLC

British Library Cataloguing in Publication Data
A catalogue record for this book is available from the British Library.

Library of Congress Cataloging-in-Publication Data
A catalog record for this book is available from the Library of Congress.

Woodhead Publishing Limited ISBN-13: 978-1-85573-960-4 (book)
Woodhead Publishing Limited ISBN-10: 1-85573-960-7 (book)
Woodhead Publishing Limited ISBN-13: 978-1-84569-140-0 (e-book)
Woodhead Publishing Limited ISBN-10: 1-84569-140-7 (e-book)
CRC Press ISBN-10: 0-8493-3437-3
CRC Press order number: WP3437

The publishers' policy is to use permanent paper from mills that operate a sustainable forestry policy, and which has been manufactured from pulp which is processed using acid-free and elementary chlorine-free practices. Furthermore, the publishers ensure that the text paper and cover board used have met acceptable environmental accreditation standards.

Project managed by Macfarlane Production Services, Dunstable, Bedfordshire
(e-mail: macfarl@aol.com)
Typeset by Godiva Publishing Services Ltd, Coventry, West Midlands
Printed by TJ International Limited, Padstow, Cornwall, England

Contents

Contributor contact details xi

Preface xv

Part I Characterisation of aroma compounds

1 The human perception of taste compounds 3
B. Bufe and W. Meyerhof, German Institute of Human Nutrition Potsdam-Rehbruecke, Germany
1.1 Introduction 3
1.2 The sense of taste 4
1.3 The molecular basis of human taste perception 5
1.4 Functional characterisation of taste receptors through calcium imaging 18
1.5 Future trends 26
1.6 Acknowledgment 31
1.7 References 31

2 Processing information about flavour 36
A. Holley, Centre Européen des Sciences du Goût, France
2.1 Introduction 36
2.2 Reception of odorants and neural processing of olfactory information 37
2.3 Reception of taste compounds and neural processing 43
2.4 Trigeminal chemosensitivity (chemesthesis) 47

2.5 Multimodal interactions and flavour integration 49
2.6 Conclusions and future trends 53
2.7 References 54

3 Sensory analysis of food flavor 62
A. C. Noble, University of California, USA, I. Lesschaeve, Brock University, Canada
3.1 Introduction 62
3.2 Current and developing techniques for sensory analysis 63
3.3 Sensory testing administration 69
3.4 Statistical analysis of data 71
3.5 Relating analytical sensory data to instrumental or consumer preference data 72
3.6 Using sensory data for business decisions 74
3.7 Conclusion 75
3.8 References 75
3.9 Appendix 80

4 Choosing the correct analytical technique in aroma analysis .. 81
G. Reineccius, University of Minnesota, USA
4.1 Introduction 81
4.2 Obtaining a complete aroma profile 82
4.3 Key components contributing to sensory properties 84
4.4 Off-notes in a food product 87
4.5 Monitor changes in aroma compounds with time 89
4.6 Using instrumental data in sensory predictions 90
4.7 Future trends 92
4.8 Sources of further information 93
4.9 References 95

5 Matching sensory and instrumental data 98
E. M. Qannari, ENITIAA/INRA, France and P. Schlich, Centre Européen des Sciences du Goût (CESG), France
5.1 Introduction 98
5.2 Investigating the structure of the data sets 99
5.3 Relating sensory to instrumental data 101
5.4 Case study 105
5.5 Conclusion 110
5.6 Acknowledgements 111
5.7 References 111

Part II Flavour retention and release from the food matrix

6 Flavour retention and release from the food matrix: an overview 117
A. Voilley, University of Bourgogne, France and I. Souchon, INRA, France
6.1 Introduction 117
6.2 Flavour properties 117
6.3 Diffusion and mass transfer 121
6.4 Main factors influencing the mobility of flavour compounds in food matrices 126
6.5 Conclusion and future trends 129
6.6 References 129

7 Lipid–flavour interactions 133
M. Ollivon, Paris Sud University, France
7.1 Introduction 133
7.2 How nature uses both types of lipids: the key role of molecular organisation 134
7.3 Classification and properties of lipids 136
7.4 Main types of lipidic structures 143
7.5 Replacing fat by structuring lipids: consequence for flavour retention and release 151
7.6 References 153

8 Emulsion–flavour interactions 156
J-P. Dumont, INRA, France
8.1 Introduction 156
8.2 Remarkable characteristics of emulsions 156
8.3 Behaviour of aroma compounds in emulsions 158
8.4 Improving flavour delivery to the consumer 162
8.5 Future trends 163
8.6 Cross-links 169
8.7 References 170

9 Protein–flavour interactions 172
A. Tromelin, I. Andriot and E. Guichard, INRA, France
9.1 Introduction 172
9.2 Protein structure in relation to flavour binding 173
9.3 Nature and strength of the interactions 182
9.4 Effect of medium on protein–flavour interactions 187
9.5 Perceptive consequences in food systems: some examples ... 191
9.6 Conclusion and future trends 193
9.7 Sources of further information 194
9.8 References 195

10 Carbohydrate–flavour interactions 208
J. Delarue and P. Giampaoli, ENSIA, France
10.1 Introduction 208
10.2 Aroma interactions with mono- and disaccharides 208
10.3 Structure of polysaccharides 214
10.4 Interaction mechanisms with aroma compounds 216
10.5 Example of retention in amylaceous matrices and their derivatives 218
10.6 Conclusion 222
10.7 References 224

11 Modelling aroma interactions in food matrices 229
K. B. de Roos, Givaudan, Nederland BV, The Netherlands
11.1 Introduction 229
11.2 Phase partitioning: aroma release under equilibrium conditions 230
11.3 Mass transfer: aroma release under non-equilibrium conditions 234
11.4 Mechanistic modelling of aroma release 239
11.5 Empirical modelling of aroma release 250
11.6 Applications 251
11.7 Future trends 252
11.8 Sources of further information 252
11.9 References 253

12 Flavour release from liquid food products 260
A. E. M. Boelrijk, G. Smit and K. G. C. Weel, NIZO Food Research, The Netherlands and J. J. Burger, Quest International Nederland BV, The Netherlands
12.1 Introduction 260
12.2 *In vivo* aroma release measurements during drinking of beverages 262
12.3 Physiological factors important for *in vivo* aroma release from beverages 265
12.4 Development of the artificial throat 266
12.5 Effect of ingredients on aroma release from beverages 271
12.6 Conclusion 277
12.7 Acknowledgements 279
12.8 References 279

Part III Influences on flavour perception

13 The process of flavour release 287
R. Linforth and A. Taylor, Nottingham University, UK
13.1 Introduction 287

13.2 Influence of the foodstuff on flavour release 288
13.3 Losses of flavour molecules after leaving the bolus 292
13.4 The transfer of volatiles to the gas phase *in vivo* 293
13.5 Aroma delivery to the upper airway and nose 295
13.6 Persistence of flavour release 301
13.7 Future trends 303
13.8 Sources of further information 304
13.9 References 304

14 Genetic influences on taste 308
J. Prescott, James Cook University, Australia
14.1 Introduction 308
14.2 Individual differences in taste perceptions 309
14.3 PTC/PROP bitterness 311
14.4 The role of taste anatomy 316
14.5 PROP taster differences and flavour perceptions 320
14.6 Conclusion 321
14.7 References 321

15 Texture–aroma interactions 327
S. Lubbers, INRA, France
15.1 Introduction 327
15.2 Influence of rheological behaviour on flavour release 329
15.3 Texture measurements and perceived intensity of aroma 336
15.4 References 341

16 Odour–taste interactions in flavour perception 345
C. Salles, INRA, France
16.1 Introduction 345
16.2 Odour–taste interactions 346
16.3 Origin of odour–taste interactions 350
16.4 Future trends 362
16.5 Source of further information 363
16.6 References 363

17 The learning of human flavour preferences 369
A. Blake, Firmenich SA, Switzerland
17.1 Introduction 369
17.2 The relationship between cooking and flavour 370
17.3 Flavours and how we experience them 372
17.4 How we learn flavour 383
17.5 Future trends 394
17.6 References 396

18 The development of flavour perception from infancy to adulthood ... 403
B. Schaal, Centre Européen des Sciences du Goût, France
18.1 Introduction ... 403
18.2 Functional value of olfaction in early development ... 404
18.3 Early functioning of olfaction ... 410
18.4 Memory and plasticity of olfactory function in early life ... 420
18.5 Conclusions and future trends ... 428
18.6 Acknowledgments ... 430
18.7 References ... 430

Index ... 437

Contributor contact details

(* = main contact)

Chapter 1

Dr Bernd Bufe and Professor Wolfgang Meyerhof*
German Institute of Human Nutrition Potsdam-Rehbruecke
Arthur-Scheunert-Allee 114-116
14558 Nuthetal
Germany

E-mail: meyerhof@mail.dife.de
E-mail: Bufe@mail.dife.de

Chapter 2

André Holley
Centre Européen des Sciences du Goût
UMR 5170 CNRS-Université de Bourgogne-Inra
15 rue Hugues Picardet
21000 Dijon
France

E-mail: holley@cesg.cnrs.fr

Chapter 3

Professor Ann Noble*
University of California, Davis
One Shields Avenue
Davis, CA 95616
USA

E-mail: acnoble@ucdavis.edu

Dr Isabelle Lesschaeve
CCOVI Biological Sciences
Brock University
500 Glenridge Avenue
St Catharines
Ontario
Canada

E-mail: ilesscha@brocku.ca

Chapter 4

Professor G. Reineccius
University of Minnesota
Food Science and Nutrition
147 F Sc N
1334 Eckles Ave
St Paul, MN 55108
USA

E-mail: greinecc@umn.edu

Chapter 5

Professor El Mostafa Qannari*
ENITIAA/INRA Unité de Sensométrie et Chimiométrie
Rue de la Géraudière BP 82 225
44322 Nantes cedex 03
France

E-mail: qannari@enitiaa-nantes.fr

Pascal Schlich
Centre Europeén des Sciences du Goût (CESG)
Laboratoire d'Interface Recherche-Industrie et de Sensométrie (LIRIS)
Campus de l'Université de Dijon
15, rue Hugues Picardet
21000 Dijon
France

E-mail: Schlich@cesg.cnrs.fr

Chapter 6

Professor Andrée Voilley*
ENSBANA – Université de Bourgogne
1 Esplanade Erasme
F 21000 Dijon
France

Tel: 03 80 39 66 59
E-mail: andree.voilley@u-bourgogne.fr

Chapter 7

Professor Michel Ollivon
Physico-Chimie des Systèmes Polyphasés
UMR CNRS 8612
University Paris-Sud
5 rue J-B. Clément
92296 Châtenay-Malabry
France

Tel: 33 1 46 83 56 29
Fax: 33 1 46 83 53 12
E-mail: michel.ollivon@cep.u-psud.fr

Chapter 8

Dr Jean-Pierre Dumont
INRA
Rue de la Géraudière BP 71627
44316 Nantes cedex 3
France

E-mail: dumont@nantes.inra.fr

Chapter 9

Dr Anne Tromelin*, Isabelle Andriot and Dr Elizabeth Guichard
UMR FLAVIC INRA-ENESAD
Flaveur, Vision et Comportement du Consommateur
INRA Dijon
17 rue Sully
BP 86150
21065 Dijon Cedex
France

Tel: 33/ (0) 3 80 69 35 12
Fax: 33/ (0) 3 80 69 32 27
E-mail: tromelin@dijon.inra.fr;
guichard@dijon.inra.fr;
Isabelle.Andriot@dijon.inra.fr

Chapter 10

Dr Julien Delarue*
Laboratoire de Perception Sensorielle et Sensométrie
Ecole Nationale Supérieure des Industries Agricoles et Alimentaires
1 avenue des Olympiades
91744 MASSY Cedex
France

E-mail: delarue@ensia.fr

Professor Pierre Giampaoli
Laboratoire de Chimie des Substances Naturelles – arômes, antioxydants, colorants
Ecole Nationale Supérieure des Industries Agricoles et Alimentaires
1 avenue des Olympiades
91744 MASSY Cedex
France

E-mail: giamp@ensia.fr

Chapter 11

Dr Kris B. de Roos
Givaudan Nederland BV
PO Box 414
3770 AK Barneveld
The Netherlands

E-mail: kris.de_roos@givaudan.com

Chapter 12

Dr Alexandra E. M. Boelrijk*, Dr Koen G. C. Weel and Professor Gerrit Smit
NIZO Food Research
Department of Flavour
PO Box 20
6710 BA Ede
The Netherlands

E-mail: alexandra.boelrijk@nizo.nl
E-mail: gsmit@nizo.nl

Dr Jack J. Burger
Food Science and Technology Centre
Quest International Nederland BV
PO Box 2, 1400 CA Bussum
The Netherlands

E-mail: Jack.Burger@questintl.com

Chapter 13

Dr R. Linforth* and Professor A. Taylor
Food Sciences
University of Nottingham
Sutton Bonington Campus
Loughborough
Leics LE12 5RD
UK

Tel: 00 44 (0)1159 516153
Fax: 00 44 (0) 1159 516142
E-mail: robert.linforth@nottingham.ac.uk; andy.taylor@nottingham.ac.uk

Chapter 14

Professor John Prescott
School of Psychology
James Cook University
PO Box 6811
Cairns QLD 4870
Australia

E-mail: John.Prescott@jcu.edu.au

Chapter 15

Dr S. Lubbers
UMR FLAVIC INRA-ENESAD
Flaveur, Vision et Comportement du Consommateur
INRA Dijon
17 rue Sully
BP 86150
21065 Dijon Cedex
France

E-mail: s.lubbers@enesad.fr; lubbers@dijon.inra.fr

Chapter 16

Dr Christian Salles
UMR INRA-ENESAD Flaveur, Vision, Comportement du consommateur (FLAVIC)
17 rue Sully
BP86510
21065 Dijon Cedex
France

E-mail: salles@dijon.inra.fr

Chapter 17

Dr Anthony Blake

E-mail: Tony@blake-nyon.com

Chapter 18

Benoist Schaal
Ethology and Sensory Psychobiology Group
Centre Européen des Sciences du Goût
UMR 5170 CNRS-Université de Bourgogne-Inra
15 rue Hugues Picardet
21000 Dijon
France

E-mail: schaal@cesg.cnrs.fr

Preface

Flavour is one of the most important characteristics of any food product. Its critical role in determining the way consumers assess food quality has made it a key area of research for the food industry. From its foundation in sensory evaluation and the isolation and analysis of flavour volatiles, flavour science has become a much broader subject aiming to provide a comprehensive understanding of flavour from its generation in food to its perception during eating. This book summarises the most important developments in flavour research and their implications for the food industry.

The first part of the book reviews the way flavour is detected and measured. The first two chapters discuss our understanding of how humans perceive and then process information about taste compounds. They provide the foundation for Chapter 3 which reviews current practice in the sensory analysis of food flavour. Complementing this chapter, Chapter 4 discusses choosing from the wide range of instrumental techniques which have been developed to identify aroma compounds. The final chapter in Part I links the preceding two chapters by discussing the complex issues in matching instrumental measurements with the results of sensory evaluation of foods.

One of the most dynamic aspects of flavour research is in understanding the way flavour compounds are retained within foods and the factors determining the way they are released. Part II reviews key research in this area. After an overview of some of the key factors influencing flavour retention and release, a group of chapters reviews the way key food components influence flavour development, retention and release. There are chapters on flavour compound interactions with lipids, emulsions, protein and carbohydrate components in food. Other chapters review modelling aroma interactions in food matrices and mechanisms of flavour retention in and release from liquid food products.

The final part of the book complements Part II by reviewing what we now know about how humans experience flavour release, together with some of the key factors influencing this process. Chapter 13 summarises the exciting research that has been done to understand the process of flavour release in the mouth. Other chapters then review the way texture–aroma and odour–taste interactions influence this process. Other chapters consider the way psychological factors, the development of flavour perception during infancy and learnt flavour preferences affect the way we perceive and evaluate flavour.

Flavour in food seeks to distil key developments in flavour science and summarise their implications for the food industry. We hope it will be a valuable reference for R&D staff, those responsible for sensory evaluation of foods and product development, as well as academics and students involved in flavour science.

Part I

Characterisation of aroma compounds

1

The human perception of taste compounds

B. Bufe and W. Meyerhof, German Institute of Human Nutrition Potsdam-Rehbruecke, Germany

1.1 Introduction

Food intake is essential for the survival of every living organism (Lindemann 2001). The failure to detect spoiled or toxic food can have lethal consequences. Therefore, it is not surprising that humans use all their five senses to analyse food quality. A first judgement about the value of a food source is made on its appearance and smell. Food that looks and smells attractive is taken into the oral cavity. Here, based on a complex sensory analysis that is not only restricted to the sense of taste but also includes smell, touch and hearing (Linden 1993, Drewnowski 1997), the final decision about ingestion or rejection of food is made. Frequently, these complex interactions between different senses is inappropriately referred to as 'taste' although it should be better called flavour perception (Linden 1993), because it uses multiple senses.

The five basic taste qualities are exclusively mediated by specialised epithelial receptor cells that are located in taste buds. Most taste buds lie within taste papillae on the human tongue, but some of them are also distributed on the palate and epiglottis (Skramlik 1926). The taste buds in the oral cavity are innervated by gustatory fibres of the VII, IX and X cranial nerve (Smith and St John 1999) Thus, the perception of the five basic taste qualities sour, salty, sweet, bitter and umami has a distinct anatomical basis.

1.2 The sense of taste

The basic taste qualities contribute differently to the assessment of the value of food (Skramlik 1926). Sweet taste is predominantly elicited by carbohydrates and indicates energy-rich food sources (Drewnowski 1995). The broth-like umami taste, that is mainly triggered by glutamate and enhanced by ribonucleotides such as inositol monophosphate (IMP), identifies protein-rich food (Bellisle 1999, Yamaguchi and Ninomiya 2000). Both taste qualities indicate valuable food components, and thus sweet and umami tastes are coupled to attractive behaviours in mammals. Salt taste is elicited by sodium chloride and other salts and contributes to electrolyte homeostasis (Lindemann 2001, Daniels and Fluharty 2004, Skramlik 1926). Consistent with this function, salt taste is attractive at low concentrations and repulsive at high concentrations (Daniels and Fluharty 2004). Strong sour taste is also repulsive and prevents the ingestion of unripe fruits and spoiled food, which often contain acids (Skramlik 1926, Lindemann 2001). Bitter taste is evoked by many compounds that belong to multiple chemical classes (Chon 1914, Delwiche *et al.* 2001). The common denominator of most bitter compounds is their pharmacological activity or toxicity (Skramlik 1926). Therefore, due to its task to avoid harmful compounds strong bitter taste is aversive (Skramlik 1926). Nevertheless humans can accept moderate bitter taste or even find it attractive. A reasonable explanation for this observation is that bitter and sour tastes should not deter us from advantageous food containing low concentrations of harmful compounds.

The five basic taste sensations are mediated by specialised epithelial cells, the taste receptor cells, that are located within the taste buds of the papillae on the surface of the tongue. These elongated taste receptors cells are deeply embedded in the surrounding epithelium and just contact the outside world in the gustatory porus of the taste buds. Thus, the porus is the place where tastants interact with the taste receptor molecules that are located at the apical site of the taste receptor cells. In contrast to obsolete textbook knowledge, humans can perceive all taste qualities on any area of the tongue that contains papillae (Häning 1901, Lindemann 2001). Only the perceived intensities of the taste qualities differ depending on the tongue region and papilla type (Häning 1901). Sweet taste saccharin for instance is highest at the tip of the tongue whereas the bitter taste of quinine is best perceived at the back of the tongue (Häning 1901). Interestingly, the anterior part of the tongue is innervated by the VII cranial nerve whereas the posterior part of the tongue is innervated by the IX cranial nerve (Smith and St John 1999). These innervations are also reflected by the distribution of the taste papilla types. The fungiform papillae are located at the anterior part of the tongue (Skramlik 1926) and thus are innervated by VII cranial nerve. In contrast, the foliate and vallate papillae that are located at the back of the tongue that is innervated by the IX cranial nerve. This nerve also innervates isolated taste buds in the palate and epiglottis. Each of three cranial nerves also carries somatosensory afferents that innervate palate taste buds and regions of the tongue neighbouring lingual taste buds. This type of innervation makes it difficult to distinguish gustatory from somatosensory information. For example, the IX nerve is, apart from its role in taste sensations, also involved in mediating the gag reflex (DeMeester *et al.* 1977). Thus, the higher sensitivity for bitter compounds at the back of the tongue might help preventing the ingestion of harmful substances by eliciting the gag reflex.

1.3 The molecular basis of human taste perception

Despite intensive efforts to identify receptor molecules that mediate the taste perceptions of humans and other species including rodents, insects or amphibians they have remained elusive for decades (Lindemann 1996). This relates to some extent to the observation that taste perception can vary significantly between different species (Lindemann 1996). This led, depending on the species studied, to some contradictory results and confusions. However, within the last five years significant progress has been made. Largely due to excellent and pioneering contributions from the laboratories of Charles Zuker and Nicolas Ryba (Zhao *et al.* 2003, Zhang *et al.* 2003, Nelson *et al.* 2001, 2002, Chandrashekar *et al.* 2000, Adler *et al.* 2000, Hoon *et al.* 1999) we now not only know receptors mediating sweet, bitter and umami taste but are also beginning to understand some of their functional properties. The molecular mechanisms of salt and sour taste in humans are presently less well understood. Their elucidation therefore represents an important objective for future work (Kim *et al.* 2004).

1.3.1 Sour taste

Sour taste is evoked by acids. Interestingly, the response to an acid is not always proportional to the pH of the substance (Ganzevles and Kroeze 1987). For example, at the same pH, acetic acid has a stronger sourness than HCl (Ganzevles and Kroeze 1987). Based on data obtained in different species various molecular mechanisms underlying sour taste have been proposed (Lindemann 1996, Lindemann 2001). Electrophysiological data obtained from mudpuppies suggest a direct blocking of potassium channels through protons present in the oral cavity (Kinnamon *et al.* 1988). In hamsters, amiloride-blockable proton currents have been recorded during acid stimulation (Gilbertson *et al.* 1992, 1993). Therefore, the amiloride-blockable epithelial sodium channel (ENaC) has been proposed to contribute to sour taste transduction (Herness and Gilbertson 1999). But also other mechanisms are candidates for playing a role in sour taste. Currently, acid sensing ion channels (ASIC) are being intensely discussed as sour taste receptors in rodents (Ugawa 2003, Ugawa *et al.* 1998, 2003). Especially ASIC2a (synonyms: MDEG1, BNC1a, BNaC1α) is considered as a promising sour taste receptor candidate (Ugawa 2003). There are several lines of evidence supporting this hypothesis. First, ASIC2a is expressed in a large proportion of rat taste receptor cells

(Ugawa *et al.* 1998). Second, electrophysiological recordings report a proton-gated sodium current upon acid stimulation within taste receptor cells (Lin *et al.* 2002). Third, the functional expression of ASIC2a in oocytes showed that this channel responds at equal pH stronger to acetic acid than to HCl, an effect that corresponds to human taste experience (Ugawa *et al.* 1998, Ganzevles and Kroeze 1987). Although at first glance, these data seem convincing, it is quite problematic that sour taste is not abolished or not even diminished in ASIC2a knock out mice (Richter *et al.* 2004, Kinamon *et al.* 2000). It remains to be seen, if the proposed coexpression and heteromerisation of ASIC2 with ASIC2b (Ugawa *et al.* 2003) can sufficiently explain the so far complete lack of any sour taste phenotype in the ASIC2a knock out mice.

Hyperpolarisation and cyclic nucleotide gated non selective cation channels of the HCN family are another family of promising sour taste receptor candidates (Stevens *et al.* 2001). Like the ASIC channels several independent lines of evidence support their role in sour taste transduction. First, the presence of HCN1 and HCN4 in a subset of taste receptor cells of the rat circumvallate papilla was shown by RT-PCR, *in situ* hybridisation and immunohistochemistry (Stevens *et al.* 2001). Second, careful electrophysiological recordings of slice preparations demonstrated that apical stimulation of taste buds with acids elicited an I_h current in a subset of taste receptor cells (Stevens *et al.* 2001). This I_h current is a hallmark for the HCN family. Third, *in vitro* functional expression of HCN1 and HCN4 in HEK293 cells showed that both channels can be activated by extracellular protons (Stevens *et al.* 2001). So far it has never been tested, if these channels show a stronger response to acetic acid than to HCl. Moreover, studies of HCN knock out animals models that would further support their role in sour taste transduction are still missing. Thus, despite large efforts of various groups our knowledge about the molecular mechanisms of sour taste perception is still insufficient.

1.3.2 Salt taste

In rodents there is reasonable evidence suggesting that the Na^+ salt taste is mediated by the epithelial sodium channel (ENaC) (Lindemann 1997). First, evidence for this assumption was based on the observation that salt taste in rodents is to a large extent amiloride-blockable, at concentrations used to block ENaC (Heck *et al.* 1984). Furthermore, in RT-PCR experiments with cDNA obtained from isolated taste buds the α, β and γ subunits of the ENaC channel could be detected in the anterior part of the tongue, whereas in the posterior part of the tongue just the α subunit was abundant (Kretz *et al.* 1999). Heterologous expression studies show that the α subunit alone is much less sensitive to amiloride than the combination of α, β and γ subunits (Benos and Stanton 1999). Therefore low expression of the two other ENaC subunits might to some extent account for the observation that salt taste is much less amiloride-blockable in the posterior part of the tongue (Doolin and Gilbertson 1996, Formaker and Hill 1991). The RT-PCR experiments are essentially confirmed by immunohistochemistry, although the majority of the ENaC protein is located inside the cell instead of the taste porus (Lin *et al.* 1999, Kretz *et al.* 1999). Moreover, electrophysiological recordings showed that amiloride-blockable sodium currents flow inward through the apical membranes of taste receptor cells (Avenet and Lindemann 1991). Another independent line of evidence is based on the influence of aldosterone on salt taste. Increasing aldosterone levels lead to a higher salt sensitivity (Herness 1992). Consistent with this observation an upregulation of ENaC protein levels in taste receptor cells have been seen (Okada *et al.* 1990).

Apart from the amiloride-sensitive salt taste that is likely mediated by ENaC there is clear evidence for a second, amiloride-insensitive, pathway (Miyamoto *et al.* 2000). Moreover, although Na^+ is the most relevant ion, this taste quality is also elicited by various other ions including LiCl, KCl, NH_4Cl and $CaCl_2$ (Miyamoto *et al.* 2000). While the ENaC channel is permeable for Na^+ and Li^+ ions it is nearly impermeable for larger ions such as K^+, Ca^{2+} and NH_4^+ (Kellenberger *et al.* 1999, 2001). This suggests the existence of a second, yet unidentified, salt taste receptor that mediates the salt taste of KCl, NH_4Cl and $CaCl_2$. This is especially evident in humans where just 20% of the salt taste is amiloride-blockable (Ossebaard and Smith 1995). Recent electrophysiological studies on mouse taste receptor cells suggested that the ENaC-independent salt taste is mediated by a non-selective cation channel that is permeable to Na^+, K^+, Ca^{2+} and NH_4^+ (Lyall *et al.* 2004, DeSimone *et al.* 2001). Based on pharmacological channel properties and recordings of taste receptor cells in vanilloid receptor 1 knock out mice a variant of the vanilloid receptor 1 channel has been proposed to mediate the amiloride insensitive salt taste (Lyall *et al.* 2004). Further studies will have to show if this holds true.

1.3.3 Sweet taste

Mice strains differ in their sensitivity towards sweet compounds (Capeless and Whitney 1995, Fuller 1974). For example, some strains have a fivefold lower sensitivity for the artificial sweetener saccharin (Capeless and Whitney 1995, Fuller 1974). Genetic studies revealed that the strain-specific differences between taster (C57BL/6J) and non-taster mice (DBA/2J) are due to a single chromosomal locus called '*sac*' (Fuller 1974) at the distal end of mouse chromosome 4 (Lush 1989, Lush *et al.* 1995). These studies in combination with data from the human genome provided a powerful tool for the discovery of the gene encoded by the *sac* locus, which was independently achieved by six research groups. Based on the assumption that the sweet taste receptor is a G-protein coupled receptor (GPCR) the six groups (Bachmanov *et al.* 2001, Nelson *et al.* 2001, Sainz *et al.* 2001, Max *et al.* 2001, Kitagawa *et al.* 2001, Montmayeur *et al.* 2001) analysed the human genome data of the distal end of human chromosome 1, which is syntenic to the distal end of mouse chromosome 4. They discovered the human *TAS1R3* gene as a new member of the putative taste receptor family *TAS1R* (Hoon *et al.* 1999) (note: *TAS1R* is the gene symbol of the human genome project nomenclature committee for the gene family previously called *T1R*s; the corresponding mouse gene symbol is *Tas1r*). The corresponding mouse gene *Tas1r3* is directly located within the *sac* locus (Bachmanov *et al.* 2001, Nelson *et al.* 2001, Sainz *et al.* 2001, Max *et al.* 2001, Kitagawa *et al.* 2001, Montmayeur *et al.* 2001). Moreover, the mouse *Tas1r3* gene shows several strain specific single nucleotide polymorphisms (Reed *et al.* 2004). Some of them correlate with the saccharin taster and non-taster status of the mice strains (Reed *et al.* 2004, Sainz *et al.* 2001, Max *et al.* 2001, Bachmanov *et al.* 2001, Kitagawa *et al.* 2001, Montmayeur *et al.* 2001, Nelson *et al.* 2001). Consistent with its proposed role as a sweet receptor, *in situ* hybridisations showed the expression of *Tas1r3* in a subset of mouse (Bachmanov *et al.* 2001, Nelson *et al.* 2001, Sainz *et al.* 2001, Max *et al.* 2001, Kitagawa *et al.* 2001, Montmayeur *et al.* 2001) and human taste receptor cells (Liao and Schultz 2003). Detailed analysis revealed that *Tas1r3* is coexpressed with its two other known family members *Tas1r2* or *Tas1r1* in two non-overlapping subsets of taste receptor cells (Max *et al.* 2001, Montmayeur *et al.* 2001, Nelson *et al.* 2001). Notably, in rodents *Tas1r2* is predominantly expressed at the posterior tongue in vallate and foliate papillae but only rarely in fungiform papillae of the anterior tongue (Hoon *et al.* 1999). In contrast, the human *TAS1R2* can be easily detected in fungiform papillae of the human tongue (Liao and Schultz 2003). In fact, this expression pattern correlates remarkably well with the known sweet sensitivities of humans and rodents. In rodents the sweet response is highest at the back of the tongue whereas humans are most sensitive at the tip of the tongue (Lindemann 1999). Thus the observed species differences in sweet taste perception can be explained by the different expression pattern of *TAS1R2*. These results led to the hypothesis that the functional sweet receptor could be a heteromer composed of *Tas1r3* and *Tas1r2* (Montmayeur *et al.* 2001). Indeed, functional expression studies in HEK293 cells proved that cells cotransfected with *Tas1r2* and *Tas1r3* responded to stimulation with various sweeteners (Li *et al.* 2002, Nelson *et al.* 2001, 2002) confirming that *Tas1r2* and *Tas1r3* form a functional sweet taste receptor. Further evidence that the *Tas1r3* gene is involved in sweet taste perception was provided by mouse models. The transgenic expression of the *Tas1r3* taster gene variant in a non-taster mouse strain rescued the taster phenotype (Nelson *et al.* 2001). In addition, behavioural experiments and neuronal recordings of *Tas1r2* and *Tas1r3* knock out mice showed that the deletion of either gene strongly reduced the nerve responses and the attractiveness of various sweeteners (Damak *et al.* 2003, Zhao *et al.* 2003) and abolished the responses to artificial sweeteners completely (Damak *et al.* 2003, Zhao *et al.* 2003). In contrast to that, the deletion of the third family member *Tas1r1* did not influence the perception of sweet compounds in mice (Zhao *et al.* 2003). Taken together there is overwhelming evidence that the rodent *Tas1r2* and *Tas1r3* and the human *TAS1R2* and *TAS1R3* are genuine sweet taste receptors. Interestingly, both the *Tas1r2* and the *Tas1r3* knock out mice kept some behavioural and nerve responses to high concentrations of the carbohydrate sweeteners glucose and sucrose (Damak *et al.* 2003, Zhao *et al.* 2003). This clearly indicates the existence of an additional low affinity transduction mechanism for natural sweeteners (Damak *et al.* 2003). It remains to be seen if this residual taste response can be completely explained through the formation of *Tas1r* monomers or homomers, which are activated *in vitro* by high sucrose concentrations (Zhao *et al.* 2003). It is also possible that *Tas1r* independent pathways contribute (Damak *et al.* 2003) such as direct activation of G-proteins by sweet tasting compounds (Naim *et al.* 1998, Peri *et al.* 2000, Naim *et al.* 1994) or of even a yet undiscovered receptor.

1.3.4 Umami taste

Umami taste strongly enhances the palatability of food (Yamaguchi and Ninomiya 2000). The umami compounds glutamic acid and 5′-ribonucleotides become enriched during the ripening processes in many foods including fruits, vegetables, cheese or meat (Yamaguchi and Ninomiya 2000). Therefore, this taste quality helps us to choose the ripest fruits and the most palatable cheese for our meal.

In humans not only glutamate but also some metabotropic glutamate receptor agonists such as ibotenate and L-AP4 elicit umami taste (Kurihara and Kashiwayanagi 2000). Moreover, various physiological and molecular studies show that the metabotropic glutamate receptors (mGluRs) are expressed in taste receptor cells of mice (Kim *et al.* 2001, Caicedo and Roper 2001, Caicedo *et al.* 2000a, 2000b, Lin and Kinnamon 1999, Toyono *et al.* 2003, Chaudhari *et al.* 1996). Both observations led to the hypothesis that metabotropic glutamate receptors contribute to umami taste. Indeed, the cDNA of an N-terminally truncated variant (mGluR4t) of the mGluR4 was isolated from rodent tongue tissue containing taste receptor cells (Chaudhari *et al.* 2000). Functional analysis of this receptor variant in a heterologous expression system showed that it responded to L-AP4 and glutamate at high concentrations, which are typically used to elicit umami taste in humans. However, IMP did not enhance the response (Chaudhari *et al.* 2000). Based on these data the truncated mGluR4 variant seemed to be a quite attractive candidate umami taste receptor, although a number of inconsistencies exist (Zhao *et al.* 2003). First, mGluR4 knock out mice show an increased preference for glutamate (Chaudhari and Roper 1998) instead of a reduced response as one would expect. Second, the expression of the mGluR4t was only detected by PCR and *in situ* hybridisation. So far, there is no evidence for the translation of the mGluR4t mRNA into a functional protein in taste receptor cells. Third, the localisation of the truncated mGluR variant on the apical cell surface of the taste receptor cells has not been shown. This is quite important because this receptor variant lacks the cell surface targeting sequence of mGluR4 (Zhao *et al.* 2003). Fourth, the normal binding site for glutamate is missing and thus the precise nature of an alternative binding site needs to be shown.

Recent studies demonstrated that receptors of the *TAS1R* family mediate umami taste. *In vitro* expression studies show that cells cotransfected with human *TAS1R1* and *TAS1R3* respond to glutamate, while cells transfected with their counterparts from mice acquired general sensitivity for glutamate and other L-amino acids (Li *et al.* 2002, Nelson *et al.* 2002). In addition 5′-ribonucleotides such as IMP strongly enhanced the responses of the transfected cells (Li *et al.* 2002, Nelson *et al.* 2002). In addition, *in situ* hybridisations clearly showed that these receptors are expressed in a subset of taste receptor cells in humans and rodents (Hoon *et al.* 1999, Liao and Schultz 2003, Max *et al.* 2001). In rodents, *Tas1r1* is expressed in the fungiform papilla at the tip of the tongue (Hoon *et al.* 1999), whereas in humans *TAS1R1* can be barely detected in the fungiform papilla (Liao and Schultz 2003). The difference in *TAS1R1* expression matches the different regional sensitivity for glutamate on the tongues of mice and humans. In rats, the response to glutamate is highest at the front of the tongue (Sako *et al.* 2000) whereas humans are most sensitive at the back tongue (Yamaguchi and Ninomiya 2000). Further evidence for a role of TAS1Rs in umami taste comes from behavioural studies and neuronal recordings in *Tas1r1* and *Tas1r3* knock out mice. The deletion of either receptor gene strongly reduced the attractiveness of umami compounds as well as the nerve response (Damak *et al.* 2003, Zhao *et al.* 2003). Taken together, a combination of genetic and physiological evidence convincingly shows that *Tas1r1* and *Tas1r3* are necessary for amino acid taste in mice. Moreover, glutamate responses from the human *TAS1R1/TAS1R3* receptor heteromer can be potentiated by monoribonucleotides, which is a hallmark of umami taste.

1.3.5 Bitter taste

In humans, the perception of different bitter compounds can vary between individuals (Blakeslee 1935). The most prominent example is the taste of the synthetic bitter compound phenylthiourea (PTC). For more than 70 years it has been known that PTC tasters and non-tasters exist (Fox 1932). PTC tasters are very sensitive and can detect micromolar concentrations of this compound, whereas the non-tasters are nearly taste-blind for PTC. Family studies subsequently revealed that the ability for PTC tasting is genetically inherited (Blakeslee 1931, 1932). Further investigations demonstrated that PTC taster status also determines the sensitivity to 6-N-propylthiouracil (PROP) and a variety of other compounds containing a N=C=S moiety (Harris and Kalmus 1949, Barnicot *et al.* 1951). Recently, genetic mapping studies in humans associated the ability to taste PTC and PROP with loci on chromosome 7 and chromosome 5, respectively (Reed *et al.* 1999, Drayna *et al.* 2003). Based on one of these studies (Reed *et al.* 1999) the analysis of the human genome project resulted in the identification of a family of putative bitter taste receptors called *TAS2R*s (Adler *et al.* 2000).

In mice, similar genetic variations for tasting bitter compounds have been observed. These include sucrose octaacetate (SOA), raffinose undedecaacetate, cycloheximide and quinine (Lush 1981, 1984, 1986, Lush and Holland 1988, Lush *et al.* 1995). Mapping studies assigned them to a locus on mouse chromosome 6 close to the *Prp* gene (Lush *et al.* 1995). Using this information, two groups independently identified a novel family of GPCRs, the *TAS2R*s, as putative bitter taste receptors (Matsunami *et al.* 2000, Adler *et al.* 2000) (note: *TAS2R* is the official gene symbol of the human genome project nomenclature committee for the gene family previously called *T2Rs* or *TRB*s; the corresponding mouse gene symbol is *Tas2r*). This newly discovered family of G-protein coupled receptors comprises, in humans, approximately 25 intact receptor genes and several pseudogenes. They are exclusively located on human chromosomes 12, 7 and 5 (Shi *et al.* 2003, Bufe *et al.* 2002, Adler *et al.* 2000). The corresponding mouse *Tas2r* receptor family consists of approximately 33 genes that are exclusively located on the mouse chromosome 6 and 15 (Shi *et al.* 2003, Adler *et al.* 2000). The chromosomal distribution of the *TAS2R* receptor family correlates well with the gene loci that determine variations in bitter perception in humans and mice (Lush 1981, 1984, 1986, Lush and Holland 1988, Lush *et al.* 1995, Reed *et al.* 1999, Kim, 2004). Moreover, *in situ* hybridisations of several mouse, rat and human *TAS2R* receptors show their expression in a subset of human and mouse taste receptor cells (Matsunami *et al.* 2000, Adler *et al.* 2000, Bufe *et al.* 2002, Behrens *et al.* 2004). Interestingly, double-label *in situ* hybridisation of two different *Tas2r* (*T2R3* and *T2R7*) receptors in rat circumvallate papillae demonstrated a coexpression in the same subset of taste receptor cells (Adler *et al.* 2000). This observation is also confirmed by *in situ* hybridisations that used mixtures of two, five and ten different *Tas2r* receptors. These studies exhibited an increase in the staining intensity of the labelled cells but no significant increase in the number of labelled cells pointing to a coexpression of the majority of *TAS2R* receptors within the same cells (Adler *et al.* 2000). Although these experiments so far do not formally prove the coexpression of all TAS2Rs in the same subset of taste receptor cells, such a localisation would explain how bitter compounds that activate different receptors can elicit the same uniform bitter taste.

Functional expression studies using recombinant receptors showed that several of the *TAS2R* receptors responded to various bitter tastants. The mouse receptor *Tas2r105* (former name *T2R5*) and the corresponding rat receptor *T2R9* respond to micromolar concentrations of cycloheximide (Bufe *et al.* 2002, Chandrashekar *et al.* 2000). The human receptor *TAS2R10* responded to strychnine. Another receptor, *TAS2R14,* was activated by multiple bitter compounds including (–)-alpha-thujone, and picrotoxinin (Behrens *et al.* 2004), whereas the human receptor *TAS2R16* responded to salicin and other bitter β-glucopyranosides (Bufe *et al.* 2002). In addition, a recent genetic study in humans identified *hTAS2R38* as a receptor for PTC and showed a good correlation between the PTC taster status and single nucleotide polymorphisms in the *TAS2R38* gene (Drayna *et al.* 2003, Kim *et al.* 2003b). There are three common variations in the *TAS2R38* gene, *P49A*, *A262V*, *V296I*. They lead to two frequently occurring haplotypes encoding the receptor variants, *TAS2R38-PAV* and *TAS2R38-AVI*, and some additional rare combinations *TAS2R38-AAI*, *-AAV*, and *-PVI* (Kim *et al.* 2003b, Wooding *et al.* 2004). Humans that are homozygous for the *AVI* form of the receptor are PTC non-tasters, whereas humans that are homo- or heterozygous for the *PAV* form are PTC tasters (Kim *et al.* 2003b). This strongly suggests that variations in the *TAS2R38* gene are responsible for inherited differences in PTC tasting. Recent functional studies of both *TAS2R38* receptor forms by our group revealed that the PAV form of the receptor responded to PTC, whereas the non-taster form does not respond to this compound (Bufe *et al.* unpublished results). Thus, there is strong molecular, genetic and functional evidence that the *TAS2R* receptors are indeed genuine bitter taste receptors. Moreover, polymorphisms in the *TAS2R* genes contribute to the inherited variations of bitter taste sensitivity within the human population.

1.3.6 Signal transduction

A first major breakthrough in the understanding of molecular mechanisms of taste transduction was due to the discovery of the G-protein α subunit gustducin by the laboratory of Robert F. Margolskee (McLaughlin *et al.* 1992). *In situ* hybridisations and immunocytochemisty revealed that α-gustducin is expressed in a subset of taste receptor cells of humans and rodents (Takami *et al.* 1994, McLaughlin *et al.* 1992). Based on its high homology to the G_i-type G-protein transducin it is thought to decrease the cAMP levels through activation of a phoshodiesterase (McLaughlin *et al.* 1994). Consistent with this assumption, quench flow assays from mouse taste tissue showed that the stimulation of cell extracts with bitter compounds resulted in decreasing cAMP levels (Yan *et al.* 2001). This cAMP breakdown could be blocked by a α-gustducin antibody (Yan *et al.* 2001). Moreover, stimulation of membrane preparations obtained from bovine taste tissue with different bitter compounds lead to an activation of α-gustducin (Ming *et al.* 1998). Further biochemical assays with membrane preparations of cells heterologously expressing the bitter taste receptor *Tas2r105* showed the activation of α-gustducin (Chandrashekar *et al.* 2000). In addition, the functional analysis of *Tas2r* receptor-coupling to chimeric G-proteins revealed a preference of the *Tas2r* receptors for gustducin-like G-protein chimeras (Ueda *et al.* 2003). Beyond that, double-label *in situ* hybridisations in rat circumvallate papillae demonstrated a nearly complete coexpression of *Tas2r* receptors with α-gustducin (Adler *et al.* 2000). Thus, many lines of evidence showed that α-gustducin is involved in the transduction of bitter taste. Surprisingly physiological and behavioural studies of α-gustducin knock out mice showed not only a reduction in bitter taste but also a reduced sweet taste (Wong *et al.* 1996). This led to the hypothesis that gustducin, apart from its role in bitter taste, might also be involved in sweet taste transduction (Wong *et al.* 1996). Indeed recent studies using single cell RT PCR and *in situ* hybridisation showed a coexpression of *Tas1r3* and α-gustducin although conflicting data about the extent of overlap exist (Max *et al.* 2001, Montmayeur *et al.* 2001, Kim *et al.* 2003a). A further hint for the involvement of α-gustducin in sweet taste signal transduction is that *Tas1Rs* couple to G_i type G-proteins in the functional expression assays (Ozeck *et al.* 2004, Li *et al.* 2002). Nevertheless, more data are needed before a sound assessment about the role of α-gustducin in sweet taste transduction can be made.

Notably, in α-gustducin gene-targeted mice the behavioural responses towards sweet and bitter compounds were reduced but not completely abolished (Wong *et al.* 1996). Therefore, either α-gustducin is not the only G-protein α subunit used in the signal transduction of sweet and bitter taste or, alternatively, in the absence of α-gustducin other G-protein α subunits are recruited as substitutes. A likely candidate in this scenario is the highly homologous G-protein α subunit transducin, which can rescue the phenotype of α-gustducin knock out mice to some degree and is also present in taste tissue (Ruiz-Avila *et al.* 1995, Perez *et al.* 2002).

Various studies showed a rise of inositoltrisphosphate (IP_3) levels upon stimulation with bitter compounds (Miwa *et al.* 1997, Nakashima and Ninomiya 1998, Spielman *et al.* 1994, 1996, Rossler *et al.* 2000, Yan *et al.* 2001). This observation cannot be explained through the activity of α-gustducin, because this molecule mediates a cAMP breakdown. Thus, apart from α-gustducin other molecules must be involved in bitter taste transduction. Quench flow assays showed that IP_3 production could be suppressed through antibodies against phospholipase Cβ2 (PLCβ2) and the G-protein γ subunit 13 (Yan *et al.* 2001). This led to the hypothesis that a $\beta/\gamma13$ complex that is associated with α-gustducin elicits the IP_3 response through activation of PLCβ2. Growing evidence supports this mechanism. Single cell RT-PCR and *in situ* hybridisations show a coexpression of the G-protein subunits $\beta1$, $\beta3$, $\gamma13$ and PLCβ2 in gustducin-positive cells (Perez *et al.* 2002, Huang *et al.* 1999). Coimmunoprecipitations revealed that $\beta1$ and $\beta3$ subunits can interact with $\gamma13$ (Blake *et al.* 2001). In addition, cell-based assays demonstrated that $\beta1/\gamma13$ and $\beta3/\gamma13$ complexes can activate PLCβ2 (Blake *et al.* 2001). Furthermore behavioural studies and nerve recording showed that PLCβ2 knock out animals lost their responses to bitter stimuli (Zhang *et al.* 2003). Notably, the PLCβ2 knock out mice also lost their sensitivity to sweet tasting molecules and to amino acids. Salt and sour responses remained unaffected (Zhang *et al.* 2003). This suggests that sweet, bitter, and umami tastes share some molecules in their signal transduction pathways. This assumption is further supported by the discovery of the ion channel TRPM5, which is involved in taste transduction (Perez *et al.* 2002). TRPM5 knock out animals essentially show the same phenotype as the PLCβ2 knock out animals. They also lost their responses to sweet, bitter, and amino acid but not to sour and salt stimuli (Zhang *et al.* 2003). Functional analysis of TRPM5 in oocytes and HEK293 cells initially produced conflicting results about the ion conductance and the gating properties of the channel (Perez *et al.* 2002, Zhang *et al.* 2003). Recent data suggest that this channel is a voltage-modulated calcium-activated channel selective for monovalent cations (Liu and Liman 2003, Prawitt *et al.* 2003, Hofmann *et al.* 2003). Based on the recent findings a preliminary model for sweet, umami (Fig. 1.1) and bitter (Fig. 1.2) taste transduction can be made. Sweet, bitter and amino acid receptors couple via Gi type G-protein α subunits to the cAMP pathway and in parallel via

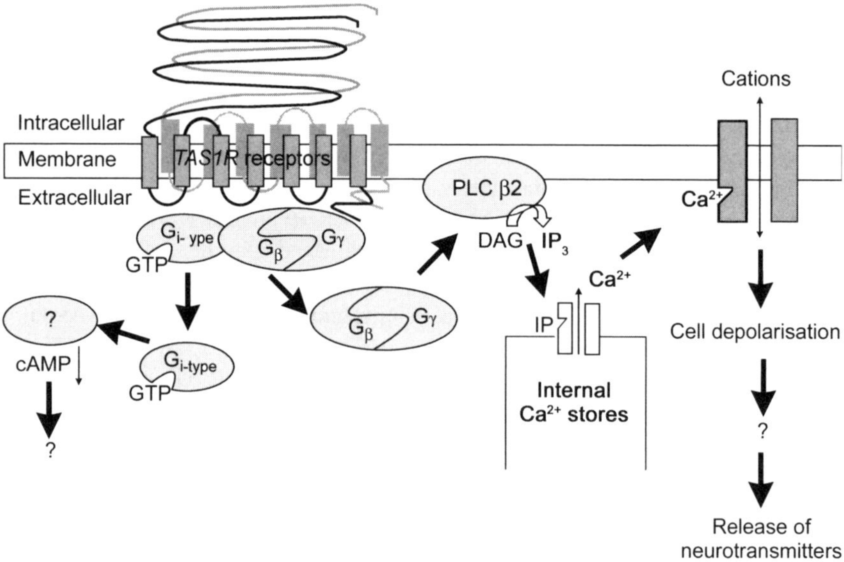

Fig. 1.1 Presumed signal transduction of *TAS1R* receptors.

G_i: Gi type G-protein α subunit; G_β: G-protein β subunit; G_γ: G-protein γ subunit; PLCβ2: phospholipase C β-2; TRPM5: TRPM5-ion channel. GDP: guonosindiphosphate; GTP: guonosintriphosphate; cAMP: cyclic adenosinmonophoshate; AMP: adenosinmonophosphate; IP_3: inositoltrisphoshphate; Ca^{2+}: calcium.

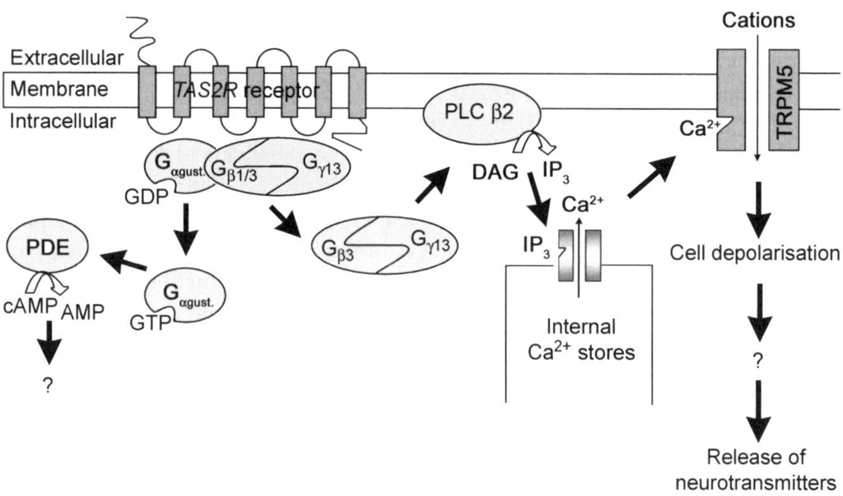

Fig. 1.2 Presumed signal transduction of *TAS2R* receptors.

$G_{\alpha\,gust.}$: G-protein α subunit gustducin; $G_{\beta 1/3}$: G-protein β subunit β-1 or β-3; $G_{\gamma 13}$: G-protein γ subunit γ-13; PLCβ2: phospholipase C β-2; PDE: phosphodiesterase; TRPM5: TRPM5-ion channel. GDP: guonosindiphosphate; GTP: guonosintriphosphate; cAMP: cyclic adenosinmonophoshate; AMP: adenosinmonophosphate; IP_3: inositoltrisphoshphate; Ca^{2+}: calcium.

a β/γ complex to PLCβ2. The activation of PLCβ2 then results in IP_3 production, which triggers a release of calcium from intracellular stores. The rapid rise of intracellular calcium levels then activates TRPM5. The resulting ion fluxes finally led to a depolarisation of the taste receptor cells and the secretion of an unknown neurotransmitter.

Although due to the discovery of gustducin and the other signal transduction components significant progress in the field of taste transduction has been made, many aspects are still elusive. For example, the role of decreasing cAMP levels and through an α-gustducin dependent activation of the phosphodiesterase is not yet understood. In addition, the release of neurotransmitters usually requires elevated calcium levels at the basolateral membrane of the taste receptor cells. As the TRP channel appears to be selective for monovalent cations, it is likely that the TRPM5-mediated depolarisation will lead to the activation of a yet unknown calcium channel. Calcium influx through this channel ultimately may trigger the release of the unknown neurotransmitter.

1.3.7 Implication for taste coding

For decades taste quality coding has been a fiercely debated field (Smith *et al.* 2000, Herness 2000). In principle, two competing models exist (Hellekant *et al.* 1998, Smith and St John 1999). The labelled line model favours a separate coding of the five basic taste qualities (Hellekant *et al.* 1998). Therefore, this model suggests the existence of specialised taste receptor cells for each taste quality, which are innervated by dedicated fibres (Hellekant *et al.* 1998). Thus, for example, a sweet stimulus such as sucrose will activate sweet taste receptor cells. Subsequently, solely sweet taste receptor cell innervating fibres will convey the signal to the brain. Consequently, in this model, the information of the taste quality is encoded at the level of the taste receptor cells.

In the competing across fibre pattern model taste receptor cells and the innervating neurons are not strictly specialised and respond to all taste stimuli (Smith *et al.* 2000). Only the strength of responses differ amongst taste modalities (Smith *et al.* 2000). In this model, for example, sweet best taste receptor cells exist, that respond stronger to sweet stimuli than to bitter, umami, sour or salt stimuli. Similarly, bitter best, salt best, sour best and amino acid best taste receptor cells would exist. As the neurons always innervate several taste receptor cells, they might, by chance, innervate more sweet best taste receptor cells. Consequently, such a neuron would be a sweet best neuron and respond stronger to sweet stimuli than to the other taste stimuli. Due to the same mechanism, neurons with a higher response for the other taste qualities would exist. Dependent on the different activation pattern, stimulation of the oral cavity by sweet, bitter, sour, salty and umami compounds would generate excitation patterns across many neurons that are decoded by the brain (Smith *et al.* 2000).

Consequently, in this model the information about the taste qualities is also encoded in higher brain centres. As both models are supported by experimental data it is not easy to decide which model is correct. Evidence that supports the across-fibre pattern model stems from various electrophysiological recordings. Nerve recordings mainly obtained from rodents and amphibians show that fibres frequently respond to stimuli of more than one taste quality, although the nerve responses to stimuli of one taste quality are especially strong. (Dahl *et al.* 1997, Yamamoto and Yuyama 1987, Pritchard and Scott 1982, Smith *et al.* 2000, Smith and St John 1999, Woolston and Erickson 1979). Moreover, recordings of taste receptor cells equally clearly demonstrate, that isolated taste receptor cells respond to stimuli of multiple taste qualities (Sato and Beidler 1997, Bealer and Smith 1975, Tonosaki and Funakoshi 1984, Herness and Gilbertson 1999, Herness 2000). In addition, non-invasive *in situ* calcium imaging of rat taste buds essentially confirmed a broad tuning of taste receptor cells (Caicedo *et al.* 2002, Caicedo and Roper 2001). In marked contrast, other data strongly argue for a labelled line model. Nerve recordings in primates revealed relatively narrowly tuned fibres for sweet, bitter and salt transduction (Danilova *et al.* 1998, 2002, Hellekant and Ninomiya 1994, Hellekant *et al.* 1998, Sato *et al.* 1994). The *Tas1r* receptors are just expressed in a subset of ~30% of the taste receptor cells (Nelson *et al.* 2001). Moreover, *Tas1r1* and *Tas1r2* do not colocalise, although they are almost always coexpressed with *Tas1r3* (Nelson *et al.* 2001, Adler *et al.* 2000, Hoon *et al.* 1999). These findings strongly argue that different subsets of taste receptor cells mediate sweet and umami taste. The observation that the *Tas2r* receptors, although expressed in ~20% of the taste receptor cells, do not colocalise with the *Tas1r* receptors (Adler *et al.* 2000, Nelson *et al.* 2001) suggests, that bitter taste is mediated by a third subpopulation of cells. In addition the salt taste receptor ENaC and the putative sour receptors HCN1 + 4 are also expressed in subsets of taste receptor cells (Kretz *et al.* 1999, Stevens *et al.* 2001). These findings are further supported by studies of taste transduction. Various signal transduction components such as gustducin, PLCβ2, and TRPM5 are expressed in subsets of taste receptor cells (McLaughlin *et al.* 1992, Zhang *et al.* 2003). Apart of a specific expression also behavioural and physiological studies of various transgenic animals also argue for specificity in taste coding. Nerve recordings and behavioural studies of PLCβ2 and TRPM5 knock out animals showed that sweet, bitter, and amino acid taste but not salt and sour taste were lost (Zhang *et al.* 2003). This demonstrates that sweet, bitter, and umami taste share some parts of their signal transduction pathways, whereas salt and sour taste use different molecules. The specificity of the signal transduction pathway is most evidently shown by a specific rescue of bitter taste. PLCβ2 knock out animals that express PLCβ2 under control of a *Tas2r* promoter selectively regain their capability to detect bitter compounds, whereas the ability to perceive sweet and umami substances is still impaired (Zhang *et al.* 2003). This strongly argues that bitter taste is mediated by *Tas2r* receptor expressing subpopulation of taste receptor cells. Analysis of *Tas1r1* and *Tas1r2* knock out mice showed that umami taste depends on the *Tas1r1* gene whereas sweet taste depends on the *Tas1r2* gene (Zhao *et al.* 2003). The *Tas1r3* knock out mice show a reduced sweet and umami taste (Zhao *et al.* 2003, Max *et al.* 2001). These findings are consistent with the specific expression pattern of the various taste receptor molecules in taste receptor cells. The expression of a receptor for the synthetic opiate spiralidone under control of the *Tas1r2* promoter resulted in a high preference of the transgenic animals for this opiate while wild type mice are indifferent to it (Zhao *et al.* 2003). This result strongly argues that sweet taste sensation is encoded by a specific subpopulation of taste receptor cells and not by the sweet receptor protein itself. Activation of any one receptor in this cell population would be perceived as sweet. Thus, a considerable amount of independent observations strongly point towards a labelled line coding of taste information in the periphery by different subtypes of taste receptor cells. This leads, of course, to the question how these results can be reconciled with the apparent broad tuning of taste receptor cells (Sato and Beidler 1997, Bealer and Smith 1975, Tonosaki and Funakoshi 1984, Herness and Gilbertson 1999, Herness 2000, Caicedo *et al.* 2002). One explanation might be based on the observation that a compound rarely has a pure taste (Skramlik 1926, Chon 1914). In many cases compounds elicit multiple taste qualities (Skramlik, 1926, Chon, 1914). For example the frequently used artificial sweetener saccharin, and other artificial sweeteners have, beside their sweet taste, a bitter aftertaste (Schiffman *et al.* 1995). Similarly, the prototype salt taste stimulus NaCl tastes sweet at low concentrations and salty at higher concentrations (Skramlik 1926). Moreover, many bitter compounds, such as the frequently used quinine, are pharmacologically active (Skramlik 1926) and might therefore influence various targets including ion channels and G-proteins in taste receptor cells. It has yet not been determined to what extent such 'unspecific' activation events occur and if such events lead to a release of neurotransmitters. Thus a certain degree of overlapping activation of taste receptor cells may be expected and does not necessarily argue against a labelled line coding. Moreover, there are clear indications that the taste receptors of different species have altered functional properties (see sections 1.3.1–1.3.5). The most drastic example might be the strong divergence of many *TAS2R* receptors in humans and rodents (Shi *et al.* 2003). This suggests strong species-specific variations in the bitter perception of humans and rodents. Such differences are also evident for salt and sweet taste (see sections 1.3.1 and 1.4.2). Thus, for some compounds, the perception in humans and rodents will drastically differ. Consequently, to some extent species differences might account for the apparent broad tuning. In addition, recent reports suggest that not all taste receptor cells might be directly connected with nerve fibres (Royer and Kinnamon 1994, Kinnamon *et al.* 1993). If this is true they must be, in an unknown fashion, connected to the so called type III cells in taste buds, which do contain synapses (Royer and Kinnamon 1994, Kinnamon *et al.* 1993). How these cells talk to each other and the implication for taste quality coding remains unknown. In summary, although much evidence argues for the labelled line model, both coding models still compete for validity. The final proof of the models depends on knowledge about the precise innervations of the cell types within taste buds and how these cells communicate. Thus transsynaptic tracing studies are necessary to elucidate how excitation of TRCs is conveyed to the brain.

1.4 Functional characterisation of taste receptors through calcium imaging

Heterologous expression of taste receptors in cell lines and calcium imaging provides a powerful tool to analyse the function of taste receptors. This system enables researchers to study the function of an isolated taste receptor that usually works on the human tongue in an artificial cell system. So far the receptors of sweet, bitter and umami taste have been studied (Behrens *et al.* 2004, Bufe *et al.* 2002, Li *et al.* 2002, Nelson *et al.* 2001, 2002, Chandrashekar *et al.* 2000, Chaudhari *et al.* 2000). Because the coexpression of the *TAS2R* receptors within the same set of taste receptor cells (Adler *et al.* 2000) prevents the measurement of single receptors, the heterologous expression of *TAS2R*s is extremely helpful to characterise the individual bitter receptors.

1.4.1 Assay principle

Plasmids containing the DNA of taste receptors are transiently transfected into HEK293 cells. Most frequently, cell lines are used that express G-proteins like $G_{\alpha 15}/G_{\alpha 16}$ that promiscuously couple with many receptors (Offermanns *et al.* 2001, Offermanns and Simon 1995). The activation of a receptor that couples to such a G-protein will result in a release of calcium from intracellular stores (Fig. 1.3). This can be monitored in real time by calcium-sensitive fluorescence dyes. Principally, two frequently used calcium imaging systems exist:

Single cell calcium imaging uses a microscope that monitors changes in intracellular calcium signals of individual cells. Based on this high resolution the system is very sensitive and due to the manual application of substances it is very flexible. Individual experiment can be performed in consecutive order, which limits the throughput to relatively few experiments a day. In addition, due to the high resolution in each experiment, typically just several dozen cells are analysed.

The fluorescence imaging plate reader (FLIPR) is the other frequently used system. Here, cells are seeded in a 96-well-plate and then analysed using a well per well resolution. In principal, up to 96 different transfections and ligand applications can be performed in parallel using a nearly fully automated system and a pipetting robot. Several dozens of 96-well-plates can be measured on the same day. Thus the main advantage is its high throughput. Moreover, the signals recorded from the wells represent the average of populations of several thousand cells. Therefore, the signals obtained in this system are very reproducible. Due to the lower resolution the system is less sensitive than single cell calcium imaging. Moreover, applied agonist cannot be removed during the experiment which limits the flexibility of the system. Thus, depending on the experimental schedule both systems are useful.

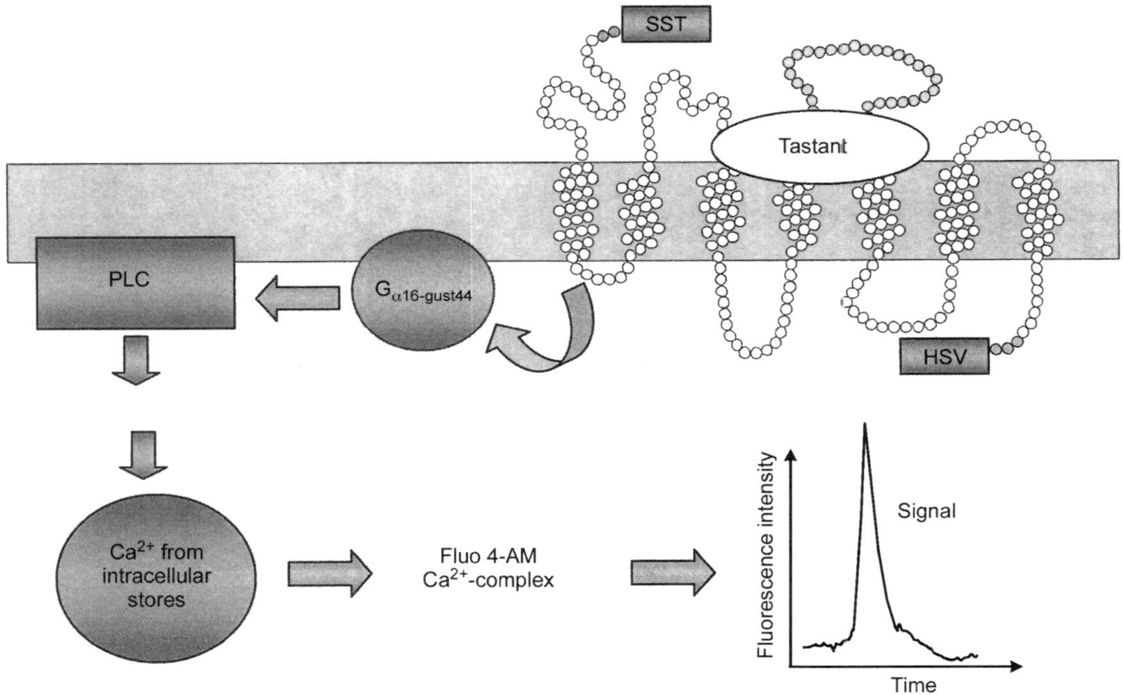

Fig. 1.3 Functional expression of *TAS2R* receptors in HEK293 G16-gust44 cells.
SST: N-terminal tag comprising the first 45 amino acid of the somatostatin receptors subtype 3 to facilitate membrane targeting. 3 HSV: C-terminal herpes simplex virus glycoprotein D tag to allow immunological detection; $G_{\alpha 16\text{-gust }44}$: Chimeric G-protein α subunit consisting of $G_{\alpha 16}$ with the last 44 amino acid are substituted by the G-protein α subunit gustducin to improve coupling of the bitter taste receptors; PLC: phospholipase C; IP_3: inositoltrisphosphate. Fluo-4-AM: calcium sensitive fluorescence dye.

1.4.2 Functional analysis of the sweet receptor *TAS1R2* and *TAS1R3*

Sweet taste is elicited by compounds of various chemical classes (Chon 1914). Natural sweeteners include sugars such as glucose and sucrose, sweet amino acids such as D-tryptophane, glycine, sweet proteins such as monellin, thaumatin, but also some other chemically quite diverse compounds including stevioside and neohesperidin dihydrochalone (Schiffman and Gatlin 1993). Most relevant sugars and sweet amino acid are low potency sweeteners (Schiffman and Gatlin 1993). This likely serves as a quantity check because only high concentrations of these compounds can indicate food sources of nutritional value (Lindemann 1996). In addition, various high potency artificial sweeteners of various chemical structures such as saccharin, cyclamate, aspartame and alitame are known (Schiffman and Gatlin 1993). Interestingly, the perception of sweet compounds varies across species. Rodents, for example, do not perceive the sweetness of the sweet proteins monellin, thaumatin, and the artificial sweeteners aspartame and cyclamate which all taste sweet to humans (Sclafani and Abrams 1986, Tonosaki *et al.* 1997, Brouwer *et al.* 1973). Based on these observations some predictions for a genuine sweet receptor can be made: first, the receptor should be activated by a multitude of chemically diverse sweeteners. Second, the receptor should show a higher affinity for artificial sweeteners than for natural sweeteners. Third, the ligand profile of the human and rodent sweet taste receptor should reflect the observed species differences. In fact, the functional analysis of the human *TAS1R2/TAS1R3* and rodent *Tas1r2/Tas1r3* heteromers showed that the receptors respond to sweeteners of multiple chemical classes (Nelson *et al.* 2001, Li *et al.* 2002). The human and the rodent receptor can be activated by the sweet mono- and disaccharides sucrose and fructose, sweet amino acids glycine and D-tryptophane, and by artificial sweeteners including saccharin and acesulfame K (Li *et al.* 2002). As predicted low concentration of the artificial sweeteners were needed for receptor activation, whereas carbohydrate sweeteners and sweet amino acids act at much higher concentrations (Nelson *et al.* 2001, Li *et al.* 2002). Notably, so far all tested compounds that are sweet to humans activate the human *TAS1R2/TAS1R3* receptor (Li *et al.* 2002). Although the number of tested compounds is still limited these results show that the *TAS1R2/TAS1R3* heteromer definitely mediates the majority of human sweet perception.

Consistent with the observed species differences the human receptor *TAS1R2/TAS1R3* but not the mouse counterpart can be activated by monellin, thaumatin, cyclamate, and aspartame (Li *et al.* 2002) which are sweet for humans but not attractive for rodents (Sclafani and Abrams 1986, Tonosaki *et al.* 1997, Brouwer *et al.* 1973). Interestingly, it could be shown that the sweet blocker lactisole abolishes the response of cells transfected with the human *TAS1R2/TAS1R3* but does not influence the response of the rat *Tas1r2/Tas1r3* to sucrose (Li *et al.* 2002). This is consistent with the observation that lactisole blocks the sweet taste in humans but not in rats (Sclafani and Perez 1997). Therefore, the observed functional differences between the human and rodent receptor support the function of *TAS1R2/TAS1R3* as a general sweet receptor.

1.4.3 Functional analysis of the umami receptor *TAS1R1* and *TAS1R3*

In humans, umami taste is elicited by glutamate but also by some agonists of metabotropic glutamate receptors like L-Apa4 and ibotenate (Kurihara and Kashiwayanagi 2000). Several 5′-ribonucleotides like inosine-5′-monophosphate (IMP) and guanosin-5′-monophosphate (GMP) strongly enhance the umami taste (Bellisle 1999, Yamaguchi and Ninomiya 2000). Therefore, these properties should be reflected by a genuine umami receptor. Indeed, functional studies showed that cells cotransfected with the human receptor *TAS1R1/TAS1R3* respond to glutamate and L-Apa4 (Li *et al.* 2002). In addition, it could be shown that IMP and GMP strongly amplify the responses of the cells to glutamate and L-Apa4 (Nelson *et al.* 2002, Li *et al.* 2002). Therefore, these studies provide the molecular explanation for the observed amplification of umami taste by 5′-ribonucleotides. They also show that the *TAS1R1/TAS1R3* receptor of humans and rodents sufficiently fulfils all functional properties that are known of umami taste. In contrast to the human *T1R1/T1R3* receptor, which is narrowly tuned to glutamate (Li *et al.* 2002), its rodent counterpart responds to a broad variety of amino acids (Nelson *et al.* 2002). It will be interesting to understand which evolutionary processes lead to different tunings of the receptors across species.

1.4.4 Functional analysis of the *TAS2R* bitter taste receptor family

Bitter tastants comprise a chemically complex family of compounds (Drewnowski 2001), belonging to a multitude of chemical classes ranging from small salts such as potassium chloride or sodium isothiocyanate to complex organic molecules such as alkaloids, polyphenols or flavonoides (Chon 1914). Thus, the perception of bitter tastants is a chemically most demanding challenge. Although humans perceive thousands of structurally very diverse compounds as bitter (Keast and Breslin 2002) the human genome only contains ~25 *TAS2R* receptors (Bufe *et al.* 2002, Shi *et al.* 2003). This raises the question whether these relatively few receptors suffice to detect all bitter tastants. The answer to this question depends on the ligand profile of these receptors. If they recognise only a few compounds, an additional family of bitter taste receptors would have to exist. Initially, functional expression and screening of various rodent and human receptors with 55 taste compounds resulted in the identification of three different receptor ligand pairs. The *Tas2r105* showed robust responses to micromolar concentrations of cycloheximide, whereas weak reactions to milimolar concentrations of denatonium benzoate were reported for *Tas2r108* and *TAS2R4* (Chandrashekar *et al.* 2000). Although these results argue for some specificity there is growing evidence that the *TAS2R* receptors are broadly tuned (Bufe *et al.* 2002, 2004, Behrens *et al.* 2004). For example, the *TAS2R14* is activated by at least eight structurally diverse agonists (Behrens *et al.* 2004). A similar broad tuning can be shown for *TAS2R38*, which is the PTC receptor (Bufe *et al.* unpublished results). So far, we identified 17 structurally diverse compounds ranging from small molecules including sodium isothiocyanate to

Table 1.1 Characterisation of the *TAS2R16* ligands

Name	Chemical structure	Threshold (mM)	EC_{50} (mM)
A: *TAS2R16* ligands with variable substitutions at the C1 position			
Naphtyl-β-D-glucoside	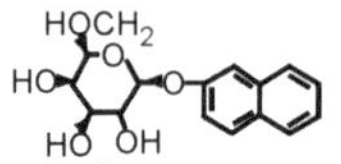	0.4 ± 0.1	1.0 ± 0.1
Phenyl-β-D-glucoside	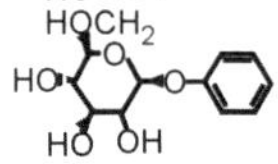	0.07 ± 0.02	1.1 ± 0.1
Salicin	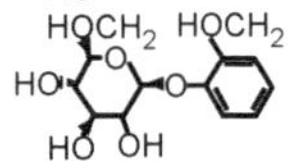	0.07 ± 0.02	1.4 ± 0.2
Helicin	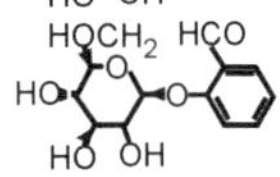	0.3 ± 0.1	2.3 ± 0.4
Arbutin	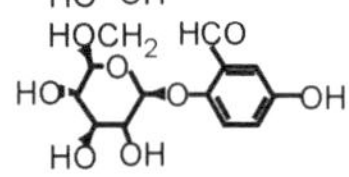	0.5 ± 0.2	5.8 ± 0.9
2-Nitrophenyl-β-D-glucoside	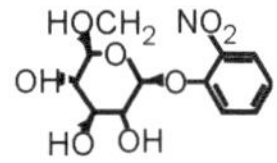	0.3–1	no data
4-Nitrophenyl-β-D-glucoside	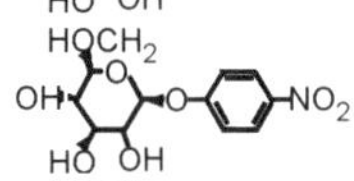	1–3	no data
Methyl-β-D-glucoside	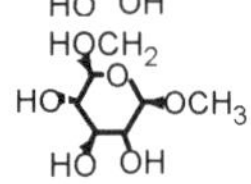	15 ± 6	no data
Esculin	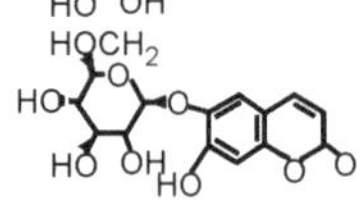	4 ± 2	no data
B: *TAS2R16* ligands with variable substitutions at the other position			
4-Nitrophenyl-β-D-thioglucoside	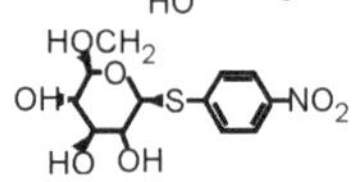	1–5	no data
4-Nitrophenyl-β-D-mannoside	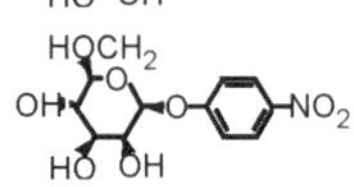	1–3	no data
Amygdalin	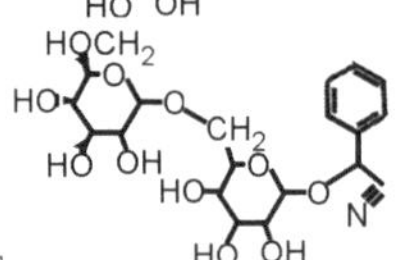	2.3 ± 0.9	20 ± 3.4
C: Structurally related compounds that cannot activate *TAS2R16*			
Phenyl-β-D-galactoside	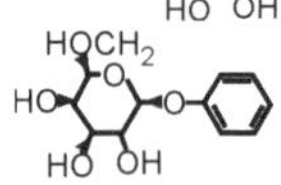	no ligand	no ligand
Phenyl-α-D-glucoside	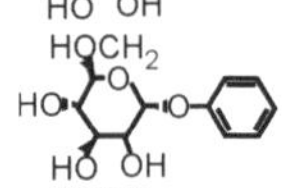	no ligand	no ligand
β-D-glucose	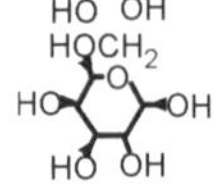	no ligand	no ligand

complex organic molecules like PROP, PTC and diphenylthiourea that can activate this receptor (Bufe *et al.* unpublished results). All agonists identified so far contain an N–C=S group as the common key structural motif and the receptor might therefore be activated by hundreds of different compounds. The receptor *TAS2R16* responded to salicin and 11 other bitter β-pyranosides (Table 1.1, (Bufe *et al.* 2002, 2004)). Comparison of the chemical structure between *TAS2R16* agonists and some chemically related compounds that do not activate the receptor allows some conclusions about the structural requirements of *TAS2R16* agonists (Fig. 1.4, (Bufe *et al.* 2002, 2004)). It is predicted that gluco- and manosides with hydrophobic substitution at C1 in the β-glycosidic configuration will be *TAS2R16* agonists. Based on this small key motive hundreds of different β-pyranosides might be able to activate the receptor. Thus, there is clear evidence for broad tuning of three human receptors. Based on these data it is reasonable to assume that this broad tuning is also typical for the other human *TAS2R* receptors. Therefore, it is possible that the ~25 different *TAS2R* receptors could detect most or even all bitter compounds. In principal, this question should be easy to address using the functional expression system. It is just necessary to express all 25 human *TAS2R* receptors and test them with compounds that humans taste bitter. If the *TAS2R*-receptors are the only bitter taste receptors, it should be possible to identify distinct receptors for all bitter tastants. Unfortunately, the functional expression of *TAS2R* receptors is still technically difficult (see also section 1.4.6) and to some extent hampered by inadequate expression and/or membrane targeting of some *TAS2R* family members (Bufe *et al.* 2004). Due to these technical problems a negative screening result with a

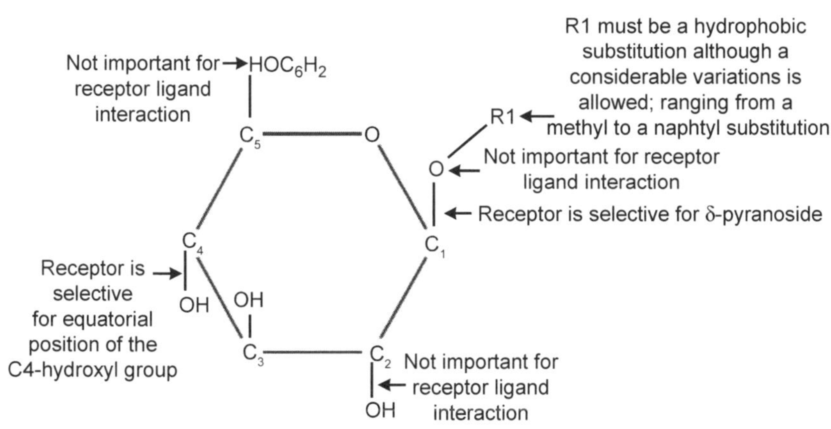

Fig. 1.4 Structural requirements of TAS2R16 agonists.
Dark grey: groups, important for receptor ligand interaction. Light grey: unimportant groups. Black: unanalysed positions.

given bitter compound does not necessarily mean that no *TAS2R* receptor for this compound exists. Therefore, the final answer to this question has to wait for improved functional assays.

1.4.5 Comparison of human taste sensations and functional expression of the taste receptors *in vitro*

The ultimate goal of taste receptor research is a better understanding of the mechanisms that lead to taste sensations in humans. Therefore, it is crucial to compare the results obtained from *in vitro* studies of taste receptors with human perception. This will help to understand which aspects of taste perception in humans are mediated at the receptor level. Some studies address these questions although currently the data are still limited to few receptors and taste compounds. Today the best examined case is the perception of salicin and other bitter β-glycosides. Nine different β-glycosides were compared for the concentration necessary to activate *TAS2R16* and for their taste threshold in humans. The observed differences were usually less than 50%. A similar good correlation was also observed between the bitter intensity and the receptor activation at higher concentrations (Bufe *et al.* 2002, 2004). This suggests that bitter taste intensities of salicin and related compounds might be encoded on the receptor level.

Interestingly, for the sweet taste receptor similar observations were made. The human threshold values for the sweeteners sucrose, D-tryptophane, aspartame and saccharin are correlated with the concentrations necessary to activate the human *TAS1R2/TAS1R3* receptor (Li *et al.* 2002). Therefore, the dose response curves obtained from the functional expression of sweet and bitter receptors might reflect the sweet and bitter intensities perceived by humans. These results appear very promising. However, due to the small set of tested compounds a fortuitous coincidence cannot be fully excluded. Further experiments with more tastants on more receptors will show if this holds true.

Human bitter taste is adaptive (Keast and Breslin 2002). This means that prolonged incubation of a bitter compound in humans results in a reduction of its bitter intensity. In addition, some bitter substances reduce the bitter intensity of other bitter compounds due to cross-adaptation. For example, the bitter tasting amino acid L-tryptophane reduces the bitter taste of L-phenylalanine (Keast and Breslin 2002). Interestingly, this cross-adaptation depended on the stimulus, because the taste of urea and quinine did not cross-adapt with that of L-tryptophan (Keast and Breslin 2002). Although the processes that lead to adaptation are yet not fully understood, it is generally believed that cross-adapting compounds share a common peripheral mechanism (Keast and Breslin 2002). The desensitisation of individual *TAS2R* receptors has been recently proposed as the molecular mechanism responsible for bitter adaptation in humans. This assumption is based on the observation that a prolonged stimulation of *TAS2R* transfected cells with cognate agonist led to a reduced response to this compound as well as to other agonists activating the same receptor (Bufe *et al.* 2002). The observed desensitisation and cross-desensitisation is well described for many other G-protein-coupled receptors (Zhang *et al.* 1997, Chuang *et al.* 1996). Here, it is usually due to a direct modification of the receptor that leads to its inactivation (Zhang *et al.* 1997, Chuang *et al.* 1996). The hypothesis that desensitisation of individual *TAS2R* receptors mediates the adaptation of humans, leads to the prediction that agonists of the same *TAS2R* receptor should show cross-adaptation in humans. In contrast, bitter compounds that do not activate the receptor should not cross-adapt. Consistent with this prediction, cross-adaptation could be shown for several *TAS2R16* agonists, whereas bitter compounds that did not activate *TAS2R16* did not cross-adapt (Bufe *et al.* 2002). Thus this result provides a basis for understanding adaptation of taste responses.

1.4.6 Current limitations of the heterologous expression system

Although heterologous expression and calcium imaging has proven to be a powerful tool to elucidate the functional properties of individual taste receptors, this system has some limitations and does not fully mimic the human tongue. On the tongue, the taste receptor cells are embedded in taste buds and present only a restricted part of their surface to the tastants (McCaughey and Scott 1998). In addition, taste receptor cells express transporters that contribute to the detoxification of compounds (Jakob *et al.* 1998). This helps to protect the cells in their native environment from osmotic pressure as well as toxic effects elicited by some tastants. In the *in vitro* expression system bath application is used and, therefore, the whole surface of the transfected cells is exposed to the tastant (Chandrashekar *et al.* 2000, Bufe *et al.* 2002, Li *et al.* 2002). The use of some tastants in the heterologous system can therefore be problematic, or even impossible, because the cells might show diminished or no responses presumably due to osmotic pressure or toxicity. Moreover, some tastants may elicit responses in mock-transfected cells in the absence of a taste receptor. Thus, more robust cell systems have to be established. Another obvious difference of the *in vitro* system is that the taste receptors on the tongue surface work in saliva, whereas in the expression system the cells are simply incubated in a physiological salt solution (Bufe *et al.* 2002, Chandrashekar *et al.* 2000). It is possible that the differences in the protein content, pH, and salt composition between saliva and the test solution influence the response of the receptors to some tastants (Schmale *et al.* 1990, 1993). Thus, the impact of the receptor environment on the receptor function needs to be determined. The expression of taste receptors in HEK293 cells also has some obstacles. In case of the *TAS2R*s, variations amongst the individual receptors with respect to expression efficacy and cell surface targeting have been observed (Bufe *et al.* 2004). Although these limitations can be compensated for to a certain extent by using high level expression cell lines such as HEK293T cells and receptor modifications that facilitate the membrane targeting (Krautwurst *et al.* 1998, Bufe *et al.* 2002), not all *TAS2R* receptors might be equally accessible for the functional analysis. In addition, the artificial signal transduction pathway that is used to couple the taste receptors to the IP_3 pathway (Fig. 1.3) might also be a problem. For the ß-adrenergic receptor and some other G-protein coupled receptors it has been shown that different G-proteins can influence the ligand specificity (Watson *et al.* 2000). Therefore, G-protein coupling might to some degree influence the functional analysis of taste receptors. The slightly different results obtained during the functional analysis of *Tas1r2/Tas1r3* by independent research groups (Nelson *et al.* 2001, Li *et al.* 2002) might be due to the different G-proteins they used (Li *et al.* 2002). Therefore, an assay that copies the native signal transduction of the taste receptors in the taste receptor cells should be established. In addition, we have to understand more about the mechanisms involved in the biosynthesis and the cells surface targeting of the *TAS2R* receptors.

1.5 Future trends

The discovery of the taste receptor genes opened several new opportunities. Comparison of the mouse and human *TAS2R* genes revealed a surprising degree of sequence divergence (Shi *et al.* 2003). The analysis of evolutionary processes, which caused this divergence will lead to a deeper understanding of the biological function of taste perception. Another aspect is the expression of taste receptors in other organs than the tongue. First studies suggested that the *TAS2R* receptors are expressed in the nasal epithelium and the gastrointestinal tract. (Wu *et al.* 2002, Finger *et al.* 2003). It will be interesting to understand the role of the *TAS2R*s in these and other tissues. In addition, the functional expression of taste receptors as well as their genetic analysis might have some impact on food design and food selection and are therefore described more in detail.

1.5.1 Taste receptor assays and their possible impact on food design

The functional expression of taste receptors is not only important for a better understanding of the molecular mechanisms of taste perception but also will enable us to predict and influence the perception of human taste compounds. Structure-activity-relations as achieved for *TAS2R16* agonists (Bufe *et al.* 2002) and molecular modelling of the ligand binding pocket of the taste receptors will be of great importance. Such models will allow rational predictions, which compounds taste bitter to humans and which do not. Beyond that, conclusions can be made as to how a compound has to be modified to avoid or reduce its bitter taste. This is potentially interesting for the pharmaceutical industry because it might help to produce less bitter medicine. In addition, the functional expression assays are also a powerful tool for the identification of new taste modifiers. Screening of substance libraries in combination with high throughput assays like the FLIPR might result in the identification of new umami or sweet taste compounds or enhancers and bitter blockers. Another promising aspect comes from the correlation of the properties of bitter and sweet taste receptors *in vitro* with human taste abilities. Notably, the threshold of activation as well as concentration response curves of *TAS2R16* showed a good correlation to the bitter perception of these compounds in humans (Bufe *et al.* 2002, 2004). A similar good correlation has been described for the activation threshold of sweeteners for *TAS1R2* and *TAS1R3* and human detection thresholds of various sweeteners (Li *et al.* 2002). If this observation holds true for other receptors, these cell-based assays might lead to the development of an artificial tongue that mimics the taste sensations of humans. This would be a useful tool for quality control in food processing.

1.5.2 Genetic variations of taste receptors and their consequences on human taste perception

PTC tasters and non-tasters differ drastically in their bitter perception of PTC but also many other N–C=S group containing compounds (Harris and Kalmus 1949, Barnicot *et al.* 1951). Some of these chemicals are natural food ingredients that occur, for example, in cruciferic vegetables such as brussel sprouts and cabbage (Drewnowski 2001) leading to rejection by the consumer. These observations support the hypothesis that food choice is influenced by individual taste sensitivity. Thus, inherited differences in taste perception could lead to different dietary habits and might therefore contribute to nutrition-related diseases. Although various research groups tried to correlate the PTC taster phenotype with differences in food liking or disliking and consumption, this issue is still far from being understood (for a review, see Prescott and Beverly 2004). To some extent these studies are hampered by technical problems (Prescott and Beverly 2004). Due to the recent discovery that variations in the bitter receptor gene *TAS2R38* mediate the PTC taster status (Kim *et al.* 2003b), a genetic tool for the determination of the PTC taster status exists. In addition, functional expression of the receptors allow the determination of a ligand profile for this receptor (Bufe *et al.* unpublished results). Based on these results prediction can be made about compounds that activate the *TAS2R38* receptor and their relevant concentrations. This could permit a better selection of foods that contain relevant concentrations of *TAS2R38* agonists. The combination of the genotypic determination of the PTC taster status and the improved knowledge about foods that might be differently perceived by PTC tasters and non-tasters will facilitate associating taster status with different dietary habits. The discovery, that polymorphisms in *TAS2R38* are responsible for the PTC taster status (Kim *et al.* 2003b) provides the first evidence that variations in taste receptors are the molecular basis for heritable variations in taste perception. Similar genetic variations have also been observed in other human *TAS2R* receptors (Ueda *et al.* 2001). Moreover, analysis of the human genome single nucleotide polymorphism data basis revealed a multitude of polymorphisms within the coding region *TAS2R* genes (Table 1.2). Many of the polymorphisms lead to amino acid exchanges, or introduce translational stop codons or frame shift mutations. This strongly suggests that there will be more cases of

Table 1.2 Single nucleotide polymorphisms in the coding region of human *TAS2R* genes

Gene name + acession number	SNP Name	Exchange		Position		Allele frequency
		Nucleotide	Amino acid	Nucleotide	Amino acid	
hTAS2R1 NM019599	rs2234231	C/T	P/L	128	43	?
	rs41469	G/A	R/H	332	111	A 0.46/G 0.54
	rs223432	G/A	C/Y	422	141	?
	rs2234233	C/T	R/W	616	206	C 0.87/T 0.13
	rs2234234	C/T	S/S	675	225	?
	rs2234235	T/C	L/L	850	284	?
hTAS2R3 NM016943	rs227009	C/T	G/G	807	369	?
hTAS2R4 NM016944	ss3181498	G/A	R/Q	8	3	?
	rs2233996	G/C	R/R	9	3	?
	rs2233997	A/C	Y/C	17	6	?
	rs2233998	T/C	F/S	20	7	?
	rs2233999	T/A	F/L	186	62	?
	rs2234000	C/T	T/M	221	74	C 0.94/T 0.56
	rs2234001	G/C	V/L	286	96	C 0.78/G 0.22
	rs2234002	G/A	S/N	512	171	A 0.78/G 0.22
	rs2234003	A/G	I/V	571	191	?
hTAS2R5 NM018980	rs2234013	G/A	G/S	58	20	?
	rs2227264	G/T	S/I	77	26	?
	rs2234014	C/T	P/L	338	113	?
	rs2234015	G/A	R/Q	638	213	?
	rs2234016	G/T	R/L	294	881	?
hTAS2R7 NM023919	rs3759251	A/T	T/S	787	263	A 0.97/T 0.03
	rs3759252	C/A	I/I	828	276	?
	rs619381	G/A	M/I	912	304	?

hTAS2R8 NM023918	ss2391467	G/A	L/L	549	183	?
	rs2537817	A/G	M/V	922	308	?
hTAS2R9 NM23917	rs3741845	T/C	V/A	560	187	C 073/T 027
	rs3944035	C/T	L/F	910	304	?
	rs2159903	C/T	P/L	926	309	?
hTAS2R10 NM23921	rs597468	C/T	T/M	467	156	?
hTAS2R13 NM23920	ss1478988	A/G	N/S	776	259	C 0.73/T 0.27
hTAS2R14 NM23922	rs3741843	G/A	R/R	375	125	A 0.97/G 0.03
hTAS2R16 NM016945	rs2233988	C/T	T/T	300	100	?
	rs2692396	G/C	V/V	303	101	?
	rs2233989	T/C	L/L	460	154	?
	rs846664	T/G	N/K	516	172	A 0.71/C 0.29
	rs860170	G/A	R/H	665	222	A 0.55/G 0.45
hTAS2R38 AF494321	rs10246939	G/A	V/I	886	296	G 0.38/A 0.62
	rs1726866	T/C	V/A	785	262	G 0.38/T 0.62
	rs713598	C/T	A/P	49	145	C 0.36/G 0.64
hTAS2R39 AF494230	–	–	–	–	–	–
hTAS2R40 AF494229	–	—	—	—	—	—
hTAS2R41 AF494232	rs1404635	A/G	T/T	189	64	?
hTAS2R42 AX097739	rs1650017	G/C	A/P	931	311	?
	rs1669411	T/C	N/N	930	310	?
	rs1669412	G/A	R/Q	875	292	?
	rs1451772	A/G	Y/C	794	265	?
	rs1669413	G/T	G/W	763	255	?
	rs1650019	A/G	L/L	561	187	?

Table 1.2 Continued

Gene name + acession number	SNP Name	Exchange		Position		Allele frequency
		Nucleotide	Amino acid	Nucleotide	Amino acid	
hTAS2R43 AF494237	rs3759246	G/C	R/T	893	298	?
hTAS2R44 AF494228	rs3759247	G/A	W/ stop	900	300	?
	rs3759246	G/C	R/T	893	298	?
hTAS2R45 AF494226	rs3759247	A/G	G/stop	900	300	?
	rs3759246	G/C	R/T	893	298	?
	rs3759245	C/T	R/C	712	238	?
	rs3759244	T/C	F/L	703	235	?
hTAS2R46 AF494227	rs2708381	G/A	W/stop	749	250	?
	rs2708380	T/A	L/M	682	228	?
	rs2598002	T/G	F/V	106	36	?
hTAS2R47 AF494233	rs2597924	G/A	R/H	920	307	?
	rs1669405	T/G	L/W	842	281	?
	rs2599404	T/G	F/L	756	252	?
	rs2600355	T/G	V/V	54	18	?
hTAS2R48 AF494234	rs1868769	T/C	L/L	418	140	?
hTAS2R50 AF494235	rs1376521	A/G	Y/C	608	203	G 0.66/ A0.34

Black: SNPs that lead to an amino acid exchange. Grey: SNPs that do not alter the amino acid composition of the *TAS2R* receptors. Underlined SNPs were observed in more that one sequence. The table refers to the NCBI Single Nucleotide Database Version 110.

genetically determined variations in taste sensitivity in the population. This conclusion is supported by the observation that individual variations for the perception of other bitter compounds exist like chloramphenicol, strychnine, quinine and cascara (Blakeslee 1935, Sugino *et al.* 2002). It appears likely that the perceptual variations of PTC tasting are just one example for many other heritable variations of bitter taste. In addition, observed differences in the perception of umami taste suggest that this could be also true for other taste qualities (Lugaz *et al.* 2002). Thus, the investigation of heritable taste variations and their impact on human diets will be a major challenge in taste research.

1.6 Acknowledgement

Work carried out in the authors' laboratory was supported by a grant from the German science foundation (DFG, Me 1024/2).

1.7 References

ADLER, E., HOON, M. A., MUELLER, K. L., CHANDRASHEKAR, J., RYBA, N. J. and ZUKER, C. S. (2000) *Cell*, 100, 693–702.

AVENET, P. and LINDEMANN, B. (1991) *J Membr Biol*, 124, 33–41.

BACHMANOV, A. A., LI, X., REED, D. R., OHMEN, J. D., LI, S., CHEN, Z., TORDOFF, M. G., DE JONG, P. J., WU, C., WEST, D. B., CHATTERJEE, A., ROSS, D. A. and BEAUCHAMP, G. K. (2001) *Chem Senses*, 26, 925–33.

BARNICOT, N. A., HARRIS, H. and KALMUS, H. (1951) *Ann. Eugen*, 16, 119–28.

BEALER, S. L. and SMITH, D. V. (1975) *Physiol Behav*, 14, 795–9.

BEHRENS, M., BROCKHOFF, A., KUHN, C., BUFE, B., WINNIG, M. and MEYERHOF, W. (2004) *Biochem Biophys Res Commun*, 319, 479–85.

BELLISLE, F. (1999) *Neurosci Biobehav Rev*, 23, 423–38.

BENOS, D. J. and STANTON, B. A. (1999) *J Physiol*, 520 Pt 3, 631–44.

BLAKE, B. L., WING, M. R., ZHOU, J. Y., LEI, Q., HILLMANN, J. R., BEHE, C. I., MORRIS, R. A., HARDEN, T. K., BAYLISS, D. A., MILLER, R. J. and SIDEROVSKI, D. P. (2001) *J Biol Chem*, 276, 49267–74.

BLAKESLEE, A. (1935) *Science*, 81, 504–7.

BLAKESLEE, A. F. (1931) *Science*, 74, 607.

BLAKESLEE, A. F. (1932) *Proc Natl Acad Sci USA*, 18, 120–30.

BROUWER, J. N., HELLEKANT, G., KASAHARA, Y., VAN DER WEL, H. and ZOTTERMAN, Y. (1973) *Acta Physiol Scand*, 89, 550–7.

BUFE, B., HOFMANN, T., KRAUTWURST, D., RAGUSE, J. D. and MEYERHOF, W. (2002) *Nat Genet*, 32, 397–401.

BUFE, B., SCHÖLEY-POHL, E., KRAUTWURST, D., HOFMANN, T. and MEYERHOF, W. (2004) *ACS Symposium Series*, 45–49.

CAICEDO, A., JAFRI, M. S. and ROPER, S. D. (2000a) *J Neurosci*, 20, 7978–85.

CAICEDO, A., KIM, K. N. and ROPER, S. D. (2000b) *J Comp Neurol*, 417, 315–24.

CAICEDO, A., KIM, K. N. and ROPER, S. D. (2002) *J Physiol*, 544, 501–9.

CAICEDO, A. and ROPER, S. D. (2001) *Science*, 291, 1557–60.

CAPELESS, C. G. and WHITNEY, G. (1995) *Chem Senses*, 20, 291–8.

CHANDRASHEKAR, J., MUELLER, K. L., HOON, M. A., ADLER, E., FENG, L., GUO, W., ZUKER, C. S. and RYBA, N. J. (2000) *Cell*, 100, 703–11.

CHAUDHARI, N., LANDIN, A. M. and ROPER, S. D. (2000) *Nat Neurosci*, 3, 113–19.

CHAUDHARI, N. and ROPER, S. D. (1998) *Ann N Y Acad Sci*, 855, 398–406.

CHAUDHARI, N., YANG, H., LAMP, C., DELAY, E., CARTFORD, C., THAN, T. and ROPER, S. (1996) *J Neurosci*, 16, 3817–26.

CHON, G. (1914) *Die Organischen Geschmacksstoffe,* Franz Seimenroth, Berlin.

CHUANG, T. T., IACOVELLI, L., SALLESE, M. and DE BLASI, A. (1996) *Trends Pharmacol Sci*, 17, 416–21.

DAHL, M., ERICKSON, R. P. and SIMON, S. A. (1997) *Brain Res*, 756, 22–34.

DAMAK, S., RONG, M., YASUMATSU, K., KOKRASHVILI, Z., VARADARAJAN, V., ZOU, S., JIANG, P., NINOMIYA, Y. and MARGOLSKEE, R. F. (2003) *Science*, 301, 850–3.

DANIELS, D. and FLUHARTY, S. J. (2004) *Physiol Behav*, 81, 319–37.

DANILOVA, V., DANILOV, Y., ROBERTS, T., TINTI, J. M., NOFRE, C. and HELLEKANT, G. (2002) *J Neurophysiol*, 88, 579–94.

DANILOVA, V., HELLEKANT, G., TINTI, J. M. and NOFRE, C. (1998) *J Neurophysiol*, 80, 2102–12.

DELWICHE, J. F., BULETIC, Z. and BRESLIN, P. A. (2001) *Percept Psychophys*, 63, 761–76.

DEMEESTER, T. R., SKINNER, D. B., EVANS, R. H. and BENSON, D. W. (1977) *Ann Thorac Surg*, 24, 278–83.

DESIMONE, J. A., LYALL, V., HECK, G. L., PHAN, T. H., ALAM, R. I., FELDMAN, G. M. and BUCH, R. M. (2001) *J Neurophysiol*, 86, 2638–41.

DOOLIN, R. E. and GILBERTSON, T. A. (1996) *J Gen Physiol*, 107, 545–54.

DRAYNA, D., COON, H., KIM, U. K., ELSNER, T., CROMER, K., OTTERUD, B., BAIRD, L., PEIFFER, A. P. and LEPPERT, M. (2003) *Hum Genet*, 112, 567–72.

DREWNOWSKI, A. (1995) *Am J Clin Nutr*, 62, 1081S–5S.

DREWNOWSKI, A. (1997) *Annu Rev Nutr*, 17, 237–53.

DREWNOWSKI, A. (2001) *Nutr Rev*, 59, 163–9.

FINGER, T. E., BOTTGER, B., HANSEN, A., ANDERSON, K. T., ALIMOHAMMADI, H. and SILVER, W. L. (2003) *Proc Natl Acad Sci USA*, 100, 8981–6.

FORMAKER, B. K. and HILL, D. L. (1991) *Physiol Behav*, 50, 765–9.

FOX, A. L. (1932) *Proc Nat Acad Sci USA*, 18, 115–20.

FULLER, J. L. (1974) *J Hered*, 65, 33–6.

GANZEVLES, P. G. and KROEZE, J. H. (1987) *Physiol Behav*, 40, 641–6.

GILBERTSON, T. A., AVENET, P., KINNAMON, S. C. and ROPER, S. D. (1992) *J Gen Physiol*, 100, 803–24.

GILBERTSON, T. A., ROPER, S. D. and KINNAMON, S. C. (1993) *Neuron*, 10, 931–42.

HäNING, D. P. (1901) *Philosophische Studien*, 17, 576–623.

HARRIS, H. and KALMUS, H. (1949) *Ann Eugen*, 15, 32–45.

HECK, G. L., MIERSON, S. and DESIMONE, J. A. (1984) *Science*, 223, 403–5.

HELLEKANT, G. and NINOMIYA, Y. (1994) *Physiol Behav*, 56, 1185–8.

HELLEKANT, G., NINOMIYA, Y. and DANILOVA, V. (1998) *Physiol Behav*, 65, 191–200.

HERNESS, M. S. (1992) *Comp Biochem Physiol Comp Physiol*, 103, 269–73.

HERNESS, M. S. and GILBERTSON, T. A. (1999) *Annu Rev Physiol*, 61, 873–900.

HERNESS, S. (2000) *Physiol Behav*, 69, 17–27.

HOFMANN, T., CHUBANOV, V., GUDERMANN, T. and MONTELL, C. (2003) *Curr Biol*, 13, 1153–8.

HOON, M. A., ADLER, E., LINDEMEIER, J., BATTEY, J. F., RYBA, N. J. and ZUKER, C. S. (1999) *Cell*, 96, 541–51.

HUANG, L., SHANKER, Y. G., DUBAUSKAITE, J., ZHENG, J. Z., YAN, W., ROSENZWEIG, S., SPIELMAN, A. I., MAX, M. and MARGOLSKEE, R. F. (1999) *Nat Neurosci*, 2, 1055–62.

JAKOB, I., HAUSER, I. A., THEVENOD, F. and LINDEMANN, B. (1998) *Am J Physiol*, 274, C182–91.

KEAST, R. S. and BRESLIN, P. A. (2002) *Chem Senses*, 27, 123–31.

KELLENBERGER, S., AUBERSON, M., GAUTSCHI, I., SCHNEEBERGER, E. and SCHILD, L. (2001) *J Gen Physiol*, 118, 679–92.

KELLENBERGER, S., GAUTSCHI, I. and SCHILD, L. (1999) *Proc Natl Acad Sci USA*, 96, 4170–5.

KIM, K. N., CAICEDO, A. and ROPER, S. D. (2001) *Neuroreport*, 12, 1715–18.

KIM, M. R., KUSAKABE, Y., MIURA, H., SHINDO, Y., NINOMIYA, Y. and HINO, A. (2003a) *Biochem Biophys Res Commun*, 312, 500–6.

KIM, U. K., JORGENSON, E., COON, H., LEPPERT, M., RISCH, N. and DRAYNA, D. (2003b) *Science*, 299, 1221–5.

KIM, U. K., BRESLIN, P. A., REED, D. and DRAYNA, D. (2004) *J Dent Res*, 83, 448–53.

KINNAMON, J. C., HENZLER, D. M. and ROYER, S. M. (1993) *Microsc Res Tech*, 26, 142–56.

KINNAMON, S. C., DIONNE, V. E. and BEAM, K. G. (1988) *Proc Natl Acad Sci USA*, 85, 7023–7.

KINNAMON, S. C., PRICE, M.P., STONE, L.M., LIN, W. and WELSH, M.J. (2000) *XII International Symposium on Olfaction and Taste (Brighton, UK)*, 104.

KITAGAWA, M., KUSAKABE, Y., MIURA, H., NINOMIYA, Y. and HINO, A. (2001) *Biochem Biophys Res Commun*, 283, 236–42.

KRAUTWURST, D., YAU, K. W. and REED, R. R. (1998) *Cell*, 95, 917–26.

KRETZ, O., BARBRY, P., BOCK, R. and LINDEMANN, B. (1999) *J Histochem Cytochem*, 47, 51–64.

KURIHARA, K. and KASHIWAYANAGI, M. (2000) *J Nutr*, 130, 931S–4S.

LI, X., STASZEWSKI, L., XU, H., DURICK, K., ZOLLER, M. and ADLER, E. (2002) *Proc Natl Acad Sci USA*, 99, 4692–6.

LIAO, J. and SCHULTZ, P. G. (2003) *Mamm Genome*, 14, 291–301.

LIN, W., FINGER, T. E., ROSSIER, B. C. and KINNAMON, S. C. (1999) *J Comp Neurol*, 405, 406–20.

LIN, W. and KINNAMON, S. C. (1999) *J Neurophysiol*, 82, 2061–9.

LIN, W., OGURA, T. and KINNAMON, S. C. (2002) *J Neurophysiol*, 88, 133–41.

LINDEMANN, B. (1996) *Physiol Rev*, 76, 718–66.

LINDEMANN, B. (1997) *Curr Opin Nephrol Hypertens*, 6, 425–9.

LINDEMANN, B. (1999) *Nat Med*, 5, 381–2.

LINDEMANN, B. (2001) *Nature*, 413, 219–25.

LINDEN, R. W. (1993) *Br Dent J*, 175, 243–53.

LIU, D. and LIMAN, E. R. (2003) *Proc Natl Acad Sci USA*, 100, 15160–5.

LUGAZ, O., PILLIAS, A. M. and FAURION, A. (2002) *Chem Senses*, 27, 105–15.

LUSH, I. E. (1981) *Genet Res*, 38, 93–5.

LUSH, I. E. (1984) *Genet Res*, 44, 151–60.

LUSH, I. E. (1986) *Genet Res*, 47, 117–23.

LUSH, I. E. (1989) *Genet Res*, 53, 95–9.

LUSH, I. E. and HOLLAND, G. (1988) *Genet Res*, 52, 207–12.

LUSH, I. E., HORNIGOLD, N., KING, P. and STOYE, J. P. (1995) *Genet Res*, 66, 167–74.

LYALL, V., HECK, G. L., VINNIKOVA, A. K., GHOSH, S., PHAN, T. H., ALAM, R. I., RUSSELL, O. F., MALIK, S. A., BIGBEE, J. W. and DESIMONE, J. A. (2004) *J Physiol*, 558, 147–59.

MATSUNAMI, H., MONTMAYEUR, J. P. and BUCK, L. B. (2000) *Nature*, 404, 601–4.

MAX, M., SHANKER, Y. G., HUANG, L., RONG, M., LIU, Z., CAMPAGNE, F., WEINSTEIN, H., DAMAK, S. and MARGOLSKEE, R. F. (2001) *Nat Genet*, 28, 58–63.

MCCAUGHEY, S. A. and SCOTT, T. R. (1998) *Neurosci Biobehav Rev*, 22, 663–76.

MCLAUGHLIN, S. K., MCKINNON, P. J. and MARGOLSKEE, R. F. (1992) *Nature*, 357, 563–9.

MCLAUGHLIN, S. K., MCKINNON, P. J., SPICKOFSKY, N., DANHO, W. and MARGOLSKEE, R. F.

(1994) *Physiol Behav*, 56, 1157–64.

MING, D., RUIZ-AVILA, L. and MARGOLSKEE, R. F. (1998) *Proc Natl Acad Sci USA*, 95, 8933–8.

MIWA, K., KANEMURA, F. and TONOSAKI, K. (1997) *J Vet Med Sci*, 59, 81–3.

MIYAMOTO, T., FUJIYAMA, R., OKADA, Y. and SATO, T. (2000) *Prog Neurobiol*, 62, 135–57.

MONTMAYEUR, J. P., LIBERLES, S. D., MATSUNAMI, H. and BUCK, L. B. (2001) *Nat Neurosci*, 4, 492–8.

NAIM, M., SEIFERT, R., NURNBERG, B., GRUNBAUM, L. and SCHULTZ, G. (1994) *Biochem J*, 297 (Pt 3), 451–4.

NAIM, M., STRIEM, B. J. and TAL, M. (1998) *Adv Food Nutr Res*, 42, 211–43.

NAKASHIMA, K. and NINOMIYA, Y. (1998) *Cell Physiol Biochem*, 8, 224–30.

NELSON, G., CHANDRASHEKAR, J., HOON, M. A., FENG, L., ZHAO, G., RYBA, N. J. and ZUKER, C. S. (2002) *Nature*, 416, 199–202.

NELSON, G., HOON, M. A., CHANDRASHEKAR, J., ZHANG, Y., RYBA, N. J. and ZUKER, C. S. (2001) *Cell*, 106, 381–90.

OFFERMANNS, S., NEGULESCU, P., HU, Y. H. and SIMON, M. I. (2001) *Naunyn Schmiedebergs Arch Pharmacol*, 364, 140–8.

OFFERMANNS, S. and SIMON, M. I. (1995) *J Biol Chem*, 270, 15175–80.

OKADA, Y., MIYAMOTO, T. and SATO, T. (1990) *Comp Biochem Physiol A*, 97, 535–6.

OSSEBAARD, C. A. and SMITH, D. V. (1995) *Chem Senses*, 20, 37–46.

OZECK, M., BRUST, P., XU, H. and SERVANT, G. (2004) *Eur J Pharmacol*, 489, 139–49.

PEREZ, C. A., HUANG, L., RONG, M., KOZAK, J. A., PREUSS, A. K., ZHANG, H., MAX, M. and MARGOLSKEE, R. F. (2002) *Nat Neurosci*, 5, 1169–76.

PERI, I., MAMRUD-BRAINS, H., RODIN, S., KRIZHANOVSKY, V., SHAI, Y., NIR, S. and NAIM, M. (2000) *Am J Physiol Cell Physiol*, 278, C17–25.

PRAWITT, D., MONTEILH-ZOLLER, M. K., BRIXEL, L., SPANGENBERG, C., ZABEL, B., FLEIG, A. and PENNER, R. (2003) *Proc Natl Acad Sci USA*, 100, 15166–71.

PRESCOTT, J. and BEVERLY, J. T. (eds) (2004) *Genetic Variations in taste sensitivity,* Marcel Dekker Inc., New York.

PRITCHARD, T. C. and SCOTT, T. R. (1982) *Brain Res*, 253, 93–104.

REED, D. R., LI, S., LI, X., HUANG, L., TORDOFF, M. G., STARLING-RONEY, R., TANIGUCHI, K., WEST, D. B., OHMEN, J. D., BEAUCHAMP, G. K. and BACHMANOV, A. A. (2004) *J Neurosci*, 24, 938–46.

REED, D. R., NANTHAKUMAR, E., NORTH, M., BELL, C., BARTOSHUK, L. M. and PRICE, R. A. (1999) *Am J Hum Genet*, 64, 1478–80.

RICHTER, T. A., DVORYANCHIKOV, G. A., ROPER, S. D. and CHAUDHARI, N. (2004) *J Neurosci*, 24, 4088–91.

ROSSLER, P., BOEKHOFF, I., TAREILUS, E., BECK, S., BREER, H. and FREITAG, J. (2000) *Chem Senses*, 25, 413–21.

ROYER, S. M. and KINNAMON, J. C. (1994) *Microsc Res Tech*, 29, 381–407.

RUIZ-AVILA, L., MCLAUGHLIN, S. K., WILDMAN, D., MCKINNON, P. J., ROBICHON, A., SPICKOFSKY, N. and MARGOLSKEE, R. F. (1995) *Nature*, 376, 80–5.

SAINZ, E., KORLEY, J. N., BATTEY, J. F. and SULLIVAN, S. L. (2001) *J Neurochem*, 77, 896–903.

SAKO, N., HARADA, S. and YAMAMOTO, T. (2000) *Physiol Behav*, 71, 193–8.

SATO, M., OGAWA, H. and YAMASHITA, S. (1994) *Chem Senses*, 19, 381–400.

SATO, T. and BEIDLER, L. M. (1997) *Chem Senses*, 22, 287–93.

SCHIFFMAN, S. S., BOOTH, B. J., LOSEE, M. L., PECORE, S. D. and WARWICK, Z. S. (1995) *Brain Res Bull*, 36, 505–13.

SCHIFFMAN, S. S. and GATLIN, C. A. (1993) *Neurosci Biobehav Rev*, 17, 313–45.

SCHMALE, H., AHLERS, C., BLAKER, M., KOCK, K. and SPIELMAN, A. I. (1993) *Ciba Found Symp*,

179, 167–80; discussion 180–5.

SCHMALE, H., HOLTGREVE-GREZ, H. and CHRISTIANSEN, H. (1990) *Nature*, 343, 366–9.

SCLAFANI, A. and ABRAMS, M. (1986) *Physiol Behav*, 37, 253–6.

SCLAFANI, A. and PEREZ, C. (1997) *Physiol Behav*, 61, 25–9.

SHI, P., ZHANG, J., YANG, H. and ZHANG, Y. P. (2003) *Mol Biol Evol*, 20, 805–14.

SKRAMLIK, E. (1926) *Handbuch der Physiologie der Niederen Sinne,* Georg Thieme Verlag, Leipzig.

SMITH, D. V., JOHN, S. J. and BOUGHTER, J. D. (2000) *Physiol Behav*, 69, 77–85.

SMITH, D. V. and ST JOHN, S. J. (1999) *Curr Opin Neurobiol*, 9, 427–35.

SPIELMAN, A. I., HUQUE, T., NAGAI, H., WHITNEY, G. and BRAND, J. G. (1994) *Physiol Behav*, 56, 1149–55.

SPIELMAN, A. I., NAGAI, H., SUNAVALA, G., DASSO, M., BREER, H., BOEKHOFF, I., HUQUE, T., WHITNEY, G. and BRAND, J. G. (1996) *Am J Physiol*, 270, C926–31.

STEVENS, D. R., SEIFERT, R., BUFE, B., MULLER, F., KREMMER, E., GAUSS, R., MEYERHOF, W., KAUPP, U. B. and LINDEMANN, B. (2001) *Nature*, 413, 631–5.

SUGINO, Y., UMEMOTO, A. and MIZUTANI, S. (2002) *Genes Genet Syst*, 77, 59–62.

TAKAMI, S., GETCHELL, T. V., MCLAUGHLIN, S. K., MARGOLSKEE, R. F. and GETCHELL, M. L. (1994) *Brain Res Mol Brain Res*, 22, 193–203.

TONOSAKI, K. and FUNAKOSHI, M. (1984) *Comp Biochem Physiol A*, 78, 651–6.

TONOSAKI, K., MIWA, K. and KANEMURA, F. (1997) *Brain Res*, 748, 234–6.

TOYONO, T., SETA, Y., KATAOKA, S., KAWANO, S., SHIGEMOTO, R. and TOYOSHIMA, K. (2003) *Cell Tissue Res*, 313, 29–35.

UEDA, T., UGAWA, S., ISHIDA, Y., SHIBATA, Y., MURAKAMI, S. and SHIMADA, S. (2001) *Biochem Biophys Res Commun*, 285, 147–51.

UEDA, T., UGAWA, S., YAMAMURA, H., IMAIZUMI, Y. and SHIMADA, S. (2003) *J Neurosci*, 23, 7376–80.

UGAWA, S. (2003) *Anat Sci Int*, 78, 205–10.

UGAWA, S., MINAMI, Y., GUO, W., SAISHIN, Y., TAKATSUJI, K., YAMAMOTO, T., TOHYAMA, M. and SHIMADA, S. (1998) *Nature*, 395, 555–6.

UGAWA, S., YAMAMOTO, T., UEDA, T., ISHIDA, Y., INAGAKI, A., NISHIGAKI, M. and SHIMADA, S. (2003) *J Neurosci*, 23, 3616–22.

WATSON, C., CHEN, G., IRVING, P., WAY, J., CHEN, W. J. and KENAKIN, T. (2000) *Mol Pharmacol*, 58, 1230–8.

WONG, G. T., GANNON, K. S. and MARGOLSKEE, R. F. (1996) *Nature*, 381, 796–800.

WOODING, S., KIM, U. K., BAMSHAD, M. J., LARSEN, J., JORDE, L. B. and DRAYNA, D. (2004) *Am J Hum Genet*, 74, 637–46.

WOOLSTON, D. C. and ERICKSON, R. P. (1979) *J Neurophysiol*, 42, 1390–409.

WU, S. V., ROZENGURT, N., YANG, M., YOUNG, S. H., SINNETT-SMITH, J. and ROZENGURT, E. (2002) *Proc Natl Acad Sci USA*, 99, 2392–7.

YAMAGUCHI, S. and NINOMIYA, K. (2000) *J Nutr*, 130, 921S–6S.

YAMAMOTO, T. and YUYAMA, N. (1987) *Brain Res*, 400, 312–20.

YAN, W., SUNAVALA, G., ROSENZWEIG, S., DASSO, M., BRAND, J. G. and SPIELMAN, A. I. (2001) *Am J Physiol Cell Physiol*, 280, C742–51.

ZHANG, J., FERGUSON, S. S., BARAK, L. S., ABER, M. J., GIROS, B., LEFKOWITZ, R. J. and CARON, M. G. (1997) *Receptors Channels*, 5, 193–9.

ZHANG, Y., HOON, M. A., CHANDRASHEKAR, J., MUELLER, K. L., COOK, B., WU, D., ZUKER, C. S. and RYBA, N. J. (2003) *Cell*, 112, 293–301.

ZHAO, G. Q., ZHANG, Y., HOON, M. A., CHANDRASHEKAR, J., ERLENBACH, I., RYBA, N. J. and ZUKER, C. S. (2003) *Cell*, 115, 255–66.

2

Processing information about flavour

A. Holley, Centre Européen des Sciences du Goût, France

2.1 Introduction

Flavours are chemical sensations elicited by a vast number of molecules released by food during eating. Several sensory systems are involved in flavour detection depending on physicochemical properties of stimulating molecules such as volatility and lipid *vs* water solubility, along with receptive properties of different detectors present in the nose and oral cavity. Three chemosensory modalities, olfactory, gustatory and trigeminal, contribute to flavour perception. They are distinct regarding their respective biological substrates but are hardly distinguishable in the consumer's experience. In everyday life, human subjects are generally not aware of the origins of chemical sensations experienced during eating, justifying the use of the global category of 'flavour' ('goût' in French). Even in experimental conditions, while olfactory and gustatory stimuli can often be accurately distinguished, it is not easy to operate the same clear-cut distinction between trigeminal and both olfactory and gustatory stimuli. While consumers do not confuse the visual image of an aliment and its acoustic properties, why do they fail to distinguish between sensations that are elicited by distinct sensory substrates in the chemical domain? The reason is the existence of multimodal interactions. Some of these interactions occur at a low level, close to the reception stage; others result from the convergence of different sensory messages to one or several common integrative areas, at higher levels in the brain. This chapter is aimed at describing neurobiological substrates of chemosensory perception. A special emphasis will be placed on those anatomical features and processes that favour interactions between sensory modalities or submodalities and promote flavour integration as experienced in food consumption and evidenced in psychophysical measurements.

2.2 Reception of odorants and neural processing of olfactory information

2.2.1 Receptor cells, receptor molecules

Whichever their route, orthonasal or retronasal, to the sensory area in the nose, volatile molecules interact with the olfactory epithelium where sensory cells have receptors to which odorants can bind reversibly as a first step towards the generation of an electrical signal. Sensory cells are neurons. A receptor neuron has a dendritic pole bearing fine cilia immersed in the nasal mucous. The ciliary membrane hosts receptor macromolecules. The cell possesses an axon that projects to the olfactory bulb and conveys electrical signals elicited by receptor activation. In the olfactory bulb, axons synapse with second order neurons that, in turn, project to the primary olfactory cortex. From there, the olfactory message is sent to many other areas in the brain.

Odorant receptors are members of a super family of G protein-coupled receptors (GPCR) (Buck and Axel 1991). All have the same general structure with seven hydrophobic membrane-spanning regions. They differ regarding their respective amino acid sequences which are especially variable in the third, fourth and fifth regions of the protein (Mombaerts 1999). It is generally thought that these 'hypervariable' regions determine an odorant-binding pocket where ligand–receptor interactions take place (Singer and Shepherd 1994). Receptor structure and properties have not been determined directly but are mostly inferred from the examination of coding genes. These genes are surprisingly numerous and diverse, and this is obviously related to the capacity of animals to discriminate among a very large number of odorants. The olfactory genes constitute the largest gene family in the vertebrate genome (Glusman *et al.* 2001). The mouse genome contains more than one thousand olfactory genes (Zhang and Firestein 2002). In humans, the number of genes is smaller and, in addition, a large fraction of them – around 60% – appear to be non-functional (Rouquier *et al.* 1998), they are 'pseudogenes'. This drastic decrease in the number of functional genes and the corresponding reduction in the number of odorant receptors is an evolutionary process that has also been observed in Old World monkeys and anthropoid primates (Rouquier *et al.* 2000, Gilad *et al.* 2004).

Olfactory transduction that is the conversion of odorant–receptor binding energy into electrical signal involves a cascade of enzymatic reactions starting with the activation of a G protein (an olfactory-specific subtype, G_{olf} (Firestein 2001)). Although two transduction cascades were originally proposed involving production of either cAMP or IP3, it seems that only one pathway, that leading to production of the cyclic nucleotide cAMP, is implied in transduction *per se* in mammals. Odorant binding to an appropriate receptor causes G protein activation that in turn activates an adenylyl cyclase (ACIII), resulting in the conversion of ATP into cAMP. The nucleotide directly controls the permeability of cyclic nucleotide-gated ionic channels (CNGC) in the cilia membrane. Cyclic AMP-increased concentration results in a depolarising influx of Na^+ and Ca^{2+}

ions through open channels. The increase in Ca^{2+} intracellular concentration induces activation of Ca^{2+}-activated Cl^- channels and a Cl^- current flows outward, contributing to further depolarisation of the cell membrane. This receptor/generator potential spreads passively to the receptor cell axon, triggering a regenerative action potential or a train of action potentials, depending on stimulation intensity (Trotier and MacLeod 1983). Because several stages of the enzymatic cascade are highly multiplicative, the whole transduction process considerably amplifies the energy transfer represented by each initial binding event. Besides, the small-sized receptor cell exhibits a high input resistance. As a consequence, a small number of odorant detection events, that is the binding of a few molecules, are liable to drive the cell membrane potential to the threshold for action potential generation (Menini *et al.* 1995). The high sensibility of the olfactory system depends on these properties.

It is notorious that olfactory responses adapt to continuous stimulation. Adaptation takes place at both peripheral and central levels. In receptor cells, the adaptation process appears as a period of reduced responsiveness following an initial response. This process involves desensitisation of cAMP-gated ionic channels by Ca^{2+}, in association with calmodulin (Kurashi and Menini 1997).

Individual responsiveness of sensory neurons has been extensively explored by means of electrophysiological recordings from individual neurons (Gesteland *et al.* 1965, Holley *et al.* 1974, Sicard and Holley 1984, Duchamp-Viret *et al.* 1999). In general, sensory cells show a low selectivity, often responding to a substantial proportion of pure odorants used as stimuli and classes of sensory cells could not be demonstrated on the basis of their response spectra. A cell can respond to many odorants and an odorant can stimulate many cells. Two alternative or associated possibilities have been proposed to explain this low selectivity. Either sensory neurons are individually equipped with several types of selective receptor or they contain a single type of receptor that have a broad selectivity. Single cell recordings could not provide a definitive answer. More recently, analysis of receptor expression in single olfactory neurons, along with *in situ* hybridisation studies allowed their authors to conclude that sensory neurons express only one type of receptors (Mombaerts 1999). However, divergent opinions have been expressed recently (Rawson *et al.* 2000, Mombaerts 2004). Individual receptor selectivity is still poorly documented. Attempts to discover what odorants activate each receptor have been rarely successful because the most appropriate methodology to explore this question, that is the expression of olfactory receptor genes in heterologous cells, revealed to be a difficult task, presumably in part because synthesised proteins are retained in the endoplasmic reticulum and are degraded (Lu *et al.* 2004). Other techniques provided results that confirmed receptor multiple sensibility (Zhao *et al.* 1998, Malnic *et al.* 1999) but the sample of receptors that could be explored is not large enough to allow firm conclusions as to the mean receptor selectivity and extent of its variations. It is possible that receptors with high selectivity coexist with receptors of much lower selectivity. Under this respect, it can be noted that selective anosmia, that is a lack of responsiveness to a limited list of odorants, could hardly be interpreted in terms of receptor absence or disruption if all odorants were exclusively detected by multiple, poorly selective receptors.

2.2.2 Odour (and aroma) representation in the olfactory epithelium

There is a consensus to see odour representation in the olfactory epithelium as a combinatorial process (Holley and Døving 1977, Malnic *et al.* 1999). Even a chemically pure odorant induces the activation of a set of several receptors of different specificity according to a unique combination. Odour identity is determined by the whole pattern of activated receptors although different odours can activate several common receptor types. This combinatorial code provides the olfactory system with its capacity of encoding a huge number of odours. The notion that each sensory neuron expresses only one receptor implies that an odour is encoded by a unique combination of excited sensory neurons, which has been expressed as the across-fibre pattern hypothesis, a notion previously elaborated to account for taste findings (see Erickson (2000) for a review). Accordingly, the task of the higher stages of the olfactory system is to process the message of the olfactory nerve in order to operate pattern discrimination and recognition. It has been assumed, in addition, that odour coding patterns had a spatial dimension because receptor neurons of different specificity were found to be heterogeneously distributed in the olfactory epithelium (Holley and Døving 1977). It was further demonstrated (Ressler *et al.* 1993, Vassar *et al.* 1993) that the olfactory epithelium was divided into four distinct, non-overlapping zones of receptor gene expression, but within a zone sensory cells expressing the same odorant receptor are randomly distributed. The functional meaning of this spatial organisation is not clear but it could prepare some functional spatial specialisation of the olfactory bulb as far as it is maintained in the olfactory nerve projection.

2.2.3 Odour maps in the olfactory bulb and subsequent cortical projections

Receptor cell axons collected by several branches of the olfactory nerve project to the olfactory bulbs (OB) in synaptically dense, discrete, spherical structures called glomeruli. These structures occupy locations which are fairly well conserved among individuals. One of the most noticeable features of the organisation of the peripheral projection is that axons of all receptor neurons expressing the same odorant receptor converge exclusively onto only one or more often two glomeruli (Ressler *et al.* 1994, Vassar *et al.* 1994). The number of glomeruli is therefore roughly twice that of receptor types, that is about 2000 in mice. Due to these selective projections, each individual glomerulus represents one receptor type and therefore one type of odour sensitivity present in the olfactory epithelium. In glomeruli, convergent axons make excitatory synapses with dendrites of a much lower number of second order projection neurons, mitral cells and tufted cells. The olfactory message transmitted to the primary olfactory cortex by mitral cell axons is shaped by the action of two main classes of local inhibitory neurons, periglomerular cells and granule cells. These cells are inserted into local feedback loops that stabilise mitral cell activity and tend to maximise the contrast between the activity levels of adjacent glomeruli through lateral inhibition. The organisation of the OB can be interpreted as based on a collection of micro modules or protocolumns that operate as functional units. Each module consists of a glomerulus, its selectively sensitive afferents, a population of output neurons and two populations of local neurons.

Several methods have been used in order to explore the neural activity elicited by olfactory stimulation in the OB: single unit electrical recordings (Chaput and Holley 1985, Mori *et al.* 1999), 2-deoxyglucose mapping of local metabolic activity (Sharp *et al.* 1975, Jourdan *et al.* 1980), voltage sensitive dyes (Cinelli and Kauer 1995) and dynamic functional magnetic resonance (Xu *et al.* 2000). As expected from the properties of receptor cells and the organisation of receptor cell projections, in the bulb the olfactory stimulation resulted in the combinatorial activation of a unique set of glomeruli. Because glomeruli have a defined and stable location, the combinatorial message can be said to have a spatial dimension.

Axons of main output neurons project to dendrites of pyramidal cells in several areas which are globally termed primary olfactory cortex (POC) (Fig. 2.1A). These areas include the piriform cortex, anterior olfactory nucleus, olfactory tubercle, entorhinal cortex and specific parts of amygdala. Anatomical and electrophysiological studies indicate that the spatial organisation of the OB is not clearly reproduced in the POC. Mitral cell axons issuing from each individual glomerular micro module remain grouped in their route to the olfactory cortex and their ramifications target several discrete sites in different parts of the cortical projection area (Buonviso *et al.* 1991). The specific sensitivity of each glomerulus seems therefore duplicated several times in the primary cortex.

Using transgenic mice in which a lectin was co-expressed only in sensory neurons equipped with a chosen odorant receptor, Zou *et al.* (2001) could visualise how neurons expressing a particular sensitivity in different locations of the OB were targeted to the POC. Expressed in sensory neurons, the lectin was first transferred to mitral cells and then to third-order cortical neurons that therefore received input from transfected sensory cells. Two different odorant receptors were used. Appropriate histochemical techniques revealed that labelled neurons carrying either sensitivity were organised in discrete clusters into several but not all projection areas. These findings seem to indicate that the specific sensitivity represented first in a subset of receptor neurons and then in a glomerulus of the olfactory bulb is duplicated in many sites of the whole cortical projection. It will be important to discover whether this projection pattern results in providing all cortical subdivisions with a homogeneous sample of all odour sensitivities. An alternative organisation could be an heterogeneous distribution leading to a relative specialisation of each subdivision (Holley 1991). In these conditions, the projections of different areas of the POC could be functionally

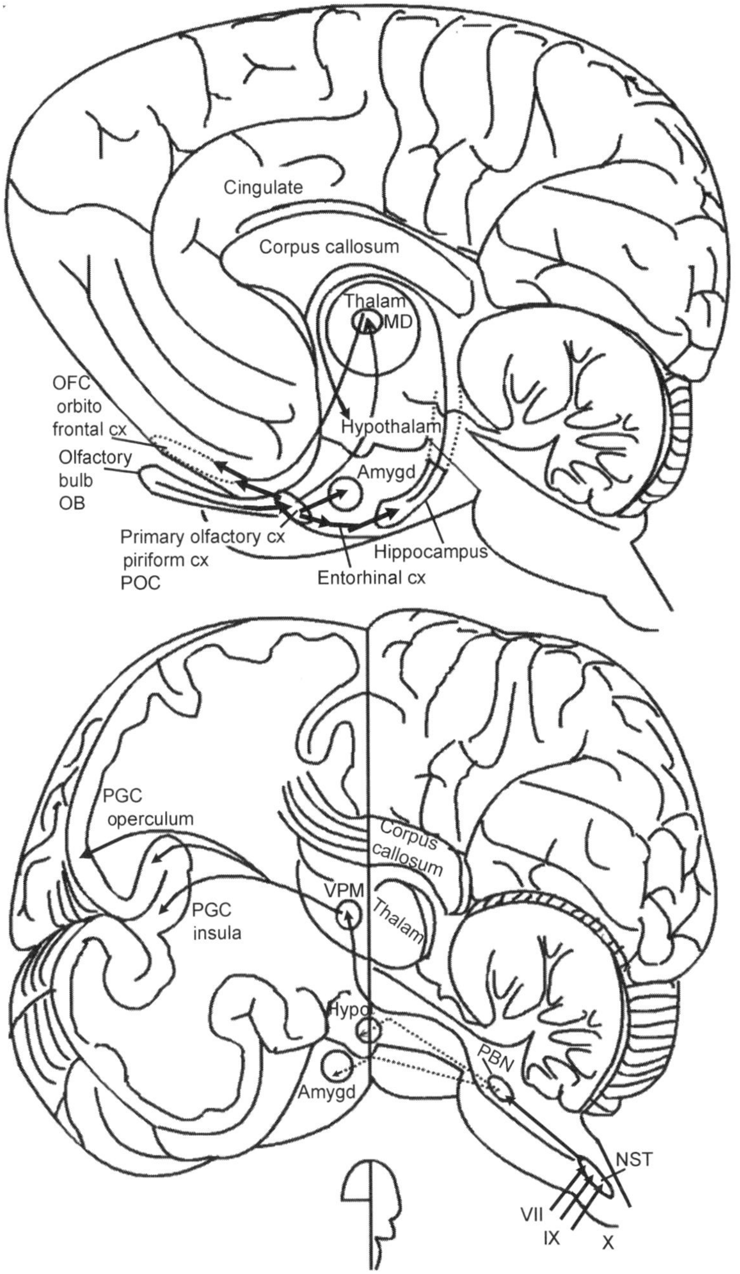

Fig. 2.1 Schematic representation of the brain with the main olfactory (A) and gustatory (B) areas and pathways. A: Olfactory system. Medial view of a hemisected brain showing olfactory structures. MD, mediodorsal nucleus of thalamus. B: Taste system. Composite view of the brain combining a medial view of the right posterior part of the brain and a frontal hemisection of the left side. NST, nucleus of the solitary tract; PBN, pontine parabrachialis nucleus; VPM, ventro-postero-medial nucleus of thalamus.

distinct, may be in relation with the behavioural role of the tertiary projection targets receiving a sample of afferent information.

2.2.4 Properties of odour (and aroma) perception in the light of peripheral processing

Several basic features of odour (and aroma) perception can be interpreted in terms of peripheral reception mechanisms. The wide range of odorant discrimination and the high number of qualities perceived by mammals and humans are direct consequences of the large number and extreme variety of receptors and depend on the combinatorial coding resting on receptor broad selectivity. Even though discriminable, odours present qualitative similarities. Qualitative distances among odours are likely to result from the extent of overlapping among populations of sensory neurons encoding their respective qualities. Intensity coding depends on changes in discharge frequency induced in individual sensory cells by variations in odorant concentration. It is also dependent on changes in the number of cells recruited by stimulation. Because odour quality is represented by a pattern of activated neurons, changing intensity that modifies this pattern may cause a change in quality as well, as observed in psychophysical studies. However, lateral inhibition by local interneurons at the OB level can be thought to limit the degree of interference between intensity and quality. Finally, mixture perception reveals qualitative and intensive interactions among components. Competitive or synergistic interactions among odorants taking place at receptor cell level are presumably involved (Duchamp-Viret *et al.* 2003).

2.2.5 Central integration of olfactory information – results from brain imaging

The primary olfactory cortex that receives organised projections from the olfactory bulb is the origin of tertiary projections to several structures (Fig. 2.1A). These projections have been first investigated in animals using anatomical and neurophysiological techniques (see Price (1985, 1987) for reviews). More recently, the central structures involved in olfactory information processing have been explored in humans, using brain imaging methods: Positon Emission Tomography scanning (PET scan) and functional Magnetic Resonance Imagery (fMRI). Both methods provide measurements of variations in local blood flow associated with cerebral/mental events such as those elicited by olfactory stimulation, with the assumption that a rise in local blood flow within a brain structure indicates an activation of this structure.

Activations following subject exposure to olfactory stimuli were systematically observed in a zone of the orbital frontal cortex (OFC) which can be considered as the secondary olfactory (neo) cortex (Zatorre *et al.* 1992). Two pathways connect the POC to this area: a direct corticocortical pathway to lateral and ventrolateral orbital areas and a transthalamic pathway through the mediodorsal (MD) nucleus of thalamus. Although the piriform area of the

POC is a relay of olfactory messages, its activation was not systematically observed (Zald and Pardo 1997, Dade *et al.* 1998, Sobel *et al.* 1998), may be as a consequence of its short duration (Poellinger *et al.* 2001). Regarding OFC activation, it presents a noticeable property which is that of not being symmetrical in right and left hemispheres (Zatorre *et al.* 1992, Small *et al.* 1997, Sobel *et al.* 1998, Royet *et al.* 1999) (see Royet and Plailly (2004) for a review). Which side is more activated depends on the type of olfactory task in which subjects are engaged, for example: detection *vs* familiarity *vs* hedonic judgements (Royet *et al.* 2001).

Many other regions of the brain were found to be activated in different brain imaging studies: the entorhinal cortex (Zald and Pardo 2000) that is an entry to the hippocampus and the memory system; the hypothalamus (Zatorre 2002) as a key structure in the control of motivated behaviour, including food intake; the amygdala (Zald and Pardo 1997, Anderson *et al.* 2003) whose role in affect, emotion and conditioning is well documented; the insula cortex that is also a projection area for taste and visceral afferents and is activated by visual perception of faces expressing disgust (Phillips *et al.* 1997); the anterior cingulate cortex (Cerf-Ducastel and Murphy 2001, Anderson *et al.* 2003) with its complex function among which one finds the elicitation of emotions and the 'guiding of decisions about whether the expected value of a reward means that it is worth acting' (Rushworth *et al.* 2004).

Again, in several of these areas which are involved in affective processing of odours, odour-induced activations manifest lateralisation. For example, in a fMRI study by Anderson *et al.* (2003) using both pleasant and unpleasant stimuli (citral and valeric acid), each at high and low concentrations, activations could be found mainly in the OFC, amygdala and rostral part of cingulate cortex. In the OFC, the pleasant stimulus (citral) elicited an activation signal in the rostral and medial part of this cortex, in the right hemisphere, while the unpleasant stimulus activated a more lateral zone of the same cortex in the left hemisphere. This suggests that two different neural circuits respectively related to stimulus acceptance and stimulus rejection are involved in odour processing. In the amygdala, the activation was correlated with intensity but not with hedonic valence, indicating that this structure responded to arousal level rather than to degree of pleasantness.

2.3 Reception of taste compounds and neural processing

2.3.1 Taste cells, taste receptors

In mammals, taste compounds released by food are detected by small bipolar cells that are densely packed in taste buds located on the tongue, palate and pharynx. In the tongue, taste buds are borne by protrusions called papillae. The apical poles of taste cells possess microvilli that are in contact with the oral cavity. Three classes of papillae are distinguished by their morphology and their location: fungiform, foliate and circumvallate. Differing from olfactory cells, taste cells are not neurons. They transfer their activation synaptically to fibres of two gustatory nerves VII bis (*chorda tympani*) and IX (glossopharyngeal nerve) that contact them in taste buds. A branch of the vagus nerve (cranial nerve X) also innervates taste buds. Taste cell microvilli are equipped with taste receptor proteins. When contacted by tastants, receptors trigger intracellular events that activate excitatory synapses and cause excitation of afferent nerve fibres. Neural discharges are transmitted to the nucleus of the solitary tract (NST), in the brain stem.

Taste reception involves two main classes of receptor processes: salt and sour tastes are elicited by ionic species detected by ionic channels whereas sweet, bitter and umami tastes involve activation of G protein-coupled receptors (GPCR) belonging to the same receptor superfamily as that comprising olfactory receptors. Salt taste is elicited by many ionic species among which Na^+ (see Lindemann (2001) for a review). The Na^+ taste involves a sodium channel which is sensitive to the channel blocker amiloride. This channel acts by providing a pathway to Na^+ ions into the cell. Na^+ influx triggers action potentials that activate synapses. For the sour taste, several mechanisms have been proposed. One group of mechanisms implies several types of H^+-gated channels; the other group includes channels that conduct an inward H^+ current when the oral space contains protons.

Identification of bitter and sweet tastes was made recently (see Montmayeur and Matsunami (2002) for a review). Studies started with the assumption that genes coding receptors for bitter-tasting compounds should be found in chromosome loci involved in the ability to detect some bitter compounds. Genetic studies had localised such loci for the ability to detect propyl-thiouracil (PROP) (Guo and Reed 2001) in man and sucrose octaacetate in mice (Capeless *et al.* 1992). A multigene family coding for candidate bitter taste receptors called T2R was identified (Adler *et al.* 2000, Matsunami *et al.* 2000). T2Rs belong to the GPCR superfamily. Twenty-six T2R genes have been identified in humans and 33 in mice. Several of them present polymorphism. T2Rs show a variety of sequences that is compatible with their ability to detect several types of bitter compounds.

Using a scanning gene strategy similar to that followed for identifying candidate bitter receptors, several groups succeeded in identifying a putative receptor for sweet taste stimuli called T1R3 (Bachmanov *et al.* 2001, Kitagawa *et al.* 2001, Max *et al.* 2001, Montmayer *et al.* 2001, Nelson *et al.* 2001, Sainz *et al.* 2001). The corresponding gene is tightly linked to the *Sac* locus that controls sensitivity to saccharin and several sweet-tasting compounds in mice. Differing from T2R, this GPCR has a long N-terminal extracellular domain that may be involved in ligand binding. In addition, several arguments indicate (Nelson *et al.* 2001) that T1R3 could form an heterodimer with a previously identified receptor T1R2 (Hoon *et al.* 1999), the function of which was unknown. The dimer can be activated by many sweet-tasting sugars and sweeteners. Another heterodimer composed of T1R3 and of another member of the T1R family, T1R1, responds to most of the common L-amino acids and to an agonist for metabotropic glutamate receptors (mGluRs). T1R1/3 could therefore function as a umami receptor. Finally, another candidate for umami taste reception is taste-mGluR4, a N-terminal truncated variant of the metabotropic glutamate receptor, mGluR4 (Chaudhari *et al.* 2000).

2.3.2 Peripheral representation of tastes

Neural representation of taste is a complex issue. It has been and remains a subject of controversy between proponents of different versions of two rival assumptions: the 'labelled line' theory and the 'across-fibre pattern' theory. These opposite views are closely related to those that oppose partisans of four (or five) discrete basic tastes to authors that conceive taste qualities as a continuum. The reason is that data brought by electrophysiological recordings from taste cells, gustatory nerve single fibres and taste relay neurons presented ambiguous features and differences according to species. The earliest electrophysiological experiments on taste fibres by Pfaffmann (1941) and subsequent studies demonstrated that peripheral axons often responded to stimuli representing more than one of the familiar qualities of salty, sweet, sour and bitter. Therefore, single fibre activity could not unambiguously represent a single quality. Pfaffmann proposed that a taste quality is represented by the relative amounts of activity across a population of cells or fibres. This notion was later popularised by Erickson as 'across-fibre pattern' theory (for a recent review, see Erickson (2000)). However, it was also observed that many afferent cells or neurons responded more strongly to a stimulus representing a given modality than to others. For example, Frank (1973) described neurons from the hamster chorda tympani that responded best to either sucrose, NaCl or HCl. Thus, although strict selectivity could not be demonstrated in mammals, subsets of afferent fibres were found to be narrowly tuned to sugars in the hamster (Frank *et al.* 1988) and to sodium in the rat (Frank *et al.* 1983).

Classifications of peripheral fibres and more central neurons according to their 'best-stimulus' were therefore proposed. It appeared that the contribution of each best-stimulus group of cells to the amount of activity evoked in response to its 'best' stimulus was prevalent over the activity evoked by other contributors to the pattern. This was rather in agreement with the labelled-line assumption. When it was possible to record stimulus-evoked intracellular activity from single gustatory cells (Kimura and Beidler 1961, Sato 1972), no elementary solution to the problem of taste discrimination emerged because taste cells also responded to more than one quality. Recent studies on receptors have not yet provided clear responses as whether a cell expresses only one receptor (as it is supposed to do in olfactory receptor cells) or several receptors. For example, divergent answers were given regarding co-expression of different T2Rs (Adler *et al.* 2000, Caicedo and Roper 2001). Taste qualities are not evenly distributed in different areas of the epithelium. For instance, in mice and rats, mRNAs of T2Rs bitter receptors are more abundant in circumvallate and foliate papillae than in fungiform papillae (Adler *et al.* 2000, Chandrashekar *et al.* 2000). However, it is not possible to strictly associate a defined area of the oral cavity with the sensitivity to any particular quality. A crude topography can be seen in the connections relating the sensory surface to the NTS and onward through thalamus to cortex but the precision of this topography is not high enough to support a spatial coding mechanism. As a conclusion, if one accepts the notion of basic tastes, it must be in a minimal fashion and without the requirement that their combination is able to create all gustatory perceptions. As formulated by Scott and Gisa (2000): 'Taste must be viewed as an integrated system rather than a federation of four to six independent channels'.

2.3.3 Central representation of tastes

Gustatory messages relay in a complex, multimodal structure, the nucleus of the solitary tract (NST). It must be emphasised that this heterogeneous nuclear complex is also a relay for visceral information coming from oral cavity, oesophagus, stomach, intestine and other viscera. In general, the gustatory and general visceral fibres do not overlap. Chorda tympani and lingual branch of glossopharyneal nerve terminate within the lateral (and anterior) subdivision of the nucleus, while the general visceral fibres terminate within the medial (and caudal) subdivision (Contreras *et al.* 1982). It is also worth noting that the gustatory subdivision of the NST receives an input from the lingual branch of the trigeminal nerve (Van Buskirk and Erickson 1977), which may offer a substrate for convergence of gustatory and tactile (or chemical) information from a same peripheral field..

The NST emits both caudal and rostral projections (Norgren and Leonard 1973). Caudally, the gustatory portion of the NST projects to areas of the reticular formation which, in turn, send connections to several motor and premotor nuclei of the caudal brain stem that participate in circuits required to produce appropriate oral movements, such as rejection or swallowing in response to gustatory stimuli (Grill and Norgren 1978), and diverse reflexes (salivation, spitting, coughing). The pattern of NST rostral projections presents some variations according to species. In humans (Fig. 2.1B), neurons from rostral NST project to the thalamus (VPM nucleus) that relays gustatory information to the cortex. In addition, NST sends visceral sensory projections to the pontine parabrachalis nucleus (PBN) which, in turn, projects to the amygdala and hypothalamus.

The primary gustatory cortex (PGC) that receives projections from the thalamic VPM is situated close to the representation of the tongue in the primary somatosensory cortex. In monkeys, this area corresponds with the region of the dorsal bank of the opercular cortex (Rolls *et al.* 1985) and the anterior insula. Taste responses can also be evoked from neurons in the posterior orbitofrontal cortex in macaques. Because this OFC area receives projections from the anterior insula, it has been considered as a secondary gustatory cortex (SGC) (Rolls 1989, Wiggins *et al.* 1987). The role of this projection will be presented later in the section devoted to flavour integration.

The pathway from rostral NST to the cortex via the thalamus has been called the lemniscal (or cognitive) gustatory pathway, while the pathways terminating in the lateral hypothalamus and basal forebrain areas form the visceral–limbic system or ventral gustatory pathway (Pfaffmann *et al.* 1977). Even in the lemniscal pathway, neurons are not exclusively tuned to taste stimuli. A rather high proportion of them in the insula/operculum and in the gustatory thalamus (Pritchard *et al.* 1989) respond to tactile stimuli or to tongue movements. At least in primates, it is generally thought that conscious sensory awareness takes place in cortical structures. Awareness of cognitive and affective properties of taste stimuli is therefore expected to result specifically from the activity in thalamic and cortical structures. However, monkeys with bilateral lesions to VPM demonstrate normal taste preferences (Reilly and Pritchard 1995) and taste awareness was not impaired following damage to insular gustatory cortex in human patients (Pritchard *et al.* 1999).

Taste perception includes two types of components: a cognitive component which is the awareness of the quality of taste and an affective component which is hedonic in nature. It seems necessary to assume that these two aspects of taste have separated neuronal representations because the cognitive dimension is expected to be essentially stable while the affective dimension is plastic, depending on learning. Separate representations could occur in some higher order structures. However, Sewards (2004) proposes a different assumption, namely that segregated populations of neurons representing sensory (cognitive) and hedonic dimensions of stimuli, respectively, coexist at each stage of the gustatory pathways. In general, neurons of the hedonic pathway seem to be located between those representing taste quality and those devoted to interoceptive visceral inputs. Studies using the fMRI method have described cortical activations induced by taste stimuli in human subjects (Faurion *et al.* 1998). Stimulation-elicited activations were regularly found in regions inside the Sylvian scissure, in the higher part of insula, frontal operculum and at the feet of precentral and postcentral gyri. These regions correspond with the location of the PGC as previously identified by electrophysiological recordings in monkeys. A posterior region of the OFC corresponding to the SGC was also labelled.

2.4 Trigeminal chemosensitivity (chemesthesis)

2.4.1 Fibres and receptors

In combination with tastes and aromas, oral and nasal irritation contributes to the flavour profile elicited by a food or a beverage. The term 'irritation' encompasses a broad range of discriminable sensations that involve the action of a cranial nerve, nerve V or trigeminal nerve, belonging to the somatosensory system and providing most of the innervation of the face, including the nasal and oral cavities along with the cornea and conjunctiva of the eye. Two branches of the trigeminal nerve, the ethmoid and nasopalatine nerves, innervate the nasal cavity. Additional branches such as the lingual branch ramify in the oral cavity.

Trigeminal fibres are associated with taste fibres in taste papillae and contribute a large part of the innervation of these papillae (Farbman and Hellekant 1978). It will be noted that fibres with trigeminal-like functions travel also in the glossopharyngeal nerve in posterior parts of the tongue. On the basis of electrophysiological studies using the active ingredient of chilli pepper, capsaicin, as a stimulus, it has been concluded that trigeminal fibres were slow conducting C and A delta fibres with relatively low conduction velocity. Many of these fibres contain neuropeptides (substance P and/or calcitonin gene-related peptide, CGRP) and ramify repeatedly, terminating as free nerve endings in the epithelium, just below the level of the apical tight junction complex. Because irritant stimuli have to penetrate through lipid membrane layers in order to reach nerve endings, lipid solubility is an important factor of their efficacy as stimuli. All fibres do not terminate as free-endings. Recently, in the rat and mouse, Finger *et al.* (2003) discovered a population of chemosensory cells that reach the surface of the nasal epithelium and form synapses with trigeminal fibres. These cells express receptors of the T2R family ('bitter' receptors) and a G protein, α-gustducin, involved in taste transduction, and they respond to bitter substances applied to the nasal epithelium.

In addition to responding to mechanical, thermal, nociceptive and proprioceptive stimuli, trigeminal fibres present in the oral cavity, nasal cavity and eye, manifest chemical sensitivity. Trigeminal receptors involved in detection of irritants inhaled into the nose seem to be polymodal nociceptors in the sense that they combine mechanical, nociceptive and/or thermal sensitivity with chemical sensitivity.

Several types of irritant-sensitive receptors have been identified. A neuronal nicotinic receptor (NnAChR) is involved in the response of the rat ethmoidal nerve to nicotine as shown by the use of blockers that decrease trigeminal responses to nicotine without affecting responses to another compound (cyclohexanone) (Alimohammadi and Silver 2000, 2003). Messenger RNA transcripts and proteins of purin receptors (P2X) – ion channels activated by ATP – have been identified in trigeminal neurons. Two subunits, P2X2 and P2X3, which can form homodimeric or heterodimeric receptors are predominantly expressed (Paul *et al.* 2002).

It is well known that some irritant chemicals, like the active component of chilli pepper, capsaicin, elicit painful thermal sensations. Insights into the mechanism underlying similar chemical and thermal sensations have come from molecular cloning of a vanilloid receptor (VR1) expressed by sensory neurons. VR1 is activated by both temperature (exceeding 43 °C) and capsaicin (Caterina *et al.* 1997, Tominaga *et al.* 1998). It is an excitatory channel of the transient receptor potential (TRP) family. Another example of chemical sensation transmitted by a thermoreceptor is provided by the identification of the molecular site of menthol action (McKemy *et al.* 2002). Long ago, Hensel and Zotterman (1951) suggested that cooling compounds elicited a cool sensation by interacting with a protein involved in cold transduction. This assumption proved to be right. The protein was identified as an excitatory ion channel, another member of the

TRP family, which is also activated by temperature in the cold-to-cool range (8–28 °C) when expressed in transfected cells. This receptor was called cold- and menthol-sensitive receptor (CMR1). The cooling agent icilin elicited strong responses from CMR1 transfected cells while eucalyptol had a smaller effect. Menthone, camphor and cyclohexanol had little or no effect and capasaicin elicited no response. The complexity of trigeminal nerve sensitivity is illustrated by the fact that a competitive inhibitor of VR1 (capsazepine) selectively decreased nerve responses of rat ethmoid nerve not only to capsaicin but also to CO_2, by 50% (Alimohammadi and Silver 2000, 2003). However, a carbonic anhydrase inhibitor (acetazolamide) diminished trigeminal nerve response to CO_2 indicating that CO_2 stimulation could be mediated by intracellular acidification.

In studies of the trigeminal system a few chemicals have been preferentially utilized, essentially nicotine, CO_2 and capsaicin along with some other components of pungent spices. However, it is generally agreed that a wide variety of compounds, including olfactory stimuli, are potentially active on trigeminal endings. Early electrophysiological recordings indicated that many compounds are active on trigeminal fibres when they are delivered at high concentration. Most of them – but not all – seem to have detection thresholds that are two or three log units higher than olfactory thresholds for the same compounds (Beidler and Tucker 1956). According to Silver and Finger (1991) who compared electrophysiological data from animals and psychophysical data from humans, there is a correlation between the human perceived intensity ratings and the magnitude of the rat maximal neural response.

2.5 Multimodal interactions and flavour integration

Several modes of interactions between different sensory modalities participating in food perception can be distinguished. Interactions can take place at the receptor level when the same stimuli activate different types of afferents. This is illustrated by the relations between the trigeminal sensitivity on the one hand and olfactory and taste sensitivities on the other hand. Interactions occur at higher levels when originally distinct flows of sensory messages converge onto the same integrative areas, as exemplified by the case of odour and taste integration.

2.5.1 Interactions between somatosensory chemesthesis and other chemical sensitivities

Olfactory-trigeminal interactions

The influence of trigeminal system activation on the function of the olfactory system may be first exerted through reflexes that participate in the protective function of trigeminal innervation against noxious stimuli (James and Daly 1969): decrease in respiratory rate, norepinephrine secretion, increase in nasal secretion, peripheral vasoconstriction, sneezing, closure of the nares, and increase in air flow resistance. It has been hypothesised that trigeminal nerve activity could affect receptor cell functioning by altering the perireceptor environment. Under antidromic stimulation of the trigeminal nerve and local application of substance P, secretory glands and cells of the olfactory mucosa show morphologic signs of activity in frogs and salamanders (Getchell *et al.* 1989) and receptor cell responses to odorants are altered (Bouvet *et al.* 1987). It is assumed that modulation of receptor cell responsiveness could occur naturally via an axon reflex, i.e. when the afferent function of a trigeminal fibre activates its efferent function or that of a collateral branch, leading to neuropeptide release. Presumably, the various trigeminal reflexes modify odour perception, especially in intensity terms.

Olfactory and trigeminal sensations combine as a result of simultaneous stimulation of both kinds of afferents by volatile compounds. As many compounds have trigeminal thresholds higher than olfactory thresholds, the trigeminal component is supposed to have little impact at low concentration. Mixtures of trigeminal stimuli that have no obvious olfactory component with olfactory stimuli which have no obvious trigeminal component produce mutual suppression of odour and pungency (Cain and Murphy 1980). Higher concentrations of the irritant (CO_2) completely inhibit responses to the odorant (amyl butyrate). As mutual suppression was still observed when stimulants were not physically mixed but presented to different nostrils, it was concluded that the interaction occurred at a central site rather at the periphery. Using a protocol of stimulus lateralisation allowing them to assess the trigeminal sensitivity of patients with olfactory impairement, Hummel *et al.* (2003) confirmed several observations showing that olfactory loss resulted in decreased trigeminal sensitivity.

Gustatory-trigeminal interactions

Irritant chemicals are found in many foods. They evoke burning or pricking sensations by activating nerve endings in the oral mucosa. As in the nasal cavity, capasaicin-sensitive or other trigeminal fibres could have an effector role and affect gustatory receptor activity via axon reflexes. This may involve changes in salivary flow, local secretions or contracting responses around taste pores, or alteration of taste transduction (Silver and Finger 1991). A few studies have explored how the presence of oral irritants modify perception of tastes. Some studies suggest that capsaicin can affect sensitivity to some bitter tastes and, in addition, that some stimuli can excite the trigeminal nerve. Green and Hayes (2003, 2004) confirmed that capsaicin (and menthol) can induce bitterness in some individuals, especially when applied to the circumvallate papillae, whereas common bitter compounds, urea and magnesium chloride, induce irritation when applied to the tongue tip. The author suggests that 'virtually all sensory irritants may be able to stimulate the gustatory system, particularly on the back of the tongue, and that some bitter tastants can stimulate "capsaicin-sensitive" somatosensory neurons, mainly on the front of the tongue'. Toda *et al.* (2002) used an *in vitro* jaw-nerve preparation, in the rat, to demonstrate that intra oral polymodal nociceptors can be excited by bitter stimuli (caffeine and quinine). In an experiment where capsaicin was applied to one side of the tongue followed by bilateral application of conventional tastants, Simons (2002) and Simons *et al.* (2003) observed that subjects assigned lower intensity ratings to sucrose-induced sweetness, quinine-induced bitterness and glutamate-induced umami on the capsaïcine-treated side. Electrophysiological recordings from neurons of the NST in rats also showed a depressing effect of lingual application of capsaicin on responses to tastant stimuli. Trigeminal ganglionectomy did not prevent response suppression by capsaicin, indicating that capsaicin acted peripherally. Moreover, in intact rats, lingual application of nicotine also suppressed tastant-evoked responses of NST units. However, in this case, bilateral trigeminal ganglionectomy prevented nicotine to exert its suppressive effect, indicating that this effect was central and depended on activation of the trigeminal pathway. It must be recalled that a trigeminal input has been seen in the NST (Van Buskirk and Erickson 1977). A large percentage of neurons in the NST were found to be bimodal, responding to cooling and to mechanical stimuli (Norgren 1984, Travers and Norgren 1995). Higher in the brain, taste and lingual somatic stimulation elicited responses in VPM nucleus of thalamus (Emmers 1966, Pritchard *et al.* 1986), these bimodal cells being located between cells responding to chemicals and cells responding to somatosensory stimulation (Scott and Mark 1986). Trigeminal modulation of gustatory NST neuron responses to tastants appeared when trigeminal nerves were either electrically stimulated or cut, in rats (Boucher *et al.* 2003). Nerve section resulted in significant reduction of tastant-evoked responses, while electrical stimulation of central cut-end of the (mandibular) nerve facilitated or suppressed taste responses, depending on NST neurons. Electrical stimulation of the peripheral cut end of the same nerve had a predominantly suppressive effect, possibly mediated by neuropeptide release in the tongue.

2.5.2 Cortical representation and flavour integration

Taste-trigeminal interactions at the cortical level

The cortical representation of both gustatory and lingual somatosensory stimulation was investigated in human using fMRI (Cerf-Ducastel *et al.* 2001). Six stimuli were applied to the tongue; four were purely gustatory in nature and two were both taste and somatosensory stimuli (HCl, pH 1.6 and aluminium potassium sulfate, called somato-gustatory stimuli). Both sets of stimuli activated the same cortical areas of the insular region, i.e. the PGC, confirming that taste and lingual somatosensory representations overlap extensively. However, the fine analysis of co-activated areas indicated discriminable patterns of activation across the sub-insular and opercular regions depending on the stimulation set.

Cortical integration of odours and tastes

In monkeys, single-neuron recording studies have identified brain areas – the insula/operculum (Scott and Plata-Salaman 1999) and orbitofrontal cortex (Rolls and Bayliss 1994, Rolls *et al.* 1996) – where neurons respond to both taste and smell stimuli. OFC neurons responding to unimodal olfactory, gustatory and even visual stimuli were found in distinct location but overlap occurred and bimodal neurons were also identified. Interestingly, bimodal cells responded to qualities which are usually associated in foods, for example sweet taste and fruit odour and not fish odour. Congruent associations also occurred for responses to visual and taste or smell stimuli. These integrative neural properties parallel perceptual experience, when an odour enhances the perceived intensity of a taste on the condition that odour and taste are congruent (Frank *et al.* 1989, Schifferstein and Verlegh 1996). Similarly, when repetitively paired with a taste, odours can acquire taste-like properties (Stevenson and Prescott 1995, Prescott 1999).

In humans, brain imaging studies have shown that independent stimulation with an odorant or a tastant elicited activation in regions of the insula (Faurion *et al.* 1998, Cerf-Ducastel *et al.* 2001, Poellinger *et al.* 2001, Gottfried *et al.* 2002), corresponding to the PGC, and OFC (Zatorre *et al.* 1992, Small *et al.* 1997, 1999, 2003, O'Doherty *et al.* 2000, Poellinger *et al.* 2001, Gottfried *et al.* 2002). The latter region is thought to contain the secondary taste cortex and secondary olfactory cortex. The amygdala (Zald and Pardo 1997, Zald *et al.* 1998, Anderson *et al.* 2003, Small *et al.* 2003) and the anterior cingulate cortex (Zald *et al.* 1998, Small *et al.* 2001, Royet *et al.* 2003, de Araujo *et al.* 2003, Small *et al.* 2003) also display activation in response to stimuli of both modalities.

Integration of aromas and tastes: the case of retronasal versus orthonasal olfaction

An interesting aspect of flavour integration is illustrated by notable differences in olfactory – gustatory interactions depending on the route followed by odorants. Whereas taste components of flavours are perceived only when the food is introduced inside the oral cavity, volatile components can reach the nasal epithelium by two routes: the 'forward' way through the nose when the food is not yet in the mouth (orthonasal olfaction) and a 'backward' way through the nasopharynx when the food is being masticated (retronasal olfaction). Both orthonasal and retronasal smelling must make use of the same receptors whichever the way followed by volatiles compounds. Accordingly, one could think that orthonasal and retro nasal smelling are nothing more and nothing less than two different routes to the same stimulation. However, ease of access to sensory areas may be different. Besides, the direction of the flow determines the order according to which olfactory (and trigeminal) receptors will be stimulated, so that different spatiotemporal patterns of stimulation can result in different neural responses (see Halpern (2004) for a discussion). Moreover, retronasal smelling is usually time-locked to taste stimulation while orthonasal smelling can be independent from it. This may result in different interactions between olfaction and taste depending on the route followed by the stimulating flow.

Several authors (Voirol and Daget 1986, Pierce and Halpern 1996) who investigated differences between orthonasal and retronasal olfaction concluded that these differences could be explained as a less efficient access of volatiles compounds to the sensory area when they follow the retronasal way. In a fMRI study (Cerf-Ducastel and Murphy 2001), brain responses to retronasally perceived odours delivered in the mouth in liquid phase were found in the same areas as those activated following orthonasal stimulation in independent studies. However, Heilmann and Hummel (2001) noted that olfactory evoked potentials differed according to the way followed by stimulation.

More directly related to the issue of flavour integration, it was reported (Small *et al.* 1997) that simultaneous presentation of tastes and orthonasal odours resulted in massive deactivation of cortical chemosensory regions. Other evidence indicate that retronasal odours are better integrated with tastes than orthonasal odours. For example, in taste aversion, potentiation by odours is more effective when odours are delivered retronasally (Slotnick *et al.* 1997). Small *et al.* (2004) recorded fMRI responses during perception of tastant/odorant liquid mixtures and compared these responses to the sum of those recorded following independent presentations of gustatory and olfactory constituents. Two flavours were utilised: one contained a familiar taste/odour pair (vanilla–sweet) defined as congruent, and the other an unfamiliar taste/odour pair (vanilla–salty) defined as incongruent. Three unimodal stimuli were also delivered: two tastes (sweet and salty) and one odour (vanilla). When compared with the sum of its constituents, the congruent flavour elicited superadditive responses in several brain areas, in particular the dorsal insula, anterior ventral insula, caudal orbitofrontal cortex and anterior cingulate cortex. A similar analysis with the incongruent flavour compared with the sum of its constituents did not reveal these regions. Finally, a direct contrast of congruent versus incongruent situations made their brain inscription again visible.

It can be concluded that insula, orbitofrontal and cingulate cortex are included in a neural network that participates in flavour integration. This integration depends on learning as evidenced by superadditive responses to familiar, congruent association of bimodal stimuli. It will also be noted that the primary gustatory cortex of insula and the primary olfactory cortex, do not play symmetrical roles in the integration process as the former area is included in the flavour network and the latter is not.

2.6 Conclusions and future trends

This brief review shows that chemosensory systems controlling food intake are particularly rich and complex. These systems allow consumers to identify many physicochemical properties of food with great precision. It also demonstrates that these sensory systems, even though distinct, interfere with each other at several anatomical and functional levels, which explains the prevalent holistic character of food perception. Interactions among sensory afferents are also responsible for integrative properties that result in either synergy or inhibition. It will be also noted that oral and visceral types of sensitivity, even though distinguishable, are closely associated anatomically, according to the principle of somatotopy implying that adjacent areas of the body have neural representations that are also close to each other.

From this review, it appears that in recent years two approaches have been particularly efficient in expanding our understanding of the biological substrate of flavour perception: studies on odorant, tastant and irritant membrane receptors on the one hand, and studies on flavour-evoked brain activation in humans, on the other hand. Works on receptors are far from being finished. It must be kept in mind that the impressive amount of data collected on receptors has been obtained from investigations on their genes and that a considerable effort remains to be achieved in order to identify their individual ligands and modellise odorant–receptor interactions. Very little is known on the processes that occur at the receptor level when different chemicals compete for the same site.

In a few years, brain imaging has demonstrated its potential to reveal functional properties of sensory systems involved in flavour processing. Brain images induced by chemosensory stimulation go beyond the stage of purely confirming anatomical localisation. They demonstrate large variations in intensity and localisation according to the task that subjects perform on sensory signals. Notoriously, flavours activate brain circuits that are much larger than those devoted to sensory modalities. As a consequence, scientists exploring olfaction, taste or their combination will be necessarily led to extend their investigations to neural structures that integrate sensory information with other brain functions such as memory, learning, affective attribution and decision to act. For this reason, future research on flavour perception based on brain exploration can be expected to considerably enrich our view on what is currently termed 'sensory information processing'.

2.7 References

ADLER E, HOON M A, MUELLER K L, CHANDRASHEKAR J, RYBA N J and ZUKER C S (2000) 'A novel family of mammalian taste receptors', *Cell*, 100, 693–702.

ALIMOHAMMADI H and SILVER W L (2000) 'Evidence for nicotinic acetylcholine receptors on nasal trigeminal nerve endings of the rat', *Chem Senses*, 25, 61–66.

ALIMOHAMMADI H and SILVER W L (2003) 'In vivo characterization of nasal trigeminal chemoreceptors', *Symposium on Trigeminal Chemoreception*, Pommersfelden, Germany (proceeding abstract).

ANDERSON A K, CHRISTOFF K, STAPPEN I, PANITZ D, GHAHREMANI D G, GLOVER G, GABRIELLI J D E and SOBEL N (2003) 'Dissociated neural representation of intensity and valence in human olfaction', *Nat Neurosci*, 6, 196–202.

BACHMANOV A A, LI X, REED D R, OHMEN J D LI S, CHEN Z, TORDOF M G, DE JONG P J, WU C, WEST D B *et al.* (2001) 'Positional cloning of the mouse saccharin preference (*Sac*) locus', *Chem Senses*, 26, 925–933.

BEIDLER L M and TUCKER D (1956) 'Olfactory and trigeminal responses to odors', *Fed Proc*, 15, 14.

BOUCHER Y, SIMONS C T, FAURION A, AZERAD J and CARSTENS E (2003) 'Trigeminal modulation of gustatory neurons in the nucleus of the solitary tract', *Brain Res*, 973, 265–274.

BOUVET J F, DELALEU, J C and HOLLEY A (1987) 'Olfactory receptor cell function is affected by trigeminal nerve activity', *Neurosci L*, 77, 181–186.

BUCK L and AXEL R (1991) 'A novel multigene family may encode odorant receptors: a molecular basis for odor recognition', *Cell*, 65, 175–187.

BUONVISO N, REVIAL M F and JOURDAN F (1991) 'The projections of mitral cells from small local regions of the olfactory bulb: an anterograde tracing study using PHA-L (*Phaseolus vulgaris* leucoagglutinin)', *Eur J Neurosci*, 3, 493–500.

CAICEDO A and ROPER S D (2001) 'Taste receptor cells that discriminate between bitter stimuli', *Science*, 291, 1557–1560.

CAIN W S and MURPHY C L (1980) 'Interactions between receptive modalities of odour and irritation', *Nature*, 284, 255–257.

CAPELESS C G, WHITNEY G and AZEN E A (1992) 'Chromosome mapping of *Soa*, a gene influencing gustatory sensitivity to sucrose octaacetate in mice', *Behav Genet*, 22, 655–663.

CATERINA M J, SCHUMACHER M A, TOMINAGA M, ROSEN T A, LEVINE J D and JULIUS D (1997) 'The capsaicin receptor: a heat-activated ion channel in the pain pathway', *Nature*, 389, 816–824.

CERF-DUCASTEL B and MURPHY C (2001) 'fMRI activation in response to odorant orally delivered in aqueous solutions', *Chem Senses*, 26, 625–637.

CERF-DUCASTEL B, VAN DE MOORTELE P F, MAC LEOD P, LE BIHAN D and FAURION A (2001) 'Interaction of gustatory and lingual somatosensory perceptions at the cortical level in the human: a functional magnetic resonance imaging study', *Chem Senses*, 26, 71–383.

CHANDRASHEKAR J, MUELLER K, HOON M, ADLER E, FENG L, GUO W, ZUKER C and RYBA N (2000) 'T2Rs function as bitter taste receptors', *Cell*, 100, 703–711.

CHAPUT M A and HOLLEY A (1985) 'Responses of olfactory bulb neurons to repeated odor stimulations in awake freely-breathing rabbits', *Physiol Behav*, 34, 249–258.

CHAUDHARI N, LANDIN A and ROPER S (2000) 'A metabotropic glutamate receptor variant functions as a taste receptor', *Nat Neurosci*, 3, 113–119.

CINELLI A R and KAUER J S (1995) 'Salamander olfactory bulb neuronal activity observed by video rate, voltage-sensitive dye imaging. II. Spatial and temporal properties of responses evoked by electrical stimulation', *J Neurophysiol*, 73, 2033–2052.

CONTRERAS R J, BECKSTEAD R M and NORGREN R (1982) 'The central projections of the trigeminal, facial, glossopharyngeal and vagus nerves: An autographic study in the rat', *J Auton Nerv Syst*, 6, 303–322.

DADE L A, JONES-GOTMAN M, ZATORRE R J and EVANS A C (1998) 'Human brain function during odor encoding and recognition. A PET activation study', *Ann N Y Acad Sci*, 855, 572–574.

DE ARAUJO E, ROLLS E T, KRINGELBACH M L, McGLONE F and PHILLIPS N (2003) 'Taste-olfactory convergence, and the representation of the pleasantness of flavour in the human brain', *Eur J Neurosci*, 18, 2059–2068.

DUCHAMP-VIRET P, CHAPUT M A and DUCHAMP A (1999) 'Odor response properties of rat olfactory receptor neurons', *Science*, 284, 2171–2174.

DUCHAMP-VIRET P, DUCHAMP A and CHAPUT M A (2003) 'Single olfactory neurons simultaneously integrate the components of an odour mixture', *Eur J Neurosci*, 18, 2690–2696.

EMMERS R (1966) 'Separate relays of tactile, pressure, thermal, and gustatory modalities in the cat thalamus', *Proc Soc Exp Biol Med*, 121, 527–531.

ERICKSON R P (2000) 'The evolution of neural coding ideas in the chemical senses', *Physiol Behav*, 69, 3–13.

FARBMAN A I and HELLEKANT G (1978) 'Quantitative analyses of the fiber population in rat chorda tympani nerves and fungiform papillae', *Am J Anat*, 153, 509–522.

FAURION A, CERF B, LE BIHAN D and PILLIAS A M (1998) 'fMRI study of taste cortical areas in humans', *An N Y Acad Sci*, 855, 535–545.

FINGER T E, BÖTTGER B, HANSEN A, ANDERSON K T, ALIMOHAMMADI H and SILVER W L (2003) 'Solitary chemoreceptor cells in the nasal cavity serve as sentinels of respiration', *Proc Natl Acad Sci USA*, 100, 8981–8986.

FIRESTEIN S (2001) 'How the olfactory system makes sense of scents', *Nature*, 413, 211–218.

FRANK M E (1973) 'An analysis of hamster afferent taste nerve response functions', *J Gen Physiol*, 61, 588–618.

FRANK M E, BIEBER S L and SMITH D V (1988) 'The organization of taste sensibilities in hamster chorda tympani nerve fibers', *J Gen Physiol*, 91, 861–896.

FRANK M E, CONTRERAS R J and HETTINGER T P (1983) 'Nerve fibers sensitive to ionic taste stimuli in chorda tympani of the rat', *J Neurophysiol*, 50, 941–960.

FRANK R A, DUCHENY K and MIZE S S (1989) 'Strawberry odor, but not red color, enhances the sweetness of sucrose solutions', *Chem Senses* 14, 371–377.

GESTELAND R C, LETTVIN J Y and PITTS W H (1965) 'Chemical transmission in the nose of the frog', *J Physiol Lond*, 181, 525–559.

GETCHELL M L, BOUVET J F, FINGER T E, HOLLEY A and GETCHELL T V (1989) 'Peptidergic regulation of secretory activity in the olfactory mucosa of the tiger salamander; immunocytochemistry and pharmacology', *Cell Tiss Res*, 256, 381–389.

GILAD Y, WIEBE V, PRZEWORSKI M, LANCET D and PÄÄBO S (2004) 'Loss of olfactory receptor genes coincides with the acquisition of full trichromatic vision in primates', *Plos Biol*, 2, 120–125.

GLUSMAN G, YANAI I, RUBIN I and LANCET D (2001) 'The complete human olfactory subgenome', *Genome Res*, 11, 685–702.

GOTTRIED J A, DEICHMANN R, WINSTON J S and DOLAN R J (2002) 'Functional heterogeneity in human olfactory cortex: an event-related functional magnetic resonance imaging study', *J Neurosci*, 22, 10819–10828.

GREEN B G and HAYES J E (2003) 'Capsaicin as a probe of the relationship between bitter taste and chemesthesis', *Physiol Behav*, 79, 811–821.

GREEN B G and HAYES J E (2004) 'Individual differences in perception of bitterness from capsaicin, piperin and zingerone', *Chem Senses*, 29, 53–60.

GRILL H J and NORGREN R (1978) 'The taste reactivity test. II. Mimetic responses to gustatory stimuli in chronic thalamic and chronic decerebrate rats', *Brain Res*, 143, 281–297.

GUO S W and REED D R (2001) 'The genetics of phenylthiocarbamide', *Ann Hum Biol*, 28, 111–142.

HALPERN B P (2004) 'Retronasal and orthonasal smelling', *Chemosense*, 6, 2–7.

HEILMANN S and HUMMEL T (2001) 'Olfactory event-related potentials to ortho- and retronasal stimulation', Achems, Sarasota, Fl.

HENSEL H and ZOTTERMAN Y (1951) 'The effects of menthol on the thermoreceptors', *Acta Physiol Scand*, 24, 27–34.

HOLLEY A (1991) 'Neural coding of olfactory information', in Getchell T V, Doty R, Bartoshuk L and Snow J B, *Smell and Taste in Health and Disease*, New York,

Raven Press, 329–343.

HOLLEY A and DØVING K B (1977) 'Receptor sensitivity, acceptor distribution, convergence and neural coding in the olfactory system', in Le Magnen J and Mac Leod P, *Olfaction and Taste VI*, Paris, 113–123.

HOLLEY A, DUCHAMP A, REVIAL M F, JUGE A and MAC LEOD P (1974), 'Qualitative and quantitative discrimination in the frog olfactory receptors: analysis from electrophysiological data', *Ann NY Acad Sci*, 237, 102–114.

HOON M A, ADLER E, LINDEMEIER J, BATTEY J F, RYBA N J and ZUKER C S (1999) 'Putative mammalian taste receptors: a class of taste-specific GPCRs with distinct topographic selectivity', *Cell*, 96, 541–551.

HUMMEL T, FUTSCHIK T, FRASNELLI J and HUTTENBRINK K B (2003) 'Effects of olfactory function, age, and gender on trigeminally mediated sensations: a study based on the lateralization of chemosensory stimuli', *Toxicol Let*, 140-141, 273–280.

JAMES J E A and DALY M DE B (1969) 'Nasal reflexes', *Proc Roy Soc Med*, 62, 1287–1293.

JOURDAN F, DUVEAU A, ASTIC L and HOLLEY A (1980) 'Spatial distribution of [14C]2-deoxyglucose uptake in the olfactory bulbs of rats stimulated with two different odours', *Brain Res*, 188, 139–154.

KIMURA K and BEIDLER L M (1961) 'Microelectrode study of taste receptors of rat and hamster', *J Cell Comp Physiol*, 58, 131–139.

KITAGAWA M, KUSAKABE Y, MIURA H, NINOMIYA Y and HINO A (2001) 'Molecular genetic identification of a candidate receptor gene for sweet taste', *Biochem Biophys Res Commun*, 283, 236–242.

KURAHASHI T and MENINI A (1997) 'Mechanism of odorant adaptation in the olfactory receptor cell', *Nature*, 385, 725–729.

LINDEMANN B (2001) 'Receptors and transduction in taste', *Nature*, 413, 219–225.

LU M, STASZEWSKI L, ECHEVERRI F, XU H and MOYER B D (2004) 'Endoplasmic reticulum degradation impedes olfactory G-protein coupled receptor functional expression', *BMC Cell Biol*, 5, 34.

MALNIC B, HIRONO J, SATO T and BUCK L B (1999) 'Combinatorial receptor codes for odors', *Cell*, 96, 713–723.

MATSUNAMI H, MONTMAYEUR J P and BUCK L B (2000) 'A family of candidate taste receptors in human and mouse', *Nature*, 404, 601–604.

MAX M, SHANKER Y G, HUANG L, RONG M, LIU Z, CAMPAGNE F, WEINSTEIN H, DAMAK S and MARGOLESKEE R F (2001) '*Tas1R3*, encoding a new candidate taste receptor, is allelic to the sweet responsiveness locus *Sac*', *Nat Genet*, 28, 58–63.

MCKEMY D D, NEUHAUSSER W M and JULIUS D (2002) 'Identification of a cold receptor reveals a general role for TRP channels in thermosensation', *Nature*, 416, 52–58.

MENINI A, PICCO C and FIRESTEIN S (1995) 'Quantal-like current fluctuations induced by odorants in olfactory receptor cells', *Nature*, 373, 435–437.

MOMBAERTS P (1999) 'Seven-transmembrane proteins as odorant and chemosensory receptors', *Science*, 286, 707–711.

MOMBAERTS P (2004) 'Odorant receptor gene choice in olfactory sensory neurons: the one receptor-one neuron hypothesis revisited', *Curr Opin Neurobiol*, 14, 31–36.

MONTMAYEUR J P, LIBERLES S D, MATSUNAMI H and BUCK L B (2001) 'A candidate taste receptor gene near a sweet taste locus', *Nat Neurosci*, 4, 492–498.

MONTMAYEUR J P and MATSUNAMI H (2002) 'Receptors for bitter and sweet taste', *Curr Opin Neurobiol*, 12, 1–8.

MORI K, NAGAO H and YOSHIHARA Y (1999) 'The olfactory bulb: coding and processing of odor molecule information', *Science*, 286, 711–715.

NELSON G, HOON M A, CHANDRASHEKAR J, ZHANG Y, RYBA N J and ZUKER C S (2001) 'Mammalian sweet taste receptors', *Cell*, 106, 381–390.

NORGREN R and LEONARD C M (1973) 'Ascending central gustatory pathways', *J Comp Neurol*, 150, 217–238.

NORGREN, R (1984) 'Central neural mechanisms of taste', in Darien-Smith I, *Handbook of Physiology*, Section 1, Vol. III. American Physiology Society, Washington, DC, 1087–1128.

O'DOHERTY J, ROLLS E T, FRANCIS S, BOWTEL R, MCGLONE F, KOBAL G, RENNER B and AHNE G (2000) 'Sensory-specific satiety-related olfactory activation of the human orbitofrontal cortex', *Neuroreport*, 11399–11403.

PAUL J, HATT H and WETZEL C H (2002) 'Electrophysiological characterization of P2X-receptors in cultured rat trigeminal neurons', *Symposium on Trigeminal Chemoreception*, Pommersfelden, Germany (proceeding abstract).

PFAFFMANN C (1941) 'Gustatory afferent impulses', *J Cell Comp Physiol*, 17, 243–258.

PFAFFMANN C, NORGREN R and GRILL H J (1977) 'Sensory affect and motivation', *Ann NY Acad. Sci*, 290, 18–34.

PHILLIPS M L, YOUNG A W, SENIOR C, BRAMMER M, ANDREW C, CALDER A J, BULLMORE E T, PERRETT D I, ROWLAND D, WILLIAMS S C, GRAY J A and DAVID A S (1997) 'A specific neural substrate for perceiving facial expressions of disgust', *Nature*, 389, 495–498.

PIERCE J and HALPERN B P (1996) 'Orthonasal and retronasal odorant identification based upon vapor phase input from common substances', *Chem Senses*, 21, 529–543.

POELLINGER A, THOMAS R, LIO P, LEE A, MAKRIS N, ROSEN B R and KWONG K (2001) 'Activation and habituation in olfaction – An fMRI study', *Neuroimage*, 13, 547–560.

PRESCOTT J (1999) 'Flavour as a psychological construct: implication for perceiving and measuring the sensory qualities of foods', *Food Qual Prefer*, 10, 349–356.

PRICE J L (1985) 'Beyond the primary olfactory cortex: Olfactory-related areas in the neocortex, thalamus and hypothalamus', *Chem Senses*, 10, 239–258.

PRICE J L (1987) 'The central olfactory and accessory olfactory systems', in Finger T E and Silver W L, *Neurobiology of Taste and Smell*, New York, John Wiley & Sons, 179–203.

PRITCHARD T C, HAMILTON R B, MORSE J R and NORGREN R (1986) 'Projections of thalamic gustatory and lingual areas in the monkey, *Macaca fascicularis*', *J Comp Neurol*, 244, 213–228.

PRITCHARD T C, HAMILTON R B and NORGREN R (1989) 'Neural coding of gustatory information in the thalamus of *Macaca mulatta*', *J Neurophysiol*, 61, 1–14.

PRITCHARD T C, MACALUSO D A and ESLINGER P J (1999) 'Taste perception in patients with insula cortex lesions', *Behav Neurosci*, 113, 663–671.

RAWSON N E, EBERWINE J, DOTSON R, JACKSON J, ULRICH P and RESTREPO D (2000) 'Expression of mRNAs encoding for two different olfactory receptors in a subset of olfactory receptor neurons', *J Neurochem*, 75, 185–195.

REILLY S and PRITCHARD T C (1995) 'The effect of thalamic lesions on primate taste preference', *Exp Neurol*, 135, 56–66.

RESSLER K J, SULLIVAN S L and BUCK L B (1993) 'A zonal organization of odorant receptor gene expression in the olfactory epithelium', *Cell*, 73, 597–609.

RESSLER K J, SULLIVAN S L and BUCK L B (1994) 'Information coding in the olfactory system: evidence for a stereotyped and highly organized epitope map in the olfactory bulb', *Cell*, 79, 1245–1255.

ROLLS E T (1989) 'Information processing in the taste system of primates', *J Exp Biol*, 146, 141–164.

ROLLS E T, VAXLEY S, SIENKIEWICZ Z and SCOTT T R (1985) 'Gustatory responses of single neurons in the orbitofrontal cortex of the macaque monkey', *Chem Senses*, 10, 443.

ROLLS E T and BAYLISS L L (1994) 'Gustatory, olfactory, and visual convergence within the primate orbitofrontal cortex', *J Neurosci*, 14, 5437–5452.

ROLLS, E T, CRITCHLEY H D and TREVES A (1996) 'Representation of olfactory information in the primate olfactory cortex', *J Neurophysiol*, 75, 1982–1996.

ROUQUIER S, BLANCHER A and GIORGI D (2000) 'The olfactory receptor gene repertoire in primates and mouse: Evidence for reduction of the functional fraction in primates', *Proc Natl Acad Sci USA*, 97, 2870–2874.

ROUQUIER S, TAVIAUX S, TRASK B J, BRAND-ARPON V, VAN DEN ENGH G, *et al.* (1998), 'Distribution of olfactory receptor genes in the human genome', *Nat Genet*, 18, 243–250.

ROYET J P, KOENIG O, GREGOIRE M C, CINOTTI L, LAVENNE F, LE BARS D, COSTES N, VIGOUROUX M, FARGET V, SICARD G, HOLLEY A, MAUGUIèRE F, COMAR D and FROMENT J C (1999) 'Functional anatomy of perceptual and semantic processing for odors', *J Cognit Neurosci*, 11, 94–109.

ROYET J P, HUDRY J, ZALD D H, GODINOT D, GREGOIRE M C, LAVENNE F, COSTES N and HOLLEY A (2001) 'Functional neuroanatomy of different olfactory judgments', *Neuroimage*, 13, 506–519.

ROYET J P, PLAILLY J, DELON-MARTIN C, KAREKEN D A and SEGEBARTH C (2003) 'fMRI of emotional responses to odors: influence of hedonic valence and judgment, handedness and gender', *Neuroimage*, 20, 713–728.

ROYET J P and PLAILLY J (2004) 'Lateralization of olfactory processes', *Chem Senses*, 29, 731–745.

RUSHWORTH M F S, WALTON M E, KENNERLEY S W and BANNERMAN D M (2004) 'Action sets and decisions in the medial frontal cortex', *Trends Cogn Sci*, 8, 410–417.

SAINZ E, LORLEY J N, BATTEY J F and SULLIVAN S L (2001) 'Identification of a novel member of the T1R family of putative taste receptors', *J Neurochem*, 77, 896–903.

SATO T (1972) 'Multiple sensitivity of single taste cells of the frog tongue to four basic taste stimuli', *J Cell Physiol*, 80, 207–218.

SCHIFFERSTEIN H N and VERLEGH P W (1996) The role of congruency and pleasantness in odor-induced taste enhancement', *Acta Psychol*, 94, 87–105.

SCOTT T R and PLATA-SALAMAN C R (1999) 'Taste in the monkey cortex', *Physiol Behav*, 67, 489–511.

SCOTT T R and MARK G P (1986) 'Feeding and taste', *Prog Neurobiol*, 27, 293–317.

SCOTT T R and GISA B K (2000) 'Issues of gustatory neural coding: where they stand today', *Physiol Behav*, 69, 65–76.

SEWARDS T V (2004) 'Dual separate pathways for sensory and hedonic aspects of taste', *Brain Res Bul*, 62, 271–283.

SHARP F R, KAUER J S and SHEPHERD G M (1975) 'Local sites of activity-related glucose metabolism in rat olfactory bulb during olfactory stimulation', *Brain Res*, 98, 596–600.

SICARD G and HOLLEY A (1984) 'Receptor cell responses to odorants: similarities and differences among odorants', *Brain Res*, 292, 283–296.

SILVER W L and FINGER T E (1991) 'The Trigeminal System', in Getchell T V, Doty R, Bartoshuk L and Snow J B, *Smell and Taste in Health and Disease*, New York, Raven Press, 97–108.

SIMONS C T (2002) 'Suppression of taste by oral irritants', *Symposium on Trigeminal Chemoreception*, Pommersfelden, Germany (proceeding abstract).

SIMONS C T, BOUCHER Y and CARSTENS E (2003) 'Suppression of central taste by oral capasaicin', *Neurosci*, 23, 978–985.

SINGER M S and SHEPHERD G M (1994) 'Molecular modelling of ligand-receptor interactions in the OR5 olfactory receptor', *Neuroreport*, 5, 1297–1300.

SLOTNICK B M, WESTBROOK F and DARLING F M C (1997) 'What the rat's nose tells the rat's mouth: Long delay aversion conditioning with aqueous odors and potentiation of taste by odors', *Anim Learn Behav*, 25, 357–369.

SMALL D M, JONES-GOTMAN M, ZATORRE R J, PETRIDES M and EVANS A C (1997) 'Flavor processing: more than the sum of its parts', *Neuroreport*, 8, 3913–3917.

SMALL D M, ZALD D H, JONES-GOTMAN M, ZATORRE R J, PARDO J V, FREY S and PETRIDES M (1999), 'Human cortical gustatory areas: a review of functional neuroimaging data', *Neuroreport*, 10, 7–14.

SMALL D M, ZATORRE R J, DAGHER A, EVANS A C and JONES-GOTMAN M (2001) 'Changes in brain activity related to eating chocolate: from pleasure to aversion', *Brain*, 124, 1720–1733.

SMALL D M, GREGORY M D, MAK Y E, GITELMAN D R, MESULAM M M and PARRISH T B (2003) 'Dissociation of neural representation of intensity and affective valuation in human gustation', *Neuron*, 39, 701–711.

SMALL D, VOSS J, MAY Y E, SIMMONS K B, PARRISH T and GITELMAN D (2004) 'Experience-dependent neural integration of taste and smell in the human brain', *J Neurophysiol*, 92, 1892–1903.

SOBEL N, PRABHAKARAN V, DESMOND J E, GLOVER G H, GOODE R L, SULLIVAN E V and GABRIELLI J D (1998) 'Sniffing and smelling; separate subsystems in the human olfactory cortex', *Nature*, 392, 282–286.

STEVENSON R J and PRESCOTT J (1995) 'The acquisition of taste properties by odors', *Learn. Motiv*, 26, 433–455.

TODA K, MIYAMOTO T, KIMOTO M, KUMEI Y and OKADA Y (2002) 'Responses of oral mucosal nociceptors to bitter solutions in rats', *Symposium on Trigeminal Chemoreception*, Pommersfelden, Germany (proceeding abstract).

TOMINAGA M, CATERINA M J, MALMBERG A B, ROSEN T A, GILBERT H, SKINNER K, RAUMANN B E, BASBAUM A I and JULIUS D (1998) 'The cloned capsaicin receptor integrates multiple pain-producing stimuli', *Neuron*, 21, 531–543.

TRAVERS S P and NORGREN R (1995) 'Organization of orosensory responses in the nucleus of the solitary tract of rat', *J Neurophysiol*, 73, 2144–2162.

TROTIER D and MAC LEOD P (1983) 'Intracellular recordings from salamander olfactory receptor cells', *Brain Res*, 268, 225–237.

VAN BUSKIRK R L and ERICKSON R P (1977) 'Odorant responses in taste neurons of the rat NTS', *Brain Res*, 135, 287–305.

VASSAR R, NGAI J and AXEL R (1993) 'Spatial segregation of odorant receptor expression in the mammalian olfactory epithelium', *Cell*, 74, 309–318.

VASSAR R, CHAO S K, SITCHERAN R, NUNEZ J M, VOSSHALL L B and AXEL R (1994) 'Topographic organization of sensory projections to the olfactory bulb', *Cell*, 79, 981–991.

VOIROL E and DAGET N (1986) 'Comparative study of nasal and retronasal olfactory perception', *Food Sci Technol*,19, 316–319.

WIGGINS L L, ROLLS E T and BAYLISS G C (1987) 'Afferent connections of the caudolateral orbitofrontal cortex taste area of the primate', *Chem Senses*, 12, 206.

XU F, KIDA I, HYDER F and SHULMAN R G (2000) 'Assessment and discrimination of odor stimuli in rat olfactory bulb by dynamic functional MRI', *Proc Natl Acad Sci USA*, 97, 10601–10606.

ZALD D H and PARDO JV (1997) 'Emotion, olfaction, and the human amygdala: amygdala activation during aversive olfactory stimulation', *Proc Natl Acad Sci USA*, 94, 4119–4124.

ZALD D H and PARDO J V (2000) 'Functional neuroimaging of the olfactory system in humans', *Int J Psychophysiol* 36, 165–181.

ZALD D H, LEE J T, FLUEGEL K W and PARDO J V (1998) 'Aversive gustatory stimulation activates limbic circuits in humans', *Brain*, 121, 1143–1154.

ZATORRE R J (2002) 'Processing of olfactory affective information: contribution of functional imaging studies', in Rouby C, Schaal B, Dubois D, Gervais R and Holley A, *Olfaction, Taste, and Cognition*, Cambridge, Cambridge University Press, 324–334.

ZATORRE R J, JONES-GOTMAN M, EVANS A C and MEYER E (1992) 'Functional localization and lateralization of human olfactory cortex', *Nature*, 360, 339–340.

ZHANG X and FIRESTEIN S (2002) 'The olfactory receptor gene superfamily of the mouse', *Nat Neurosci*, 5, 124–133.

ZHAO H, IVIC L, OTAKI J M, HASHIMOTO M, MIKOSHIBA K and FIRESTEIN S (1998) 'Functional expression of a mammalian odorant receptor', *Science*, 279, 237–242.

ZOU Z, HOROWITZ L F, MONTMAYEUR J P, SNAPPER S and BUCK L B (2001) 'Genetic tracing reveals a stereotyped sensory map in the olfactory cortex', *Nature*, 414, 173–179.

3

Sensory analysis of food flavor

A. C. Noble, University of California, USA and I. Lesschaeve, Brock University, Canada

3.1 Introduction

Despite advances in instrumental analysis the flavor sensations perceived by humans can be measured only by sensory tests. Analytical sensory tests determine if there is a perceptible difference among products and describe and quantify the nature of the differences, providing very detailed information about the sensory properties. The effect of any flavor differences on consumer preference can only be assessed by subjective sensory tests in which consumers rate preference or acceptance.

Two key issues affect the success of the sensory evaluation efforts. First, many sensory managers and personnel have no formal training in this area. Proper sensory testing ideally requires an awareness of the psychological and physiological factors which influence sensory perception. In addition, an understanding of the correct methods of statistical analysis of sensory and consumer data, and the limitations of these methods, is critical. Second, there is a lack of communication within most organizations at the management level between the sensory group and the engineering, production and marketing units. Products are often sent to the sensory team for difference tests or profiling without any background information. For example, a sensory group is asked to determine whether a formulation change produces a significant aroma difference. This can be done very sensitively using simple difference tests with trained panelists, but is this difference detectable by or relevant to the consumer? The initial step in any sensory study is to determine the purpose for the analysis so the appropriate tests can be conducted. The global purpose of the test in the company context has to be understood by the sensory group so that the appropriate analytical sensory tests are used and appropriate recommendations made.

The purpose of this chapter is to introduce the basic types of sensory tests and the standard sensory practices which provide guidelines for their administration. In addition, factors influencing the conclusions drawn from such tests will be presented. Excellent references for standard sensory methods are available (Carpenter *et al.* 2000, Lawless and Heymann 1998, Stone and Sidel 1993, Meilgaard *et al.* 1999).

3.2 Current and developing techniques for sensory analysis

3.2.1 Discrimination tests

Difference tests

When the products vary by differences that are too small to be described, the two most common difference tests are the duo-trio test and the triangle test. When the difference can be defined, pair tests (also known as two alternative forced choice tests or 2AFC) are used to ask which sample is stronger in a specified attribute (Lawless and Heymann 1998). For example, to determine if a modification in a cake formulation produced a difference in the sensory properties, duo-trio tests or triangle tests can be used to determine if overall differences can be detected at a predetermined level of significance. If the change in formulation appeared only to affect one attribute such as sweetness or intensity of vanilla aroma, then a pair test asking which of two cakes is sweeter or has the more intense vanilla aroma is employed. The three alternative forced choice test (3AFC) is similar to the pair or 2AFC test, except that three samples (two of one product and one of the other) are presented and panelists are asked to select which sample(s) are higher or lower in sweetness.

Threshold tests

The threshold value is the concentration of a compound at which a detectable difference in aroma or taste is found (detection threshold) or at which the characteristic odor or taste can be recognized (recognition threshold). Typically threshold values are determined by difference tests to estimate the compound's contribution to flavor or determine the sensitivity of panelists to a specific compound. The average threshold of a group of panelists is often measured to estimate the contribution of specific compounds to the aroma of a product. For example, the thresholds of potent volatiles associated with off-flavors in wines, such as ethyl mercaptan (1.1 μg/l) or dimethyl sulfide (25 μg/l) (Goniak and Noble 1987) are much lower than for acetic acid (800–1300 mg/l) (Corison and Ough 1979). Even more potent is that for trichloroanisole (TCA, 4 ng/l) (Amon *et al.* 1989) which produces a moldy smell in water and wines. Although a detection threshold indicates the likelihood that an off-flavor can be detected 50% of the time, at this level a 'tainted' milk or wine might be still acceptable to a consumer. To address this problem, a method has been proposed to determine the consumer rejection threshold of a given flavor or off-flavor in a product (Prescott *et al.* 2005). Thresholds are also determined by gas chromatography-olfactometry. It should be noted that threshold values are of limited use since the value applies only to the tested product under specific testing conditions.

In tests for specific compounds or spoilage notes such as TCA, determination of the threshold value of individual subjects identifies panelists who are sensitive to the specific compounds. Determination of thresholds is also done to study factors that influence individual differences, such as age, gender or disease. An alternative application of threshold testing is in shelf-life studies. For example, Chardonnay wines were stored at 40 °C for 5 to 9 days before a difference in aroma was perceived (de la Presa Owens and Noble 1997), whereas Chardonnay wines in clear and green glass bottles developed a detectable change in aroma when exposed to fluorescent light for 4 and 18 hours, respectively (Dozon and Noble 1989).

3.2.2 Analytical intensity rating tests

Types of scales

Several scaling procedures and types of scales are used to rate intensity of attributes. Category scales or unstructured line (graphic) scales have been used most frequently. Magnitude estimation (ratio scaling) in which the intensity of a sensation is rated relative to the intensity of a reference is often used in psychophysical studies. Magnitude matching (or cross modal matching) is a variation in which the intensity of taste, smell or mouthfeel is rated relative to a standard sound or light (Marks *et al.* 1988). Recently, the labeled magnitude scale (LMS), which is a combination of the ratio and category scales, was developed (Green *et al.* 1993). In contrast to category and graphic scales that are anchored at the ends by the terms 'low' and 'high', the LMS is anchored by 'barely detectable' and 'strongest you can imagine'.

When the sensory properties of most products are being compared, the unstructured line scales or category scales are ideal. For a normal range of intensities, the same results are obtained using ratio scaling as those found with category or graphic scales, yet the latter are simpler to use than magnitude estimation methods (Giovanni and Pangborn 1983). For descriptive analysis in which many attributes are rated, estimating each on a scale anchored by 'strongest you can imagine' seems difficult and use of the LMS scale has not been attempted.

However, the LMS scale or cross-modal matching techniques are more appropriate for comparing responses of different groups of subjects (such as those classified by age, culture, salivary flow status, or by sensitivity to compounds such as propyl thiouracil (PROP). Use of either method, removes the variability in use of scale among individuals. As a consequence, the LMS scale is more sensitive for measuring the strength of the sensation perceived by individuals who perceive stimuli very differently because of inherent physiological differences (Prutkin *et al.* 2000).

Descriptive analysis

Descriptive sensory analysis (DA) is a highly sophisticated technique that is used to obtain complete sensory descriptions of products. These quantitative ratings or profiles of flavor attributes can be used to monitor products produced by competitors, identify underlying ingredient and process variables, in shelf-life studies and in product development, as described elsewhere (Heymann *et al.* 1993, Issanchou *et al.* 1997). In addition, data from descriptive analyses can be related to preference ratings (see section 3.5.2) or to instrumental results (see section 3.5.1 and Chapter 5 (Qannari and Schlich 2006)).

The different methodologies for doing DA are compared in detail elsewhere (Murray *et al.* 2001). They include Quantitative Descriptive Analysis (QDA)® (Stone *et al.* 1974, Stone and Sidel 1993), Sensory Spectrum® (Meilgaard *et al.* 1999) and Free Choice Profiling (FCP) (Williams and Langron 1984, Williams and Arnold 1985, Beal and Mottram 1993). In QDA®, each panel develops a consensus list of terms to describe differences among products, in contrast to the Spectrum® approach in which a standardized lexicon of terms is used in all evaluations of a class of products along with an 'universal measurement scale'. In FCP, each panelist develops his/her own unique list of terms. Although this technique requires the least training, it is the most difficult to interpret given the idiosyncratic descriptions of flavors produced by untrained panelists (Williams and Langron 1984). Generic descriptive analysis, which follows the general guidelines of the methods above, is used in industry and academic research. The specific methods for descriptive analysis vary in the manner in which terms are generated and standardized and panelists are trained. Despite this, in general, they yield similar results and are subject to the same limitations (European Sensory Network 1996).

There are two primary requirements for DA to yield complete flavor profiles. First, the vocabulary or list of terms derived must fully describe the differences in flavor among the products, using specific concrete terms that are not redundant. Second, extensive training of panelists is required for consistent, reproducible rating of each attribute. Untrained consumers cannot be used as panelists, unless they undergo the same rigorous training as trained descriptive panel members. If these requirements are not met, then any attempt to relate the sensory DA data to consumer or instrumental data will be severely limited or flawed.

In the first sessions, samples are presented which have large flavor differences. The panel proposes terms to describe the differences, which are then defined by physical reference standards prepared by the panel leader. At initial sessions, samples and reference standards are presented, and the panel discusses the appropriateness of the terms for rating the flavor of the experimental products. Over time, the panelists generate a list of attributes for which there is consensus. Subsequent training sessions focus on training panelists to consistently use the terms. After the panelists are trained, the intensity of each term is rated in practice sessions on a range of products. Analysis of replicated data from these sessions permits evaluation of panelist performance. See section 3.4.2.

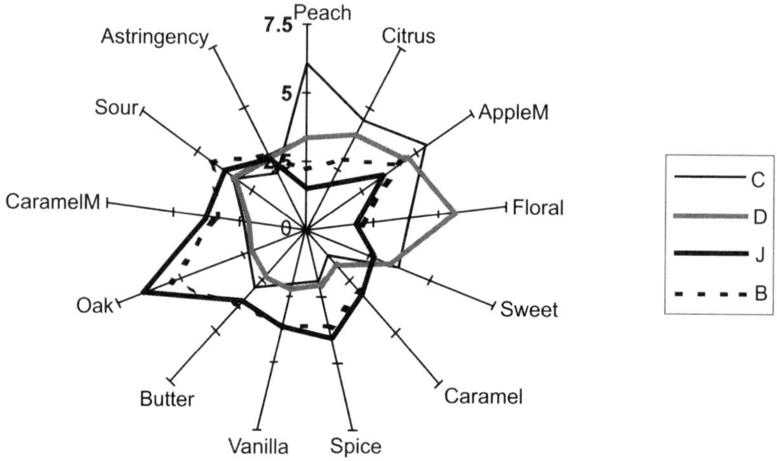

Fig. 3.1 Mean intensity of aroma and flavor by mouth terms for four Chardonnay wines. At the origin intensity = 0; at the perimeter, high intensity = 7 (Yegge, 2001).

The flavor profiles of the samples are revealed by plotting the average ratings from the descriptive analysis. For example, ten Chardonnay wines were evaluated by descriptive analysis. In Fig. 3.1, the intensity ratings are shown for four of the most different wines. C and D are high in peach, citrus and floral aroma, and low in the attributes associated with aging wine in oak barrels (oak, vanilla, spice, and caramel). In contrast, J and B are high in the oak notes, but very low in the fruity and floral characteristics. To show the overall relationship of the flavors of the ten wines, a principal component analysis (PCA) of these data was performed. A PCA represents most of the information contained in the descriptive terms (13 in this example) in two (or three) principal components (PC) or factors, as described elsewhere (Jolliffe 1986, Qannari and Schlich 2006). It is an excellent exploratory method to have an overview of the relationship of the samples and their overall differences in the descriptive terms. The relationship of the wines and the attributes can then be seen by their projection on these few PCs as shown in Fig. 3.2. Here, the loadings for the sensory terms are shown as vectors and the factor scores of the wines as points for the first two principal components. The first PC (which accounts for 72.6% of the total variation) contrasts wines (on the left) which are high in peach/apricot, citrus, sweetness and floral *vs* those wines on the right which are low in these terms, but high in oak, spicy, vanilla and butter notes (Yegge 2001, Yegge and Noble 2001).

Training panelists to perform descriptive analysis on flavor attributes is challenging due to the nature of the flavor compounds imparting the sensation and their dynamic interaction with other compounds. Moreover, there is no standardized or universal vocabulary for description of aromas, in contrast to color, taste or some texture attributes such as soft/hard. Therefore, untrained panelists use words that are familiar and descriptive of their individual experiences. Several techniques have been developed to align panelist's individual

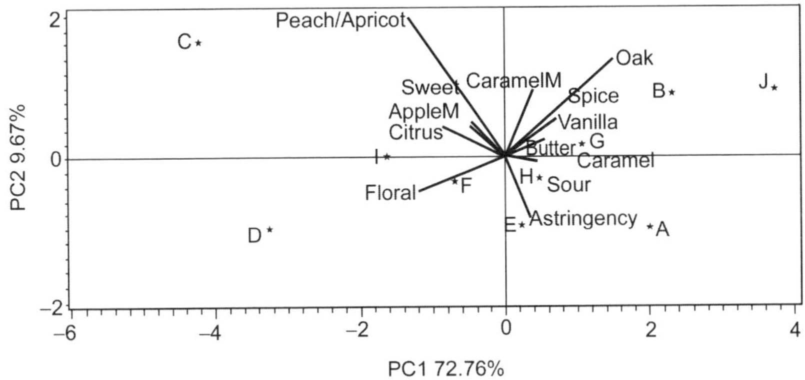

Fig. 3.2 Principal component analysis of descriptive analysis data. The wines are shown as letters and attributes are shown as vectors for the first two principal components (Yegge and Noble, 2001). (Copyright © 2001. Reprinted by permission of the American Society for Enology and Viticulture.)

words into one consensus concept. All of them rely on reference standards, such as those for cheese (Murray and Delahunty 2000) and wine (Noble *et al.* 1987). Reference standards are very useful to initiate the training of novice descriptive panelists (Rainey 1986), but using actual test products is a more efficient method for concept alignment than relying on reference standards (Sulmont *et al.* 1999). Calibration of the descriptive panel implies the alignment of descriptive attributes among panelists as well as in the use of the rating scale. For the latter, giving feedback to panelists on their rating performance is critical (Kuesten and McLellan 1994).The use of visual and immediate feedback has been shown to reduce the training time by 50% for panels working on a complex product such as wine (Findlay *et al.* 2004).

Time–intensity (T–I)

Time–intensity procedures have been used to characterize persistent sensations and to monitor perceived intensity as flavor is released during ingestion and mastication. For example, T–I can determine the manner in which variation in composition or the food matrix affects the temporal patterns of flavor. It has also been used in tandem with analysis of volatiles in the oral or nasal cavities, see Chapter 13 (Linforth and Taylor 2006). T–I is a very time consuming process since rating intensity continuously is a complex task. Even with extensive training, individual panelists have characteristic curves with idiosyncratic, yet reproducible, patterns. The way in which panelists move their tongues, the rate at which they chew and their salivary flow rates influence their perception of intensity and persistence of the sensation. Despite the difference in temporal patterns, panelists can be trained to give reproducible responses that are consistent across samples.

The time to maximum intensity, decay rate and total duration of sweetness of sucrose and artificial sweeteners vary, so T–I is a useful technique to determine the equal sweetness at maximum intensity when sweeteners are substituted for sucrose (Matysiak and Noble 1991, Noble 1995). However, there is no method to measure overall sweetness. Is it determined on the first impression, as the sensation after swallowing or as an integrated response? Do individuals do this in the same way? Development of a methodology to address this and similar issues is needed.

In many products, changing the composition affects the temporal patterns of many flavor notes. For example, this occurs in ice-cream or yogurt when the fat content is altered (Chung *et al.* 2003). T–I is appropriate to characterize the flavors of the modified product *vs* the full fat one, but T–I is far too time consuming since each attribute is rated separately. A better method in this case is a dynamic time intensity method that focuses on the dominant attributes, rather than rating intensity of one attribute. In the Temporal Dominance of Sensations technique (TDS), the monitor displays all attributes and panelists continuously record the intensity of the dominant sensation. The TDS method has permitted better discrimination among products than standard T–I (Pineau *et al.* 2004). TDS has potential application for the flavor industry in formulation of flavors. It permits modification of the flavor essence so that the temporal profile of the low-fat product more closely resembles the original product.

3.2.3 Consumer tests

For all preference or acceptance testing, it is crucial to identify and recruit target consumers of the products or flavors to be evaluated. Data from analytical sensory tests do not predict consumer liking. Similarly, preferences of personnel at a work site or 'experts' do not represent the preferences of either frequent or infrequent consumers of a product. Focus groups can be useful to provide guidance in identifying factors that influence the consumer's preference (Casey and Krueger 1994). However, it should be recognized that these data are only qualitative and may not reflect the actual behavior of the consumers in normal consumption situations. A comprehensive review of methods for consumer research during product development is available (van Kleef *et al.* 2005).

Most frequently, pair preference tests are used or samples are ranked or rated for preference on hedonic scales anchored by terms such as 'dislike extremely' to 'like extremely'. An alternative approach is based on auction methods. The experimenter assigns a price to the products, and the consumers indicate which products they wish to be given at the end of the test using a defined total budget (Lange *et al.* 1999). Regardless of the method, a large number of target consumers must be tested to obtain valid results because of the tremendous variation in preferences. For a general reference on consumer testing, see Resurreccion (1998).

Preference and purchase decisions are affected by price, brand, label and product information in addition to flavor as discussed elsewhere (Tuorila *et al.* 1994, Bower *et al.* 2003). As a consequence, this information is often presented to the consumer during testing to allow them to make a purchase decision based on flavor and external factors in contrast to analytical tests in which samples are tasted blind to remove any bias or expectation.

Consumers cannot consistently describe why they like or dislike samples, because they do not have the vocabulary to describe flavor notes accurately. Therefore, to identify the flavors of highly preferred products, preference ratings are related to descriptive analysis data using multivariate techniques called preference mapping (see section 3.5.2). Another approach to understanding flavor preference provides insights into consumer language by comparing the use of consumers' descriptive words with the descriptive terms used by the trained panel on the same set of products (Carr *et al.* 2001). For example, the terms 'dry', 'bad', 'bitter', 'astringent' and 'sour' were used interchangeably to describe sour or bitter or astringent products (Lesschaeve 2003). An alternative approach for eliciting and analyzing words from consumers which can be related to analytical sensory profiles is repertory grid analysis which is described elsewhere (McEwan and Thomson 1989).

3.3 Sensory testing administration

3.3.1 Standard sensory practices

Perception of trained panelists is influenced by psychological and physiological factors that can cause additional sources of 'experimental noise'. Consequently, analytical sensory experiments are conducted under conditions designed to minimize or exclude bias and distraction. For example, samples are always coded and presented uniformly. For evaluation of flavor, differences in appearance are masked by serving samples in opaque containers or under red light. The environment in which sensory tests are conducted should be temperature-controlled, quiet and odor-free and have controlled lighting as described by Eggert and Zook (1986). In contrast to analytical tests, consumer preference tests are done at home or a central location more often than in a lab environment. For a good discussion of principles of 'good practice', see Lawless and Heymann (1998).

In most analytical sensory tests, samples are expectorated rather than being swallowed to reduce any change in response due to satiety, fatigue or, in the case of alcoholic beverages, increase in blood alcohol. To reduce fatigue and remove the previous sample, subjects should breathe fresh air, sniff or rinse with water between samples. The specific inter-stimulus protocol varies with each product. Warm water is helpful for rinsing between oily or fat samples, while crackers or rinses of hydrocolloid gums (Brannan *et al.* 2001) or pectin (followed by thorough rinses with water) help reduce bitterness or astringency of beverages such as tea or wine (Colonna *et al.* 2004).

Many trigeminal compounds (which are perceived by pain and touch receptors) show both sensitization (increased perception of intensity) and desensitization, depending on the time between presentation of samples. Pungency and burn of compounds such as capsaicin initially increase when samples are presented at one to five minute intervals. When the sample is presented again after a break of 15 minutes, desensitization occurs and the intensity of burn is much lower (Green 1989). Thus evaluation of 'hot' products must be carefully made for valid results.

3.3.2 Experimental design

When there are too many products to be evaluated in one session without fatiguing the panelists, experimental designs should be used to randomly assign samples to different sessions. In addition, both the context of presentation and the sequence in which samples are presented affect perception, thus samples should be served in randomized orders within each session. For example, shifts in intensity due to context were observed in sweetness ratings of a fruit beverage. Mid-range juices were less intense in the context of stronger items and more intense in the context of weaker items (Diamond and Lawless 2001). For experimental designs, see texts such as Cochran and Cox (1957); designs for difference tests are available as well (Stone and Sidel 1993, Meilgaard *et al.* 1999). Sophisticated software is available as well to assist in the design of experiments. In analytical tests and preference studies sequence has a large effect. Samples presented in the first position are usually rated higher and are more preferred, than subsequent samples irrespective of sample identity. In an inter-laboratory study of preference, eight coffees were presented in a Latin square design. The preference score for the first position, regardless of coffee identity, was significantly higher than the other positions (European Sensory Network 1996). To eliminate the effect of the first sample being rated differently, a 'warm-up' can be presented first and its scores discarded (Stone and Sidel 1993).

In evaluation of some products, time-order or carry-over effects also occur (Amerine *et al.* 1965). In the evaluation of astringent or bitter products, the second sample is almost always perceived as higher in intensity. Thus, the design must randomize the order of presentation as well as balance the number of times each specimen is presented in a specific sequence. This can be done using William's Latin squares designs which are balanced for first order carry-over effects (MacFie and Bratchell 1989, Schlich 1993b). In some cases, the carry-over effect is compound specific. In a study in which bitter compounds were presented in specific sequences, no matter what compound was presented first, all bitter compounds in the study increased in bitterness when presented after another bitter stimuli. Caffeine increased bitterness of all compounds that followed it by the largest amount, although bitterness of caffeine was least affected by other compounds. In contrast the increase in bitterness of quinine was higher than the other bitter compounds in the study (Cubero-Castillo and Noble 2001).

3.3.3 Panelist selection

Initially, panelists should be selected on the basis of availability and motivation. The specific requirements for selecting panelists vary with the type of test. If a test involves detection of a specific off-odor, such as the moldy or musty notes of geosmin in water or trichloroanisole in wine corks, select panelists who are very sensitive to these compounds. To determine the sensitivity of individuals, a series of pair tests can be presented to the panelist in which a blank is compared with low concentrations of the specific chemical, or a formal threshold test can be done. For more general studies, panelist performance should be examined. Retain subjects who perform reproducibly and are consistent with other panelists, as discussed in section 3.4.2.

3.4 Statistical analysis of data

3.4.1 Determining statistical significance of results

Tables for binomial probability are available for determining the significance of difference tests (Lawless and Heymann 1998, O'Mahony 1986). Data from simple scaling tests, descriptive analysis, or time–intensity studies are evaluated by analysis of variance (ANOVA) to determine whether the samples were significantly different in an attribute and if the panelists were consistent and reproducible in rating this attribute. When significant differences among samples are found for an attribute by ANOVA, tests such as Duncan's multiple range test or Fischer's least significant difference (LSD) identify which products differed significantly from each other (O'Mahony 1986). Multivariate analysis of variance (MANOVA) is also used to determine whether products differ significantly over all attributes. A good introduction to MANOVA is found in Lawless and Heymann (1998).

3.4.2 Evaluating panelist performance

Keeping records on panelist performance permits selection of panelists who are consistent, sensitive, and reproducible. We define performance as the ability of a panel to make valid and reliable product and attribute assessments. For difference tests, the cumulative number of correct judgements can be tracked (Amerine *et al.* 1965). For rating tests or descriptive measurements, performance is characterized by the ability of a panel to detect an attribute, use attributes in the same way as other panels or panelists, discriminate between products, use a scale properly, repeat their own results, and reproduce results from other panels or panelists. Many techniques have been proposed to visualize or track descriptive panelists' performance, using graphical, univariate or multivariate statistics.

Most frequently, ANOVAs of replicated data for individual panelists for each term are used to determine if panelists were reproducible and able to differentiate products according to attribute intensity; an ANOVA of all panelists on a term reveals whether they were consistent with each other in rating the attribute (i.e. rated the term in the same manner). However, if panelists were inconsistent in their rating of a term, determining which panelists contributed significantly to this error can be tedious. Several graphical methods for comparing panelist performance based on analysis of variance of each attribute have been proposed (Baardseth *et al.* 1994, Brockhoff 2001, Hirst and Naes 1994; Schlich,1994) (see also Schlich (1996)). Generalized procrustes analysis (GPA) compares spaces derived by principal component analyses of each panelist to determine whether there is a significant fit or consensus among the panelists (Arnold and Williams 1986, Dijksterhuis and Punter 1990, Dijksterhuis and Gower 1991, McEwan *et al.* 2002). A significant fit among the panelists indicates their consistency in use of terms. Similarly, GPA has been used to compare results of different sensory panels (Martin *et al.* 2000).

3.5 Relating analytical sensory data to instrumental or consumer preference data

3.5.1 Sensory–instrumental

Because of the complexity of both sensory data and instrumental information, multivariate statistical methods are used for interpreting their relationships. For a detailed discussion of methods for analysis of sensory instrumental relationships, see Chapter 5 (Qannari and Schlich 2006). Many of the techniques for relating sensory and instrumental data compare the configurations of sensory and instrumental data matrices. Calculation of the RV coefficient has been used as a measure of similarity between the matrices of sensory data and instrumental data (Schlich and Guichard 1989, Schlich *et al.* 1987). Generalized procrustes analysis has also been used to compare the relationship of instrumental to sensory data (Le Fur *et al.* 2003). The method of partial least squares (PLS) analysis of latent variables (Martens and Martens 1986) is used to uncover relationships between sensory and instrumental data sets, design variables or sample composition. PLS indicates how well variables in one data set predict or model the variation among variables in the second. RV, GPA and PLS have been shown to yield similar results for interpreting instrumental–sensory relations (Noble and Ebeler 2002).

3.5.2 Sensory–consumer

Generic and proprietary preference mapping programs have been used to relate consumers' preferences to the sensory descriptive analysis data. Internal preference analysis looks at the configuration of preferences of each individual and by inference from sensory profiles reveals qualitative information about the basis for the preferences. In contrast, external preference mapping (prefmap) relates preference data to sensory profiles. The data for each consumer can be used, but more commonly cluster analysis of the preference data is done first to identify market segments of consumers with similar preferences. The average preferences of the segments are then regressed onto the sensory space. Different models, such as linear, circular, elliptical and quadratic, are used to best model consumer preferences (Schlich 1995). The models developed by external pre-mapping characterize the sensory attributes driving consumer preference and identify consumer segments based on their product preferences (Lesschaeve *et al.* 2001).

In addition, relating consumer data to sensory profiling information permits development of a product which maximizes consumer preferences (Lesschaeve and Findlay 2004). Multivariate techniques such as 'reverse engineering' (Moskowitz 1994) or the PrefMax approach (Schlich *et al.* 2003) define the optimal sensory attribute intensity for a given consumer (or consumer segment) that would maximize his/her liking score for the product considered. Using this approach, sensory specifications for an 'ideal' product can be generated for the product development team. For more information of preference mapping see (McEwan 1996, Risvik *et al.* 1994, Schlich 1995, Greenhoff and MacFie 1994).

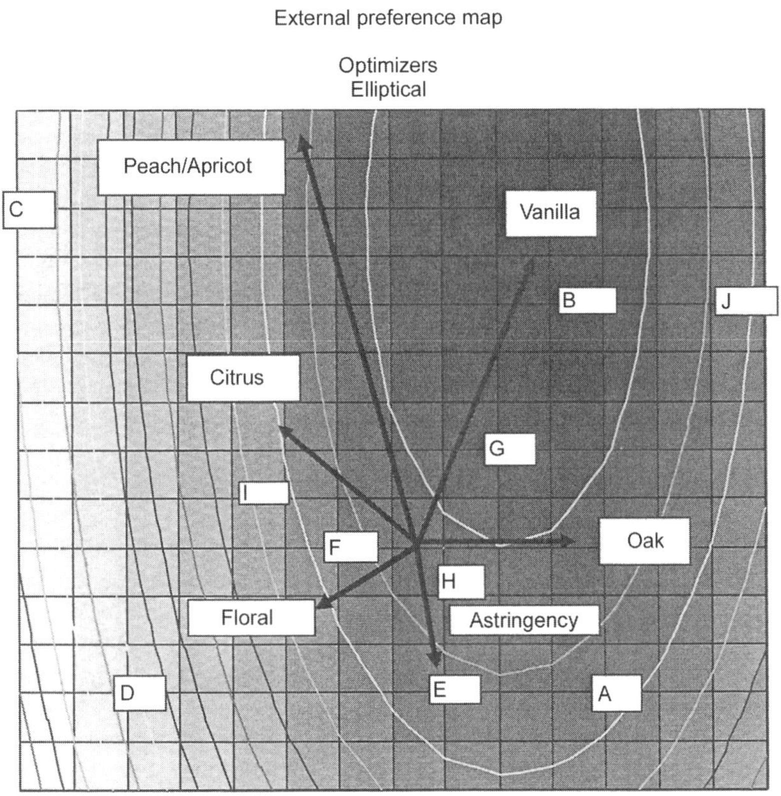

Fig. 3.3 Two-dimensional external preference map for Chardonnay wines by 'optimizer' segment. Preference contours are superimposed on PC1 and PC2 of Principal Component Analysis shown in Fig. 3.2. Vectors indicate sensory attributes which account for the most variation in PC1 and PC2. Wines are shown as letters. Center contour equals highest preference rating (5.0) (Yegge and Noble, 2001). (Copyright © 2001. Reprinted by permission of the American Society for Enology and Viticulture.)

To illustrate preference mapping results, data from a Chardonnay study are presented. Consumers rated preferences for the ten Chardonnays for which descriptive analysis results were provided in Figs 3.1 and 3.2. Cluster analysis performed on these preference scores revealed five market segments. The average preference ratings for each segment were regressed onto the descriptive analysis sensory space. Two of the three segments which were modeled significantly were linear. Segment one liked C and D best but would have rated even fruitier wines even higher in preference. Segment two most preferred wines J, B, and G, which were high in oak notes and relatively low in fruitiness. A third segment was fit by an elliptical model and represents 'optimizers'. As shown in the external preference map which modeled this cluster (Fig. 3.3), their most preferred wine was G, followed closely by B. Wine J which was higher in the oak-related notes was less preferred. Similarly, wines F and I which were higher in fruitiness and lower in oak notes were liked less than B and G (Yegge and Noble 2001).

3.6 Using sensory data for business decisions

Sensory data to be used for making business decisions can only be meaningful if the sensory unit is informed about the global purpose of the test. Often the sensory team is isolated from other corporate functions. The sensory results are fed into a pipeline, with no further communication with management. Results of sensory tests can be used for making meaningful business decisions only when the sensory group is integrated with other teams at the management level. With this integrated approach, the experiment can be designed to provide the optimal information and permit the sensory group to make actionable recommendations based on the test results. Below, we give two examples of the need for interpreting sensory data in an integrated context.

3.6.1 Significance of results

What level of significance should be used in interpreting the results? This depends on the importance of the attribute. For example, a change in formulation is made in which a different sweetener is substituted. Pair tests or rating tests using trained panelists in booths show that a difference in sweetness is perceptible at a 5% significance level or value of alpha. This means there is a probability of 5% ($p = 0.05$) that the results could have occurred by chance alone or conversely we are 95% confident that there is a difference. Is this difference large enough to justify the cost of implementing a new process or using a new ingredient?

In contrast to this scenario, if the testing is to determine whether a defect or negative characteristic is detectable, a probability of 5% indicates a difference that is of concern. If a very small number of consumers can detect this off-note, then the product will be rejected. In this situation, it is important that the test has high power. Power is the probability that a real difference is found or that we did not fail to find a real difference. Power is determined by three factors: the number of panelists and replications, the magnitude of the differences among the samples (effect size) and the alpha level chosen. Power can be increased by using more panelists and conducting more replications. Trained, sensitive panelists can perceive very small differences and are more reproducible, thus increasing power over untrained panelists. Testing of power is seldom done, but it should be performed when decisions are made based on the sensory tests. For more information about power, see Cohen (1988), Lipsey (1990), Schlich (1993a) or Lawless and Heymann (1998).

3.6.2 Relevance of analytical results to consumer perception

Usually, the very small difference that is detected by analytical panelists in a booth will not be perceived by consumers during normal consumption or use. But if the attribute is important or relevant to the consumer, even a small difference, for example at $p < 0.05$, may be detected by consumers under normal conditions of consumption. Paradoxically, consumers who could not detect differences between two waters demonstrated a significant preference for one of them (MacRae and Falahee 1995). A similar result has been found using authenticity tests. Consumers who could not discriminate among milks (Wolf Frandsen *et al.* 2003) or beers (Köster 1998) were able to identify the brand which they usually consumed when it was presented with other samples. A sensory method for determining how big a difference must be for it to be meaningful (i.e. perceptible to a consumer) needs to be developed.

3.7 Conclusion

Sensory analysis can provide objective quantitative information about the sensory properties of flavors when trained sensory professionals conduct properly designed tests, using trained panelists and appropriate sensory protocols. The existing sensory methods provide valuable information when used rigorously, but full value of these sensory tests is only fully realized when they are conducted with knowledge of the ultimate purpose for the test. In many cases, projects should be started at the consumer end, initially assessing what factors are important in exploratory focus groups or even with preference studies of existing products. Sensory tests can be designed, interpreted and used for making meaningful business decisions only by full integration of the sensory unit with other company sections at the management level.

3.8 References

AMERINE M A, PANGBORN R M and ROESSLER E B (1965) *Principles of Sensory Evaluation of Foods*, New York, Academic Press.

AMON J M, VANDEPEER J M and SIMPSON R F (1989) 'Compounds responsible for cork taint in wine', *Wine Ind*, 62–69.

ARNOLD G M and WILLIAMS A A (1986) 'Use of generalised procrustes techniques in sensory analysis', in Piggott J R, *Statistical Procedures in Food Research*, 233–254.

BAARDSETH P, HIRST D and NAES T (1994) 'Using cumulative ranks to detect individual differences in sensory profiling', *J Sensory Studies*, 9, 87–99.

BEAL A D and MOTTRAM D S (1993) 'An evaluation of the aroma characteristics of malted barley by free-choice profiling', *J Sci Food Agric*, 61, 17–22.

BOWER J A, SAADAT M A and WHITTEN C (2003) 'Effect of liking, information and consumer characteristics on purchase intention and willingness to pay more for a fat spread with a proven health benefit', *Food Qual Pref*, 14, 65–74.

BRANNAN G D, SETSER C S and KEMP K E (2001) 'Effectiveness of rinses in alleviating bitterness and astringency residuals in model solutions', *J Sensory Studies*, 16, 261–275.

BROCKHOFF P B (2001) 'Sensory profile average data: combining mixed model ANOVA with measurement error methodology', *Food Qual Pref*, 12, 413–426.

CARPENTER R P, LYON D H and HASDELL T A (2000) *Guidelines for Sensory Analysis in Food Product Development and Quality Control*, Gauthersburg, MD, Aspen Publishers.

CARR B T, CRAIG-PETSINGER D and HADLICH S (2001) 'A case study in relating sensory descriptive data to product concept fit and consumer vocabulary', *Food Qual Pref*, 12, 407–412.

CASEY M A and KRUEGER R A (1994) 'Focus Group Interviewing', in MacFie H J H and Thomson D M H, *Measurement of Food Preferences*, Glasgow, Blackie Academic, 77–98.

CHUNG S J, HEYMANN H and GRUN L U (2003) 'Temporal release of flavor compounds from low-fat and high-fat ice cream during eating', *J Food Sci*, 68, 2150–2156.

COCHRAN W G and COX G M (1957) *Experimental Designs*, New York, Wiley.

COHEN J (1988) *Statistical Power Analysis for the Behavioral Sciences*, Hillsdale, NJ, Lawrence Erlbaum Associates.

COLONNA A, ADAMS D O and NOBLE A C (2004) 'Comparison of procedures for reducing astringency carry-over effects in evaluation of red wines', *Australian J Grape Wine Res*, 10, 26–31.

CORISON C and OUGH C S (1979) 'Must acetic acid and ethyl acetate as mold and rot indicators in grapes.', *Am J Enol Vitic*, 30, 130–134.

CUBERO-CASTILLO E M and NOBLE A C (2001) 'Effect of compound sequence on bitterness enhancement', *Chem Senses*, 26, 105–116.

DE LA PRESA OWENS C and NOBLE A C (1997) 'Effect of storage at elevated temperatures on aroma of Chardonnay wines', *Am J Enol Vitic*, 48, 310–316.

DIAMOND J and LAWLESS H T (2001) 'Context effects and reference standards with magnitude estimation and the labeled magnitude scale', *J Sensory Studies*, 16, 1–10.

DIJKSTERHUIS G B and GOWER J C (1991) 'Interpretation of generalized procrustes analysis and allied methods', *Food Qual Pref*, 3, 67–87.

DIJKSTERHUIS G B and PUNTER P H (1990) 'Interpreting generalized Procrustes analysis "Analysis of Variance" tables', *Food Qual Pref*, 2, 255–265.

DOZON N M and NOBLE A C (1989) 'Investigation of the effect of fluorescent light on the aroma of sparkling wine and its base wine', *Am J Enol Vitic*, 40, 265–271.

EGGERT J and ZOOK K (1986) *Physical Requirement Guidelines for Sensory Evaluation laboratories*, Philadelphia, PA, American Society Testing Materials.

EUROPEAN SENSORY NETWORK (1996) *European Sensory and Consumer Study. A case study on coffee*, Chipping Camden, England, Campden & Chorleywood Food Research Association.

FINDLAY C, CASTURA J, SCHLICH P and LESSCHAEVE I (2004) 'Optimizing the proficiency of wine panels trained using feedback calibration', In *Proceedings 7th Sensometrics Conference*, Davis, CA, 39.

GIOVANNI M E and PANGBORN R M (1983) 'Measurement of taste intensity and degree of liking of beverages by graphic scales and magnitude estimation', *J Food Sci*, 48, 1175–1182.

GONIAK O and NOBLE A C (1987) 'Sensory study of selected sulfur volatile compounds in white wine', *Am J Enol Vitic*, 38, 223–227.

GREEN B G (1989) 'Capsaicin sensitization and desensitization on the tongue produced by brief exposures to a low concentration', *Neurosci Lett*, 107, 173–178.

GREEN B G, SHAFFER S and GILMORE M (1993) 'Derivation and evaluation of a semantic scale of oral sensation magnitude with apparent ratio properties', *Chem Senses*, 18, 683–702.

GREENHOFF K and MACFIE H J H (1994) 'Preference mapping in practice', in MacFie H J H and Thomson D M H, *Measurement of Food Preferences*, Glasgow, Blackie Academic, 137–166.

HEYMANN H, HOLT D L and CLIFF M A (1993) 'Measurement of flavor by sensory descriptive techniques', in Ho C-T and Manley C H, *Flavor Measurement*, New York, Marcel Dekker, 113–132.

HIRST D and NAES T (1994) 'A graphical technique for assessing differences among a set of rankings', *J Chemometrics*, 8, 81–93.

ISSANCHOU S, SCHLICH P and LESSCHAEVE I (1997) 'Sensory analysis: Methodological aspects relevant to the study of cheese', *Lait*, 77, 5–12.

JOLLIFFE I T (1986) *Principal Component Analysis*, London, Springer-Verlag.

KÖSTER E P (1998) 'Épreuves Hédoniques', in *Évaluation sensorielle: Manuel Méthodologique*, Paris, Lavoisier, 182–207.

KUESTEN C L and MCLELLAN M R (1994) 'Strategies Toward the Development of an Expert-System For Sensory Evaluation Testing and Database-Management', *Behav Res Methods Instruments & Computers*, 26, 112–116.

LANGE C, ROUSSEAU F and ISSANCHOU S (1999) 'Expectation, liking and purchase behavior under economical constraint', *Food Qual Pref*, 10, 31–39.

LAWLESS H T and HEYMANN H (1998) *Sensory Evaluation of Food. Principles and Practices*, New York, Chapman & Hall.

LE FUR Y, MERCURIO V, MOIO L, BLANQUET J and MEUNIER J M (2003) 'A new approach to examine the relationships between sensory and gas chromatography-olfactometry data using generalized procrustes analysis applied to Six French Chardonnay wines', *J Food Agric Chem*, 51, 443–452.

LESSCHAEVE I (2003) 'Communicating wine sensory benefits to consumers', in *Proceedings 54th Am Soc Enol Vitic Ann Meeting*, Reno, NV, ASEV.

LESSCHAEVE I and FINDLAY C (2004) 'Development of a wine style guided by consumer research', in *Proceedings 12th Australian Wine Industry Technical Conference*, AWRI, Melbourne, Australia.

LESSCHAEVE I, NORRIS L N and LEE T H (2001) 'Defining and targeting consumer preferences, in *Proceedings 11th Australian Wine Industry Technical Conference*, AWRI, Adelaide, Australia.

LINFORTH R and TAYLOR A (2006) 'The processing of flavour release', in Voilley A and

Etiévant P (eds), *Flavour in Food*, Cambridge, Woodhead, 287–307.

LIPSEY M W (1990) *Design Sensitivity: Statistical Power for Experimental Research*, Newbury Park, CA, Sage Publications.

MACFIE H and BRATCHELL N (1989) 'Designs to balance the effect of order of presentation and first-order carry-over effects in Hall tests', *J Sensory Studies*, 4, 129–148.

MACRAE A W and FALAHEE M (1995) 'Theoretical note on a practical problem: effective screening of drinking water for taints', *Food Qual Pref*, 6, 69–74.

MARKS L E, STEVENS J C, BARTOSHUK L M, GENT J F, RIFKING B and STONE V K (1988) 'Magnitude-matching: The measurement of taste and smell', *Chem Senses*, 13 (1), 63–87.

MARTENS M and MARTENS H (1986) 'Partial least squares regression', in Piggot J R, *Statistical Procedures in Food Research*, London, Elsevier Applied Science, 292–359.

MARTIN N, MOLIMARD P, SPINNLER H E and SCHLICH P (2000) 'Comparison of odour sensory profiles performed by two independent trained panels following the same descriptive analysis procedures', *Food Qual Pref*, 11, 487–495.

MATYSIAK N L and NOBLE A C (1991) 'Comparison of temporal perception of fruitiness in model systems sweetened with aspartame, an aspartame+acesulfame K blend, or sucrose', *J Food Sci*, 56, 823–826.

MCEWAN J, HUNTER T, VAN GEMERT L J and LEA P (2002) 'Proficiency testing for sensory profile panels: measuring panel performance', *Food Qual Pref*, 13, 163–172.

MCEWAN J A (1996) 'Preference mapping for product optimization', in Naes T and Risvik E, *Multivariate Analysis of Data in Sensory Science*, New York, Elsevier Science, 71–102.

MCEWAN J A and THOMSON D M H (1989) 'The repertory grid method and preference mapping in market research: a case study on chocolate confectionery', *Food Qual Pref*, 1, 59–68.

MEILGAARD M, CIVILLE G V and CARR B T (1999) *Sensory Evaluation Techniques*, Boca Raton, FL, CRC Press.

MOSKOWITZ H R (1994) *Food Concepts and Products: Just in Time Development*, Trumbull, Food & Nutrition Press.

MURRAY J M and DELAHUNTY C M (2000) 'Selection of standards to reference terms in a cheddar-type cheese flavour language', *Journal of Sensory Studies*, 15, 179–199.

MURRAY J M, DELAHUNTY C M and BAXTER I A (2001) 'Descriptive sensory analysis: past, present and future', *Food Res Int*, 34, 461–471.

NOBLE A C (1995) 'Application of time-intensity methodology for evaluation of taste and mouthfeel', *Am J Enol Vitic*, 46, 128–133.

NOBLE A C and EBELER S E (2002) 'Use of multivariate statistics in understanding wine flavour', *Food Rev Int*, 18, 1–21.

NOBLE A C, ARNOLD R A, BUECHSENSTEIN J, LEACH E J, SCHMIDT J O and STERN P M (1987) 'Modification of a standardized system of wine aroma technology', *Am J Enrol Viticult*, 38.

O'MAHONY M (1986) *Sensory Evaluation of Food: Statistical Methods and Procedures*, New York, Marcel Dekker.

PINEAU, N, CORDELLE S, IMBERT A, ROGEAUX M and SCHLICH P (2004) 'Temporal dominance of sensations: Comparison with conventional profiling and time-intensity', in *Proceedings 7th Sensometrics Conference*, Davis, CA, 40.

PRESCOTT J, NORRIS L N, KUNST M and KIM S (2005) 'An estimateof the "consumer rejection threshold" for TCA in white wine', *Food Qual Pref*, 16, 345–349.

PRUTKIN J, DUFFY V B, ETTER L, GARDNER E, LUCCHINA L A, SNYDER D J, TIE K, WEIFFENBACH J and BARTOSHUK L M (2000) 'Genetic variation and inferences about perceived taste intensity in mice and men', *Physiol Behav*, 69, 161–173.

QANNARI M and SCHLICH P (2006) 'Matching sensory and instrumental data', in Voilley A and Etiévant P, *Flavour in Food*, Cambridge, Woodhead, 98–114.

RAINEY B A (1986) 'Importance of reference standards in training panelists', *J Sensory Studies*, 1, 149–154.

RESURRECCION A V A (1998) *Consumer Testing for Product Development*, Gaithersburg, MD, Aspen.

RISVIK E, McEWAN J A, COLWILL J, ROGERS R and LYON D H (1994) 'Projective mapping: A tool for sensory analysis and consumer research', *Food Qual Pref*, 5, 263–269.

SCHLICH P (1993a) 'Risk tables for discrimination tests', *Food Qual Pref*, 4, 141–151.

SCHLICH P (1993b) 'Uses of Change-over designs and repeated measurements in sensory and consumer studies', *Food Qual Pref*, 4, 223–235.

SCHLICH P (1994) 'Grapes: A method and a SAS program for graphical representations of assessor performances', *J Sensory Studies*, 9, 157–169.

SCHLICH P (1995) 'Preference Mapping: Relating consumer preferences to sensory or instrumental measurements', in Etievant P and Schreier P, *Bioflavour '95*, Paris, INRA Editions, 135–150.

SCHLICH P (1996) 'Defining and validating assessor compromises about product distances and attribute correlations', in Naes T and Risvik E, *Multivariate Analysis of Data in Sensory Science*, New York, Elsevier Science, 259–306.

SCHLICH P and GUICHARD E (1989) 'Selection and classification of volatile compounds of apricot using the RV coefficient', *J Agric Food Chem*, 37, 142–150.

SCHLICH P, ISSANCHOU S, GUICHARD E, ETIEVANT P and ADDA J (1987) 'RV coefficient: A new approach to select variables in PCA and get correlations between sensory and insturmental data', in Martens M, Dalen G A and Russwurm H, *Flavour Science and Technology*, New York, John Wiley & Sons, 469–473.

SCHLICH P, LESPINASSE N, NAVEZ B and SCANDELLA D (2003) 'Preference mapping of tomato varieties from the French market. A PrefMaX consumer segmentation validated by a two-year study', in *Proceedings 5th Pangborn Sensory Science Symposium*, Boston, MA, Elsevier, London.

STONE H and SIDEL J (1993) *Sensory Evaluation Practices*, Orlando, FL, Academic Press.

STONE H, SIDEL J, OLIVER S, WOOLSEY A and SINGLETON R C (1974) 'Sensory evaluation by quantitative descriptive analysis', *Food Technol*, 28, 24–34.

SULMONT C, LESSCHAEVE I, SAUVAGEOT F and ISSANCHOU S (1999) 'Comparative training procedures to learn odor descriptors: Effects on profiling performance', *J Sensory Studies*, 14, 467–490.

TUORILA H, MEISELMAN H, BELL R, CARDELLO A and JOHNSON W (1994) 'Role of sensory and cognitive information in the enhancement of certainty and liking for novel and familiar foods', *Appetite*, 23, 231–246.

VAN KLEEF I, VAN TRIJP H and LUNING P (2005) 'Consumer research in the early stages of new product development: a critical review of methods and techniques', *Food Qual Pref*, 16, 181–201.

WILLIAMS A A and ARNOLD G M (1985) 'Comparison of the aromas of six coffees characterised by conventional profiling, free-choice profiling and similarity scaling methods', *J Sci Food Agric*, 36, 204–214.

WILLIAMS A A and LANGRON S P (1984) 'The use of free-choice profiling for the evaluation of commercial Ports', *J Sci Food Agric*, 35, 558–568.

WOLF FRANDSEN L, DIJKSTERHUIS G, BROCKHOFF P, HOLM NIELSEN J and MARTENS M (2003) 'Subtle differences in milk: comparison of an analytical and an affective test', *Food Qual Pref*, 14, 515–526.

YEGGE J M (2001) *Influence of sensory and non-sensory attributes of Chardonnay wine on acceptance and purchase intent*, Ph.D. Dissertation, University of California.

YEGGE J M and NOBLE A C (2001) 'Identification of sensory and non-sensory attributes of Californian Chardonnay wines that influence acceptance and purchase intent for differing segments of consumers', in *Proceedings of the ASEV 50th Anniversary Annual Meeting*, Vol. 51, *Am. J. Enol. Vitic.*, 28–31.

3.9 Appendix

Sensory resources

To keep abreast of developments in the sensory field and in statistical analysis of sensory data, the biennial international symposiums, the Pangborn Sensory Science Symposium and the Meeting of the Sensometrics Society (www.sensometrics.org) are valuable resources. Sensory training is offered at the University level in Europe and North America, but is more limited elsewhere. Professional associations include the sensory evaluation division of the Institute of Food Technology (http://www.ift.org/divisions/sensory/index.html), and sensory analysis technical committees of the International Organization for Standardization (ISO) (TC 34/SC 12) (www.iso.org) and American Society for Testing Materials (ASTM E18) (www.astm.org). Internet sensory user groups are also very active and have specific forums on sensory methodology and sensory software (www.sensory.org). In Europe, there are many multinational sensory projects sponsored by the EU such as FLAIR and Healthsense. Other organizations focus on basic sensory research. They include European Chemoreception Research Organisation (ECRO) and Association for Chemoreception Science (AchemS) the Australasian Association for ChemoSensory Science (AACSS) and Japanese Association for the study of taste and smell (JASTS).

4

Choosing the correct analytical technique in aroma analysis

G. Reineccius, University of Minnesota, USA

4.1 Introduction

Historically, aroma was considered the major component of flavor and, thus, flavor chemists focused on the study of volatiles in food while other sensory inputs were neglected. As knowledge related to sensory perception has grown and closer relationships developed with sensory scientists, flavor chemists have come to appreciate the multimodal nature of human perception (Taylor and Roberts 2004). Thus, there has been an increasing effort placed on the chemistry of non-olfactive sensory inputs and their interactions to form flavor perception (e.g. taste and chemesthesis) (Hollowood *et al.* 2002, Ashima and Nakai 1991, Engel *et al.* 2000, 2001, Hofmann and Schieberle 2002). Unfortunately, our understanding how non-volatiles influence flavor perception is very incomplete and thus this chapter will reflect knowledge in the field and focus on volatiles (Pickenhagen *et al.* 1996, Hollowood 2002).

There are numerous analytical methods that one might use in approaching a flavor problem (Marsili 2002, Mussinan and Morello 1998, Schreier 1993, Reineccius 2002, 2005). One may simply couple a static headspace aroma isolation method with gas chromatography, or be very complicated using detailed, tedious aroma extractions and distillations coupled to mass spectrometry or olfactometry. Sample considerations (e.g. sample composition, and aroma compound concentrations and properties) will influence the choice of analytical method. For example, a solvent extraction of a food will not be of great value if the food contains lipid for one extracts not only volatiles but also lipid. The presence of lipid dictates the use of a method to further isolate volatiles from the lipid before analysis can proceed. Headspace analyses are not useful if the volatiles of interest are present in only trace amounts or have very low vapor pressures.

Furthermore, each method offers differing strengths/weaknesses quite separate from the ability to isolate aroma compounds. Automated methods offer precision and permit the analysis of a greater number of samples. The majority of variability in final results comes from the aroma isolation step. Whenever this process can be automated, variability decreases improving the value of the results. Sample handling capacity also differs greatly with methodology. Automated methods make 24-hour operation possible (perhaps 40 or more samples/day) while tedious, labor intensive methods such as simultaneous distillation extraction (SDE) are limited to 1–2 samples/day. Analysis time is critical in most projects. Stability of the method over time is critical in storage studies. Adsorbents used in solid phase microextraction (SPME) or Stir-bar methods may deteriorate over time and use thereby changing the volatile profile obtained in isolation. Purge and trap, solvent extraction and SDE are stable over time and may be better suited to storage studies.

It is important to consider the sample, volatiles of interest, analysis time and objectives of the analytical study in choosing analytical method. In terms of analytical objectives, one may wish to:

1. Obtain a 'complete' aroma isolate to accurately identify and quantify every aroma constituent in a food.
2. Identify only key components of an aroma profile, that is, those components that are responsible for the characteristic food aroma.
3. Identify an off-note in a food product.
4. Monitor aroma changes with time.
5. Predict sensory attribute(s).
6. Determine if a food flavoring is adulterated.

Each of these tasks imposes different requirements on the methodology. These considerations will be discussed in the following sections.

4.2 Obtaining a complete aroma profile

One of the most challenging analytical tasks is to obtain a complete and accurate analytical profile of the aroma components present in a food. The difficulties start with choosing a sample for study. Sample selection is a major problem in this analysis since the methods employed are so tedious that only very few samples can be analyzed. One cannot do an appropriate statistical sampling/analysis scheme but must make a judicious selection of a very limited number of samples for analysis.

Some of the sampling issues can be illustrated best by example. If we wish to obtain an aroma profile of Cheddar cheese, for example, which sample, or limited number of samples of Cheddar cheese is characteristic of all Cheddar cheeses? Quite obviously, no individual sample or even limited number of samples is characteristic of all Cheddar cheeses, thus, one must limit the study to one particular location, factory, or type of Cheddar cheese. Ideally one will choose Cheddar cheeses that are characteristic of that cheese and free of defects.

Again, who says that a sample of cheese is 'typical' or free of defects? Everyone has a different expectation or preference for a given food. This problem is normally addressed by having an individual, or group of individuals, that are experienced in the flavor of a given food make sample selection. While experts often disagree as broadly as consumers, there is little alternative; one must choose samples for study and there is no 'perfect' way to do this. So one is going to analyze a sample, or group of samples, that hopefully is (are) characteristic of the product they wish to characterize.

The next step in the study is to obtain an accurate aroma isolate from the chosen samples. Unfortunately, no individual isolation technique will yield an accurate analytical profile. Every method of aroma isolation will provide a biased and unique view of the volatiles in a food (Leahy and Reineccius 1984, Werkhoff *et al.* 2002, Reineccius 1993). Thus, if one wishes to have a complete view of the aroma compounds in a food, one must use several isolation techniques in combination. A good combination might be using a static headspace method to obtain a profile of the most volatile and most abundant aroma compounds. One can follow this method with a purge and trap (or 'Stir bar') methodology to obtain data on the less volatile and abundant constituents (no solvent to 'cover' early eluting components of interest). The task might end with a solvent extraction (if there are no lipids in the food) or simultaneous distillation/extraction (SDE) method to obtain a profile of the least volatile aroma components. While each technique will yield a unique aroma isolate, this combination of techniques would yield a reasonably complete qualitative view of the aroma profile of a food (for example, see Qian (2000)).

The next question is what information is needed about this aroma profile? If one is interested in obtaining quantitative data on this profile, there are several approaches available. For example, one might use multiple internal standards, standard addition or, isotopic dilution analysis. Each of these methods can yield good quantitative data if used properly.

In the case of using an internal standard method, one must choose the internal standards to represent the compounds of interest as closely as possible, i.e. have similar physical and chemical properties as the aroma compounds of interest. Thus, one typically uses multiple internal standards with each standard representing some subclass of the overall volatile profile (Qian 2000). Once having chosen the internal standards, one must get relative recovery and GC response data for all compounds present in an aroma profile relative to their corresponding standards in an appropriate matrix. This is tedious.

The use of a standard addition method is equally tedious. In this approach, one obtains an aroma profile (perhaps using 2–3 different methods), adds a known amount of each aroma compound present in the sample, and then reanalyzes the spiked sample (Uematsu *et al.* 1994, Spitz 1977). The amount of each volatile initially present in a food can be estimated by the increase in GC peak area found after adding a known quantity of this volatile to the food. This approach requires that one has pure reference compounds available for every volatile compound present in the food. That is highly unlikely.

The preferred method for quantification is Isotope Dilution Assay. In this method one adds a known quantity of isotopically labeled analog to the sample for each compound one wishes to quantify (Fay *et al.* 2000, Blank *et al.* 1999). Obviously, in studies where one wishes to obtain complete, accurate quantitative data on all of the volatiles present in a food, one would need to have stable isotope standards of every volatile in the food. The synthesis of this number of isotopically labeled compounds is not practical so this method is used only when a limited number of volatiles are of interest, for example, in studies where 'key' aroma compounds are of interest as opposed to complete profiles. Thus, this method will be discussed in greater detail in the following section of this chapter.

Traditionally, flavor chemists have been less than rigorous in the quantification of aroma compounds in foods. It has long been the practice to simply report GC area percentage in tabulations of aroma compounds and their quantities found in foods. The data obtained and reported in this manner are typically grossly in error due to the many biases in aroma isolation and GC responses. Occasionally a researcher will add an internal standard to the sample and report quantitative data in terms of the internal standard. Unless one does very rigorous recovery and response studies for each compound being quantified, this approach offers little or no improvement over GC peak area. Obtaining accurate quantitative data is a very formidable task that many researchers choose to short-cut.

4.3 Key components contributing to sensory properties

The task of identifying only the compounds which have sensory significance may appear to be less complicated than obtaining an accurate analytical view of all of the aroma compounds in a food because fewer compounds are involved. Unfortunately, this task has other obstacles that make it more complicated. In this task, one has to isolate all of the volatiles from a food and then determine which of these make a sensory contribution. Determining which compounds make a sensory contribution has been the subject of considerable research over the last 20 or more years and despite this effort, has met with very limited success.

The initial problem is again deciding on an appropriate sample for study (as discussed in section 4.2 above). If one wishes to determine the characterizing aroma compounds in Parmesan cheese, for example, one cannot analyze all of the Parmesan cheeses on the market to develop a chemical description that embraces all of the Parmesan cheeses of the world. One typically selects two or three products that are determined to be characteristic of Parmesan cheese and they serve as the models for study. As noted earlier, the magnitude of the work involved in this task precludes rigorous statistical sampling and analysis of multiple samples. In publishing results, one has to be careful to note that the data obtained are valid only to describe the limited number of cheeses examined, not all Parmesan cheeses. However, assuming judicious choices were made in sample selection, one can generally expect that the data obtained are reasonably valid in characterizing the product overall and that other products have only subtle differences in sensory character.

Once products have been selected for study, then one turns to preparing aroma isolates for characterization. One must obtain aroma isolates that include all aroma compounds in the food so each compound may be evaluated for potential sensory contribution. Due to problems in obtaining aroma isolates, there is always the possibility that one does not obtain aroma isolates that truly represent the sample even when multiple aroma isolation procedures are used. An aroma compound may not readily be isolated due to its chemical or physical characteristics, or perhaps, chemical stability. There is no guarantee that one has isolated and delivered to the gas chromatographic-olfactometric (GC-O) port every aroma compound of sensory significance. Some researchers do a sensory evaluation of each aroma isolate. One can sniff each isolate and determine if it contains the sensory character of the sample. This is, of course, a very crude evaluation for one would not be able to distinguish if an aroma isolate contains any more than the major sensory notes of a food: the approach would not pick up subtle nuances in character. Also, every aroma isolate will be biased in its composition and thus, no individual aroma isolate is representative of the food. The existence of this problem is often evident only at the end of the study when results do not yield success. This is manifest by the sensory recombinations based on analytical data not reproducing the aroma character of the product.

At this point we will assume we have made a judicious choice in samples for study, selected appropriate combinations of aroma isolation methods, and thereby isolated *all* aroma compounds from our samples in adequate quantities to meet our objectives. The next step is to determine which aroma compounds in the aroma isolates *define* the aroma component of the food. Ideally, one would identify and quantify all aroma compounds in the samples and then conduct sensory studies to determine which are important. The large number of volatiles in foods make it impossible to obtain and sensorially evaluate all of them. Thus, one must narrow this list down to some characterizing subset of the total volatile profile. (We are fairly safe in assuming that not all volatiles in a food are present at concentrations that make a sensory contribution so we can work with some subset of the complete aroma profile.) This selection of a subset of aroma compounds that truly characterize the food is another major problem (Grosch 2001, Leland *et al.* 2001).

The selection of characterizing volatiles from the total volatile profile involves GC-O methods. This is where we link people to instruments. The GC is used to provide individual aroma compounds isolated from a food to human assessors. Human assessors make some judgments on each of these aroma substances that are used to determine likely importance of that compound to the sensory character of that food. Since GC-O methods are described in numerous other references (e.g. Acree *et al.* 1984, Grosch 2001), there will be no detailed discussion of GC-O methodology here. However, it is important to recognize the weaknesses of the GC-O methods in this application (Abbott *et al.* 1993, Piggott 1990, Frijters 1978, Mistry *et al.* 1996).

In all GC-O methods, aroma compounds are presented *individually* to the human assessors in an 'atypical' form. In eating we take the combined input from all olfactory stimuli *together* and make a judgment as to sensory character and quality. This allows for sensory interaction between stimuli that are not accounted for by evaluating aroma compounds individually. Also, the fact that each individual stimulus (aroma compound) is evaluated in air (GC column effluent) without the benefit of all of the supporting taste, texture and contextual aspects of eating further weakens the predictive quality of these human assessments.

A further complication is that while GC-O methodologies will indicate which volatile compounds in an aroma isolate have an odor, they are limited in ability to suggest the relative importance of each aroma compound. Some GC-O methods involve sequential sample dilution and determination of ability to smell a compound across dilutions (Ullrich and Grosch 1987, Acree *et al.* 1984). Others take the methodology further permitting the calculation of a number that reflects the 'odor value' or 'Odor Activity Value' of a given compound (Grosch 2001). However, each of these methods violates psychophysical sensory laws and thus, has reduced value for our purposes. The OSME method has the soundest theoretical base (Miranda Lopez *et al.* 1992).

A final limitation of these methodologies in general is evident from the recent thesis research of (Atanasova *et al.* 2004). To be selected as a key odorant by GC-O methodology, all of the GC-O methods require that an aroma compound has a detectable odor at the sniff port and relative importance amongst odorants is based on compound intensity (or dilution) in GC-O. Basically, all GC-O methodologies used to select important odorants reject compounds with no, or little, odor from selection. Atanasova *et al.* (2004) has shown that volatiles present in wine (woody notes) *below their sensory detection thresholds* can mask the sensory perception of other volatiles (fruity notes). This finding suggests that odorants present in a GC-O run having no (or weak) odor may be important to the sensory character of a product. None of the existing GC-O methods have a mechanism to select aroma compounds that fit this phenomenon.

The 'bottom line' is that it is impossible to accurately assess the contribution of an aroma compound using any of the GC-O methodologies. These methods can be used only to suggest which aroma compounds may be important and even this ability is severely compromised by constraints in methodology. The only way to determine what compounds characterize the aroma of a food is through sensory evaluation.

The application of sensory evaluation in this context is very complicated. One can never simply take analytical data and successfully recreate an aroma that cannot be distinguished from the real food using rigorous sensory methodologies. (Often weak or inappropriate sensory methods are applied that suggest 'success'.) We must use sensory methods to validate the list of characterizing aroma compounds (too many or too few) or adjust their concentrations in the aroma formulation. Unfortunately, we have no sensory methods to optimize concentrations of 20–40 variables (aroma compounds) in mixtures. Simply determining whether an odorant makes a sensory contribution can be problematic. Often omission testing is used for this purpose (Grosch 2001, Engel *et al.* 2000, 2001). In omission testing, a complete aroma formulation is prepared and is compared by sensory methods to the formulation minus one component. If the panel cannot determine a difference between the complete formulation and the formulation minus one, the omitted component is viewed as not making a contribution to aroma. Yet omission testing has weaknesses. For example, one may find that compounds A and B may individually be omitted from a mixture without perceptibly changing the sensory character of the mixture. However, the deletion of both, may result in a change. Individual, minor effects too small to be statistically significant can become significant when added with those of other minor compounds.

The numerous problems discussed in this section have resulted in limited success in chemically characterizing the aroma component of foods. Many studies have identified and quantified the major chemical contributors to aroma of given foods. Often the major sensory properties of foods can be recreated by synthetic formulation. However, in no case has the aroma of any food been adequately characterized that:

1. *all contributing* aroma compounds are known; and
2. their ranges in concentration that give an indistinguishable sensory character have been defined.

We still have some distance to go in chemically defining the aroma component of our foods.

4.4 Off-notes in a food product

In this type of study, it is critical that the most intensely and yet characteristic off-flavored samples are selected for study (Kilcast 2003). Once selected, the off-flavored and control samples are subjected to some aroma isolation/concentration process. Fortunately, the requirements imposed upon aroma isolation methodology for this task are much less stringent than those imposed by the tasks discussed above. Virtually any isolation method can be used that yields an aroma isolate *containing the off-note*. Selection of the isolation method can be initially guided by experience. For example, off-notes that have a generically 'solvent-like' character (e.g. food packaging contamination or printing inks), can be initially approached using a purge and trap or even perhaps a static headspace method. Off-notes that are 'heavier' in sensory character (e.g. earthy, musty, cooked or burnt) may require more rigorous methods such as steam distillation or high vacuum distillation. If the analyst has no experience in methods selection, one tries a range of methods determining which yields the best isolate (highest concentration of off-note judged by sensory evaluation of the isolate) (Marsili 1997).

It is essential that the aroma isolate(s) be produced and then sensorially evaluated to ensure that the chemical constituent(s) responsible for the off-notes have actually been isolated from the food. Method validation can be done in different ways. For example, if the isolate is a liquid, a blotter can be dipped in it, the solvent evaporated, and the blotter smelled for the off-note. Alternatively, an isolate obtained using a SPME, Stir bar or purge and trap can be thermally desorbed into a cold trap, the trap warmed and then smelled. Often the analyst just injects the aroma isolate to a GC without preliminary sensory evaluation and then sniffs the effluent of the GC for the off-note. If the off-note is not evident on sniffing, then either the method did not effectively isolate the component responsible for the off-flavor or the off-flavor was unstable during GC analysis. It is a good habit to use on-column injection (or a temperature programmed injection port), deactivated injection port liners and transfer lines to minimize the possibility of compound decomposition in the GC.

The analytical approach for finding the odorant(s) causing an off-flavor then involves analyzing the isolates from the control and tainted samples by GC while smelling the column effluent (GC-O) to locate the chemical constituent corresponding to the characteristic off-note. One must recognize that any food aroma isolate likely contains odorants that are 'unpleasant'. However, the presence of unpleasant odorants in the GC effluent may only be due to their concentration in the effluent: they may not be a source of off-flavor in the food itself. One focuses further attention only on GC peaks *characteristic* of the off-flavor in the food.

It should be obvious that the individuals doing the GC-O work must be sensitive to the off-note and reasonably trained in this task. Deibler has presented substantial evidence to show that specific anosmias are common in the population (Deibler and Delwiche 2004). Thus, GC-O assessors must be tested for their ability to detect the off-note in the tainted product. Also, training of the assessors is helpful. There is no question that one's ability to do GC-O work improves with experience (although training does not correct for anosmia – if a panelist cannot smell a substance, repeated evaluation will not necessarily improve detection ability).

Ideally, the aroma isolation procedure should yield an isolate sufficiently concentrated for GC-MS identification of the tainting compound. If there is no GC peak in the chromatogram where the offending aroma is smelled (or too little compound is present for a good mass spectrum), then the aroma isolation procedure needs to be changed, or sample size increased, so larger quantities of the tainting chemical are obtained. In some cases mass spectral data may not be required for 'identification'. Identification is in quotation marks since one can occasionally identify a compound simply by odor character and GC retention time. While this may be adequate to know what is causing the off-note, it would not be an adequate identification for a scientific publication or to stand up in a court of law. In these latter cases, solid MS data are required in support of the chromatographic data.

Once the offending compound has been identified, one must obtain a pure reference of it, add it to a control sample of food product at the determined concentration, and do sensory evaluation to determine whether the addition of the pure reference compound gives the food the same sensory off-note as the tainted product. This may be done informally by the researcher or formally using rigorous sensory protocols depending upon the end use of the data.

In most cases the analytical chemist is successful in identifying the chemical compounds giving the off-note and thus, identifying the mechanism off-flavour development. However, the probability of success depends upon the character and intensity of the off-note. If the off-note is not very intense and individuals disagree on the character or even the presence of the off-note, the probability of success goes down rapidly. The more unique (out of place) the odorant is and/or intense, the more likely the human sniffer is to detect it in the GC effluent. The task also becomes more problematic when the defect is caused by volatiles normally in the food product for then there is no unique odor to key on, only a change in intensity of a given sensory note.

4.5 Monitor changes in aroma compounds with time

There are many situations where the flavor chemist wishes to monitor changes in food aroma over time. For example, one may wish to analytically monitor flavor losses from a food product (e.g. instant coffee), the formation of desirable flavors (e.g. wine or cheese aging), or the appearance of off-flavors (e.g. lipid oxidation) during storage or aging. For these types of problems, one has to consider at least four method requirements:

1. Does the aroma isolation method provide data on the aroma compounds of interest?
2. Is the method sufficiently robust to be stable over time?
3. Is there adequate precision in the method to see the anticipated variation?
4. Is the method rapid enough to be used in the study?

The importance of each requirement depends on the research objectives. The first requirement is self evident. One must know what chemicals will contribute to the aroma of a food if one wishes to monitor changes over time that are of sensory significance. This knowledge permits selection of the proper methodology, i.e. it must isolate the desired volatile components in a manner that permits the collection of reproducible, quantitative data.

The idea of stability of a method over time has been discussed briefly in the introduction to this chapter. If one is using an absorption method, the absorbent may deteriorate (or break) thereby changing its ability to absorb volatiles. Thus, some portion of the changes observed over time may relate to changes in the absorbent, not the sample. The changing of an absorbent fiber due to breakage during a shelf-life study can be devastating.

Precision, typically, is obtained through the use of automated methods and/or analytical standards. Automated headspace methods often have a coefficient of variation (CV) of only 2–3% (compound dependent) while a distillation method may have a CV ranging from 10–50%. Many of our analytical methods in flavor research suffer from poor reproducibility. A lack of reproducibility may mask the appearance of true changes in the sample.

Analysis time is also an issue. For example, a storage study with 10–20 samples to be analyzed each day (or even each week) will preclude any method that is very time consuming (distillation methods may be problematic). One needs to use an automated headspace (static or dynamic) method when possible. Ultimately, the approach taken in this type of a study is too wide in scope to be discussed here. Each study must be designed with the aforementioned considerations in mind.

4.6 Using instrumental data in sensory predictions

While there is no question that sensory evaluation is the only true measure of sensory quality, there are numerous reasons why one would like to use instrumental data to screen or supplement sensory testing. Reasons include the cost, availability of time and manpower required in sensory testing, variability associated with sensory testing, and the inability of people to determine the stimuli (e.g. individual chemicals) providing a given sensory response. Thus, supplementing sensory testing with instrumentally-based support is very attractive.

The primary problem with using instrumental data for this purpose is the complexity of stimuli involved in determining perception. If one chemical defined a flavor, predictions would be easy, however, this is never the situation. We find that perception results from the complex cognitive integration of numerous stimuli provided by several sensory systems (e.g. smell, taste, touch, pain, hearing, and sight). Along with this recognition is the application of complex instrumental techniques that generate potentially thousands of data points to be used in predicting perception. Our ability to analyze and interpret the shear number of data points (variables) one can now gather from a given GC run, for example, has been outpaced. Thus, we are turning increasingly to the use of statistical methods (chemometrics) for data analysis and interpretation (Ashima and Nakai 1991). Since chemometrics is a tool for dealing with numerous variables, we could include a discussion of all sensory stimuli in this section, however, the principles are the same so this discussion will be limited to olfaction (the ability to use data on volatile substances in foods to predict sensory aroma judgments).

As with all other tasks discussed thus far, this task also starts with the isolation of volatiles from a food. However, unlike other tasks, a complete or accurate aroma profile is generally not required. The goal is to be able to obtain a reproducible set of instrumental data on the volatiles in a food (precision aids prediction reliability) and contains components related to the property one wishes to predict. For example, one might use this approach to detect adulteration of a food, its region of production, its plant origin, or some sensory attribute. If the objective is to predict some sensory attribute, the volatile constituents used in prediction may or may not be not causative, however, they must be statistically correlated to the sensory attribute of interest. This task as it appears in the scientific literature is nearly always done in research settings, but is intended for production or a quality control setting so time, ease, and reliability are additional factors influencing method choice.

'Direct' isolation techniques are well suited for this task. Methods such as direct thermal desorption and equilibrium (or non-equilibrium) headspace analysis are commonly used. For example, headspace analysis is routinely used in the quality control of fats and oils (analysis of hexanal as an indicator of rancidity) (Barrefors *et al.* 1995, Harayama *et al.* 1991, Brewer *et al.* 1999). In some cases, these direct methods may be replaced by a purge and trap or a SPME/Stir bar method to yield more information (greater sensitivity) for obtaining valid statistical models/classifications (Marsili 1997). These methods can be automated to reduce analysis variability and improve on sample numbers (statistical considerations). Sensitivity may not be as great an issue as one might expect because there is no need to measure causative substances, only substances related to the causative substances. Often one can find volatiles in a food in much greater quantity to measure that are predictive as opposed to causative.

A second difference in approach used for this task is that a wide range of instrumental data can be used. One may use mass spectral data with or without GC separation (e.g. atmospheric pressure ionization or electron impact), GC profiles, HPLC profiles, FT-IR, etc. Basically, any instrument that provides a 'view' of the volatiles in a food may be used as an input for chemometrics. The instrumental technique chosen will influence the method of sample preparation/ analysis.

This task requires the gathering of other data (non-instrumental) to permit developing statistical correlations. In some cases the data are analyzed using an unsupervised method, i.e. one puts all of the data into a statistical program and the program searchers for relationships, trends or grouping of the samples. The method is exploratory in nature. The major statistical tools used are principal component analysis (PCA), factor analysis (FA), cluster analysis and multidimensional scaling (MDS, sensory applications) (Ashima and Nakai 1991).

In other cases, a supervised method is used. In supervised methods, the analyst has *a priori* knowledge of the grouping of samples. Thus, one can use sample knowledge in the development of optimal classification rules. Multiple regression analysis (MRA), canonical correlation analysis (CCA), partial least squares (PLS), linear discriminant analysis (LDA), and linear learning machine (LLM) methods are commonly used in this classification (Ashima and Nakai 1991). The *a priori* knowledge may be geographical classification of samples, grape used in wine production, storage time, use of an antioxidant, etc. For example, one may analyze the volatiles in olive oils from France, Greece and Morocco and then attempt to predict the growing region based on volatile profile. Alternatively, sensory data may be gathered on samples to permit a prediction of some sensory attribute or overall quality.

Ultimately, the task becomes choosing a statistical method to establish relationships between the analytical and sensory data. The choice in chemometric method depends upon the project objectives. For example, one may wish to categorize the data into groups. The groups may be as simple as 'accept' or 'reject', or be more definitive such as predicting the level (e.g. none, slight, mild, intense) of an off-note such as 'cardboardy'. The statistical method may be directed toward predicting some continuous property such as cooked flavor in thermally processed foods or oxidized notes in stored snack foods. The task is to develop statistical relationships between instrumental data and a continuous sensory judgment. This approach can also be used to establish relationships between one data set (e.g. a sensory profile) and volatile compounds. While it was said that the instrumental data used do not have to be causative, often they are and thus determining the compounds that predict a response may, in fact, point out causative substances. Thus, one may gain insight into causes of some sensory attribute. The application of chemometrics to problem solving is becoming more popular in the flavor field for the reasons discussed above. Due to the complexity of the application of chemometrics, flavor researchers benefit greatly from collaborative relationships with those in the discipline.

Reviews on advancements in chemometrics are periodically available in *Analytical Chemistry* (e.g. Lavine and Workman 2002) and a comprehensive review specifically focused on flavor applications has been provided by Aishima and Nakai (1991).

4.7 Future trends

Despite more than 40 years passing since modern flavor chemistry was born, we are still left with numerous gaps in the knowledge needed to solve many of the issues facing the food industry today (Teranishi *et al.* 1999). The most challenging is an understanding of how stimuli provided by a food and the eating process are translated into overall flavor perception by the human being. We are slowly gaining an understanding of the basic mechanisms of the olfactory, gustatory and somato senses in the human (Taylor and Roberts 2004). We are also gaining an understanding of the stimuli provided by a food and the eating process. Work on the chemical characterization of food aroma and then its release during eating has progressed greatly with the advent of real-time breath analysis and model mouth simulators (Grab and Gfeller 2000, Linforth and Taylor 1993, Taylor *et al.* 2000, Deibler *et al.* 2001, Roberts 1995). While neither of these areas is well understood, unparalleled progress has been made in the last 10 years. A major unknown is the cognitive processing of all sensory input and ultimately the formation of flavor perception. We find that flavor chemists can no longer work within their own discipline (or common disciplines) but must reach out to psychologists, physiologists, and geneticists for collaborative help. It is only in this manner that true progress will be made.

We need progress in sensory science as well. As mentioned earlier in this chapter, our only means of determining if a chemical makes a contribution to the flavor of a food is through sensory evaluation of mixtures in the appropriate food system. Currently we cannot manage working with 20–40 variables (aroma compounds) in sensory studies and need new approaches to this task.

Finally, we have only recently begun to study the role non-volatiles can serve in determining the flavor of foods. Historically we have treated olfaction, taste and chemesthesis as separate entities, not recognizing how each sensory stimulant interacts and influences the other (e.g. taste and aroma). This is fertile ground for future work.

4.8 Sources of further information

4.8.1 Scientific journals

While flavor research appears in many scientific journals, journals that are particularly well respected and focus on flavor include the *Journal of Agricultural and Food Chemistry* (American Chemical Society), *Flavour and Fragrance Journal* (John Wiley and Sons) and *Perfumer and Flavourist* (Allured Pub. Corp.). As noted, many other journals publish flavor research but these journals are typically general in scope (e.g. *Food Chemistry*, *Z. Lebens. Unters. Forschung*, *Journal of Science Food Agriculture*, and *Journal of Food Science*) or focus on a commodity/discipline (e.g. *Food Engineering*, *Cereal Chemistry*, *Journal of Dairy Science*, and *Lipid Chemistry*) where flavor may also be relevant. The majority of these journals are now online so retrieval of articles is simple if one's library/company has a subscription to this service.

Today trade journals abound and flavor is a common topic (e.g. *Snack Food & Wholesale Bakery*, *New Products*, *Prepared Foods*, *Food Processing*, *Food Technology*, and *Cereal Foods World*). There is little published in these journals that one would consider being of a research nature but often interesting information is presented on new products or market trends. A special category of trade journals is produced by flavor companies to show off expertise or inform potential customers of new products. Examples of flavor companies that produce a publication for the trade include Givaudan Inc. *Inspire*, Sigma-Aldrich *Fine Chemicals Quarterly*, Hasegawa USA Inc. *Hasagawa Letter*, and a newsletter from Symrise.

4.8.2 Internet

The internet has brought us both convenience and a greatly expanded world of information directly to our desktop. One can find information related to virtually all aspects of flavor research via the internet. For example, if one wishes to find information about the industry overall, Leffingwell & Associates (http://www.leffingwell.com/) is an excellent site. This site provides information on current events in the industry, what new chemicals have come on the market or are approved (current FEMA GRAS list is posted), where to buy chemicals and software, checking international legislation, and finding contacts and consultants. The site also has links to many flavor companies.

The greatest value of the internet is the access to the literature. Often there is merit in doing very general searches to get a perspective on what is happening across all disciplines. Perhaps one is interested in encapsulation. A search on this word provided 456,000 hits! That is clearly unmanageable but crossed all fields. A cursory browsing of 50–100 is manageable and may introduce the reader to something he/she had never thought of.

One may choose to limit the topic area and focus on a specific aspect of encapsulation, e.g. cyclodextrins. A search on cyclodextrins brings the number of hits down to 89,000 – still a bit unmanageable. But again one can browse the top 50–100 hits and get an overview of the broad activity in cyclodextrins or narrow the topic more by putting in qualifying key words. Simply tying cyclodextrins to 'flavor' brought the number down to 367. That is very manageable. It is worth noting that there are many search engines available and each offers unique search mechanisms/criteria.

Direct access to the scientific literature is unquestionably one of the most useful services of the internet. There are several ways one can access the literature through the internet. The National Agriculture Library (http://agricola.nal.usda.gov/) provides free search and retrieval accesses and 'for fee' articles can be ordered and received by internet or by mail. Most libraries have sophisticated searching routines that become available at the desktop. If the library has a subscription to an online subscription, very often entire journal articles can be downloaded. If the library does not have an online subscription, an abstract can likely be downloaded. A second approach to getting literature by internet is that most organizations permit members to access their journals if they are subscribers. For example, the *Journal of Agricultural and Food Chemistry* is available through ACS at http://pubs.acs.org/journals/jafcau/index.html. If one has a subscription to the journal online, the articles can be downloaded in HTML or pdf formats.

A final approach is to contact the author directly by email and request the article be returned as an email attachment. Sending an article by email is relatively easy. One simply clicks on 'reply' and attaches the file for the article. This is much easier than finding the article, putting it in an envelope, addressing the envelope, adding postage and finally mailing it. This ease of providing an article greatly enhances the probability of an author honoring a request for a reprint of his/her article and it is much quicker.

Retrieving patents, either domestic (http://www.uspto.gov/patft/index.html) or international (http://ep.espacenet.com/espacenet/ep/en/e_net.htm), has been made easy via the internet. It is possible to search and download complete patents at no cost. In the past, obtaining domestic patents was a problem, and international patents were nearly impossible to get in a timely manner. Leffingwell & Associates have provided a compilation of patents (downloadable) at their site for the major flavor companies that have been issued patents since 1976. The patents listed are directly linked to the selected patent at the USPTO (http://www.leffingwell.com/patents/patents1976_2001.htm).

It should go without saying that finding books, publishers and ordering online is again simple. Virtually all book publishers have websites to display their products and facilitate ordering.

4.8.3 Conferences

Conferences have traditionally been a major outlet for publications in the flavor area. Conferences focused primarily on flavor chemistry (limited sensory aspects) are offered in Europe (e.g. Weurman and Wartburg) and the USA (American Chemical Society). Sensory Science is more diverse and meetings such as the Pangborn Sensory Science Symposium, Associates for Chemoreception Sciences (AChemS), ASTM Committee E18 on Sensory Evaluation, The Sensometrics Meeting, ECRO Biennial Congress, and Australasian Association for Chemosensory Science focus on various aspects of sensory science. A brief summary of the organizations listed as sources of sensory information is provided at http://ssc.fst.ohio-state.edu/conferences.asp. More detailed information can be found on each of these meetings by viewing their specific websites.

4.9 References

ABBOTT, N, ETIEVANT, P, ISSANCHOU, S and LANGLOIS, D (1993) 'Critical evaluation of two commonly used techniques for the treatment of data from extract dilution sniffing analysis', *J Agric Food Chem,* 41, 1698–1703.

ACREE, T E, BARNARD, J and CUNNINGHAM, D G (1984) 'A procedure for the sensory analysis of gas chromatographic effluents', *Food Chem,* 14, 273–286.

ASHIMA, T and NAKAI, S (1991) 'Chemometrics in flavour research', *Food Reviews Int,* 7, 33–101.

ATANASOVA, B, THOMAS-DANGUIN, T, LANGLOIS, D, NICKLAUS, S and ETIEVANT, P (2004) 'Perceptual interactions between fruity and woody notes of wine', *Flavour Fragrance J,* 19, 476–482.

ATANASOVA, B, THOMAS-DANGUIN, T, CHABANET, C, LANGLOIS, D, NICKLAUS, S and ETIEVANT, P (2005) 'Perceptual interactions in odour mixtures: odour quality in binary mixtures of woody and fruity wine odorants', *Chem Senses*, 30(3), 209–217.

BARREFORS, P, GRANELLI, K, APPELQVIST, L A and BJOERCK, L (1995) 'Chemical characterization of raw milk samples with and without oxidative off-flavour', *J Dairy Sci,* 78, 2691–2699.

BLANK, I, MILO, C, LIN, J and FAY, L B (1999) 'Quantification of aroma-impact components by isotope dilution assay-recent developments' in Teranishi R, Wick E L and

Hornstein I, *Flavour Chemistry: Thirty Years of Progress*, New York, Kluwer Acad/Plenum Publ, 63–74.

BREWER, M S, VEGA, J D and PERKINS, E G (1999) 'Volatile compounds and sensory characteristics of frying fats', *J Food Lipids*, 6, 47–61.

DEIBLER, K and DELWICHE, J F (2004) *Handbook of Flavour Characterization: Sensory Analysis, Chemistry, and Physiology*, New York, Marcel Dekker, 283.

DEIBLER, K D, LAVIN, E H, LINFORTH, R S T, TAYLOR, A J and ACREE, T E (2001) 'Verification of a mouth simulator by in vivo measurements', *J Agric Food Chem,* 49, 1388–1393.

ENGEL, E, NICKLAUS, S, SEPTIER, C, SALLES, C and LE QUERE, J L (2000) 'Determination of the taste active compounds of a goat cheese water-soluble extract', in Schieberle, P and Engel, K-H, *Frontiers of Flavour Science*, Garching, Deutsche Forschung Lebensmit, 84–88.

ENGEL, E, SEPTIER, C, LECONTE, N, SALLES, C and LE QUERE, J-L (2001) 'Determination of taste-active compounds of a bitter Camembert cheese by omission tests', *J Dairy Res,* 68, 675–688.

FAY, L B, METAIRON, S, LIN, J and BLANK, I (2000) 'Stable isotope dilution assay mass spectrometry in flavour research: internal standard and calibration issues', in Schieberle, P and Engel, K-H, *Frontiers of Flavour Science*, Garching, Deutsche Forschung Lebensmit, 125–131.

FRIJTERS, J E R (1978) 'A critical analysis of the odour unit number and its use', *Chem Senses Flavour,* 3, 227–233.

GRAB, W and GFELLER, H (2000) 'Flavourspace – a new technology for the measurement of fast dynamic changes of the flavour release during eating', in Schieberle, P and Engel, K-H, *Frontiers of Flavour Science*, Garching, Deutsche Forschung Lebensmit, 261–270.

GROSCH, W (2001) 'Evaluation of the key odorants of foods by dilution experiments, aroma models and omission', *Chem Senses,* 26, 533–545.

HARAYAMA, K, HAYASE, F and KATO, H (1991) 'Evaluation by multivariate analysis of off-flavour in headspace volatiles formed during storage of beer', *Agric Biol Chem*, 55(2), 393–398.

HOFMANN, T and SCHIEBERLE, P (2002) 'Chemical interactions between odor-active thiols and melanoidins involved in the aroma staling of coffee beverages', *J Agric Food Chem,* 50, 319–326.

HOLLOWOOD, T A, DAVIDSON, J M, DEGROOT, L, LINFORTH, R S T and TAYLOR, A J (2002) 'Taste release and its effect on overall flavour perception', in *Chemistry of Taste*, Washington DC, Amer Chem Soc, 166–178.

KILCAST, D (2003) 'Sensory analytical methods in detecting taints and off-flavours in food', in Saxby, M J, *Taints and Off-Flavours in Food*, Cambridge, Woodhead, 5–30.

LAVINE, B-K and WORKMAN, J, JR (2002) 'Chemometrics', *Anal Chem,* 74, 2763–2769.

LEAHY, M M and REINECCIUS, G A (1984) 'Comparison of methods for the isolation of volatile compounds from aqueous model systems', in Schreier P, *Analysis Volatiles: Methods Applications,* New York, de Gruyter, 19–47.

LELAND, J V, SCHIEBERLE, P, BUETTNER, A and ACREE, T E (eds) (2001) *Gas Chromatography-Olfactometry*, Washington DC, Amer Chem Soc, 235.

LINFORTH, R S T and TAYLOR, A J (1993) 'Measurement of volatile release in the mouth', *Food Chem,* 48, 115–120.

MARSILI, R (1997) 'Off-flavours and malodors in foods: mechanisms of formation and analytical techniques', in Marsili, R, *Techniques for Analyzing Food Aroma*, New

York, Marcel Dekker, 237–264.

MARSILI, R. (ed.) (2002) *Flavour, Fragrance, and Odor Analysis*, New York, Marcel Dekker.

MIRANDA LOPEZ, R, LIBBEY, L M, WATSON, B T and McDANIEL, M R (1992) 'Identification of additional odor-active compounds in Pinot noir wines', *Amer J Enol Viticul,* 43, 90–92.

MISTRY, B S, REINECCIUS, T A and OLSON, L K (1996) 'Gas chromatography-olfactometry for the determination of key odorants in foods' in Marsili, R, *Food Aroma Analysis: Science and Technology*, New York, Marcel Dekker, 265–292.

MUSSINAN, C and MORELLO, M J (eds) (1998) *Flavour Analysis*, Washington DC, Amer Chem Soc, 389.

PICKENHAGEN, W, HO, C-T and SPANIER, A M (eds) (1996) *Contribution of Low- and Nonvolatile Materials to the Flavour of Foods*, Washington DC, Amer Chem Soc, 242.

PIGGOTT, J R (1990) 'Relating sensory and chemical data to understand flavour', *J Sensory Studies,* 4, 261–272.

QIAN, M (2000) *Study of Parmesan cheese aroma*, PhD Thesis, University of Minnesota, Minneapolis.

REINECCIUS, G A (1993) 'Biases in analytical profiles introduced by isolation method', in Ho C-T and Manley C H, *Flavour Measurement*, New York, Marcel Dekker, 61–76.

REINECCIUS, G A (2002) 'Instrumental methods of analysis', in Taylor A J, *Food Flavour Technology*, Sheffield, Sheffield Acad Press, 212–253.

REINECCIUS, G A (2005) *Flavor Chemistry and Technology*, 2nd edn, Boca Raton, FL, CRC Press, 450.

ROBERTS, D D and ACREE, T E (1995) 'Simulation of retronasal aroma using a modified headspace technique: investigating the effects of saliva, temperature, shearing, and oil on flavour release', *J Agric Food Chem,* 43, 2179–2186.

SCHREIER, P W (ed.) (1993) *Progress in Flavour Precursor Studies: Analysis, Generation, Biotechnology*, Carol Stream, IL, Allured Publ Corp, 507.

SPITZ, H D (1977) 'Flavour analysis by objective analytical techniques', *Int Flavours Food Add,* 8, 248–249.

TAYLOR, A J, LINFORTH, R S T, HARVEY, B A and BLAKE, A (2000) 'Atmospheric pressure chemical ionisation mass spectrometry for in vivo analysis of volatile flavour release', *Food Chem,* 71, 327–338.

TAYLOR, A J and ROBERTS, D D (eds) (2004) *Flavour Perception*, Oxford, Blackwell.

TERANISHI, R, WICK, E L and HORNSTEIN, I (eds) (1999) *Flavour Chemistry: Thirty Years of Progress*, New York, Kluwer Acad/Plenum Publ, 439.

UEMATSU, Y, HATOOKA, Y, HIRATA, K, HIROKADO, M and KAZAMA, M (1994) 'Determination of residual organic solvents in flavour by standard addition head-space gas chromatography', *J Food Hygienic Soc Japan [Shokuhin Eiseigaku Zasshi],* 35, 645–651.

ULLRICH, F and GROSCH, W (1987) 'Identification of the most intense volatile flavour compounds formed during autoxidation of linoleic acid', *Zeitschrift Lebensmit Untersuch Forschung*, 184, 277–82.

WERKHOFF, P, BRENNECKE, S, BRETSCHNEIDER, W and BERTRAM, H J (2002) 'Modern methods for isolating and quantifying volatile flavour and fragrance compounds', in Marsili R, *Flavour, Fragrance, and Odor Analysis*, New York, Marcel Dekker, 139–204.

5

Matching sensory and instrumental data

E. M. Qannari, ENITIAA/INRA, France and P. Schlich, Centre Européen des Sciences du Goût (CESG), France

5.1 Introduction

The relation of sensory to instrumental data is manifold and involves (i) understanding the mechanisms by which chemical and physical properties of foods act to produce specific sensations (MacFie and Hedderley 1993); (ii) assessing how a change of some ingredients is reflected in the sensory properties; (iii) establishing a relationship between sensory and instrumental data, thus allowing the detection of those variables that change at the same time; (iv) predicting sensory from instrumental measurements, considering that sensory data are generally time consuming and the assessors are not easily available (Naes and Kowalski 1989).

For the past three decades, the investigation of this issue has been (and still is) a challenge to the statisticians and sensory analysts. Increasingly, there is a need not only to predict one data set from another (black-box modeling) but also to enhance the understanding of the mechanisms that govern the underlying phenomena. A model is a simplification of a system and several models, more or less complex, may be proposed to describe a particular situation. However, a good model is required to predict accurately and to be easily interpretable in addition to being stable and economical. In other words, a model should reflect the systematic variations inherent to the system and not to the sampling errors. As will be discussed subsequently, these features will disqualify some usual and popular methods such as multiple linear regression.

The current practice in relating sensory to instrumental data is to use successively several statistical methods in order to shed light from different angles and gather new facts as the various perspectives of the problem are thus illuminated. Lee *et al.* (1999) compared sensory and instrumental data using correlation analysis, principal component analysis, cluster analysis and partial least squares regression analysis. In a study on ice creams, Chung *et al.* (2003) also performed various multivariate statistical procedures to investigate the relationship between sensory and instrumental flavor profiles. In a first stage, they undergo principal component analysis and canonical variate analysis separately on the chemical and sensory data tables to explore the structure of each set. In a second stage, the impact of flavor volatiles on the sensory profile of ice cream was investigated by means of general Procrustes analysis and partial least squares regression. Moreover, the effect of a log-transformation of chemical variables was evaluated within each statistical method.

In the subsequent sections, we start by recommending that each data set should be investigated separately in order to get insight into their structure. Thereafter, we outline some useful methods to relate sensory to instrumental data. Although we ultimately recommend the use of partial least squares regression as one of the most appropriate tools in this context, we believe that users should have a general overview of the methods and how they are related to each other. This makes it possible to better understand on which grounds a strategy of analysis can be preferred to another and the benefit and the limitations of a method that they may come across, for instance when reading a paper.

5.2 Investigating the structure of the data sets

It is highly recommended that before undertaking the investigation of the relationships between sensory and instrumental data, each kind of data should be analyzed separately in order to explore their underlying structure. The benefits from this preliminary study are manifold and involve (i) detection of the presence of outliers which if not discarded might have a great impact on the analysis; (ii) in the case of sensory evaluation, assessing the performance of the panel and the panelists; (iii) assessing the extent to which the variables within each set of data are related to each other. Of particular interest is the investigation of the structure of the instrumental data. Indeed, as these variables are generally used as independent variables (or predictors) in several models, the presence of colinearity among them may be very harmful as it results in unstable models (Belsley 1991, Fox 1991). The colinearity among the instrumental variables may be easily diagnosed from the outcomes of a principal component analysis (PCA) performed on these variables. More precisely, principal components with small variances should arouse suspicion as they may reflect noise only. The variances of the principal components are given by the so-called eigenvalues of the variance–covariance matrix (or the correlation matrix if the instrumental variables are standardized). The ratio of the largest eigenvalue to the smallest eigenvalue is referred to as the conditioning index and is usually used to flag the presence of high colinearity. Belsley (1991) advocates that a condition index over 30 indicates a serious colinearity problem. Therefore, particular care should be taken as to which statistical method should be considered. This aspect is subsequently discussed in more details.

In sensory analysis, appearance, odor, flavor and texture are usually evaluated by trained assessors. Owing to differences in sensitivity to chemical stimuli, to different ways of using the scoring scale and, possibly, differences in the interpretation and the understanding of the sensory attributes, variations among individuals are inevitable and can blur the characterization of the products if not alleviated. The reduction of the sources of variations among assessors is above all a matter of a careful selection and training of the judges (Labbe *et al.* 2004). Care should also be taken in designing sensory experiments in order to counteract flaws that may result, for instance, from an inappropriate order of presentation as this is known to have an important effect (Muir and Hunter 1991, Hunter 1996).

The investigation of the performance of the panel and the reduction of some sources of variations among the assessors is a major domain of research in sensometrics (McEwan *et al.* 2003, Brockhoff 2003). In the following, we shall restrict ourselves to sensory profiling data obtained by means of a descriptive sensory analysis such as Quantitative Descriptive Analysis (QDA). More precisely, we consider the setting where different products are assessed by a panel of assessors using the same descriptors (fixed vocabulary profiling). The simplest and more popular way to deal with this kind of data is to average the data tables over assessors. The data table thus obtained is subsequently submitted to PCA in order to depict the relationships among the products or used to relate sensory to instrumental or preference data. MacFie and Hedderley (1993) recommend investigating and overcoming individual differences before averaging the data over assessors. This can be done by means of ANOVA (Naes and Kowalski 1989, Brockhoff 2003, Kuti *et al.* 2004). Some pre-treatments performed on the data can also help to eliminate some sources of variations. Column centering of the data corrects for the variations among assessors in the overall level of the scores they give. It consists in subtracting from each entry of each data set the average of the corresponding column. The scaling of the individual data tables entails standardization in order to adjust for variations among assessors in the range of scoring. This problem is generally solved by multiplying each assessor's data table by a positive scalar in order to set them to the same total variance (Kunert and Qannari 1999). In order to compensate for different use and interpretation of the descriptors, more sophisticated methods such as generalized Procrustes analysis (Arnold and Williams 1986, Dijksterhuis 1996) or STATIS (Schlich 1996) can be considered. Multivariate analysis of variance which is a natural extension of ANOVA to a multivariate setting and canonical variate analysis are gaining ground and becoming more accepted by sensory analysts. In particular, we draw the reader's attention to the interest in canonical variate analysis. This method is to a certain extent similar to PCA performed on the average data set, but it makes it possible to take uncertainty and error correlations into account (Porcherot and Schlich 2000, Schlich 2004).

In the following sections, we shall assume that we dispose of a data set which gives a description of the products from the sensory point of view. This data set is generally obtained as the average of the (pre-processed) data tables associated with the assessors. Alternatively, this data table can be the group average configuration obtained by means of generalized Procrustes analysis or STATIS method or the average over assessors of the canonical variates obtained by means of canonical variate analysis.

5.3 Relating sensory to instrumental data

The investigation of the relationships between sensory and instrumental data is motivated by the desire to achieve an accurate prediction and enhance the understanding of the latent phenomenon that underlies these relationships. As a matter of fact, these two points (predicting and understanding) are the two sides of the coin and should be achieved by the same model. Moreover, an analyst feels more confident with a model that he or she can interpret using the contextual background knowledge (Martens and Martens 2001).

5.3.1 Multiple regression analysis and regression on principal components

Multiple regression analysis is by far the most popular statistical method used to relate a (dependent) variable to a set of (independent) variables. In the present context, each sensory variable is, in turn, regressed upon the instrumental variables thus leading to a model that expresses the sensory variable under consideration as a linear combination of the instrumental variables. A first shortcoming of this procedure of analysis emerges. Indeed, the approach has a propensity to generate a large number of estimated parameters which makes the interpretation of the outcomes rather tedious. A second and more serious shortcoming is that the models thus obtained are likely to be unstable as a consequence of the high colinearity among the instrumental variables. This is reflected by unacceptable uncertainty (large variance) in regression coefficient estimates. Furthermore, the coefficients can change drastically depending on which individuals are in or out of the model or because of small errors in the data.

Selection of variables in linear regression is a common solution to circumvent the problem of multicolinearity. There are various strategies to select variables that should be included in the model. These strategies encompass backward, forward and stepwise selection together with other criteria that have been developed to assist the user in finding a good subset of variables. As a matter of fact, the problem of selection of variables is more complex than it appears and is still the focus of a lot of research in several areas of applications. It is common knowledge that it is a satisfactory strategy for improving the prediction performance but has some drawbacks insofar as the understanding of the underlying concept is concerned. This is a direct consequence of the automatic nature of the selection process which entails that key variables for the comprehension of the phenomenon may be discarded and replaced by other variables whose meaning is less obvious.

Another popular alternative to multiple linear regression is regression on principal components. We have already advocated performing PCA on instrumental data as a preliminary analysis in order to investigate their structure and reveal the linear relationships among the instrumental variables. The outcomes of this analysis can subsequently be used in order to investigate how the underlying dimensions to the instrumental measurements (i.e. principal components) are related to the sensory variables. This can be done by, first, discarding the last principal components of the instrumental data and, second, performing multiple linear regression of the sensory variables on the retained principal components. Graphical summaries of the method of analysis showing how the sensory variables are related to the principal components and, consequently, to the instrumental variables themselves are a compelling feature of the method of analysis. This consists in displaying the correlations of the sensory and instrumental variables with the most relevant instrumental principal components. MacFie and Hedderley (1993) refer to this strategy of analysis as 'principal component correlation'.

A serious pitfall of principal regression analysis is related to the choice and the number of principal components to be included in the model. Considering that the derivation and the selection of the principal components are based on their ability to recover the variability in the instrumental data, it may occur that some of these components are not relevant to explain the variability in the sensory data whereas some of the discarded principal components may contain relevant information connected to sensory data. This is one reason why we turn our attention to multivariate statistical methods where sensory variables have an active role in the determination of the latent components derived from the instrumental data.

5.3.2 Canonical correlation analysis

Canonical correlation analysis (CCA) is a multivariate linear statistical analysis, first introduced by Hotelling (1936). It can be seen as an extension of multiple linear regression in the sense that it investigates the relationships between two (multivariate) data tables (Bruce 1984). Its interest also lies in its fecund conceptual framework which encompasses, besides multiple linear regression, various other statistical methods such as correspondence analysis and canonical variate analysis. If applied to sensory and instrumental data, it makes it possible to highlight the most correlated variables from both data sets and detect those variables that change simultaneously. From a technical point of view, CCA proceeds in a stepwise manner. In a first stage, it seeks a latent variable called canonical variate from the sensory space (i.e. a linear combination of sensory attributes) and a canonical variate from the instrumental space such that the correlation between these two latent variables is as large as possible. In subsequent stages, CCA computes other pairs of such canonical variates corresponding to relations of decreasing strength. The canonical variates in each space are constrained to be orthogonal (uncorrelated). If the correlation coefficient (called canonical correlation coefficient) between two associated canonical variates is close to 1, this means that we have identified two patterns, one from each data set, that mirror each other. As usual in multivariate statistical methods, the interpretation of each canonical variate is done by considering the loadings (coefficients, weights) associated with the sensory or instrumental variables as the case applies. Alternatively, and preferably, they can be interpreted in terms of their correlations with the original sensory and/or instrumental variables. However, two pitfalls can hinder the use of CCA in many applications. First, as it is clear from the brief description of CCA, the role played by sensory and instrumental data is symmetric whereas the purpose of the study is intrinsically not symmetric as we aim to predict sensory data from instrumental data. Nevertheless, this is not a serious objection insofar as CCA can be seen as a means to capture patterns which are relevant to both kinds of data. In a subsequent stage, sensory variables can be predicted by means of regression upon canonical variates associated with the instrumental variables. The second and more serious pitfall of CCA is its vulnerability to multicolinearity among either sensory or instrumental variables. Similarly to multiple linear regression, CCA may lead to unstable and difficult-to-interpret models. Furthermore, the common patterns reflected by the first pairs of canonical variates may consist of minor aspects in both sensory and instrumental data. By minor aspects, we mean directions of the sensory or instrumental spaces which carry only a small amount of total variance and which may be connected to noise only. For these reasons, we recommend the use of redundancy analysis or partial least squares regression, which will be outlined below, instead of CCA.

5.3.3 Redundancy analysis

Whereas it may occur that the first canonical variates of CCA reveal minor aspects which are common to sensory and instrumental analysis, redundancy analysis (RA) is more clearly directed towards recovering the variability in sensory data from latent components (linear combinations) derived from the instrumental measurements. This feature makes RA a promising alternative to CCA.

RA, also called PCA with instrumental variables (Rao 1964), can be defined as a PCA performed on the sensory data but the principal components are constrained to be linear combinations of the instrumental variables. It can also be viewed as multiple linear regression of the sensory variables upon orthogonal (non-correlated) linear combinations of the instrumental variables which are successively computed so as to maximize the proportion of total variance in the sensory variables that can be accounted for by these linear combinations (van den Wollenberg 1977). Therefore, it appears that the rationale behind RA is very compelling in the context discussed herein because it precisely meets the purpose of predicting sensory from instrumental data. Unfortunately, like multiple linear regression and canonical correlation analysis, RA is numerically vulnerable to multicolinearity among predictor variables. Therefore, users of this method of analysis should be cautioned against accepting the outcomes of the analysis uncritically. We will discuss, on the basis of a case study, a way to deal with this problem. Alternatively, one may consider performing partial least squares regression which is based on the same principle as RA but counteracts the problem of multicolinearity in an efficient way.

5.3.4 Partial least squares regression

PLS regression (PLSR) is nowadays one of the most popular methods for relating two data sets. This widespread use is justified by the fact that it combines simplicity and efficiency. It was first proposed by Wold *et al.* (1983). From a conceptual point of view, PLSR bears some similarity to canonical correlation analysis as it seeks pairs of directions, one in the sensory space and another in the instrumental space that are highly related. But, whereas canonical correlation analysis uses the correlation coefficient to measure the strength of the linear relation between the canonical variates, PLSR uses the covariance to measure the strength of the linear relation between the components. This is a small change in the optimality criterion but it has a great and positive impact on the stability of the prediction model. To understand the rationale behind this criterion, one should bear in mind that the covariance between two components t and u (say) is the product of their correlation coefficient and their standard deviations. Therefore, it turns out that by maximizing the covariance, we realize a compromise between seeking a common structure to both sensory and instrumental data (correlation coefficient of t and u) and recovering the variability in both data sets (standard deviations of t and u). By taking account of the major latencies present in both sensory and instrumental spaces, we prevent minor aspects which may be related to noise only to emerge as the first dimensions. Another important difference between PLSR and canonical correlation analysis is that the former analysis is not symmetric in terms of the roles played by the sensory and instrumental data sets, thus highlighting the fact that one set of variables (sensory) is being predicted from the other set (instrumental). This is done by a clever device which consists, at each stage of the procedure, of regressing the sensory variables on the PLS components from the instrumental space and which have been previously computed. Thereafter, the residuals of these regressions are considered in the subsequent study instead of the original variables in order to seek new directions in the instrumental space that explain (part of) the remaining information in the sensory variables which has not been explained so far. As a matter of fact, the same strategy (i.e. regression upon the instrumental components and computation of the residuals) is also applied on the instrumental variables in order to enforce the orthogonality (non-correlation) of the PLS components computed at the various stages of the algorithm, thus making their interpretation easier.

PLSR also bears similarities to redundancy analysis in the sense that it aims at recovering the variability in the sensory variables from the instrumental measurements. But, PLSR achieves a trade-off between the stability of the model and the recovery of the variation in the sensory variables by favoring those directions in the instrumental space that account for a large amount of total variance.

5.4 Case study

We investigate the relationships between forty aroma volatile compounds in tomatoes (Table 5.1) and eight flavor attributes: earthy, green tomato, herb, flavor intensity, floral, lemon, sweet flavor and tomato leaf. In the following, the emphasis is put on the outcomes of the statistical methods rather than why and how the volatile compounds affect the sensory characteristics. For more details regarding these latter aspects, we refer to papers which have been published on the subject (Langlois *et al.* 1996, Tandon *et al.* 2000, Krumbein *et al.* 2004). The products consist of several varieties of tomatoes some of which were harvested in two successive years (1992 and 1993). In total, we dispose of 20 products on which forty volatile compounds were acquired and eight sensory variables were measured.

The data table (20 rows and 40 columns) associated with the volatile compounds was analyzed by PCA in order to investigate the structure of the variables

Table 5.1 List of the volatile compounds recorded by gas-chromatography

Volatile compound (abbreviation)	Volatile compound (abbreviation)
3-methyl-butanal (3butanal)	isobutylthiazole (isobutyl)
2-methyl-butanal (2butanal)	butanolactone/heptadienal (buta-hep)
pent-1-en-3-one (pentenon)	3-methyl-thio-propan-1-ol (thioprol)
pentanal (pentanal)	2-methyl-butanolactone (butanola)
pentan-3-one (pentanon)	2-phenyl-ethanal (phethaal)
2-methylbut-2-enal (2butenal)	nonanal (nonanal)
2- et 3-methylbutan-1-ol (23butaol)	phenyl-methanol (phmethol)
pent-2-enal (pentenal)	2-methoxy-phenol (oxphenol)
hexanal (hexanal)	2-phenyl-ethanol (phethaol)
(Z)-hex-3-enal (Zhexenal)	decanal (decanal)
3-methylpentan-1-ol (3pentaol)	methyl salicylate (salicylat)
(E)-hex-2-enal (Ehexenal)	2,3-epoxy-geranial (epoxyger)
hex-3-en-1-ol (hexenol)	benzothiazole (benzothi)
hexan-1-ol (hexanol)	neral (neral)
2-methyl-thio-ethanol (thiothol)	geranial (geranial)
3-methyl-thio-propanal (thiopral)	3(1,1-dimethyl-ethyl)phenol (etphenol)
hept-2-enal (heptenal)	geranylacetone (geranyla)
6-methylhept-5-ene-2-one (heptenon)	b-ionone (b-ionone)
benzaldehyde (benzalde)	5,6-epoxy-b-ionone (epionone)
6 methylhept-5-ene-2-ol (heptenol)	

which will be used subsequently as predictor variables. The two first principal components which are nearly of equal importance account for 45.7% of the total variance. Up to six principal components are needed in order to recover 80% of the total variance. Some eigenvalues of the correlation matrix (variances of the principal components) are very small and, not surprisingly, the eigenvalues from rank 20 (number of products) upwards are equal to zero. This is a consequence of the fact that the number of variables exceeds the number of individuals (products). In this case, the correlation matrix is said to be singular and several statistical methods such as multiple linear regression, canonical correlation analysis and redundancy analysis cannot be performed, at least in their standard conceptual framework. This issue is discussed below in more details.

Figure 5.1 depicts the correlation of the instrumental variables with the first two principal components and Fig. 5.2 depicts the correlation of sensory attributes with the same principal components. This makes it possible to visualize which variables (instrumental or sensory) are positively correlated (arrows pointing in the same direction), negatively correlated (arrows pointing in the opposite directions) or uncorrelated (perpendicular arrows). This visual inspection is particularly true for those variables which are close to the circle.

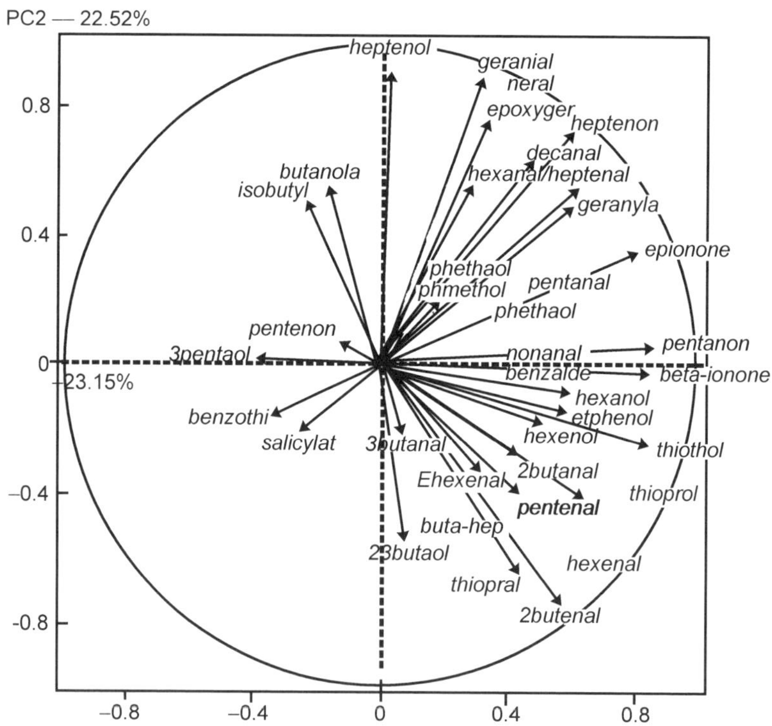

Fig. 5.1 Correlation of the instrumental variables with the first two principal components from instrumental data.

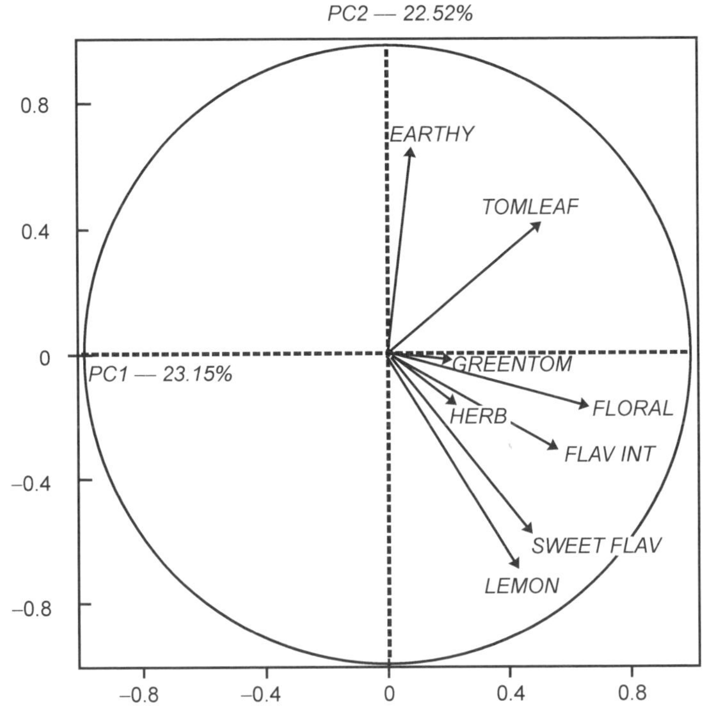

Fig. 5.2 Correlation of the sensory variables with the first two principal components from instrumental data.

The representation of the products on the basis of the first two principal components from the instrumental variables is given in Fig. 5.3. It can be seen that the varieties Cerise2 (Cherry, harvested in 1992) and Cerise3 (harvested in 1993), for instance, are characterized by large values of 2-methylbut-2-enal (2butenal), (Z)-hex-3-enal (Zhexenal) .... These tomatoes were assessed by the sensory panel as having a relatively high sweet flavour and high intensity of lemon. Contrariwise, varieties Elena2, Daniela3 and Cencara3 have larger values for volatile compounds 6-methylhept-5-ene-2-ol (heptenol) and geranial, among others. Accordingly, they were assessed by the sensory panel as having a larger value than average for the attribute 'earthy' and small intensities for sweet flavour and lemon.

As we have mentioned before, a technical problem arises from the fact that the correlation matrix is singular. Methods such as linear regression, canonical correlation analysis and redundancy analysis cannot be directly performed. A way to deal with this problem is to select a subset of instrumental variables as will be illustrated within the framework of redundancy analysis. Within this context, the algorithm used by Schlich *et al.* (1987) has proven to be very effective for relating sensory to volatile compounds quantified by gas chromatography. The rationale behind this algorithm is to select step-by-step

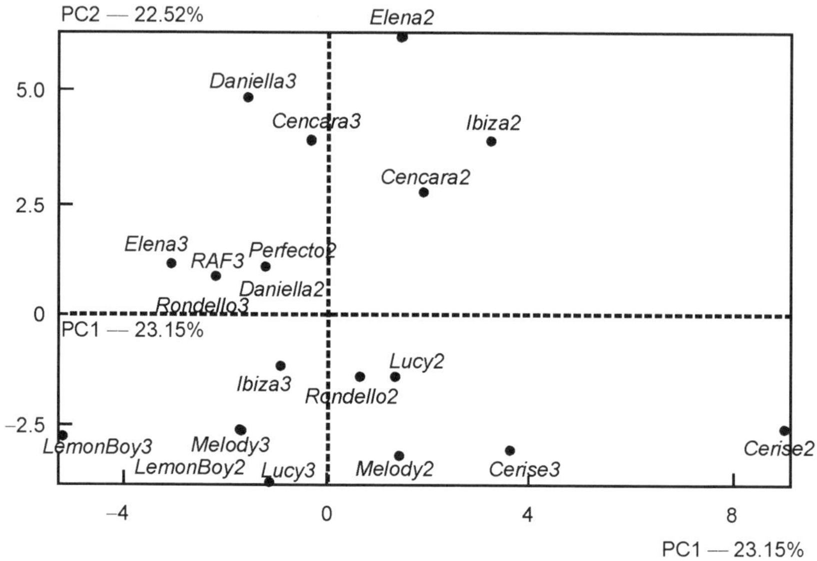

Fig. 5.3 Representation of the products on the basis of the first two principal components from instrumental data.

instrumental variables that recover as much as possible the variability in the sensory data. However, as mentioned above, selection of variables by maximizing a criterion that reflects the extent to which the selected instrumental variables predict the sensory variables may lead to discarding key variables and selecting, instead, variables of minor interest or those which are difficult to interpret. A solution to this problem is given by cluster analysis of variables (Schlich and Guichard 1989, Vigneau and Qannari 2003). This consists in defining homogeneous clusters of variables (i.e. clusters formed of highly correlated variables). Thereafter, one variable from each cluster can be selected taking account, besides statistical criteria, of considerations pertaining to interpretation, relevance, etc.

In order to select a good subset of instrumental variables for the purpose of predicting the sensory attributes by means of redundancy analysis, we run the algorithm used by Schlich *et al.* (1987). This led to the selection of eight volatile compounds. The outcomes of redundancy analysis on the basis of these selected variables are shown in Fig. 5.4 which gives a clear picture of how the retained instrumental variables are related to the sensory attributes and how the varieties of tomatoes are characterized from the sensory and instrumental points of view.

We also performed PLS regression in order to relate the sensory attributes to the forty volatile compounds. The first two PLS components from the instrumental space explain 47.2% of the variation in the sensory data. This percentage gradually increases when more PLS components are introduced in the model. It

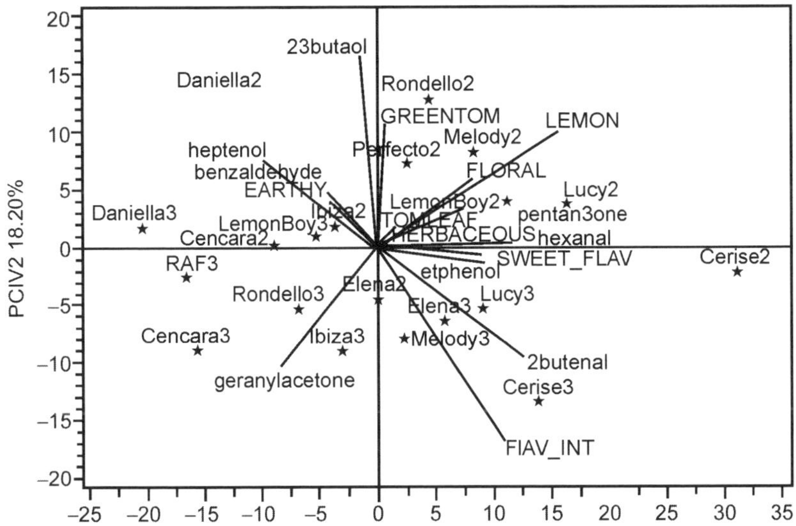

Fig. 5.4 Representation of the products on the basis of the first two redundancy analysis components and correlations of the sensory and the selected volatile compounds with these components.

is also of interest to consider the percentage of total variance explained by the same components obtained by a cross-validation procedure. This is a more objective criterion for the choice of the number of PLS components to be introduced in the model. For a given number of components A (say), cross-validation consists in setting, in turn, each individual (product) aside, elaborating a PLS model with the A components and predicting the sensory characteristics for the individual that was set aside. A cross-validated percentage of total variance explained by the A components is computed by comparing the actual and predicted sensory values. Not surprisingly, this criterion was much smaller than the usual percentage of total variance explained. It increased when passing from one component to two components, reached a plateau and decreased when more than four components were introduced in the model indicating that these components are irrelevant for the prediction of the sensory variables.

Figure 5.5 shows the loadings associated with the first two PLS components from both sensory and instrumental variables. This makes it possible to investigate which variables are associated. In parallel, it is also possible to depict the representation of the products on the basis of the first PLS components from the instrument data. This graphical display (not shown here) made it possible to draw conclusions which are, to a large extent, similar to those drawn from Fig. 5.3, which shows the representation of the products on the basis of the first two principal components of the instrumental data.

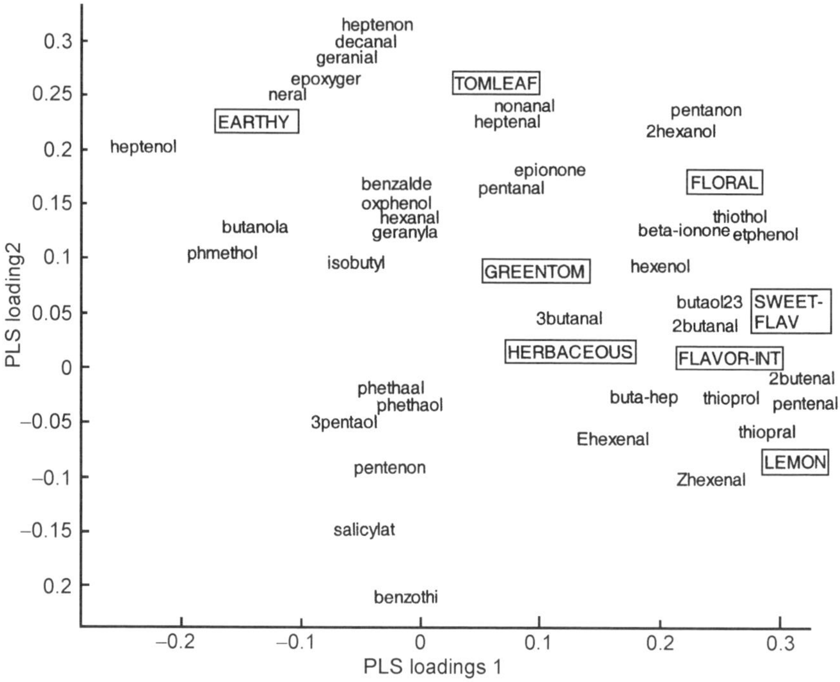

Fig. 5.5 Loadings associated with the first two PLS components from both the sensory and the volatile compounds variables.

5.5 Conclusion

The appealing feature of most of the multivariate techniques discussed herein is that they provide visualization tools which can be helpful for researchers to unveil hidden patterns and relationships among variables. Therefore, the portrayal of the relationship between sensory and instrumental data makes an important contribution to understanding how physical and chemical properties act to produce specific sensations. This exploratory data analysis inherent to multivariate statistical techniques is not shared by predictions methods based on neural networks which pertain to black-box modeling. However, these latter tools are gaining ground in sensory analysis because of their ability to take account of non-linear relationships between input and output variables (Bardot *et al.* 1994, Wilkinson and Yuksel 1997, Donahue *et al.* 2000).

Another important issue that was not sufficiently addressed in this paper concerns the validation of the models. This is a key part of the modeling process which deserves increased attention if systematic progress is to be made in the development of predictive models. It encompasses the assessment of the stability of the model and the uncertainty about the estimated parameters. It also consists of setting up a hypothesis testing framework in order to evaluate the significance of some parameters, the relevance of some patterns revealed by the method of analysis or, more generally, to make inferential decisions about the models. The sampling theory of most of the techniques outlined herein requires multivariate normality which is generally not met in the context of sensory and instrumental data. Nonetheless, with the advent of computer technology, new paradigms for model validations have emerged which render the requirement of normality obsolete. These tools range from resampling methods such as jackknife or bootstrapping techniques (Efron and Tibshirani 1993, Mooney and Duval 1993) to randomization tests (Good 1994, Edgington 1995). A description of these techniques is beyond the scope of this chapter but users should be aware of their interest and their potential.

The advent of technology also helped the statistical software companies to design menu driven interfaces to make their software even more user-friendly and more interactive. However, sensory analysts should not be discouraged by the more complex packages because they offer greater flexibility and power of statistical inference and make it possible to share (generally for free) programs with other users via the internet.

5.6 Acknowledgments

Thanks are due to Dominique Langlois from INRA–FLAVIC, Dijon for providing the data and for the time he spent looking through his files and translating some technical words.

5.7 References

ARNOLD G M and WILLIAMS A A (1986) The use of Generalized Procrustes analysis in sensory analysis. In Piggott J R (ed.) *Statistical Procedures in Food Research*. Elsevier, London.

BARDOT I, BOCHEREAU L, MARTIN N and PALAGOS B (1994) Sensory-instrumental correlations by combining data analysis and neural network techniques. *Food Quality and Preference*, 5, 1/2, 159–166.

BELSLEY D A (1991) *Conditioning Diagnostics: Colinearity and Weak Data in Regression*. Wiley, New York.

BROCKHOFF P B (2003) Statistical testing of individual differences in sensory profiling. *Food Quality and Preference*, 14, 5/6, 425–434.

BRUCE T (1984) *Canonical Correlation Analysis: Uses and Interpretation*. Sage Publications in Quantitative Applications in the Social Sciences, 47.

CHUNG S J, HEYMANN H and GRÜN I U (2003) Application of GPA and PLSR in correlating sensory and chemical data sets. *Food Quality and Preference*, 14, 5/6, 485–495.

DIJKSTERHUIS G B (1996) Procrustes Analysis in sensory research. In: Næs T and Risvik E (eds), *Multivariate Analysis of Data in Sensory Science*. Elsevier Science, London.

DONAHUE D W, BENOIT P W, LAGASSE B J and BUSS W R (2000) Consumer and instrumental evaluation of Maine wild blueberries for the fresh pack market. *Postharvest Biology and Technology*, 19, 3, 221–228.

EDGINGTON E S (1995) *Randomization Tests*, 3rd edn. Marcel Dekker, New York.

EFRON B and TIBSHIRANI R (1993) *An Introduction to the Bootstrap*. Chapman & Hall, London.

FOX J (1991) *Regression Diagnostics: An Introduction*. Sage Publications in Quantitative Applications in the Social Sciences, 79.

GOOD P (1994) *Permutation Tests: A Practical Guide to Resampling Methods for Testing Hypotheses*. Spring-Verlag, New York.

HOTELLING H (1936) Relations between two sets of variates. *Biometrica*, 28, 321–377.

HUNTER E A (1996) Experimental design. In Næs T and Risvik E (eds), *Multivariate Analysis of Data in Sensory Science*. Elsevier Science, London.

KRUMBEIN A, PETERS P and BRÜCKNER B (2004) Flavour compounds and a quantitative descriptive analysis of tomatoes (Lycopersicon esculentum Mill.) of different cultivars in short-term storage. *Postharvest Biology and Technology*, 32, 1, 15–28.

KUNERT J and QANNARI E M (1999) A simple alternative to Generalized Procrustes Analysis. Application to sensory profiling. *Journal of Sensory Studies*, 14, 197–208.

KUTI T, HEGYI A and KEMÉNY S (2004) Analysis of sensory data of different food products by ANOVA. *Chemometrics and Intelligent Laboratory Systems*, 72, 2, 253–257.

LABBE D, RYTZ A and HUGI A (2004) Training is a critical step to obtain reliable product profiles in a real food industry context. *Food Quality and Preference*, 15, 4, 341–348.

LANGLOIS D, ETIEVANT P X, PIERRON P and JORROT A (1996) Sensory and instrumental characterisation of commercial tomato varieties. *Zeitsch Lebens Unters Forsch*, 203, 534–540.

LEE S Y, LUNA-GUZMÁN I, CHANG S, BARRETT D M and GUINARD J X (1999) Relating descriptive analysis and instrumental texture data of processed diced tomatoes. *Food Quality and Preference*, 10, 6, 447–455.

MACFIE H J H and HEDDERLEY D (1993) Current practice in relating sensory perception to instrumental measurements. *Food Quality and Preference*, 4, 41–49.

MARTENS H and MARTENS M (2001) *Multivariate Analysis of Quality: An Introduction*. Wiley, New York.

MCEWAN J A, HEINIÖ R L, HUNTER E A and LEA P (2003) Proficiency testing for sensory ranking panels: measuring panel performance. *Food Quality and Preference*, 14, 3, 247–256.

MOONEY C Z and DUVAL R D (1993) *Bootstrapping: A Nonparametric Approach to Statistical Inference*. Sage Publications in Quantitative Applications in the Social Sciences, 95.

MUIR D D and HUNTER E A (1991) Sensory evaluation of Cheddar cheese: order of tasting and carryover effects. *Food Quality and Preference*, 3, 3, 141–145.

NAES T and KOWALSKI B (1989) Predicting sensory profiles from external instrumental measurements. *Food Quality and Preference*, 4/5, 135–147.

PORCHEROT C and SCHLICH P (2000) FLASH table and canonical mapping of potato varieties. *Food Quality and Preference*, 11, 163–165.

RAO C R (1964) The use and interpretation of principal component analysis in applied research. *Sankhya A*, 26, 329–358.

SCHLICH P (1996) Defining and validating assessor compromises about product distances and attribute correlations. In Næs T and Risvik E (eds), *Multivariate Analysis of Data in Sensory Science*. Elsevier Science, London.

SCHLICH P (2004) L'analyse en variables canoniques des données de profils sensoriels.

Applications pour le contrôle des performances des sujets et pour la comparaison des produits. *8èmes Journées Agro-Industries et Méthodes Statistiques,* Rennes, 33–43.

SCHLICH P and GUICHARD E (1989) Selection and classification of volatile compounds of apricot using the RV coefficient. *J. Agr. Food Chem.*, 37, 142–150.

SCHLICH P, ISSANCHOU S, GUICHARD E, ETIEVANT P and ADDA J (1987) RV coefficient: A new approach to select variables in PCA and to get correlations between sensory and instrumental data. In *Flavour Science and Technology*. Chichester, Wiley & Sons, 469–474.

TANDON K S, BALDWIN E A and SHEWFELT R L (2000) Aroma perception of individual volatile compounds in fresh tomatoes (Lycopersicon esculentum, Mill.) as affected by the medium of evaluation. *Postharvest Biology and Technology*, 20, 3, 261–268.

VIGNEAU E and QANNARI E M (2003) Clustering of variables around latent components. *Communications in Statistics – Simulation and Computation*, 32, 4, 1131–1150.

WILKINSON C and YUKSEL D (1997) Using artificial neural networks to develop prediction models for sensory-instrumental relationships; an overview. *Food Quality and Preference*, 8, 5/6, 439–445.

WOLD S, MARTENS H and WOLD H (1983) The multivariate calibration problem in chemistry solved by the PLS method. In *Proc. Conf. Matrix Pencils*, Ruhe.

VAN DEN WOLLENBERG A L (1977) Redundancy Analysis: an Alternative to Canonical Correlation Analysis. *Psychometrika*, 42, 207–219.

Part II

Flavour retention and release from the food matrix

6

Flavour retention and release from the food matrix: an overview

A. Voilley, University of Bourgogne, France and
I. Souchon, INRA, France

6.1 Introduction

Flavour release depends on the mechanisms that influence the volatilisation of aroma compounds from foods in many situations such as formulation and processing, during product storage or consumption. In all cases, the same series of unit operations with mass and heat transfers, diffusion or reactions as in food engineering have to be considered.

Thermodynamic and kinetic parameters control the flavour release from foods. They depend not only on the composition but also on the physical state of the matrix: liquid systems (water or lipids) are relatively simple to study and are useful to understand physical partitioning and release; when moving to viscous or solid systems, more parameters are involved such as diffusion; real foods are often even more complex, with heterogeneous composition and structure and must be studied at different levels.

The objectives of this overview are to give some physico-chemical characteristics of flavour, the fundamentals of diffusion and mass transfers, and to discuss some of the main factors influencing flavour release or retention, in order to introduce the following chapters which concern all these aspects.

6.2 Flavour properties

Flavours are volatile and odorous organic compounds at atmospheric pressure; their retention is a function of the availability of the flavour compounds in the gas phase and, therefore, of the affinity of these compounds for the food matrix.

The concentration of a volatile compound, very often low, in one or between several phases, depends not only on the characteristics of the pure compound (molecular size, functional groups, shape) but also of thermodynamic parameters at the macroscopic scale (vapour pressure, solubility, partition coefficients and activity coefficients). All these colligative properties can be derived by related equations (Atkins 1996; Bruin 1999). For example, the gas–liquid partition coefficient (K_i) or the volatility of a compound i is expressed at a given temperature:

$$K_i = \frac{Y_i}{X_i}$$

where Y_i is the concentration of i (in general, expressed in molar fraction) in the vapour phase; X_i is the concentration of i in the liquid phase; and K_i can be related to both the activity coefficient of i at infinite dilution (γ_i^∞), representing interactions between aroma compounds and the matrix, and to the saturated vapour pressure (P_i^S) at the same given temperature:

$$K_i = \frac{\gamma_i^\infty P_i^S}{P_T}$$

where P_T is atmospheric pressure.

Experimental data of saturated vapour pressures (volatilities of pure compounds) in literature are scarce (Covarrubias-Cervantes *et al.* 2004); reliable methods for estimating this parameter are of increasing importance as a tool in predicting the behaviour of aroma compounds in different media. Numerous equations and correlations are presented in the literature (Reid *et al.* 1987) but nevertheless some differences are sometimes obtained (Table 6.1). These differences have an effect in the calculation of other thermodynamic constants related to the vapour pressure as activity coefficients (γ_i^∞) which differ significantly from one (Table 6.2), meaning that there are physico-chemical interactions between components and many techniques are available based on headspace measurement; however, interlaboratory work done during the European Cost Action 96 on 'interactions of food matrix with small ligands influencing flavour and texture' demonstrates that, in any case, much caution

Table 6.1 Saturated vapour pressures (experimental and estimated values) of some aroma compounds at 25 °C

Aroma compound	Experimental (Pa)	Reid *et al.* (1987)	Gomez-Nieto and Thodos (1978)
d-linalool	27	14 (−48)	31 (+15)
2-nonanone	59	80 (+36)	65 (+10)
d-limonene	200	173 (−14)	279 (+40)
Isoamyl acetate	733	802 (+09)	804 (+10)

() % error = [(estimated − experimental value)/experimental value] × 100

Table 6.2 Activity coefficient of aroma compounds calculated using experimental and estimated values of saturated vapour pressure (25 °C)

Aroma compound	γ_i		
		Used value of P_i^S	
	Experimental	Reid *et al.* (1987)	Gomez-Nieto and Thodos (1978)
d-linalool	8740	16 805 (+92)	7569 (−13)
2-nonanone	58 036	42 800 (−26)	52 483 (−10)
d-limonene	99 307	114 672 (+15)	66 536 (−33)
Isoamyl acetate	5348	4888 (−09)	4878 (−09)

() % error = [(estimated − experimental value)/experimental value] × 100

should be taken (Voilley 2000). An overview of the developments in phase equilibria in aqueous solutions is given by Le Maguer (1992).

Food matrix components, forming part of the bulk of the food, such as proteins, polysaccharides and lipids are known to interact with flavour components (Solms 1986, Bakker 1995, Taylor 1999, Guichard 2002). Examples are given in Table 6.3:

- Volatility of aroma compounds at infinite dilution. In water, it increases with the chain length (on the contrary, the volatility – P_i^S – of the pure components decreases).
- Selective interactions with proteins and lipids. One of the most important

Table 6.3 Vapour–liquid coefficient (K_i^∞) and retention (r%) of aroma compounds in complex systems ($T = 25$ °C)

Aroma compound	Water	5 g.L^{-1} sodium caseinate		triolein	
	$K_{iw}^\infty \times 10^3$ γ_i^∞	$K_{im}^\infty \times 10^3$ γ_i^∞	r (%)	$K_{im}^\infty \times 10^3$ $\gamma_i^\infty \times 10^4$	r (%)
Ethyl butyrate ($P^S = 12$ mm Hg) log $P^{**} = 1.7$	13.5 *988*	12.5^{NS} *760*	7.4	0.22* *8*	98.4
Ethyl hexanoate ($P^S = 1$ mm Hg) log $P^{**} = 2.8$	34.0 *26 428*	22.0* *16 720*	35.3	0.03* *65*	99.9

Source: Landy *et al.* (1998)
NS not significant ($P > 0.05$)
* $P \leq 0.05$
** calculated from Rekker method (1977)

$$r = \frac{K_{iw} - K_{im}}{K_{iw}}$$

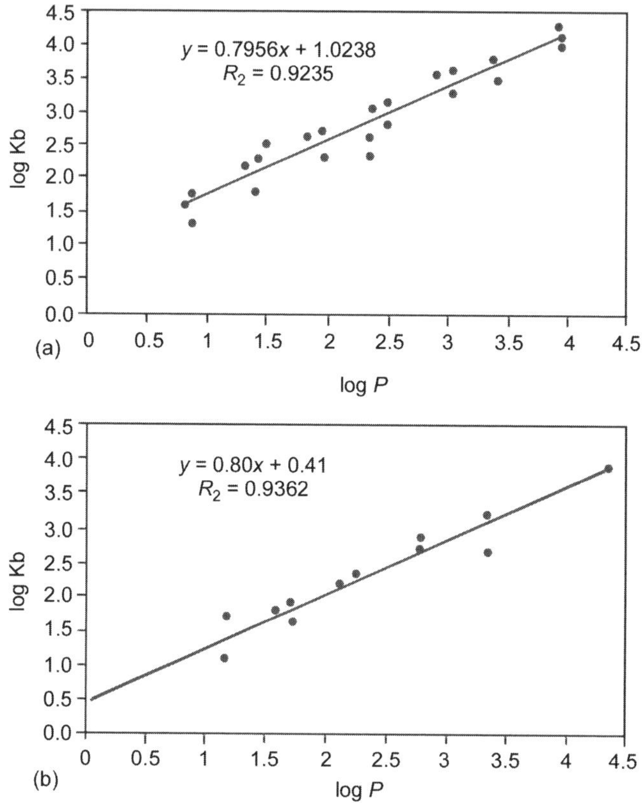

Fig. 6.1 Interactions between β-lactoglobulin and (a) methyl ketones, alcohols, aldehydes, and (b) esters, relation between the logarithm of the binding constant (log kb) and the hydrophobicity (log p) of the flavour compound (Guichard 2000).

parameters which play a role is the hydrophobicity quantified by log P (experimental or estimated value). With β-lactoglobulin, a good linear correlation is found between the logarithm of the binding constant measured by affinity chromatography and the hydrophobicity of aroma compounds (Fig. 6.1).

The research on the effect of other molecules on flavour release is not so often developped (Bakker 1995). Only, influence of small molecules as salts (Voilley *et al.* 1977) or ethanol (Conner *et al.* 1994, Pittia *et al.* 1998) have been intensively studied.

The behaviour of aroma compounds depends also on their solubilities (expressed in mole fraction), which is usually very low in water and inversely related to the activity coefficient. For example, the limit of solubility of ethyl acetate in water is higher than in sucrose solutions, by a factor around 2 at 25 °C up to a factor 4 at 0 °C depending on the sucrose concentration. Moreover, the

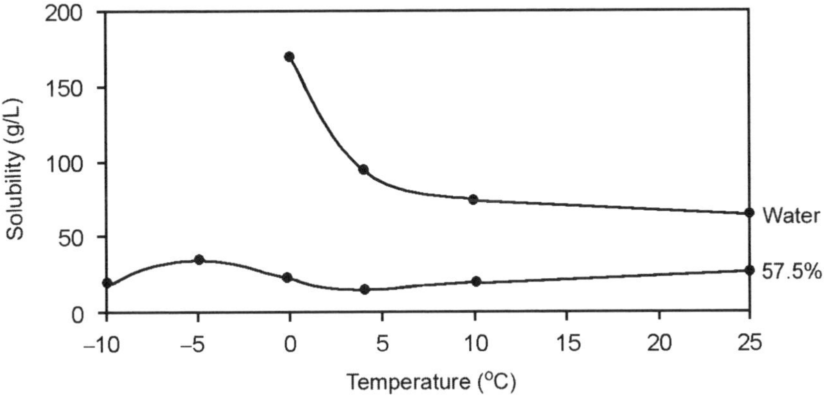

Fig. 6.2 Solubility of ethyl acetate in aqueous sucrose solutions (Covarrubias-Cervantes *et al.* 2004).

temperature effect on solubility in both solvents shows opposite trends: in water, as temperature decreases, solubility increases (Fig. 6.2).

A number of different interactions have been proposed to explain association of flavour with other food components and these are discussed in the following chapters.

6.3 Diffusion and mass transfer

There are two major factors that control the rate of aroma release from foods, namely the volatility of the flavour compounds and the resistance to mass transfer between phases and especially from the product to air.

In other words, the two factors that influence the rate of mass transfer are a driving force to move materials and the resistance to their flow. When considering dissolved substances in liquids, the driving force is a difference in the substance concentration, whereas for gases or vapours, it is a difference in partial pressure or vapour pressure; in all cases, it is a chemical potential difference. The resistance arises from the medium and the interactions between the materials and the medium.

6.3.1 Fundamentals

Unit operations in the food industry most often consist of the mass, heat, and momentum transfers and are often limited by such transport. Similar fundamentals exist for flavour release from food. Mechanisms of transfer are detailed well in literature (Charm 1971, Leniger and Beverloo 1975, Loncin and Merson 1979, Treybal 1980, Crank 1986, Gekas 1992, Perry and Green 1997, Fellows 2002 amongst others). Only the basic principles are given in this chapter concerning mass transfer (diffusion and convection).

Diffusion coefficient (D)
Mass transfer in immobile systems is classified as a 'diffusional' process if this transfer due to the movement of groups of molecules is negligible compared with the movement of individual molecules; a system is called 'isotropic' when its properties are independent of the direction considered.

For immobile, isotropic and permanent systems, Fick's first law is expressed as:

$$\frac{dm}{dt} = -A \cdot D \frac{dC}{dx}$$

where m is the quantity of diffusing molecules (kg) passing normally across the element (dA, unit: m^2) in a time (dt, unit: s); D is the diffusion coefficient (m^2s^{-1}); dC/dx is the concentration gradient with C ($kg.m^{-3}$) and x, the distance (m).

For non-permanent systems, Fick's second law is expressed as:

$$\frac{\partial C}{\partial t} = D\left(\frac{\partial^2 C}{\partial x^2} + \frac{\partial^2 C}{\partial y^2} + \frac{\partial^2 C}{\partial z^2}\right)$$

It represents the variation of concentration at a given point as a function of time.

The resolution of these equations depends on experimental system, initial and boundary conditions (Crank 1986).

Mass transfer coefficient (k)
In a biphasic system with a moving fluid, if a mass transfer between the two phases occurs, there are (Fig. 6.3a):

- at the interface a regular boundary layer with a constant thickness, Δx, in which the flow is laminar;
- beyond the boundary layer, the mass transfer is infinitely rapid and a mean concentration $\bar{C}_1$ exists throughout the turbulent core.

In this case, the equation of the mass flux from the boundary layer to the fluid

$$\frac{dm}{dt} = -A \cdot D \frac{dC}{dx}$$

can be written

$$\frac{dm}{dt} = -A \cdot D \frac{\bar{C}_1 - C_2}{\Delta x}$$

The mass flux expressed with the mass transfer coefficient

$$k = \frac{D}{\Delta x} (m.s^{-1})$$

is equal to

$$\frac{dm}{dt} = A \cdot k(C_2 - \bar{C}_1)$$

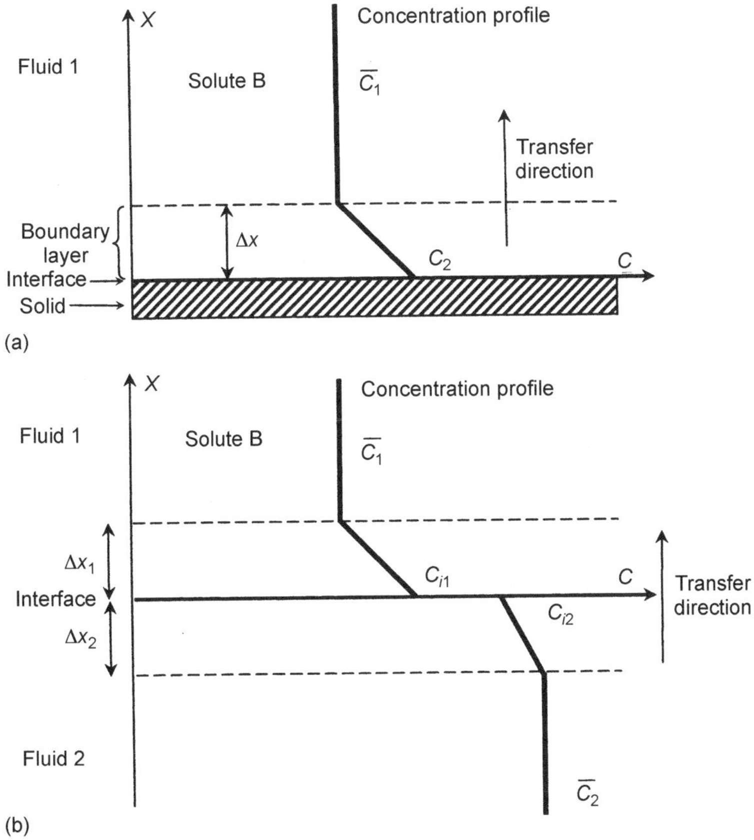

Fig. 6.3 Determination of mass transfer coefficient.

At equilibrium, the concentration at the interface C_2 tends towards the limit reached by the concentration $\bar{C}_1$ in the fluid and

$$\frac{dm}{dt} = A \cdot k(C_1^* - \bar{C}_1)$$

This reasoning can be applied to mass transfer between two phases in turbulent flow (Fig. 6.3b). In the upper phase, the fluid beyond the boundary layer possesses a concentration $\bar{C}_1$ of solute B; the concentration at the interface is C_{i1} and the thickness of the boundary layer is ΔX_1. For the lower phase, the concentration of solute B is $\bar{C}_2$, beyond the boundary layer of thickness ΔX_2 and C_{i2} at the interface. If the phases are in equilibrium at the interface, we have:

$$\frac{C_{i1}}{C_{i2}} = K_c$$

the partition coefficient of solute B between the two phases, in general, is not unity.

If we consider that all of the resistance to transport is found in the boundary layers, the rate of transfer is:

$$\frac{dm}{dt} = A \cdot k_{\text{global}}(K_c\bar{C}_2 - \bar{C}_1)$$

If $K_c\bar{C}_2 = C_1^*$,

$$\frac{dm}{dt} = A \cdot k_{\text{global}}(C_1^* - \bar{C}_1)$$

with

$$\frac{1}{k_{\text{global}}} = \frac{1}{k_1} + \frac{K_c}{k_2}$$

This global mass transfer coefficient, k_{global} (m.s^{-1}), is a function of all the characteristics of the system and is always determined by the similarity methods (dimensional analysis).

6.3.2 Flavour diffusion in the products

Not so many values of aroma diffusion coefficient, especially in food products are avalaible in the literature.

Methods of diffusion measurements

One of the aims of the COST 90bis Project on the Physical Properties of Foods (1982–1987) was the determination of diffusion coefficients. Two types of methods were used to measure *translational* diffusivities of aroma compounds:

Direct measurements

Measurement in a Stokes' cell could be performed only with solutions having a relatively low viscosity.

$$D = \frac{1}{\beta \cdot t}\text{Ln}\left(\frac{\Delta C_i}{\Delta C_t}\right)$$

ΔC_i and ΔC_t are the differences in the concentrations of aroma between the two compartments, respectively at the beginning of the experiment and after time t; D is the diffusivity (m^2.s^{-1}); β is the characteristic constant of the cell.

In a second method, the tested solution was gelified by a low concentration of a gelling agent and was poured into a tube. One extremity of this tube was immersed in a medium with a constant aroma concentration (C_1). After a time t, the cylinder was cut into thin slices. The diffusivity was calculated from the experimentally measured concentration profile in the tube by using a solution of Fick's second law:

$$\frac{C_1 - C_{(x,t)}}{C_1 - C_0} = \text{erf}\left(\frac{x}{2[D \cdot T]^{1/2}}\right)$$

where erf is the error function, x is the distance inward from the exposed surface,

$C_{(x,t)}$ is the aroma concentration at the distance x and time t, C_0 is the initial uniform concentration in the tube.

Indirect measurements, from the results of drying experiments
The aroma solution was air-dried as a slab, in which the remaining aroma concentration was measured as a function of time. Diffusivity was calculated by using the solution given by Crank (1986) for a slab:

$$\frac{C_1}{C_0} = \frac{8}{\pi^2} \sum_{n=1}^{\infty} \frac{1}{(2n-1)} \exp\left(-(2n-1)^2 \pi^2 \frac{D_A t}{L_t^2} \right)$$

D_A, the diffusivity obtained here, was called apparent diffusivity, taking into account that it may be different from the values obtained in direct measurements because of possible structure heterogeneity in the sample due to the drying process and/or influence of the aroma volatility on the rate of evaporation. C_0, C_t are respectively the initial concentration and the remaining concentration of aroma in the slab after the drying time t. L_t is the thickness of the slab, varying with t.

Other methods such as fluorescence recovery after photobleaching (FRAP) (Champion *et al.* 1995) or diffusion ordered spectroscopy (RMN Dosy) (Gostan *et al.* 2004). The observation scale is the micrometre.

Another spectroscopic technique, to determine *rotational* diffusivity (electron spin resonance) can be used; there is a relationship between the probe diffusion coefficients (translational and rotational) in different sugar solutions (Le Meste and Voilley 1988).

Diffusion coefficients of aroma compounds in different matrices
Examples of experimental values found in literature are given in Fig. 6.4. In the same matrix, diffusion coefficient of flavour compounds are of the same order.

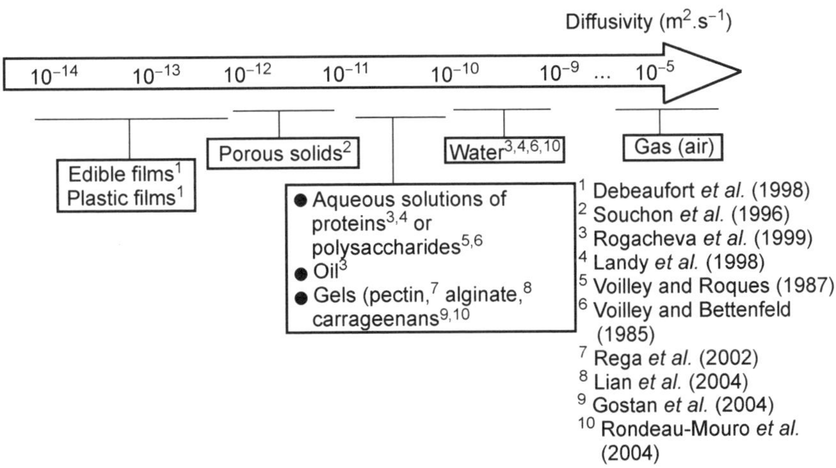

Fig. 6.4 Diffusion coefficient of flavour compounds in different matrices.

Table 6.4 Diffusivity of volatiles in glucose solution (50%, 25 °C) (experimental and estimated values)

Aroma compound	D ($\times 10^{10}$ m^2.s^{-1})			
	Experimental	Wilke and Chang (1955)	Perkins and Geankopolis (1969)	Voilley and Roques (1987)
Acetone	1.21	0.84 (−30)	0.95 (−21)	1.27 (+6)
Diacetyl	0.80	0.72 (−10)	0.81 (−1)	0.62 (−22)
n-hexanol	0.41	0.56 (−36)	0.63 (−54)	0.29 (−29)

() , relative error in reference to the experimental values (%)

A comparison with estimated values is presented in Table 6.4. The relatively simple model takes into account some intrinsic parameters and temperature but, sometimes, the relative error is not negligeable.

6.3.3 Mass transfer at the interfaces

Many quantitative approaches have been developed to study flavour release from the matrix to the headspace. Depending on experimental conditions, mass transfer coefficient can vary between 10^{-4} and 10^{-9} m.s^{-1} (Table 6.5). Foods are often multiphasic, like emulsion. For example, we developed a method to determine the resistances at the lipid–water interfaces, 'the rotating diffusion cell' (Harvey *et al.* 1995, Voilley *et al.* 2000). The addition of sodium caseinate modifies the aroma transfers (Table 6.6).

6.4 Main factors influencing the mobility of flavour compounds in food matrices

6.4.1 Composition: interactions with other constituents

It is clear that a change in volatility of the flavour compound, due to the interactions with the matrix modifies the release from the matrix to the air phase. Not only the nature of the matrix but also the dry matter content plays an important role especially on diffusion.

6.4.2 The physical state and structure of the matrix

Diffusivity and mass transfer coefficient of volatile compounds are affected by these two parameters. It is well known that the change of the structure and the physical state in relation to glass transition of the matrix during processes such as drying, freezing and storage modify flavour release (To and Flink 1978a,b,c, Levi and Karel 1995, Druaux and Voilley 1997, Seuvre *et al.* 2000, de Roos 2003).

Table 6.5 Vapour–matrix mass transfer coefficient of flavour compounds in different food matrices

Matrix	Flavour compounds	Mass transfer coefficient ($\times 10^5 m.s^{-1}$)	Experimental conditions	Reference
Water	Mix of five flavour compounds	10–100	Dynamic headspace dilution, 25 °C	Marin *et al.* (1999)
	Diacetyl	0.04	Stirring rate: 1.67 rps, 37 °C	Bakker *et al.* (1998)
	Ethyl hexanoate	9.2×10^{-4}	Dynamic cumulative headspace, 25 °C	Juteau *et al.* (2004)
		7.6×10^{-4}	Static headspace, 25 °C	
Aqueous solutions of gelatine (2.5–20% w/w)	Diacetyl	0.21–0.04	Stirring rate: 1.67 rps, 37 °C	Bakker *et al.* (1998)
Aqueous solutions of gelatine (15% w/w)		0.02–0.2	Stirring rate: 1.67 7.33 rps, 37 °C	Bakker *et al.* (1998)
Aqueous solutions of β-lactoglobulin	Methyl ketones (C_7, C_8, C_9)	2–5	Stirring, 25 °C	Andriot *et al.* (2000)
Sucrose solutions (5–60% w/w)	Esters and aldehydes	$5\text{–}12.5 \times 10^{-3}$	Gas bubbling through the sample	Nahon *et al.* (2000)
Oil in water emulsion (2% fat)	Ethyl hexanoate	0.1	Dilution by a stream of gas, 22 °C	Doyen *et al.* (2000)
Starch gel	Ethyl hexanoate	6.9×10^{-4}	Dynamic cumulative headspace, 25 °C	Juteau *et al.* (2004)

Table 6.6 Relative percentages of each of the resistances, R_{aq}, R_{oil} and R_i, to the overall mass transfer of the flavour compounds through a miglyol layer[a]

Solute	Sodium caseinate (%, m m^{-1})	R_{aq} (%)	R_{oil} (%)	R_I (%)
Ethyl acetate	0	35.0	27.4	37.6
	5	59.2	16.6	24.2
Ethyl butanoate	0	86.2	6.7	7.1
	5	78.0	5.3	16.7
Ethyl hexanoate	0	74.4	0.4	25.2
	5	58.8	0.3	40.9

Source: Landy *et al.* (1998)
[a] Percentages were calculated by considering that at $\omega^{-1/2} = 0.87\,s^{1/2}$, $1/k$ corresponded to 100% of the resistances.

6.4.3 Examples of flavour release from complex dairy products

A lot of work has been done with simple or model systems. This physicochemical approach is, of course, very important to explain the different phenomena. Nowadays, more complex systems are studied (Charles *et al.* 2000, Paci *et al.* 2004, Nongonierma *et al.* 2005).

For example, sensory evaluation of processed cheeses flavoured with 2-nonanone shows that the intensity of aroma perception depends on the structure of the matrix. The relative effects of all studied factors on flavour release in specific conditions are given in Fig. 6.5.

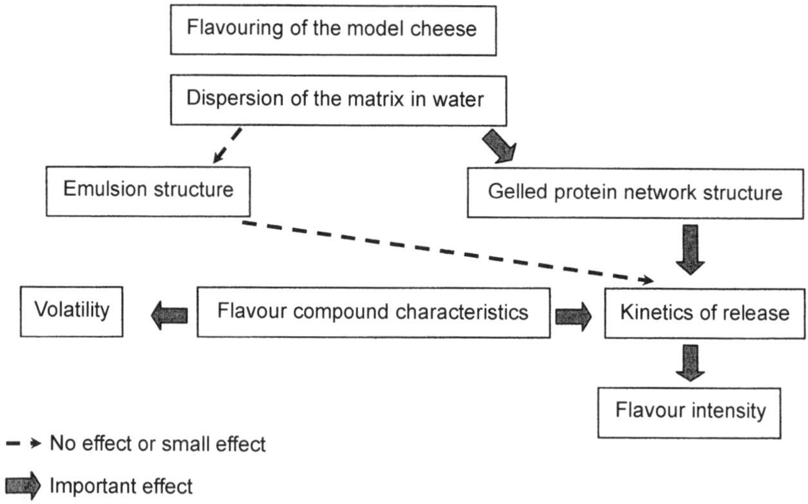

Fig. 6.5 Influence of the physicochemical characteristics of model cheeses upon their texture and flavour.

6.5 Conclusion and future trends

In the literature, much scientific research has been published on flavour release from food. The physico-chemical properties of aroma compounds must be quantified correctly and their behaviour (partition coefficients and mass transfer coefficients) must be known as a function of the space (multiphasic systems) and time.

At present, it is difficult to compare the different studies and the impact of the various factors on flavour release need to be better classified. For this last reason, a mechanistic approach has to be developed to better understand the limiting steps of mass transfer and so to better control flavour release during processes and consumption. Of course, in each case, it would also be interesting to know much more about perception.

6.6 References

ANDRIOT, I., HARRISON, M., FOURNIER, N. and GUICHARD, E. (2000) Interactions between methyl ketones and β-lactoglobulin: sensory, analysis, headspace analysis, and mathematical modeling. *J. Agric. Food Chem.* **48**: 4246–4251.

ATKINS, P. W. (1996) *The Elements of Physical Chemistry*. Oxford University Press, Oxford, England.

BAKKER, J., BOUDAUD, N. and HARRISON, M. (1998) Dynamic release of diacetyl from liquid gelatin in the headspace. *J. Agric. Food Chem.* **46**: 2714–2720.

BAKKER, J. (1995) Flavor interactions with the food matrix and their effects on perception. In *Ingredient Interactions. Effect of Food Quality*, Gaonkar, A.G. (ed.) Marcel Dekker Inc., New York, 411–439.

BRUIN, S. (1999) Phase equilibria for food product and process design. *Fluid Ph. Equil.* **158–160:** 657–671.

CHAMPION, D., HERVET, H., BLOND, G. and SIMATOS, D. (1995) Comparison between two methods to measure translational diffusion of a small molecule at subzero temperature. *J. Agric. Food Chem.* **43**: 2887–2891.

CHARLES, M., ROSSELIN, V., BECK, L., SAUVAGEOT, F. and GUINCHARD, E. (2000) Flavor release from salad dressings: sensory and physicochemical approach in relation with structure. *J. Agric. Food Chem.* **48**: 1810–1816.

CHARM, S. (1971) *Fundamentals of Food Engineering*. AVI, Westport.

CONNER, J. M., PATERSON, A. and PIGGOTT, J. R. (1994) Agglomeration of ethyl esters in model spirit solutions and malt whiskies. *J. Sci. Food Agric.* **64**: 45–53.

COVARRUBIAS-CERVANTES, M., MOKBEL, I., CHAMPION, D., JOSE, J. and VOILLEY, A. (2004) Satured vapor pressure of aroma compounds at various temperatures. *Food Chem.* **85**(2): 221–229.

CRANK, J. (1986) *The Mathematics of Diffusion*, Clarendon Press, Oxford, England.

CRANK, J., McFARLANE, N. R., NEWBY, J. C., PATERSON, G. D. and PEDLEY, J. B. (1981) *Diffusion Processes in Environmental Systems*. The Macmillan Press, London, England.

DE ROOS, K.B. (2003) Effect of texture and microstructure on flavour retention and release. *Int. Dairy J.* **13**: 593–605.

DEBEAUFORT, F., QUEZADA-GALLO, A. and VOILLEY, A. (1998) Edible films and coatings: Tomorrow's packagings. A Review. *Crit. Rev. Food Sci. Nut.* **38**: 299–313.

DOYEN, K., CAREY, M., LINFORTH, R. S. T., MARIN, M. and TAYLOR, A. J. (2000) Volatile release from an emulsion: headspace and in-mouth studies. *J. Agric. Food Chem.* **49**: 804–810.

DRUAUX, C. and VOILLEY, A. (1997) Effect of food composition and microstructure on volatile flavor release. *Trends Food Sci. Technol.* **8**: 364–368.

FELLOWS, P. J. (2002) *Food Processing Technology. Principles and Practice*, 2nd edn, Woodhead Publishing Limited and CRC Press LLC, Cambridge, England.

GEKAS, W. (1992) *Transport Phenomena of Foods and Biological Materials*. CRC Press, Boka Raton.

GOMEZ-NIETO, M. and THODOS, G. (1978) Generalized vapour pressure equation for nonpolar substances. *Ind. Eng. Chem. Fundam.* **17**(1): 45–51.

GOSTAN, T., MOREAU, C., JUTEAU, A., GUICHARD, E. and DELSUC, M.-A. (2004) Measurement of aroma compounds self-diffusion in food models by DOSY. *Magn. Reson. Chem.* **42**: 496–499.

GUICHARD, E. (2000) Interactions between β-lactoglobulin and flavour compounds. *Food Sci. Tech.* COST 96, interactions of food matrix with small ligands influencing flavour and texture **7**: 76–81.

GUICHARD, E. (2002) Interactions between flavor compounds and food ingredients and their influence on flavor perception. *Food Rev. Int.* **18**(1): 49–70.

HARVEY, B. A., DRUAUX, C. and VOILLEY, A. (1995) Effect of proteins on the retention and transfer of aroma compounds at the lipid–water interface. In *Food Macromolecules and Colloids*. Dickinson, E. and Lorient, D. (eds) The Royal Society of Chemistry, Cambridge, 154–163.

JUTEAU, A., CAYOT, N., CHABANET, C., DOUBLIER, J. L. and GUICHARD, E. (2004) Flavour release from polysaccharide gels: different approaches for the determination of kinetic parameters. *Trends Food Sci. Technol.* **15**: 394–402.

LANDY, P., DRUAUX, C. and VOILLEY, A. (1998) Retention of aroma compounds by proteins in aqueous solution. *Food Chem.* **54**: 387–392.

LE MAGUER, M. (1992) in: Schwartzberg, H.G. and Hartel, R.W. (Eds) *Physical Chemistry of Foods*, Marcel Dekker, New York, USA, 155–215.

LE MESTE, M. and VOILLEY, A. (1988) Influence of hydration on rotational diffusivity of solutes in model systems. *J. Phys. Chem.* **92**(6): 1612–1616.

LENIGER, H. A. and BEVERLOO, W. A. (1975) *Food Process Engineering*, Reidel, D. (ed.), Kluwer, Dordrecht.

LEVI, G. and KAREL, M. (1995) The effect of phase transition on release of *n*-propanol entrapped in carbohydrate glasses. *J. Food Eng.* **24**: 1–13.

LIAN, G., MALONE, M. E., HOMAN, J. E. and NORTON, I. T. (2004) A mathematical model of volatile release in mouth from the dispersion of gelled emulsion particles. *J. Contr. Rel.* **98**: 139–155.

LONCIN, M. and MERSON, R. L. (1979) Principles and selected applications. *Food Eng.* Academic Press, New York, USA.

MARIN, M., BAEK, I. and TAYLOR, A. J. (1999) Volatile release from aqueous solutions under dynamic headspace dilution conditions. *J. Agric. Food Chem.* **47**: 4750–4755.

NAHON, D. F., HARRISON, M. and ROOZEN, J. P. (2000) Modelling flavor release from aqueous sucrose solutions, using mass transfer and partition coefficients. *J. Agric. Food Chem.* **48**: 1278–1284.

NONGONIERMA, A., SPRINGETT, M., LE QUÉRÉ, J. L., CAYOT, P. and VOILLEY, A. (2005) Flavor release at gas/matrix interfaces of stirred yoghurt models. *Int. Dairy J.*, in press.

PAÇI, E., ENGEL, E., LATRILLE, E., MARTIN, N. and MARIN, M. (2004) Composition rather than viscosity modifies the aroma compounds retention of flavored stirred yogurt. *J. Agric. Food Chem.* **52**: 3048–3056.

PERKINS, L. R. and GEANKOPOLIS, C. J. (1969) Molecular diffusion in a ternary liquid system with the diffusing component dilute hem. *Eng. Sc.*, 1035–1042.

PERRY, R. H. and GREEN, D. W. (1997) *Perry's Chemical Engineers' Handbook*, 7th edn. McGraw-Hill, New York, USA.

PITTIA, P., NICOLI, M. C. and LERICI, C. R. (1998) Ethanol liquid–vapor partition in model systems. *Food Sci. Techn.*: COST 96, interactions of food matrix with small ligands influencing flavour and texture. **2**: 59–67.

REGA, B., GUICHARD, E. and VOILLEY, A. (2002) Flavour release from pectin gels: effects of texture, molecular interactions and aroma compounds diffusion. *Sci. Alim.* **22**: 235–248.

REKKER, R. F. (1977) The hydrophobic fragment constant. In *Pharmacochemistry Library,* Nauta W. and Rekker R. F. (eds), Elsevier Scientific, Amsterdam.

REID, R. C., PRAUSNITZ, J. M. and POLING, B. E. (1987) *The Properties of Gases and Liquids.* McGraw-Hill Book Company, New York, USA.

ROGACHEVA, S., ESPINOSA-DIAZ, M. A. and VOILLEY, A. (1999) Transfer of aroma compounds in water-lipid systems: binding tendency of β-lactoglobulin. *J. Agric. Food Chem.* **47**: 259–263.

RONDEAU-MOURO, C., ZYKWINSKA, A., DURAND, S., DOUBLIER, J. L. and BULÉON, A. (2004) NMR investigations of the 4-ethyl guaicol self-diffusion in iota (ι)-carrageenan gels. *Carb. Pol.* **57**: 459–468.

SEUVRE, A. M., ESPINOZA-DIAZ, M. A. and VOILLEY A. (2000) Influence of matrix structure on retention of aroma compounds. *J. Agric. Food Chem.* **48**, 4296–4300.

SOLMS, J. (1986) Interactions of non-volatile and volatile substances in foods. In *Interactions of Food Components*, Birch, C. G. and Lindley, M.G. (eds), pp. 189–210.

SOUCHON, I., ROJAS, J. A., VOILLEY, A. and GREVILLOT, G. (1996) Trapping of aroma compounds by adsorption on hydrophobic sorbents. *Sep. Sci. Technol.* **31**: 2473–2491.

TAYLOR, A. J. (1999) Flavor matrix interactions. In *Current Topics in Flavor and Fragances.* Swift, K. A. D. (ed.), Kluwer Academic Publishers, Dordrecht, 123–138.

TO, E.C. and FLINK, J.M. (1978a) 'Collapse', a structural transition in freeze dried carbohydrates. I. Evaluation of analytical methods. *J. Food. Technol.* **13**: 551–565.

TO, E.C. and FLINK, J.M. (1978b) 'Collapse', a structural transition in freeze dried carbohydrates. II. Effect of solute composition. *J. Food. Technol.* **13**: 567–581.

TO, E.C. and FLINK, J.M. (1978c) 'Collapse', a structural transition in freeze dried carbohydrates. III. Prerequisite of recrystallization. *J. Food. Technol.* **13**: 583–594.

TREYBALL, R.E. (1980) *Mass Transfer Operations.* McGraw-Hill, New York.

VOILLEY, A. (2000) Kinetic and thermodynamic constants: The work of cost 96 SUBGROUP. *Food Sci. Tech.* COST 96, interactions of food matrix with small ligands influencing flavour and texture. **7**: 33-39.

VOILLEY, A. and BETTENFELD, M. L. (1985) Diffusivities of volatiles in concentrated solutions. *J. Food Eng.* **4**: 313–323.

VOILLEY, A. and ROQUES, M. A. (1987) Diffusivity of volatiles in water in the presence of a third substance. In *COST 90bis physical properties of foods. Final Seminar Proceedings*. R. Jowitt, F. Escher, M. Kent, B. McKenna and M. Roques (eds), London, England and New York, USA, 109–121.

VOILLEY, A., SIMATOS, D. and LONCIN, M. (1977) Retention of volatile trace components in freeze-drying model solutions. *Lebensm.- Wiss. U. Technol.* **10**: 285–289.

VOILLEY, A., ESPINOZA DIAZ, M.A., DRUAUX, C. and LANDY, P. (2000) Flavor release from emulsions and complex media. In *Flavor Release*, Deborah, D. R. and Taylor, A. J., American Chemical Society, Washington, 142–152.

WILKE, C. R. and CHANG, P. (1955) Correlation and diffusion coefficients in dilute solutions. *AIChE J.*, **1**, 264–270.

7

Lipid–flavour interactions

M. Ollivon, Paris Sud University, France

7.1 Introduction

Lipids play a key role in the storage and release of aroma molecules which has been recognised for a long time (Forss 1969, Maier 1970, Haahr *et al.* 2000, Drewnowski 1992, de Roos 1997). It is reasonable to think that flavour entrapment in lipids had been observed soon after the discovery of soap, more than ten thousand years ago by the simple mixing of the soap paste with flower petals – the starting point of a succesful industry, still active today. The parallel evolution of food flavouring, and possibly an even older process, using aromatic oil for cooking is also difficult to date. For both cases of aroma entrapment, both storage and release were achieved. Soap perfectly achieves the combination of both. In soap, aromatic molecules are stored, like mechanical energy in a spring, entrapped for a very long time and almost immediately released upon the addition of water. While for aromatic oils the release occurs more slowly during cooking and serving. In fact, the two main points to respect in the design of a successful encapsulation system, whatever the domain of activity, food, pharmacy, cosmetics, agrochemistry and so on are *stable storage* and *controlled release*.

Indeed, lipids are carriers and release modulators of aroma but they are also flavour precursors. Lipid oxydation as well as lipolysis generates numerous short chain compounds to which we are extremely sensitive because of their low levels of olfactory detection. Such reactions which are especially important in the dairy industry are perfectly controlled in cheese-making processes.

In addition, lipids play a key role in modifying our organoleptic perception of food, including mouth-feel, appearance (gloss, colour, opacity), structure (texture, consistency), heat transfer (melting profile), and non-sensory effects (satiety). They influence qualitative, quantitative, and temporal perception of flavour in products through structure modifications and specific flavour-ingredient interactions (Allaneau *et al.* 1981, Plug and Haring, 1993). Indeed, the flavour quality of foods which are mostly emulsions, is largely dependent on flavour release within the different phases including the gas phase at each step of food preparation and consumption. Partitioning and diffusion of the solutes into the lipidic (oil and interface) and aqueous phases influence the flavour release in emulsions through both thermodynamically and kinetically-controlled processes. The importance of food composition and structure on the flavour release together with the impact of the interfacial mass transfer and methods to evaluate the respective influences of these factors have been developed recently (Druaux and Voilley 1997, Pothakamury and Barbosa-Gnovas 1995, de Roos 2003).

Like most of the molecules used for flavour entrapment, polar lipids display amphiphilic property. Amphiphile molecules display a double affinity for water and oil that results in the juxtaposition in the same molecule of hydrophilic and hydrophobic moieties. In lipids, these moieties self-assemble to give supramolecular assemblies displaying large multimolecular hydrophilic and hydrophobic domains. Self-organisation of both domains simultaneously results in water–oil interface and aggregate formations. Depending on criteria that will be examined below, the shape of this interface is planar or curved and the supramolecular assemblies are either spontaneously formed or not. Lipids may also display a unique characteristic, that of making hydrophobic reservoirs allowing permanent storage of aromatic molecules. Both properties, *interface* and *reservoir* formations are not possessed by the same lipids and a lipid classification is necessary. This classification of lipids is generally made with respect to their affinity for the main solvent, *water*. Depending on their class, lipids exhibit more or less affinity for water. Although more classes of lipid–water interactions can be distinguished, let us start below with two classes, polar lipids which are not necessarily charged but frequently are like fatty acid molecules in soap, and weakly polar lipids that are frequently neutral.

7.2 How nature uses both types of lipids: the key role of molecular organisation

Formation of compartments and delimitation of functions is mainly achieved in living systems by polar lipids. These compartments are made in eucaryotic cells by the self-assembly of polar lipids, mainly phospholipids that form single or bilayered membranes. A phospholipid molecule is shown in Fig. 7.1 together with its symbolic drawing. Most of the compartments formed are closed and more or less of spherical shape. Most frequently, bilayers, obtained by the juxtaposition of two monolayers of lipids oriented in opposite directions, delimit water compartment from a water continuum. The case of single-layer membrane delimiting oil compartments also exists and will be examined below. Such bilayers delimit all compartments of living systems from organels to cytoplasm

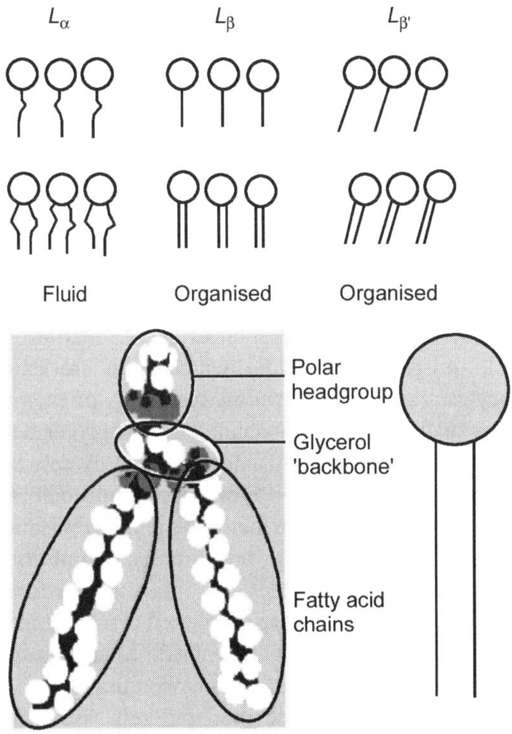

Fig. 7.1 Schematic drawings of lipid molecules in different organisations of the interfacial lamellar phase (adapted from Luzzati classification). Top and middle are single and double chain molecules, $L\alpha$ with chain(s) in the fluid state, $L\beta$ with chain(s) organised in a solid state, untilted, $L\beta'$ with chain(s) organised in a solid tilted state, respectively. Bottom shows typical phospholipid molecule with its schematic representation (the two lines represent the two chains).

membranes. Indeed, proteins and others molecules contribute to the properties of these membranes but as active materials while lipids are mainly passive components. In fact, polar lipids are the walls of the cell building while proteins are its windows and doors. The proteins offer communication routes through the bilayers while polar lipids act as wall builders ensuring their impermeability to molecules. In this way, only the molecules selected by the proteins are allowed to cross the membrane. This allows the formation of molecule or ion concentration gradients from one side to the other of each membrane.

However, unlike the walls of most of our buildings which are made of solid and immobile blocks of concrete unable to resist to the mechanical waves of an earthquake because they are too rigid, polar lipids self-assemble in *soft* liquid-crystalline organisations that also self-repair, thanks to the lateral mobility of molecules in such structures. The fluidity which is associated with the different organisations made by lipids, is the key to their stability. Thanks to lipids, living bodies are soft and deformable and are thus able to resist many stresses against which rigid crystalline systems fail. The main type of self-organisation found in nature is vesicles. Such vesicles are made by a bilayer of phospholipids closed onto itself in a single walled object. Unilamellar liposomes are perfect models for cell organisations. Vesicles are the smallest biological reservoirs known. The properties of such aggregates will be discussed below.

A second type of lipid is found in living systems, its main function is energy storage. These molecules such as glycerol esters of fatty acids are generally neutral, very weak polar lipids. The most abundant class is triglycerides. As polar lipids they are never found pure in living systems but as complex mixtures. Triglycerides also self-assemble but in a different way to polar lipids since they tend to isolate from water by forming oil droplets. Such droplets are found in large amounts in oil-rich fruits from which they are extracted by simple pressure. Adipocytes are the corresponding reservoirs of energy of mammelians. Milk, which is an oil in water O/W emulsion of triglycerides, containing also proteins and lactose in the serum, is a form of energy reservoir provided to a young mammal for its growth. In all cases, the oil droplet is surrounded by a monolayer of amphiphilic molecules that allows its stabilisation. Depending on the temperature, oils might transform into fats by partial crystallisation of the more saturated and longer triglycerides present in the mixtures.

Indeed, both types of aggregates are privileged carriers of aromatic molecules depending on their properties. We only considered above spherical aggregate shapes resulting from the curvature of one or two quasiplanar interface(s) which is the emerged part of iceberg since the lipid self-assembly property is also capable of generating infinite combinations of interface shapes and tri-dimensional structures. Many other uses and locations of lipids exist in nature, the description of which is beyond the scope of this chapter.

This chapter will be organised in such a way that we describe first the relation between the molecules properties and interface shapes. We will then describe the formation conditions and properties of the main aggregates regarding encapsulation and release of flavours starting from pure lipid spontaneous self-assembly and moving to induced and unstable ones. We will discuss advantages and inconveniences of the various aggregates for encapsulation whatever their types, including those made from mesophases.

7.3 Classification and properties of lipids

The amphiphilic character of a lipidic substance is frequently expressed using HLB, (hydrophile, lipophile balance) a value which characterises the affinity of lipid for respectively the water and oil compartments of a two phase system (Small 1986). Briefly, emulsifiers having a low HLB are suitable for stabilisation of water-in-oil (W/O) emulsions and the ones having high HLB are for oil-in-water (O/W) emulsions. Although, this value is always a guide to select an emulsifier for a given application this single rough number cannot cover all the parameters that characterise an emulsifier/surfactant. Moreover, emulsifiers and surfactants that are currently analysed using this parameter are frequently mixtures of tenths or hundredths of different compounds that do not exhibit individually the same characteristics and, for instance, do not show the same partition coefficient between oil and water phases. Therefore, more parameters are needed to characterise the properties of such products.

7.3.1 Affinity for water

In more refined classification of lipids, these molecules are divided into four different main classes depending on their affinity for, and solubility in, water and propensity to spread at air–water interface (Small 1986, Larsson 1994). Affinity for water depends on the polarity of polar groups present in the lipid molecule. Ionisation further reinforces interactions. The more polar or ionised the function the stronger the affinity for water. Solubility is the second criterion. For a given head group, the shorter the chain the higher the solubility. The water solubility of amphiphilic compounds such as detergent and emulsifiers is expressed as critical micellar concentration (CMC, see below). The affinity for water also depends on the number of chains linked to polar group(s). The affinity decreases with increasing number of chains. In fact, affinity for water results of a balance between the affinity of polar group(s) for water and that of chain(s) as HLB translate it into number.

However, affinity for water is not the only criterion for the classification of lipids and of superstructures formed by lipids, self-assembly also obeys other parameter considerations among which geometrical considerations are important ones.

7.3.2 Molecular shape, critical packing parameter

Polar lipids are not necessarily ionisable molecules bearing an electrical charge, some are electrically uncharged or neutral (zwitterionic) but all are lyotropic. This means that they modify their molecular self-organisation as a function of solvent, which is generally water. Thus, change of water content modifies the lipid head group hydration. This polar head group hydration in turn induces changes in the packing unit of self-assemblying molecules and the whole interface shape. A very simple but efficient model of the self-organisation of lipidic molecules at the interface has been proposed by Israelachvilii (Israelachvilii *et al.* 1976). Figure 7.2 shows the main self-assembly cases as a function of molecule geometry as well as the expected interface geometry for various types of molecules. The molecule shape, assimilated to a cone, a truncated cone or a cylinder, is assumed to determine the interface curvature through the determination of a packing parameter $p = v/al$ in which v is the volume of the hydrophobic moiety of the molecule, the chain(s) moiety, l is the average chain(s) length and a is the molecular area at the interface. Straightforward geometrical considerations lead to the maximum values listed in Fig. 7.2 for the different packings. Volume and surface calculations for all these types of

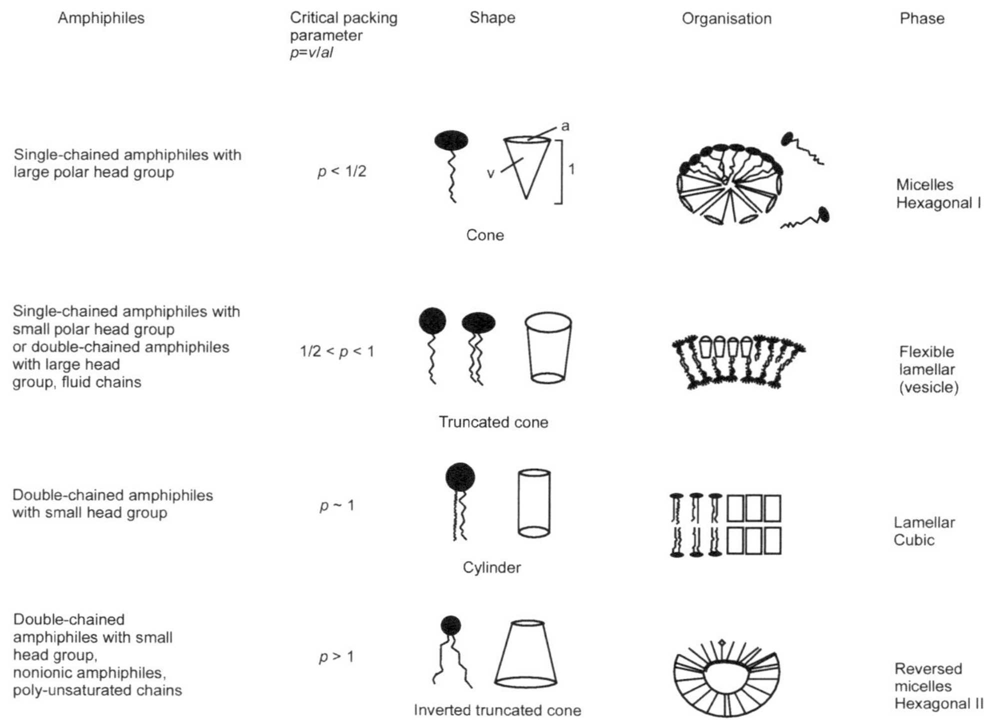

Fig. 7.2 Critical packing parameter p of the main lipidic organisations and main types of molecules and molecules shapes associated with these organisations (adapted from Israelachvilii *et al.* 1976).

supramolecular assemblies imply that sphere, cylinder, plane and reverse micelles necessarily display $p \leq 1/3$, $p \leq 1/2$, $p \leq 1$, $p > 1$, respectively. Tanford and Small (Tanford 1980, Small 1986) reported useful values for the calculation of v and l in the liquid state as a function of the number of methylene groups of the chains n_{CH_2}:

$$v \cong (27.4 + 26.9 \cdot n_{CH_2})\ \text{Å}^3$$

$$l = (1.5 + 1.265 \cdot n_{CH_2})\ \text{Å}^3$$

The surface occupied by each polar group is a function of interactions between them. Such interactions depend on the type of polar group, its net charge and the charge distribution which are functions of the pH and ionic strength of the medium. At low ion concentration, the larger the repulsive force between ionised groups the larger the interface curvature and conversely. As a consequence, an ionic strength increase decreases the interface curvature by decreasing either their repulsion or attraction.

However, the volume occupied by chains depends on the chain state, length and unsaturation. The volume occupied by chain methylene in the solid state is reduced by 25–30%. As a consequence, important volume contractions are observed upon lipid crystallisation.

Origins of the interface curvatures

The lipids with a single short chain with a large and ionised head group ($p \leq 1/3$) (e.g. short chain anionic detergent such as sodium dodecyl sulfate – SDS) are preferentially forming direct micelles (the term 'direct' means that the continuous phase in which the aggregate is formed is water contrary to the term 'reverse' for which oil is the continuum) in water. Such micelles display an aggregation number (average number of molecules in the aggregate) in the range of 20 to 500 for medium chain length ($8 \leq n \leq 16$). The solubility in water as monomers of such molecules is pretty high so a dynamic equilibrium exists between micelle and monomer. The monomer–micelle exchange rate depends on the chain length. For a medium chain length ($n = 12$) the exchange rate is in the order of microseconds, while the whole replacement of micelle molecules is in the millisecond range. The system reaches equilibrium by itself and no external energy is needed to obtain the micellar solution. A domain exists in the lipid–water phase diagram where the micellar solution is the stable state. The system is thermodynamically stable since it is dynamically stabilised.

On the other hand, the lipids with two or more long chains with a small and neutral or poorly ionised head group ($p \geq 1$) (e.g. dioleylphosphatidylethanolamine) are preferentially forming reverse aggregates like micelles (e.g. H_{II} structure) in an oil continuum. Except for some hexagonal phases, their structures are not well documented.

In between, the molecules with a cylindrical shape ($p \sim 1$) such as phosphatidylcholine (the main constituent of lecithin) are preferentially forming either mono or bilayers stabilising 'planar' aggregates such as emulsions or liposomes respectively or lamellar phases. Such aggregates are not thermodynamically stable unlike micelles. They display an aggregation number in the range of 500 to 10^7. They are formed only with long chain compounds ($n \geq 12$). The solubility in water as monomers of such molecules is pretty low (in the range of 10^{-10} mol/l) so no dynamic equilibrium exists between aggregate and monomer because of the quasi absence of monomer. The system does not reach equilibrium and aggregates are only formed by dissipation (input) of energy (of mechanical, chemical or biological origin). When aggregates are formed they are unstable, they aggregate/floculate, fuse/precipitate/cream and form layers of lamellar crystalline phase next to the water (or oil in the case of emulsions) in excess. The stable state is the lamellar phase separating by gravity from the excess water. No micelles are formed in such systems, the aggregate size of which can be theoretically infinite in the stable state. However, the aggregates formed can be stable for years if properly stabilised. For instance, charged (bi or mono) layers are more stable in low ionic strength solutions. While aggregates are unstable objects, molecule rotation and diffusion exist within layers and between layers. Molecular rotation and position exchange within the layer takes nano to microseconds and milliseconds, respectively depending on the state of the layer (gel or liquid-crystalline) while flip-flop, molecule exchange between layers, may take minutes or weeks. As a consequence of the metastability of such aggregates, no domain can be found in the lipid–water phase diagram where they exist in a stable state. A system formed of such aggregates is thermodynamically unstable while the corresponding system with separated phases is stable.

By only using very simple geometrical packing considerations, the above model is unable to take into account specific deformations of the interface involving its gaussian curvature and the formation of triply periodic minimal surfaces such as those found in cubic phases formed, for instance, by monoglycerides. Discussion of such interface deformation is beyond the scope of this chapter (Garstecki and Holyst 2002a,b, Schwarz and Gompper 2000). Also, mixture of interfacial lipid with other amphiphilic molecules, including other lipids, may lead to a variety of curvature(s) combinations or local phase separations that further complexify the interface topology (Seddon and Cevè 1993). Angelov addressed the change of packing parameter as a function of lipid composition (Angelov *et al.* 1999).

7.3.3 Aggregate and mesophase formation

The supramolecular structures made by lyotropic lipids are called mesophases since they are neither perfectly organised as crystals nor desorganised as liquids. They frequently form liquid-crystalline structures in which chains are either in the liquid state or poorly organised. Micellar and hexagonal phases, bicontinuous microemulsions and cubic phases which are made by lipid molecules bearing chains of short or medium length (or frequently, a single chain), are thermodynamically stable phases formed spontaneously and rapidly by simple mixing of constituents. The lipids with long and double chains form unstable aggregates which frequently reach equilibrium slowly after the energy dissipation used to get the mixture homogeneous.

Number of phases coexisting in the system and partition coefficient

While lipid-containing systems are not necessarily at equilibrium (Bruin 1999), the number of phases that coexist in a system at a certain time is limited and in principle might always be determined. Such determination and identification of the number of phases coexisting, at equilibrium or not, is of prime importance regarding the distribution of flavour molecules within the various compartments of the system considered. This is also of importance regarding the distribution of molecules not directly interacting with aromas but rather influencing this distribution in an indirect way through the repartition of aromas between phases and then their release.

The stability of the structures formed in lipidic systems is such that the dynamic of the molecular diffusion within an aggregate is fast compared to that of equilibration of the lipidic system at the level of supramolecular organisations and structures. Apparent molecular equilibrium is reached in some metastable systems like emulsions and liposomes (see below). Thus, in these metastable systems when both equilibration times differ by several orders of magnitude, it is still possible to determine significant/apparent partition coefficients. The main difficulty resides in the determination of the exact number of phases and their properties.

The interface phase: an 'interphase'

A majority of food products are complex multiphase systems resulting from emulsification in high shear conditions of oils by solutions containing water, proteins and sugars. To increase emulsion stability and resistance to the severe conditions sometimes encountered during long shelf life as temperature-induced changes, droplet size has to be reduced to a minimum, then emulsifier concentration has to be increased. At constant oil concentration in an emulsion, the reduction of its mean droplet size increases the interface specific surface and the need for interface coverage by amphiphilic products such as stabilisers or emulsifiers (note: the limit of size reduction corresponds to the obtention of a microemulsion). Such processes have been applied recently to many dairy emulsions to increase shelf life, improve reactivity by increasing interface or flavour release. In this respect, it is useful to remember that amphiphilic molecules such as surfactants and emulsifiers do not display the same properties regarding interface stabilisation. Single short chain surfactants like sodium dodecyl sulfate are not emulsifiers but detergent molecules since their solubility in water (cmc) is pretty high compared with that of emulsifiers. The solutions they form in water cannot be considered as biphasic. Then, depending on their nature and their concentrations, amphiphiles added to emulsions for their stabilisation should be considered as forming a new phase or not directly in relation to their water or oil solubility or not. For instance, emulsifiers are lyotropic molecules that differ from surfactants by not forming micelles but rather acting in statically stabilising the oil–water interface. They frequently display very low CMCs depending on chain length and number they might be soluble either in water or in oil compartments when not in both for some of them.

Lecithins are a perfect example of long chain amphiphilic lipids quasi-insoluble in water, that do form a separate phase in most food products. Although they might be extracted from oilseeds, cereal germs, egg yolk, or fish, industrial phospholipids are almost entirely obtained from soybeans with phosphatidylcholine (PC, lecithin), phosphatidylethanolamine (PE, cephalin), and phosphatidylinositol (PI) as major constituents of the phospholipid fraction. Phospholipids are used in the food industry as emulsifiers and stabilisers, for example, in baked products, chocolate, instant products, margarine, mayonnaise, shortenings, and as fat replacers in low fat products. In excess water, lecithins swell forming lamellar phases that can be mono or bilayered depending on the surrounding phases (oil and water or water and water, respectively). In both environments, the number of associated water molecules is pretty high, from 10 to 30 per phospholipid molecule depending mainly on lecithin composition and temperature. In excess water (oil in water emulsions for instance), the 'interface phase' should be considered as constituted by mono, bi or multilayers of fully hydrated phospholipid. On the other hand, lecithin in chocolate should be considered as a phase containing few water molecules eventually shared with molecules of other components that they are supposed to coat: sucrose, starches, cocoa butter crystals and liquid phases. This aspect has to be taken into account when using lipids and fats as edible films (Debeaufort *et al.* 2000).

The interface phase plays a determining role in the flavour molecule distribution between compartments. Landy *et al.* (1998) have shown that the resistance to the transfer increases for ethyl butanoate and ethyl hexanoate in the presence of sodium caseinate. The presence of the protein in modifying the interface between oil and water plays a role in transport of flavour, kinetic release and then perception. This illustrates the role played by the interface phase that should be considered as a distinct phase. The protein layer as well as that made by emulsifiers eventually mixed with proteins does not play the same role as the surrounding solution (Landy *et al.* 1998). The role played by the different phases, including interface, has been discussed by Druaux and Voilley. They examined the transfer between different phases and pointed out the importance of the food composition and structure on the flavour release together with the impact of the interfacial mass transfer. Lipids affect the partitioning of flavour compounds between product and vapour phase. The partition coefficient between the product and vapour phases describes the potential extent of the release of flavour compounds (Druaux and Voilley 1997).

In fact, once formed the interface phase is quite stable. The formation of emulsion droplets usually by shearing processes leads to a mixture of emulsion droplets and aggregates corresponding to interface phase in excess. This aggregate excess can be isolated. It corresponds to single- or multi-walled vesicles, vesicles making single or multiple coatings of the emulsion droplets or even in specific cases mixed aggregates formed by the assembly of an emulsion droplet with a vesicle. On emulsion storage, as the mean size of emulsion droplets is expected to increase, the number of emulsion droplets is decreasing and the proportion of non associated phase is increasing.

7.4 Main types of lipidic structures

Table 7.1 lists main aggregate and mesophase types as well as some of their characteristics and properties, typical or estimated in order to get rough typical values of sizes, lifetimes and possible applications in mixtures with other ingredients. Figure 7.3 shows schematic drawings for some of these organisations. The full description of these aggregates and of their properties is beyond the scope of this chapter. Only some of the main structures and aggregates will be described below.

7.4.1 Micelles and related aggregates

Some emulsifiers and polar short chain compounds (including aromatic molecules) may form micelles providing their lipidic chains are in the liquid state and their molecular solubility in water is sufficiently high. Surfactant

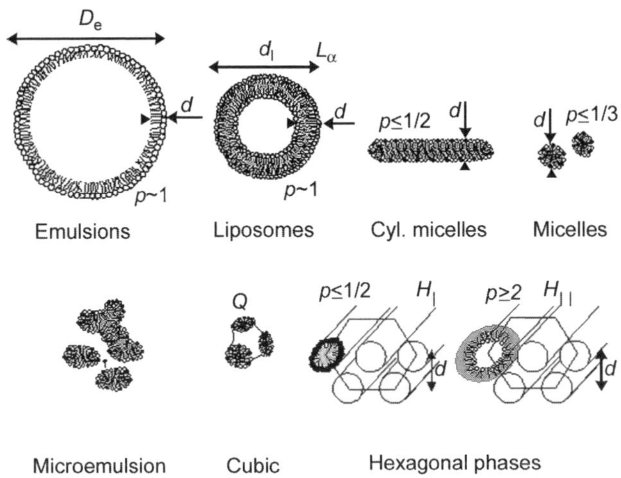

Fig. 7.3 Main aggregate (top) and mesophase (bottom) lipidic organisations (shown as cuts) drawn irrespective of their dimensions in order to show the molecular interfacial organisations. *L, H, Q* mean Lamellar, Hexagonal and Cubic organisations respectively (conventions as shown Fig. 7.1), I and II meaning direct and reverse. Associated packing parameters *p* are shown, *D* and *d* represent aggregate and interface dimensions (see Table 7.1). Cuts of microemulsion, cubic and hexagonal phases show elementary organisation. Cubic and hexagonal phases show periodic organisations with defined local curvatures while microemulsion are locally fluctuating.

Table 7.1 Characteristics of the main lipid aggregates and mesophases

Type	Subtype	Size or diameter De (nm)[a]	Lipid layer(s) thickness or distances d (nm)	Formation[c]	Stability[e] (aggregate lifetime)	Stability upon dilution (in eq. medium)[d,e]
Emulsion	direct	$100–10^5$	3	Energy input	hours up to years	5–10
Emulsion	inverse	$100–10^5$	3	Energy input	hours up to years	5–10
Liposomes	direct	$25–10^4$	6	Energy input	hours up to years	5–10
Lamellar phase	direct	25[f]–∞	6	Spontaneous	hours up to years	5–10
Micelles	direct	4–15	5–10	Spontaneous	msec	~1[e]
Micelles	inverse	2–10	3–10	Spontaneous	> sec	~1[e]
Cylindrical micelles	direct	20–∞	5–10	Spontaneous	–	~1[e]
Cylindrical micelles	inverse	20–∞	5–10	Spontaneous	–	~1[e]
Microemulsion	bicontinuous	∞[b]	5–20[b]	Spontaneous	msec	~0.5[e]
Microemulsion	direct	25–100	6–25	Energy input	> sec to hours	~1–2[e]
Microemulsion	inverse	25–100	6–25	Energy input	> sec to hours	~1–2[e]
Multiple emulsion	direct or inverse	$100–>10^5$	5–10	Energy input	hours up to years	3–4
Cubic phases	bicontinuous	5–15[b]	5–10	Spontaneous	hours up to years	–
Hexagonal phases	direct	∞[b]	5–20	Spontaneous	–	~0.5[e]
Hexagonal phases	inverse	∞[b]	5–20	Spontaneous	–	~0.5[e]

[a] Size (diameter of objects or length), external diameter.
[b] Characteristic dimension (lipid layer or pore thicknesses).
[c] Formation requires energy (metastable organisation) or is spontaneous (thermodynamical equilibrium).
[d] Magnitude of possible dilution in equilibrated medium without major change of aggregate, e.g. 5 means the aggregate can be diluted 10^5 time by the continous medium (oil or water solutions).
[e] Estimated.
[f] Discs a finite size to infinite lamellar phases.

solubility mainly depends on the affinity of its polar group for water. However, as discussed above, this solubility also depends on chain number and length. The solubility in water of long chain compounds decreases with increasing chain length. As a rule of thumb, increasing the chain length of a single chain surfactant by two carbons decreases its water solubility by an order of magnitude. At molecule concentrations larger than the limit of molecular solubility as monomer, micelles that are aggregates of molecules form. This concentration limit is called critical micellar concentration (CMC). The CMC value depends on temperature. At concentrations > CMC, dynamic equilibrium is established between micelle and monomer solution thanks to very fast exchange of molecules between the micelle and surrounding medium. The aggregation number which is the average number of molecules per micelle is low (frequently < 200) and depends on lipid chain length as well as molecule head group ionisation. As micellisation can be considered as a certain form of molecular solubility, micellar solutions are considered as true solutions containing a single phase for the Gibbs phase rule.

There are limitations to this hypothesis. For instance, increasing the phospholipid content of a micellar solution will result in the formation of larger elongated (cylindrical) micelles that might interconnect forming a new phase or when water is no longer in excess, the formation of a hexagonal phase. Such phases, inverted or direct, are present in pure emulsifier (Tween 80) and persist over dilution depending on the water concentration. They have also been observed in mixtures of bile salts with phospholipids. For both phases the structural unit is a cylinder as shown in Fig. 7.3.

More complex structures are those built by unsaturated monoglycerides. In excess water, they develop cubic phases which are bicontinuous structures meaning that they are built from two interwoven continuous domains. The period size is in the 10 nm range. Such cubic phases can be dispersed in the form of cubosomes, the size of which ranges from a few tenths of nm to micronic sizes. These nanoscale 'container' units, also termed 'nanostructured hierarchical fluids' can give rise to the design of new liquid devices of three-dimensional (3D) architecture involving networks of nanochannels and nanocompartments. These particles which may encapsulate proteins or many types of biological substances such as aromas or aroma precursors, are interesting open structures. Many applications are in a process of development in order to utilize the important structural and chemical advantages of the lipid cubic structures such as periodic organisation of aqueous nanochannels, huge interface area per unit volume (400 m^2/g), and biocompatible environment (Angelova *et al.* 2005). As discussed above, cubic phases as well as cubosomes might be used as a reservoir for flavours. While monoglycerides are known to make complexes with starch, one might imagine that aroma release could be produced upon heating when complexes form at T > gelatinisation temperature of starch, i.e. between 65 and 100 °C, depending on water content (Le Bail *et al.* 1999).

7.4.2 Liposomes

Liposome entrapment which is considered as one of the most promising techniques for encapsulation has recently been the subject of numerous studies in the domain of food science although not in the aroma field (Gouin 2004, Lin *et al.* 2004). However, liposomes are not a single class of vesicles with similar shapes. Many types of liposomes exist and schematic drawings of some of them are shown in Fig. 7.4.

First of all, liposomes are neither necessarily spherical nor unilamellar as generally considered. Some liposomes are elongated, twisted or onion shaped. Moreover, liposomes display a large size range from diameters of 25 nm to several tenths of micrometres (see Table 7.1). An average size value easily obtained by homogenisation, extrusion or soft sonication techniques is 100–500 nm (Lasic 1993). Figure 7.4 shows different types of liposomes (drawn spherical) as a function of their increasing stability, from left to right. Single walled liposomes are less stable than multilamellar vesicles (MLVs). Large unilamellar vesicles (LUVs) are more stable than small unilamellar vesicles (SUVs).

The main advantage of using liposome for encapsulation is the protection that it offers against chemical (e.g. pH) and biological (e.g. enzymes) degradations. Encapsulated substances such as vitamins and aromas can be protected for months at neutral pH and at room temperature or below. However, when departing from neutral pH, hydrolysis of phospholipids is increased with increased release of compounds entrapped in vesicles. Then, sphingolipids can also be used to prevent hydrolysis especially at pH values far from neutrality. For flavouring use of liposomes, the stability of the system should be overcome

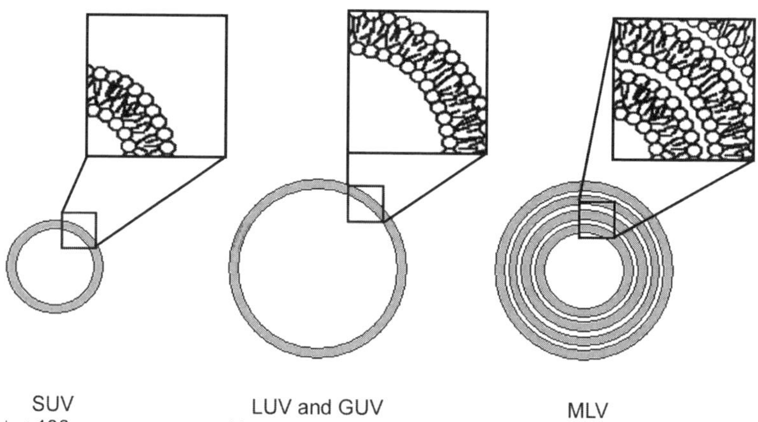

Fig. 7.4 Schematic drawings of small, large or giant and multilamellar, vesicles (SUV, LUV or GUV and MLV, respectively). Multi-walled liposomes of MLV show the alternance of lipidic bilayers and water layers, a hydrated bilayer is about 6 nm. Frames represent enlargements of the bilayers. While SUV exhibit strong local curvature, increasing diameter from LUV to GUV results in more and more neglectable curvatures (in GUVs the local curvatures are only due to thermal fluctuations).

to get the flavour release. Taking the advantage of such an inconvenience, one can use the slow process of hydrolysis to get the aromatic release from vesicles for instance by combining processes of acidification by lactic fermentation and that of phospholipid hydrolysis. Because of the lysolecithin formation through this process and thank to the vesicle–micelle transition one can expect delayed flavour release that will operate upon package opening, for instance in yoghourt, providing the liposome-encapsulated flavours are mixed with the other ingredients before closure of packaging (Ollivon *et al.* 2000).

While it is possible to use pure phospholipids or phospholipid classes (e.g. a mixture of phosphatidylcholines (PC) with different chain lengths) to form liposomes in the study of food science model, the use of such compounds in food production is prohibited by cost considerations. Food grade lecithins which are a byproduct in the refining of vegetable oils, are mixtures of phospholipids and glycolipids, containing some triglycerides, carbohydrates, traces of sterols, free fatty acids and carotenoids. During the usual batch degumming process, the crude oil is heated to about 70 °C, mixed with a few per cent water and subjected to thorough stirring for about half-an-hour to an hour. Addition of water to the oil results in the swelling and insolubilisation of polar lipids. Centrifugation of the resulting lecithin sludge allows separation from the oil. Dried mixtures of high viscosity crude lecithin are commercialy available as standard fluid lecithins. Upon rehydration with large amounts of water and submission to high shear, such vegetable lecithin mixtures form liposomes and are the most used products. Extraction with acetone yields de-oiled lecithin qualities with a residual content of only 2–3% neutral lipids while ethanol or ethanol–water mixtures allow separation of phospholipid classes (e.g. separation of PC from phosphatidylethanolamine (PE), phosphatidylinositol (PI) and phosphatidic acid (PA)). De-oiled products are predominantly for 'instant' food applications and/or for food applications requiring defined phospholipid patterns. Providing phospholipids form a single phase which is frequently the case especially when chains are in the liquid state, liposome suspensions are considered to be formed by two phases, that of hydrated lipids and water. When substances are encapsulated in one of the compartments, external or internal, systems are made of three phases. Cholesterol addition (e.g. 10–50% *vs* PC) improves the impermeability of the bilayer to water soluble substances.

As cost reduction is a determining factor in liposome use, the replacement of rather expensive phospholipids for a mixture of high and low HLB emulsifiers is interesting (Roux *et al.* 1996). Liposome-like structures have been obtained at about a hundredth of the cost of usual phospholipids using a mixture of glyceride-based hydrophobic emulsifiers, mono- and di-glycerides or lactate, acetate or citrate esters of monoglycerides (HLB between 3 and 7) and hydrophilic emulsifiers, sucroesters and/or stearoyl lactilates (HLB between 8 and 15).

Another major advantage of liposomes and emulsions over thermodynamically stable preparations such as micelles and bicontinuous microemulsions is the fact these aggregates can be diluted with water by several orders of magnitude (e.g. up to 10^6 times for egg PC vesicles) not changing their structures and properties (see Table 7.1). This property is especially useful when dilution of the preparation has to be obtained without release of encapsulated material. Indeed, for liposomes this only applies to water soluble compounds entrapped in the vesicles, since compounds located in the external water compartment are diluted readily as the suspension. Great attention should be drawn to molecules that are amphiphilic like most of the flavouring compounds, since these molecules will partition between lipidic and water compartments. Then as a consequence of such a high partition coefficient, the increase of the external volume resulting from dilution might lead to a flavour release. For amphiphilic compounds there is no way to prevent release using liposomes upon dilution but the number of barriers to delay the process can be increased. This is achieved using multilamellar vesicles which are obtained by simple mixing of hydrated phospholipids at high shear. In this respect, it is worth noting that the production of liposomes is no longer obtained using sonication, solvents or detergents. Recent progress in technology improved the techniques used for production which is not limited to pilot scale. Liposomes can be produced on a large scale by continuous techniques such as homogenisation/microfluidisation and at a very moderated cost. So their use is no longer limited to products with high added value such as those employed in pharmacy or in some cosmetic preparations.

Another unique property of liposomes is the possibility of targeted delivery of their content in specific parts of the foodstuff (Gouin 2004). For instance, it has been shown that liposome encapsulated enzymes concentrate preferably in the curd during cheese formation while non-encapsulated enzymes are usually distributed evenly in the whole milk mixture, which leads to very low (2–4%) retention of the flavour-producing enzymes in the curd (Kheadr *et al.* 2000). Development of more sophisticated targeting involving high affinity recognition sites are currently under way for pharmaceutical applications that could be transposed in the future to food delivery.

7.4.3 Emulsions

As emulsion will be the focus of another chapter of this book, we decided to reduce this point to a minimum in this chapter. Only specific points like the influence of lipid crystallisation on flavour retention and release, influence of structure on lipolysis and oxydation as well as the role played by interface are considered elsewhere in this chapter.

Influence of temperature on lipid structures

Below a certain temperature lipidic chains are not anymore in the liquid state. This temperature depends on the chain type (saturated or not) and length, number of aliphatic or ramified chains, of the polar head group, etc. and time because of the slow processes of molecular organisation and diffusion in relation with the viscosity of the medium. Then temperature and time have a determining influence on the structures formed especially when the lipidic system allows the formation of metastable structures (very frequently). Then, several types of lateral packings of chain leading to more or less compact structures from liquid crystalline state to solid crystalline state can be formed. The formation of such or such lateral packing again depends on the same parameter as those listed above for temperature. Depending on the crystallisation conditions, most of the lipid molecules display several polymorphic forms, some, like the major triglycerides of cocoa butter (the fat of chocolate), exhibit six identified varieties and even more. Both types of lipids, polar and lyotropic as well as weakly polar mainly displaying oily character, and in both conditions, anhydrous or hydrated, do undergo polymorphic transitions on cooling or on heating, making the understanding of the lipidic structure very complex at low temperature.

On another hand, many fat-containing foods require refrigeration or freezing for preservation. Then, understanding of the polymorphic evolutions of lipids occurring upon cooling is of prime importance to understand mechanisms of flavour diffusion and release as well as texture changes, the latter influencing the former. For instance, dairy products are very sensitive to biological degradation and then are frequently kept at low temperature. The influence of cooling rate and storage conditions has been studied recently using time-resolved synchrotron radiation and DSC (Lopez *et al.* 2000, 2001, 2002) on cream that is emulsified milk fat. Four types of crystals were found in equilibrium with a liquid within micronic globules of cream while up to six different types of molecular packings have been identified in milk fat. This is not surprising regarding the broad diversity of fatty acids and triglycerides (TG) present in milk fat (>500 and >10^6, respectively). By affecting drastically the oil content (OC is complementary of the solid fat content, SFC), OC (%) = 100 − SFC (%), which is the phase expected to retain flavours more efficiently, especially lipidic ones, such processes might modify the flavour partition coefficient between liquid and solid as well as between oil and water compartments and the release processes. This influence is directly in line with the increase of aroma retention observed in cheese upon increasing of triolein content (Piraprez *et al.* 1998).

The situation is even more crucial for ice-cream where most of the phases, including a large moiety of the lipids, are frozen and in which air represents about 50% of the volume (see below). The design of low fat ice-cream in which fat replacers are supposed to mimic the texture given by the presence of cream exacerbates the problem of flavouring, as shown below.

Reactivity in lipid structured food: influence of structure on oxidation and lipolysis

Both oxydation and lipolysis are processes occuring in most of the food products containing fat. Both are degradative processes which depends on molecular diffusion that are modified by the compartmentalisation of foodsuff. Both processes are complex and not fully understood even in the simplest cases.

While lipid oxydation is perceived as oil degradation and is not accepted by consumers, the lipolysis which is also a form of degradation is perceived in some cases as a benefit for tasting depending of the product and local customs and cultures. For instance, the taste of cheese of some North American chocolates, which is an added value there, is considered off-flavour in the European market.

Lipid oxydation depends on many factors including both oxygen and double bond presences, temperature, traces of catalysts, etc. as well as texture of food and fat distribution within. While the description of mechanisms of lipid oxydation in bulk has been the subject of many reports, lipid oxidation is still a major cause of quality deterioration in food emulsions (Coupland and MacClements 1996). The physicochemical mechanisms of lipid oxidation for emulsified lipids differs from that of bulk lipids and are complexified by the presence of the droplet membrane, the interactions between the ingredients, and the partitioning of ingredients between the oil, aqueous and interfacial regions. Understanding such mechanisms, developing methods of preventing, or at least retarding, lipid oxidation in emulsified products is a major concern for food manufacturers because it leads to the development of undesirable 'off-flavours' (rancidity) and potentially toxic reaction products.

Lipolysis is either a chemical process catalysed by pH deviations from neutrality or a biochemical mechanism. Both result in the hydrolysis of lipids which in turn generate flavours. While lipolysis is perceived positively in dairy products, it should be a controlled process. It has been found that cheddar cheese proteolysis and lipolysis could be accelerated using liposome-encapsulated enzymatic cocktails (Kheadr *et al.* 2000). Neutral bacterial protease, acid fungal protease and a lipase were individually entrapped in liposomes and added to cheese milk prior to renneting while the resulting cheeses were chemically, rheologically and organoleptically evaluated during three months of ripening at 8 °C. Strong typical cheddar flavour was found by the second month while the cheese did not exhibit any off-flavour when ripening was extended for a further month.

7.4.4 Nano and microencapsulation: from micro to nanoparticles and nanocapsules

A variety of techniques of micro and nanoencapsulation using lipids as an encapsulating agent has been proposed recently. Temperature jumps or quenchings as well as use of supercritical fluids (using mainly carbon dioxide) and pressure have been used to obtain encapsulation of sensitive materials such as flavours in hydrophobic solid particles or core-shell particles (Müller *et al.* 2002, Benoit *et al.* 1996).

Many types of lipid nanoparticle-based devices can be imagined to get release triggered by internal or external stimulation. For instance, mixing, heating or cooling processes can be used to trigger release whatever timing is anticipated, as in the case described above. Such release can be also applied to emulsions. In this respect, the release can be site-specific, stage-specific or signalled by changes in pH, temperature, irradiation or osmotic shock. In the food industry, the most common method is by solvent-activated release. The addition of water to dry soups, dry beverages or cake mixes is an example (Gibbs *et al.* 1999). Liposomes have been applied in cheese making, and its use in the preparation of food emulsions such as spreads, margarine and mayonnaise is a developing area.

In this respect, the great advantage of liposomes and emulsions over other microencapsulation technologies is the stability that liposomes impart in food applications with high water activity or high water content. Others techniques of encapsulation, producing dried and solid powders, e.g. by spray drying or fluidised beds, also impart great stability to food ingredients but shows the inconvenience of the dry products' advantage not to resist in high water activity with immediate release in water.

7.5 Replacing fat by structuring lipids: consequence for flavour retention and release

The sharply reduced cost of vegetable oils has been the main reason for increased fat consumption worldwide. Although, fat-rich food consumption is still associated with highest incomes in a given country. On the other hand, reduction of the food fat level is continually being advised by health authorities in an attempt to reduce the diseases and mortality linked to high fat consumption (Drewnowski 2003). In developed countries, the tendency is towards both the reduction of the fat and cholesterol-containing food, in spite of the fact its metabolism is not understood. Many companies are making efforts to develop fat-free and cholesterol-free products.

As expressed above, the implication of fat on flavour release from food matrices has been known for a long time. Briefly, lipid phases act as a reservoir for the aromas resulting in delayed release and perception. Then, generally, flavour release decreases with increasing lipid level in the food matrix, with the exception of hydrophilic compounds possessing log P values near or below zero (Guichard 2002). However, the generally used octanol/water partition coefficient should not be considered as a reliable parameter for describing the polarity of all lipids. In fact, individual experimental rather than theoretical partition coefficients for different lipids seem to be necessary for the precise description of flavour partitioning in emulsions (Rabe *et al.* 2004). As a result of the reduction of fat in food, fat-reduced or fat-free food matrices are characterised by a significantly changed aroma profile and a generally disliked transient flavour burst.

In dispersed systems, the release is much more complex and depends on many more parameters than in bulk phases. For an oil in water emulsion, at fixed oil and water content, the smaller the fat droplet the larger the surface of exchange. Then, in principle, the faster the external compartment is saturated with aroma. The situation is more complex since the interface plays a role in the aroma transfer. Moreover, at room temperature the fat is not necessarily melted. Furthermore, both fat and interface lipids might be solid phases influencing aroma transfer to water and air compartments (Kalnin *et al.* 2004). Precise localisation of solid fat crystals and emulsifiers in dispersed systems is not known in spite of the progress made by confocal microscopy and X-ray diffraction. Are these crystals forming a shell or not at interface? Answering this question would help in understanding flavour release in complex matrixes such as cheese, yoghourt, frozen dessert and ice-cream (Kheadr *et al.* 2000, Brauss *et al.* 1999a,b, de Roos, 1997, 2003).

Recent studies by confocal microscopy helped in understanding the distribution of fat within complex food matrices. In cooked and pressed cheese, like Emmental, the fat is found in very large domains of undefined shape, fat globules are almost intact in soft cheese, like Camembert. The globular size is efficiently reduced in whipped cheese and soft cheese processed with ultrafiltration technology (UT) (Lopez 2005, Michalski *et al.* 2004). However, the distribution of fat crystals within these structures is still unknown as well as its consequences on flavour development. The effect of globule size reduction using UT is a promising field regarding the possibility of modulation of fat globule reactivity for technological or biological processes.

Fat removal from vanilla ice cream was found to result in a drastic flavour profile distortion and loss in vanillin intensity during storage. Encapsulation was proposed to improve the stability and to restore the flavour balance in locally reproducing the full-fat product in low-fat ice cream. The flavouring was encapsulated in a low-melting fat (mp < 35 °C) to create a microenvironment for the flavour compounds that mimic the full-fat base. During consumption, the fat is melting and releases the vanilla flavour. According to some authorities, the encapsulation not only restored most of the original flavour of the full-fat product, but it also provided a lingering sensation during consumption and a higher stability during storage (de Roos 2003).

7.5.1 Methods for inducing/monitoring flavour release using lipids

Release of flavour compounds is an essential parameter to control in the design of food. The physicochemical properties of the flavour compounds as well as that of the food matrix determine the release before ingestion as well as during mastication. As explained in the first part of this chapter lipids by modifying the number of compartments and the matrix structure of many food products greatly influence their physical properties and are determining factors of release (Haahr *et al.* 2000, Druaux and Voilley 1997). Lipids also indirectly influence aroma development through oxydation or lipolyse reactions. Progress of both reactions depends on diffusion rates, interface area and dispersion of fats. In these reactions lipids act as flavour precursors, flavour storage compartments and, in addition, as flavour release modulators.

On the other hand, as a paradox, lipid molecules by their numerous possibilities of self-assembly provide unique and perfectly safe ways for encapsulation of flavours, flavour precursors, and flavour developing molecules in new food matrixes in which the amount of fat is reduced. Some suggestions for new encapsulation systems have been given in the text above. However, the infinite combinations of lipids to built up structures have far from all been explored and constitute a reservoir for future encapsulation and release devices, exactly as fats act as a reservoir for flavours.

7.6 References

ALLANEAU, B., NICOLAS, J., MONTENAY-GARESTIER, T. and DRAPRON, R. (1981) Effect of lipids on the retention of volatile compounds. Experimental determination of activity coefficients. *Sciences des Aliments* 1, 447–465.

ANGELOVA, A., ANGELOV, B., PAPAHADJOPOULOS-STERNBERG, B., OLLIVON M. and BOURGAUX C. (2005) Proteocubosomes: Nanoporous vehicles with tertiary organized fluid interfaces. *Langmuir* 21, 4138–4143.

ANGELOV, B., OLLIVON, M. and ANGELOVA, A. (1999) X-ray diffraction study of the effect of the detergent octylglucoside on the structures of lamellar and non-lamellar lipid/water phases of use for membrane protein reconstitution. *Langmuir* 15, 8225–8234.

BENOIT, J. P., ROLLAND, H., THIES, C. and VANDEVELDE, V. (1996) Method of coating particles, PCT WO 96/11055.

BRAUSS, M. S., LINFORTH, R. S. T., CAYEUX, I., HARVEY, B. and TAYLOR, A. J. (1999a) Altering the fat content affects flavor release in a model yogurt system. *J. Agric. Food Chem.* 47, 2055–2059.

BRAUSS, M. S., BALDERS, B., LINFORTH, R. S. T., AVISON, S. and TAYLOR, A. J. (1999b) Fat content, baking time, hydration and temperature affect flavor release from biscuits in model-mouth and real systems. *Flavour and Fragrance J.* 14(6), 351–357.

BRUIN, S. (1999) Phase equilibria for food product and process design. *Fluid Phase Equilibria* 158–160, 657–671.

COUPLAND, J. N. and MacCLEMENTS, D. J. (1996) Lipid oxydation in food emulsions. *Trends in Food Sci. and Technol.* 71, 83–91.

DEBEAUFORT, F., QUEZADA-GALLO, J.-A., DELPORTE, B. and VOILLEY, A. (2000) Lipid hydrophobicity and physical state effects on the properties of bilayer edible films. *J. of Membrane Science* 180, 47–55

DREWNOWSKI, A. (1992) Sensory properties of fats and fat replacements. *Nutr. Rev.* 50 (4), 17–20.

DREWNOWSKI, A. (2000) Nutrition transition and global dietary trends. *Nutrition* 16 (7–8), 486–487.

DRUAUX, C. and VOILLEY, A. (1997) Effect of food composition and microstructure on volatile flavour release. *Trends in Food Sc. and Technol.* 8, 364–368.

FORSS, D. A. (1969) Role of lipids in flavors. *J. Agric. Food Chem.* 17, 681–685.

GARSTECKI, P. and HOLYST, R. (2002a) Scattering patterns of self-assembled cubic phases. 1. The model. *Langmuir* 18(7), 2519–2528.

GARSTECKI, P. and HOLYST, R. (2002b) Scattering patterns of self-assembled cubic phases. 2. Analysis of the experimental spectra. *Langmuir* 18(7), 2529–2537.

GIBBS, B. F., KERMASHA, S., ALLI, I. and MULLIGAN, C. N. (1999) Encapsulation in the food industry: a review. *Int. J. of Food Sci. and Nutrition* 50, 213–224.

GOUIN, S. (2004) Microencapsulation: industrial appraisal of existing technologies and trends. *Trends in Food Science & Technology* 15, 330–347.

GUICHARD, E. (2002) Interactions between flavor compounds and food ingredients and their influence on flavor perception. *Food Rev. Int.* 18, 49–70.

HAAHR, A.-M., BREDIE, W. L. P., STAHNKE, L. H., JENSEN, B. and REFSGAARD, H. H. F. (2000) Flavour release of aldehydes and diacetyl in oil/water systems. *Food Chemistry* 71, 355–362.

ISRAELACHVILI, J. N., MITCHELL, D. J. and NINHAM, B. W. (1977) Theory of self-assembly of lipid bilayers and vesicles. *Biochim. Biophys. Acta* 470, 185–201.

KALNIN, D., SCHAFER, O., AMENITSCH, H. and OLLIVON, M. (2004) Fat crystallization in emulsion: Influence of emulsifier concentration on triacylglycerol crystal growth and polymorphism. *Crystal Growth & Design* 4(6), 1283–1293.

KHEADR, E. E., VUILLEMARD, J.-C. and EL DEEB, S. A. (2000) Accelerated Cheddar cheese ripening with encapsulated proteinases. *Int.J. of Food Sci. and Technol.* 35, 483–495.

LANDY, P., ROGACHEVA, S., LORIENT, D. and VOILLEY, A. (1998) Thermodynamic and kinetic aspects of the transport of small molecules in dispersed systems. *Colloids and Surfaces B: Biointerfaces* 12, 57–65.

LARSSON, K. (1994) *Lipids-Molecular Organization Physical Functions and Technical Applications*. Oily Press, Dundee, Scotland.

LASIC, D. D. (1993) *Liposomes: From physics to applications*. London: Elsevier.

LE BAIL, P., BIZOT, H., OLLIVON, M., KELLER, G., BOURGAUX, C. and BULEON, A. (1999) Monitoring the crystallization of amylose-lipid complexes during maize starch melting by synchrotron X-ray diffraction. *Biopolymers* 50, 99–110.

LIN, J., LESER, M.E., LÖLIGER, J. and BLANK, I. (2004) A new insight into the formation of odor active carbonyls by thermally-induced degradation of phospholipids in self-assembly structures. *J. Agric. Food Chem.* 52, 581–586.

LOPEZ, C. (2005) Focus on the supramolecular structure of milk fat in dairy products. *Reproduction, Nutrition, Development*, 45(4), 497–511.

LOPEZ, C., LESIEUR, P., BOURGAUX, C. and OLLIVON, M. (2000) Thermal and structural behaviour of milk fat: 1. Unstable species of cream. *J. Colloid Interf. Sci.* 229, 62–71.

LOPEZ, C., LESIEUR, P., BOURGAUX, C., KELLER, G. and OLLIVON, M. (2001) Thermal and structural behavior of milk fat: 2. Crystalline forms obtained by slow cooling of cream. *J. Colloid Interf. Sci.* 240, 150–161.

LOPEZ, C., BOURGAUX, C., LESIEUR, P., BERNADOU, S., KELLER, G. and OLLIVON M. (2002) Thermal and structural behavior of milk fat: 3. Influence of cream cooling rate and droplet size. *J. Colloid Interf. Sci.* 254, 64–78.

MAIER, H.G. (1970) Volatile flavoring substances in foodstuffs. *Angew. Chem., Int. Ed.* 9, 917–926.

MICHALSKI, M.C., OLLIVON, M., BRIARD, V., LECONTE, N. and LOPEZ, C. (2004) Native fat globules of different sizes selected from raw milk: thermal and structural behavior. *Chem. Phys. Lipids* 132, 247–261.

MÜLLER, R.H., RADTKE, M. and WISSING, S.A. (2002) Nanostructured lipid matrices for improved microencapsulation of drugs. *Int. J. Pharm.* 242 (1–2), 121–128.

OLLIVON, M., PATERNOSTRE, M., LESIEUR, S. and GRABIELLE-MADELMONT C. (2000) Vesicle reconstitution from lipid–detergent mixed micelles, special issue of BBA Biomembranes, guest editors, F. Goñi and A. Alonso. *Biochim. Biophys. Acta* 1508, 34–50.

PIRAPREZ, G., HERENT, M.-F. and COLLIN, S. (1998) Flavor retention by lipids measured in a fresh cheese matrix. *Food Chemistry* 61, 119–125.

PLUG H. and HARING P. (1993) The role of ingredient-flavour interactions in the development of fat-free foods. *Trends in Food Science & Technology* 4, 150–152.

POTHAKAMURY, U.R. and BARBOSA-GNOVAS, G.V. (1995) Fundamental aspects of controlled release in foods. *Trends in Food Science & Technology* 6, 397–406.

RABE, S., KRINGS, U. and BERGER, R.G. (2004) Dynamic flavour release from Miglyol/water emulsions: modelling and validation. *Food Chemistry* 84, 117–125.

DE ROOS, K.B. (1997) How lipids influence food flavor. *Food Technol.* 51, 60–62.

DE ROOS, K.B. (2003) Effect of texture and microstructure on flavour retention and release, *Int. Dairy J.* 13, 593–605.

ROUX, D., DEGERT, C. and LAVERSANNE, R. (1996) Microencapsulation of food ingredients using surfactants, FR 2735658 A1.

SCHWARZ, U. S. and GOMPPER, G. (2000) Stability of inverse bicontinuous cubic phases in lipid–water mixtures. *Physical Review Letters* 85(7), 1472–1475.

SEDDON, J. M. and CEVÈ, G. (1993) Lipid polymorphism: Structure and stability of lyotropic mesophases of phospholipids. In *Phospholipids Handbook*, Cevc, G. (ed.), Marcel Dekker: New York, 403–454.

SMALL, D. M. (1986) *The Physical Chemistry of Lipids. From Alkanes to Phospholipids*. Plenum Press, New York.

TANFORD, C. (1980) *The Hydrophobic Effect: Formation of Micelles and Biological Membranes*, 2nd edn, 233 pp.

8

Emulsion–flavour interactions

J. P. Dumont, INRA, France

8.1 Introduction

Emulsions consist in binary arrangements of non-miscible fluid phases. Phases are distributed in such a manner that emulsions look and can be handled as one single system. One phase is contained, scattered (a dispersed phase) within another (continuous phase). Thermodynamically, the system is not stable. The dispersed phase is bound to recover into one single continuous phase. From one emulsion to the other, it may take a few minutes or a few months before the dispersed phase achieves total recovery.

Edible emulsions are basic constituents of the human diet. Dairy cream and butter have been prepared since the Dark Ages. Most dressings, particularly mayonnaise and salad dressing and sauces, namely white sauce, are emulsified systems. Emulsion pictures a unique structural pattern. Water, fat and the solutes they contain are intimately mixed. As such, emulsion is a suitable carrier for aroma components and seasonings and is widely used as a support for elaborated sauces. Emulsion makes food tastier; in the meantime, it also proves relevant for granting palatability to a large share of dietary fats. Odour, taste and texture are the sensory characteristics more exposed to changes induced by this processing.

8.2 Remarkable characteristics of emulsions

8.2.1 Interface: crucial for stability and mass transfer

Physical stability of emulsified systems (i.e. keeping with the starting distribution of droplet size and the even repartition of the dispersed phase) relies on conservation of physical interactions in the tripod (dispersed phase; interface;

continuous phase). Roughly speaking, that means droplets of the dispersed phase must be kept surrounded by films of the continuous phase to prevent coalescence. The interface accommodates tensioactive molecules showing affinity for either phase. These compounds may have pre-existed in one or the other liquid phase or be added deliberately (emulsifiers) prior to emulsification. Whatever the origin, they help to anchor the non-miscible phases together and thus counteracts droplet aggregation and extends the life span of the system. In some instances, these emulsifying aids have been suspected of affinity towards sensory compounds (Land 1996). Oil–water emulsions involve water either as the continuous phase (oil-in-water, o/w) or as the dispersed phase (water-in-oil, w/o). In real life, most edible emulsions are of the oil-in-water type.

8.2.2 Aroma components

In emulsions, aroma compounds are solutes partitioned between water and fat. Balance of partial pressure between the liquid phases and the headspace is controlled by established laws of thermodynamics (possible connection with Chapter 11, see page 229). The driving trend for components is the move towards pressure equality in adjacent compartments. Liquid phases depict an intricate patchwork of droplets and surrounding films, and it may take time to achieve a resting system before coming to perfect equilibration. In real life, food emulsions face major disturbances when taken into the mouth. Dilution of the aqueous phase with saliva can produce, in some places, a severe lowering of the concentration of aroma compounds. In turn, kinetics of mass transfer between neighbouring compartments are exposed to deep modification. In such a situation, information gained on systems at rest (by far, the larger share of published data) is of little help for inferring the practical impact of events.

8.2.3 Taste components

Taste compounds demonstrate a different situation. Most of them are not soluble in lipids. Very often they have very low, even negligible, saturated vapour pressure at ambient temperature. Transportation of active components towards taste receptors, through the continuous aqueous phase, is the main concern to deal with. This is achieved more or less readily, depending on the involved components. Water-in-oil emulsions (butter, margarine) deserve specific consideration regarding taste components. The major part is bound to gather into the water phase and find itself trapped in the dispersed droplets. Low partial pressures developed by these components do not enable them to break through the oil film while access of extracting saliva is denied by the wrapping lipid phase. Under such conditions, extraction of the taste material must be slow. Surprisingly enough, Bakker and Mela (1996) evidenced no significant difference on sensory perception of taste compounds in w/o and o/w emulsions (oil volumic fraction: 0.5). The authors pointed that reversion of the water-in-oil into an oil-in-water emulsion may occur, particularly when small volumes are involved. They suggested that, in this watched experiment, 'both emulsion types may have shared a common physical structure within the mouth'.

8.3 Behaviour of aroma compounds in emulsions

8.3.1 Most aroma compounds readily dissolve in lipids

Aroma components, on the contrary to taste stimuli, partition between oil and water liquid phases. Free mobile molecules, located in regions where overall internal pressure is high, are inclined to move, whenever possible, towards regions bearing lower internal pressures. Pressure developed, when single, by a solute is directly proportional to concentration, saturated vapour pressure and activity coefficient in the solvent. Most volatile organic compounds have a much higher affinity (i.e. lower activity) for fat than for water. This results in frequent high scores for experimental oil–water partition coefficient (Rabe *et al.* 2004) and positive values of the theoretical Hansch's factor (log [*n*-octanol/water partition] or log *P*) (Rekker 1977). Accordingly, internal pressure developed by a fixed mass of a volatile organic solute is used to be lower in oil than in water. In turn, achievement of pressure equilibrium at the border of the two liquid phases implies prolonged neighbourhood of uneven concentrations of the solute. This kind of distribution for solutes concentration is consistent with emulsified systems that favour extensive conjunction of non-miscible phases. Summation of the mass of solute contained by each liquid phase makes clear that, in most cases, the major share of flavour compounds is crowding in the fat fraction.

Further their apparent homogeneous looking, emulsions show a fine micro-structure supported by a more or less stable arrangement of non-miscible solvents. Confined to the close range, exchange of solutes can still carry on long after the overall system has come apparently to a standstill. Frequently, equality of internal pressure is assumed completed in the system as soon as external variation of concentration cannot be anymore detected. The postulate is daring, as chance for instability breaking out in the deep of the system is not totally ruled out. Standing from the point of instrumental monitoring everything looks quiet but, as soon as the sample is taken into the mouth, underlying features will burst at the surface.

Apparent concentration

Apparent concentration of individual solutes (calculated as the ratio of mass of contained component to total volume of emulsion) is readily available. The numerical value is often published as an intrinsic characteristic of the system (Rabe *et al.* 2004), or even more, involved as an objective variable against which variations of sensory characteristics have been monitored (Van Ruth *et al.* 2000). Unfortunately, convenience does not necessarily suit pertinence. Apparent concentration is a pure abstraction bound to fit reality in one unique configuration: equality of solute partition in oil and water. Actually, sensory grading deals with taste compounds concentration in the aqueous phase and odorants concentration in the air or in the water phase (assuming equilibration within the mouthspace achieved). In the mouth, total volume of the water phase is abruptly increased by dilution with saliva. Transfer of a proportion of aroma components from the oil to the water phase is necessary to restore the proper balance of concentration. At the level of the individual compound, the proportion of the stored-in-oil molecules due to shift into water is as much higher as the dealt oil–water partition coefficient is lower. On the contrary, the concept of apparent concentration does not make any qualitative difference between components: any increase of the total volume results in a same apparent decrease of all components in the system, proportional to the dilution factor. Extensive recovery of the solutes concentration in the water phase is fully consistent with experimental data gained on the sensory impact of dilution of an oil-in-water emulsion. In experiments carried out using middle range hydrophobic odorant (ethyl butanoate), Brossard *et al.* (2000) evidenced a very limited impact of dilution on odour intensity rated by the sensory panel. Estimation of active flavour concentration in the water phase, on spotting intensity notes on the (intensity vs concentration) Stevens plot, yielded values well above those predicted on the account of the dilution factor. Doyen *et al.* (2001) came exactly to the same observation, using GC monitoring of volatiles concentration in the headspace. The recovery mechanism is likely to be involved in flavour release from mouthpieces, under the only provision that the timescale for completing transfer is of the same order as the mouth stay. Further support is given, although indirectly, by Roberts *et al.* (2003) and Rabe *et al.* (2004), who reported, at constant apparent concentration of hydrophobic compounds in the emulsion, that proportion of odorants released in the air decreases as the oil volumic fraction is risen. Assuming concentration in the air to be pictured by concentration in the continuous phase, the report sounds perfectly relevant: fewer volatiles detected in the air refers to less material in the water phase as normally produced on equilibrating the aqueous phase against raised volumes of oil that contain fixed amounts of the solutes. This is basically the same mechanism as involved in the restoration of solute concentration in the water phase after sample dilution with saliva. Either produced by water dilution or deliberate lowering of the oil content, decrease of the oil-to-water ratio in the system, at constant overall flavour content, raises the proportion of hydrophobic molecules readily available for release in the air.

Regarding hydrophilic compounds, Rabe *et al.* (2004) have stated little (diacetyl) or none (2,3-dimethyl pyrazine), effect of oil volumic fraction on volatiles release. Under the experimental conditions used (oil volumic fraction set from 0 up to 0.2), concentration of solutes prone to gather in the water phase rises within a narrow range (relatively to pure water, the maximum possible increase amounts 25%) while wandering hydrophobic components in the aqueous phase go down from 100% to traces. According to the scheme we assume, variations imparted to the vapour contents of the related hydrophilic odorants must stay relatively small.

At the moment, recorded experimental data and predictions relying on principles presently forwarded are in good agreement. It can be concluded that dilution of aqueous continuous phase modifies concentration of each individual aroma components to a particular extent in water and oil. Observed patterns are quite different from one compound to the other, but there is a clear cut between hydrophilic and hydrophobic components. To finish with rebuttal of apparent concentration, Land (1996) reported that higher concentrations of volatiles are consistently released in the headspace by biphasic oil/water systems as the primitive two layers configuration is turned into an emulsion. Sensible estimation of flavour potential of foods must rely on the array of partial pressures developed in the liquid phases by the solutes rather than just stick to the listing of apparent concentrations.

Emulsifiers
It has been pointed out that, in emulsions at rest, solutes show differences in concentration across the interface of the liquid phases. Nevertheless, exchange of solutes at the oil–water interface is kept at a standstill as long as a driving force is not generated. Molecule transfer usually starts with the advent of an internal pressure gap. Emulsifiers, accommodated by the interface, may interfere in the mass transfer through physical hindrance or chemical interaction with migrant solutes coming across the border. Hydrophobic bonding is often suspected for possible transfer disturbance, particularly if protein is used as the emulsifier. In fact, owing to their high molecular weight, true (mass/mass) capacity of macromolecules for ligands accommodation is limited. Calculated from the data reported by Hansen and Booker (1996), it appears to rate well below 1%. Moreover, protein docking at the interface probably leaves a small number of available sites in front of the bulk of molecules fat is able to supply. No conspicuous effect is expected, except if a large amount of residual emulsifier is contained by the water phase and meets emulsifier involvement postulated by Land (1996). Experiments by Roberts *et al.* (2003) have uncovered no impact from various emulsifiers on the milk/water static balance for some selected aroma compounds. However, the paper reports on balanced systems and does not totally exclude the event of transient restraining of migrants. Wondering whether, at the time scale of the mouth stay, emulsifiers can slow solute transfer down is still a question that is worth asking.

8.3.2 Some other important features

Droplet size
The smaller the droplets, the easier solutes can escape the dispersed phase. At constant volumic fraction of the dispersed phase, smaller droplets develop a larger interfacial area that in turn allows a richer flow of volatiles outgoing into the swallow breath (Carey *et al.* 2003). In addition, practical impact of solute diffusion inside the droplets can be waived in many cases as the diameter is less than twice the thickness of the limiting layer. Diffusion is assumed to account for a lag in the order of one second on 10 μm oil droplets (Weel *et al.*, 2004), but

a smaller average diameter is most commonly reported in emulsions with low oil volumic fractions.

Viscosity of the continuous phase

High viscosity of the continuous phase is viewed as a potential hindering factor for mass transfer. Addition of thickening agents to the water phase has been used to delay droplet aggregation and result in more stable systems (McClemens 1999). The principle working with droplets of the dispersed phase is assumed to extend (possibly) to molecular solutes. According to accepted theory, migration of molecules in a fluid matrix is bound to change its pace when facing a tougher resistance. This is likely to have an impact on the release of volatiles. Experimental evidence has been reported, *in vitro*, by Roberts *et al.* (1996). However, viscosity is often disregarded in studies on model systems as many of them involve fluid samples. In most instances, recorded measurements deal with emulsion viscosity whereas theory on flavour transfer is developed from systems containing one single continuous phase. In the light of the data reported by Rabe *et al.* (2004) viscosity of emulsions made of pure water and miglyol volumic fraction up to 0.2, does not rate much different from water. Equations proposed by Banarara *et al.* (2002) incline to expect the marginal impact of such a small variation of viscosity on the liquid–gas phase transition and insignificant effect on conscious sensory perception. The point becomes less trivial with food emulsions, like sauces, that are in a much higher viscosity range. Visual, uneasy dripping mentally suggests extensive hindrance of all types of mass transfer.

Headspace equilibration normally relies on free access and secured supply of volatiles to the air–liquid interface. The interfacial area offered for exchanges is quite small comparatively to the huge area developed by a liquid interface. On the way to aroma balance, a limited area of exchange can prove a bottleneck. Hindrance to solute transfer in the continuous phase may also deny free access of volatiles to the gas phase. We must be aware that when volatiles transfer from an emulsion to the air they have to face different possible limiting factors. None is bound to prevail and eventual application of one or the other depends on the characteristics of the food system.

Flavour fraction

Most frequently, food flavours arise from mixtures of sensory active components. At the level of the individual compound, a question is pending on the economy of the exchanges between fat, water and air. In the process ending in equilibration of internal pressures at the junction of the liquid phases, are exchanges balanced at the level of each individual volatile or controlled globally on the pool of components co-migrating in the same area? In the absence of any pumping effect at the interface, related, for example, to chemical affinity of the emulsifier or the receiving solvent for a particular solute, all molecules merely have the same chance of crossing. If so, instant composition of the transferred fraction is determined by the pressure gap across the interface and the components' relative abundance on the leasing side of the border.

8.4 Improving flavour delivery to the consumer

8.4.1 Making of the mouthspace

Dilution with saliva

When food emulsion is taken into the mouth, hydrophobic continuous phases resist sample soaking with saliva and restrain odorant extraction to the outer layer of the bowl. If oil is the dispersed phase, saliva can readily mix with the aqueous continuous phase. This results in a transient dilution of solutes followed by subsequent extraction from the oil droplets.

Pushing to the gas phase

Overall residence time of bites in the mouth is too short to enable complete mobilisation of the bulk of odour and taste active components contained in the food. Bowl dilution by saliva, subsequent extraction of flavour components by the aqueous phase and movement of aroma compounds in the mouthspace is managed on a local scale. On real foods, as stated by Land (1996) 'unite solute cells or particles are very small and diffusion distances will be of the order of micrometers'. Eventually, merged individual contributions from a variety of local sites produce the vapour fraction carried towards the upper airways. It is sensible to assume that the higher the local pressure, the larger the amount of released vapours and the heavier the contribution of the corresponding region to the overall mouthspace. On swallowing, the 'swallow breath' is expelled towards the nasal cavity (Buettner *et al.* 2001). Subsequent restoration of the mouthspace takes place at the expense of the liquid phase left in the throat (Normand *et al.* 2004). As already noted on thickened solutions (Baines and Morris 1989, Taylor *et al.* 2003), the fluid emulsions, with viscosity spotting beyond a critical point, certainly oppose the other (thickened) ones. The former can be assumed to grossly fit the models proposed for fluid drinks whereas even distribution of the aroma fraction is questionable in the latter. Nosespace is fed on successive swallow breaths until the next cycle is started by taking a new mouthpiece.

Guess work

Winner molecules (those that succeeded in docking to an olfactory receptor) are necessarily withdrawn from the air circulating in the nasal cavity and subsequently conveyed through the mucus layer. Nosespace analysis, probably the most suitable instrumental strategy available at the moment, relies on the volatile content of expired air. So, a question is raised: can we reliably deduce what has been trapped from the composition of what is left? The negative option is clearly undermining what has been carried out up to now but the positive answer implies, more or less, that what has been trapped is an aliquot of the original collection. 'Maybe' looks wiser without being of much help. However, besides the modelling by Keyhani *et al.* (1997), the sensible way to take into account mucus affinity, geometry of the air circulation or time span of receptors activation, in their real life involvement, is still a pending question.

8.4.2 Acknowledged perception is more than straight translation of molecular signals

Model food systems flavoured similarly with iso-amyl acetate, but given different textures, produced a close atmospheric pressure ionisation (API) pattern but deserved different banana intensity ratings (Cook *et al.* 2003). Moreover, changes in flavour release monitored through atmospheric pressure ionisation coupled with mass spectroscopy (API-MS), evidenced on varying fat content of milk emulsions, do not strictly parallel modifications in flavour perception, assessed by time–intensity (TI) (Miettinen *et al.* 2004). Good agreement between related variations of TI data and monitored concentrations, was evidenced for highly hydrophobic odorants only. Linalool, added to model systems of different fat contents, displayed a stable sensory pattern with intensity of the perception scaled proportionally to the measured concentrations. Volatility of these hydrophobic components, whether monitored at the first swallow breath or the subsequent breaths (also named persistence), was shown severely impacted by the fat content of the emulsion (Carey *et al.* 2003). Rise in the fat content accounts for definite, but opposite, effects on intensity (decrease) and persistence (increase) of the nosespace. As such, it appears able to control odorant release in the system while volatile concentration in the air accounts for the right variable to be monitored. Then, it is sensible to expect close connection of (API-MS/TI) data. On other instances or sensory attributes, regarding hydrophilic or medium hydrophobic odorants, conjunction of experimental (API-MS/TI) data is loose. This suggests that other factors other than impact of fat content are not negligible. One must be aware that the overall impression acknowledged by the consumer is actually the closing event of a complex course of actions. It involves matching and integration of a number of factors out of which the bulk of flavour material available for stimulation, although pregnant, is not decisive. This makes all the more difficult, anticipation of effects aroused at the consumer's level by deliberate changes introduced in samples processing.

8.5 Future trends

8.5.1 Some expectations in the mood

Making the more of active agents (aroma, salt, sugar) contained by food systems

There are different reasons for aiming at getting as high a sensory gratification as possible from the available bulk of flavour components. For a long time, efforts have been merely spent on cutting production costs. For instance, care was taken to make parsimonious use of expensive flavourings. Nowadays, the leading trend is towards a healthy diet. Sucrose and common salt are widely involved in food elaboration as taste compounds whereas they are currently blamed for a negative contribution to nutritional status. Apart from turning towards alternative taste compounds, the challenge is to maintain taste while cutting down on quantities. Regarding food emulsions, a single strategy is unlikely to care for simultaneous savings of both the hydrophobic odorants and the hydrophilic taste compounds. Clues and possible pathways for investigations are discussed further in this chapter.

Lower fat content

It is commonplace to stress that the share of lipids in the diet of people in the Western hemisphere is too high. This is alleged to produce widespread obesity and to expose people to increased risk of cardiovascular diseases. Although advisable on the one hand for the sake of better health, lowering fat content of processed foods deals, on the other hand, with various technical implications that cannot be disregarded. The concern is probably less with taste, as most active components are contained by the water phase. Less fat in the emulsion would not affect the capacity of the system to accommodate them nor impart extensive modification in their availability. However, changes in texture characteristics of the system may result from cuts in fat content and show limited indirect impact on mass transfer of taste components and time scale of their docking to receptors. On the contrary, the pattern of available aroma is prone to be heavily modified as fat content is manipulated. In any case, less fat at the consumer's level means forced shrinking of the storage capacity of hydrophobic molecules, especially aroma components. Different effects can be anticipated; depending on the manner flavour material is fed to the system. If flavouring is food-borne in the course of processing (fermented foods, heated foods developing Maillard reaction), instant trapping of volatiles is made more difficult and overall recovery of produced flavour material rates lower. If formulated flavour is provided from outside the system, two alternatives have to be taken into account: keeping with total amount of fed flavour or keeping with flavour concentration of fat. It is clear from Miettinen *et al.* (2004) that keeping with quantity added to milk emulsions results, at lower fat content, in enhanced intensity of the more hydrophobic molecules whereas hydrophilic compounds are little affected. On the other hand, keeping with concentration in the oil phase 'freezes' the balance between hydrophilic and hydrophobic odorant samples and affords good preservation of original sensory characteristics of the emulsion. As discussed earlier in section 8.3.1, dilution of food emulsions by saliva has limited effect on the concentration of hydrophobic odorants available to the consumer and the practical impact on instant perception is usually low. The latter strategy is apparently more advisable despite the limited capacity of flavour storage inherent in low fat levels, but is at risk of falling short on counterbalancing dilution. However, lowering fat content in a regular food is bound to involve changes in intensity, persistence, or balance of its flavour pattern. In addition, practical effects are heavily dependent on the aroma component itself (because of the huge range of oil/water partition coefficient), making all the more uncertain any forecast of sensory change at the level of the whole system.

Ways to better stability

Fats and oils are known to be materials sensitive to oxidation. Although, in some instances and at an early stage, oxidation may look desirable to promote flavour formation in the system, much more often it is viewed as a detrimental occurrence. Commonly, oxidation is feared as it imparts off-flavours and is considered as a real concern to good keeping quality of food. Many manufactured dressings and flavourings consist of emulsified food systems (mainly oil-in-water) from which quick delivery of flavouring material is aimed. Increasing interfacial area on decreasing droplet size is the layman's recipe commonly advised to master molecule exchanges in complex systems. These practices, liable to help in the migration of solutes from fat, improve in turn accessibility to the oil phase. Oxygen dissolved in the aqueous phase gains easier access to the fat globule as reported by Genot *et al.* (2003) that evidenced, on highly dispersed lipid phase, shortening of the time lag before oxidation starts. One may wonder whether there is any chance that an optimised distribution of droplet size could support simultaneously excellent physical stability of the dispersed phase, active flavour release from fat and acceptable delaying of oxidation. The point deserves discussion and further investigation.

8.5.2 Driving principles applied in the elaboration of strategies

A large proportion of flavour components contained by food emulsions may look underemployed. In systems operating normally, the larger share of aroma compounds is not even given a chance to escape the bowl at any moment throughout the mouth stay. However, these molecules, although not carried forth in the nosespace, are not idle material as they find themselves, at one time or the other, in a position to influence overall balance of partial pressures. Oil-in-water emulsion is a remarkable model system that accommodates each one of the three physical presentations of volatile components in a different phase. Stored in the fat globules, molecules move rather freely in the water phase and maintain a vapour in the air. Whereas the contribution of the latter presentation is readily acknowledged and studied, the stay in the liquid phases has been overlooked. People do not commonly wonder how many molecules are needed on a specified spot to enable one single molecule to escape in the air.

Taste components are not bound to skip from one phase to the other, as most of them are strictly soluble in water. Anyway, on coming in the receptors' vicinity, the bulk of molecules also partitions into a trapped and an unrecorded share. Maximized food, of which not a single flavour molecule would have gone unnoticed in the swallowed mouthpiece, is just a matter of dreaming.

Upgrading the management of mobile material in the mouthpiece

It is pointed that odorants invest separate compartments of the system. Achievement of a balanced partial pressure in the whole system requires proper distribution of volatiles between the air and liquid phases. Partition then assigns to the liquid phases a given proportion of the bulk, which is denied access to olfactory receptors. Selecting the right conditions to prevent excess reservation of odorants and making the most of the disposable portion are immediate challenges for technical studies.

Increased efficiency of the flavour release is intended to speed up overall mass transfer while keeping driving force constant. Alternatively, maintaining the same flow rate of exchanged material under a smaller gap of pressure would open the gate to a lowering of solute concentration in oil and of oil volumic fraction in the emulsion. Anyone aiming at optimised flavour yield must consider two different strategies.

First, easing the solute's migration towards the air phase. End concentration of the volatile's components in the air is bound to reach a fixed value strictly determined by thermodynamics. Assuming pressure balance achieved in the whole system, concentration in air can be readily calculated from oil volumic fraction of the emulsion, oil–water partition coefficient of the solute and apparent (w/v) concentration of the solute. Under static conditions, the main structural features involved are liquid interfaces and the continuous phase. They possibly impact on the span of time needed for transfer completion but, on the average, the system looks locked and little freedom is left for modulating odour perception. As a mouthpiece is being chewed, the system enters a dynamic evolution. Then, the rate of release makes much more sense than the theoretical end concentration of flavour (de Roos *et al.* 1994) and the model that accounted reliably for static conditions deserves upgraded tailoring. The time scale of events joins listed factors (interface, the solute's release from the continuous phase) as a key feature. Kinetics of solute availability from the dispersed phase and impact of rheological characteristics of fluids have to be considered and pertinent related variables have to be dug out. At the first sight, viscosity looks like the immediate candidate, but it must be remembered that, earlier in this text, viscosity has been identified as a frequently ill-assigned and misused characteristic. If it is intended to grade food properties by a sensory panel, monitoring of system apparent viscosity is no doubt advisable as evidenced by Brossard *et al.* (2002) on relating fluidity (sensory) to viscosity (instrumental). If the focus is on the release of sensory active material in the system, the light should be better directed on viscosity of the continuous water phase. It must be remembered that increase of emulsion viscosity, concomitant to a rise in oil volumic fraction, is bound to have only a small impact compared with the disturbance of the mass balance caused by the supply of pure refined fat. Care has to be taken to avoid the rash attribution of facts to a conspicuous factor.

In the mouth, moving of the food between cheeks and tongue, deformation by mechanical stress and dilution with saliva are involved simultaneously. The original structure collapses and gives birth to a number of secondary patches, available for subsequent dilution with saliva. Actually, after a few seconds, the mouth of the taster contains a variety of scattered pieces of emulsions showing different characteristics. One may wonder which phase stands out as the more relevant token of the disintegration: whole emulsion or the continuous phase? At oil volumic fractions up to 0.2, numerical data reported by Rabe *et al.* (2004) on viscosity incline to rebut significant restraint in the mixing of the liquid phases. Even dilution with saliva is assumed and the mouthpiece ends in a homogeneous mass. It is acceptable that further speculation on flavour optimisation is based on a single liquid phase containing evenly distributed solute, at the concentration of the water phase.

At high oil volumic fraction, viscosities of the emulsion and the continuous phase are variables that clearly spot on different orders of magnitude. Smooth mixing of the liquid phases is less than warranted. Charles *et al.* (2000) have reported that the smaller oil droplet size, the higher the viscosity of o/w emulsions of high oil volumic fraction. This gives a hint of the diversity of situations possibly involved in mixing high fat emulsions with saliva and conveys the implicit conclusion that viscosity of the system pictures internal resistance to mass transfer. Regarding practical aspects of flavour availability to the consumer, pockets of undiluted emulsion may well survive being taken into the mouth. Consequently, a proportion of oil droplets, wrapped in a companion water phase, are able to contain solutes at a concentration close to the initial level. This sustains relatively higher internal pressure in some local regions. After swallowing, the emulsion film lining the throat may contain patches of higher internal pressure that release related higher concentrations of aroma components in the expired air. Brossard *et al.* (2002) have evidenced a higher (flavour intensity/benzaldehyde concentration in water) ratio as the miglyol volumic fraction rose from 0.3 to 0.6. Moreover, rising in oil volumic fraction lowers the content in continuous phase of the system. Dispersion of oil is made more difficult and average droplet size inclines to increase. The solute's diffusion inside the larger droplets may slow down odorant release from the film lining the throat and contribute to extended persistence in the breath. According to the model proposed by Weel *et al.* (2004), diffusion inside droplets of the Brossard *et al.* (2002) experiment (14 μm; $\Phi = 0.6$) induces a time lag shorter than the time span (3 s) allotted to *in vivo* aroma release. In such a case, benzaldehyde is made available progressively with no real hindrance of the maximum intensity and strengthened persistence. This looks contradictory to earlier alleged principles (Overbosch *et al.* 1991) on the connection of intensity to viscosity. Actually, it is concluded that viscosity must not be viewed as a systematic hindrance to the burst of the emulsion's aroma. Good keeping of perceived intensity is achieved when volatile materials are kept undisturbed in one phase until the 'swallow breath' and are subsequently exposed to repeated extraction by air in the throat.

Second, the interface may hide a bottleneck for molecule migration. In an oil-in-water emulsion, going through the air–water interface is merely a matter of bridging an energy gap while crossing the oil–water interface may prove otherwise entangling. Besides the requested overcoming of the energy barrier, components possibly face transfer hindrance caused by molecular interactions between migrant and emulsifier molecules. Presently, documented information on aroma–emulsifier interaction is scarce and ruling principles are not clear. As pointed out earlier in the text, the impact of irreversible interactions is likely to be marginal while the lengthened residence time on crossing the interface cannot be ruled out. Odorant exchange into the throat takes place on a short time scale, suggesting a learned selection of the emulsifier may open ways on the control of sensory perception. To date, a study (unpublished) using proteins as the tensioactive agent, has shown the rate of volatile transition at the interface to be affected to a very different extent by β-lactoglobulin and bovine serum albumin.

Raising the driving force at no risk of increasing resistance

Levelling up internal pressure, at constant apparent concentration, is not easy. In most instances, pressure push on the oil side of the liquid interface affords its theoretical maximum. The only track left open is towards the increase of activity coefficient of solutes. Increasing the proportion of crystalline fat in the lipid phase looks like the immediate sensible proposal as solutes, excluded from the solid fraction, are intended to crowd into the liquid fraction. The predicted effect is consistent with data recorded by Relkin *et al.* (2004) on fats of different origin, added with hydrophobic solutes, that contained different proportions of solids at the same temperature. However, Roberts *et al.* (2003) gave evidence that, on the average, the effect on release is usually moderate unless the proportion of solid fat scores very high. Keeping within the scope of acceptable food commodities, one may wonder whether workable changes in the fat phase of emulsions can have a sensible impact on molecule availability or significantly affect the impression gained by taste panellists.

On the same line as the influence of high viscosity of the emulsion (when linked to high oil volumic fraction), a stimulating question is raised on the effect of rigidification of the continuous phase. In the literature, an implicit ruling principle assumes uniformity of the solute concentration in the dispersed phase, on starting. But, what if the whole flavour material happened to be fed, before rigidification of the system, incorporated within a portion of the total count of droplets? An uneven flavour distribution would be produced in the dispersed phase that may not necessarily achieve perfect balance in the sample. On being taken into the mouth, flavour release would take place as previously stated with viscous emulsions and intensity of perceived aroma would possibly outrange intensity gained on emulsion fed with odorants evenly distributed in the dispersed phase.

Investigating the potential of co-occurring stimulations

When monitored on a local scale, mass transfer of solutes crossing the oil–water interface looks impacted by the total pressure developed by the bulk of the fellow migrants. It has been shown, on a model system consisting of resting layers of oil and water (Brossard *et al.* 2003), that the time required to complete equilibration of a given individual component varies in conjunction with volatility of its companion molecules browsing at the interface. Initially contained in the oil phase with volatile esters, 1 Octen-3-ol equilibrates readily in the water phase whereas it does not when associated with a series of secondary alcohols. This hints that, if a particular sensory effect is aimed at, it would be possible to work out the composition of odorant mixtures along two dimensions at right angles. In connection with an established selection of flavouring material on the grounds of quality and mutual balance, shrewd choice of odourless volatile companions would open perspectives on rational management of the release of key components.

Better knowledge on the interactions, at the sensory level, between aroma, taste and texture stimulations would be appreciated by those who intend to broaden the scope of perceptions aroused by foods. At the moment, it is accepted that acknowledged sensory impact does not arise straight from recorded availability of a bundle of selected components. Factors such as the microstructure of the food, viscosity of liquid phases, delivery pattern of active and inactive components or strategy of mastication are also involved (LeThuaut *et al.* 2005). They influence perception at the integration stage in the consumer's mind. Obviously, it is necessary to come to an integrated view of the whole process starting with the collection of signals on food and ending in the acknowledged perception. This is not possible at the moment as most of the available information does not sustain generalisation. In the future, special attention must be paid in extending the scope of validity of the investigations.

8.6 Cross-links

Emulsion proves a very flexible and reasonably handy system for field investigation on molecular exchanges at the liquid interface. In addition, the scope allowed for variation of imparted textural characteristics and distribution of flavour components is fairly wide. Unlike solid samples, emulsion is not crushed by mastication. In many cases, deformation takes place during the mouth stay without extensive disintegration. This certainly helps to restrain dispersion of experimental data and lowers the risk of incidental generation of unreliable samples.

Emulsions are not so different from other types of food system. From the standpoint of structure, they can be classified as a dispersion of a liquid phase into a fluid or solid network. This is far from being an uncommon pattern in foodstuffs. Emulsions and gels do not appear much different regarding release of water-soluble taste components into the mouth and after swallowing. In any case, the concern is for the mass transfer of solutes in an aqueous solvent concomitantly diluted with saliva. On the other hand, aroma compounds show differences in availability of the volatile material and the diverging impacts of firmness and viscosity. Variation, relating to differences in firmness of gels samples, is generally caused by chewing. On emulsion being taken into the mouth, different schemes can be activated depending on rheological characteristics of the system. It is noticeable that principles involved in solute exchange in high viscosity emulsion (Brossard *et al.* 2002), also power the machinery of the composite model system (superimposed resting layers of air, residual diluted food and mucus), proposed by Normand *et al.* (2004) for modelling the kinetics of flavour release during drinking. Taste panellists do not usually acknowledge evaluation of sensory characteristics of emulsion samples as the most pleasant task they have to deal with. This should be kept in mind as people may feel a little uncomfortable when rating samples. Despite some inherent flaws, experiments carried out to show the effect of texture characteristics on flavour quality or perceived intensity are essential for uncovering connections between objective measurements and subject assumed realities.

8.7 References

BAINES Z V and MORRIS (1989) 'Suppression of perceived flavour and taste by food hydrocolloids', in Bee R D, Richmond P and Mingens J, *Food Colloids*, Cambridge, Royal Society of Chemistry, 184–192.

BAKKER J and MELA D J (1996) 'Effect of emulsion structure on flavor release and taste perception', in McGorrin R J and Leland J V, *Flavor–Food Interactions*, ACS Symposium series 633, Washington, ACS, 36–47.

BANARAVA D S, RABE S, KRINGS U and BERGER R G (2002) 'Modeling dynamic flavor release from water', *J Agric Food Chem*, 50, 6448–6452.

BROSSARD C, MEYNIER A and DUMONT J P (2003) 'Partition and tranfers of aroma mixtures between water, oil and air phases: alcohols, esters and their mixtures show disjunct cases', in Le Quéré J L and Etiévant P X, *Flavour Research at the Dawn of the Twenty-first Century*, proceedings of the 10th Weurman flavour research symposium, Paris, Lavoisier, 55–58.

BROSSARD C, ROUSSEAU F and DUMONT J P (2000) 'Odour intensity of o/w emulsions: Looking for the keys to prediction', in Schieberle P and Engel K H, *Frontiers of Flavour Science*, proceedings of the 9th Weurman flavour research symposium, Garching, Deutsche Forschunganstalt für Lebensmittelchemie, 304–310.

BROSSARD C, ROUSSEAU F and DUMONT J P (2002) 'Perception of odor and flavor of liquid emulsions', in Anton M, *Food Emulsions and Dispersions*, Trivandrum, Research Signpost, 83–91.

BUETTNER A, BEER A, HANNIG C and SETTLES M (2001) 'Observation of the swallowing process by application of videofluroscopy and real time magnetic resonance imaging-consequences for retronasal aroma stimulation', *Chem Senses*, 26, 1211–1219.

CAREY M, LINFORTH R and TAYLOR A (2003) 'Factors affecting dynamic flavour release from emulsions', in Le Quéré J L and Etiévant P X, *Flavour Research at the Dawn of the Twenty-first Century*, proceedings of the 10th Weurman flavour research symposium, Paris, Lavoisier, 212–216.

CHARLES M, ROSSELIN V, SAUVAGEOT F, BECK L and GUICHARD E (2000) 'Sensory and physico-chemical approaches to flavour release from salad dressings in relation to the structure', in Schieberle P and Engel K H, *Frontiers of Flavour Science, proceedings of the 9th Weurman flavour research symposium*, Garching, Deutsche Forschunganstalt für Lebensmittelchemie, 320–324.

COOK D J, HOLLOWOOD T A, LINFORTH R S T and TAYLOR A J (2003) 'Oral shear stress predicts flavour perception in viscous solutions', *Chemical Senses*, 28, 11–23.

DE ROOS K B and WOLKSWINKEL K (1994) 'Non-equilibrium partition model for predicting flavour release in the mouth', in Maarse H and Van der Heij D G, *Trends in Flavour Research*, proceedings of the 7th Weurman flavour research symposium, Amsterdam, Elsevier Science, 15–32.

DOYEN K, CAREY M, LINFORTH R S T, MARIN M and TAYLOR A J (2001) 'Volatile release from an emulsion: headspace and in-mouth studies', *J Agric Food Chem*, 49, 804–810.

GENOT C, MEYNIER A and RIAUBLANC A (2003) 'Lipid oxidation in emulsions', in Kamal-

Eldin A, *Lipid Oxidation Pathways*, AOCS Press, Champaign.

HANSEN A P and BOOKER D C (1996) 'Flavor interaction with casein and whey protein', in McGorrin R J and Leland J V, *Flavor–Food Interactions*, ACS Symposium series 633, Washington, ACS 75–89.

KEYHANI K, SCHERER P W and MOZELL M W (1997) 'A numerical model of nasal odorant transport for the analysis of human olfaction', *J Theor Biol*, 186, 279–301.

LAND D G (1996) 'Effects of interactions on flavor perception', in McGorrin R J and Leland J V, *Flavor–Food Interactions*, ACS Symposium series 633, 2–11.

LETHUAUT L, BROSSARD C, MEYNIER A, ROUSSEAU F, LLAMAS G, BOUSSEAU B and GENOT C (2005) 'Sweetness and aroma perceptions in dairy desserts varying in sucrose and aroma levels and in textural agent', *Int Dairy J*, 15(5), 485–493.

McCLEMENS D J (1999) *Food emulsions: Principles, practice and techniques*, Boca Raton, CRC Press.

MIETTINEN S M, HYVONEN L, LINFORTH R S T, TAYLOR A J and TUORILA H (2004) 'Temporal aroma delivery from milk systems at 0–5% fat levels observed by free choice profiling, time-intensity and API-MS techniques', *J Agric Food Chem*, 52(26), 8111–8118.

NORMAND V, AVISON S and PARKER A (2004) 'Modeling the kinetics of flavour release during drinking', *Chem Senses*, 29, (3), 235–245.

OVERBOSCH P, AGTEROF W G M and HARING P G M (1991) 'Flavor release in the mouth', *Food Rev Int*, 7, 137–184.

RABE S, KRINGS U D and BERGER R G (2004) 'Dynamic flavour release from Miglyol/water emulsions: modelling and validation', *Food Chem*, 84, 117–125.

REKKER R F (1977) 'The hydrophobic fragmental constant', in Nauta W and Rekker R F, *Pharmacochemistry Library*, Amsterdam, Elsevier Science.

RELKIN P, FABRE M and GUICHARD E (2004) 'Effect of fat nature and aroma compound hydrophobicity on flavor release from complex food emulsions', *J Agric Food Chem*, 52, 6257–6263.

ROBERTS D D, ELMORE J S, LANGLEY K R and BAKKER J (1996) 'Effect of sucrose, guar gum and carboxymethylcellulose on the release of flavor compounds under dynamic conditions', *J Agric Food Chem*, 44, 1321–1326.

ROBERTS D D, POLLIEN P and WATZKE B (2003) 'Flavour release from milk-based liquid emulsions: key influence of fat phase', in Le Quéré J L and Etiévant P X, *Flavour Research at the Dawn of the Twenty-first Century*, proceedings of the 10th Weurman flavour research symposium, Paris, Lavoisier, 21–26.

TAYLOR A, HOLLOWOOD T, DAVIDSON J, COOK R and LINFORTH R (2003) 'Relating taste, aroma and mouthfeel stimuli to overall flavour perception', in Le Quéré J L and Etiévant P X, *Flavour Research at the Dawn of the Twenty-first Century*, proceedings of the 10th Weurman flavour research symposium, Paris, Lavoisier, 194–199.

VAN RUTH S M, ROOZEN J P and JANSEN F J H M (2000) 'Release of odour active compounds from oxidized sunflower oil and its oil-in-water emulsion', in Schieberle P and Engel K H, *Frontiers of Flavour Science*, proceedings of the 9th Weurman flavour research symposium, Garching, Deutsche Forschunganstalt für Lebensmittelchemie, 292–299.

WEEL K G C, BOELRIJK A E M, BURGER J J, JACOBS M A, GRUPPEN H, VORAGEN A G J and SMIT G (2004) 'Effect of emulsion properties on release of esters under static headspace, in vivo, and artificial throat conditions in relation to sensory intensity emulsions', *J Agric Food Chem*, 52, 6572–6577.

9

Protein–flavour interactions

A. Tromelin, I. Andriot and E. Guichard, INRA, France

9.1 Introduction

Flavour is considered one of the most important attributes determining the acceptance of food by the consumer. Flavour perception is known to be influenced not only by the composition of odorant molecules in the food matrix but also by the amount of molecules to be released in the mouth during food consumption. Food matrix ingredients, among them food proteins, have little flavour of their own, but are known to bind and trap aroma compounds. In function the nature and the strength of the binding, the release of aroma compounds in the gas phase will be more or less decreased and this will have a significant impact on the overall aroma perception, due to changes in the aromatic balance (Fischer and Widder 1997, Guichard 2002, Kinsella 1990). However, protein ingredients not only reduce perceived impact of desirable flavours but they can also produce undesirable off-flavours (O'Neill 1996).

This chapter reviews the main results obtained up to now on protein–flavour binding. The first section deals with the molecular structure of some of the main proteins susceptible to bind flavour compounds, and underlines the types of structure which are more likely to be involved in flavour binding. The localisation of the more probable binding sites is presented, together with the effect of the medium on the conformation of the proteins.

The second section compares the different methods used for the determination of the binding constants and for the localisation of the binding sites, focusing on more recent developments.

The third section points out the effects of the medium changes, on the strength of the binding to food proteins, for example pH, temperature, salt and co-solvent, underlining the direct effects and the indirect effects due to conformational changes of the proteins.

The last section will present a selection of relevant examples showing the impact of protein–flavour binding on flavour release and flavour perception, in model systems and real food systems.

9.2 Protein structure in relation to flavour binding

Most investigation which aimed to study the mechanism of flavour binding showed the role of proteins structure as well as the type of flavour compound (aldehyde, alcohol, ketone, ester) in binding process (Heng *et al.* 2004, Semenova *et al.* 2002a, 2002b, 2002c). Globular proteins are used as food ingredients for different applications such as gelation, thickening, emulsification and foaming. The structures adopted by a globular protein under a particular environment result from an equilibrium between physicochemical parameters including hydrophobic interactions, electrostatic interactions, hydrogen bonding, van der Waals forces, and configurational entropy (McClements 2002, and cited references). The system tends to reduce the contact area between non-polar groups and water, and for this reason the main driving force stabilising the compact structure of globular proteins is related to hydrophobic effect.

The functional properties of globular proteins strongly depend on the molecular structure and dynamics of the proteins under the environmental conditions. In response to a change in environment during a process (heating, mechanical stress, etc.), a globular protein may undergo change from a 'native' state to a 'denatured' state.

In this way, it is of particular interest to examine the structure of some food proteins, such as milk and vegetable proteins. We focus in this paper on the structure of the most represented and studied proteins, such as β-lactoglobulin and soy proteins.

9.2.1 β-lactoglobulin

Among the different food proteins, β-lactoglobulin is one of the best characterised and studied proteins (Batt *et al.* 1994, Brownlow *et al.* 1997, Kuwata *et al.* 1999, Papiz *et al.* 1986, Sawyer *et al.* 1998, Sawyer and Kontopidis 2000). The β-lactoglobulin belongs to the super-family of lipocalins (Akerstrom *et al.*, 2000; Flower, 2000), together with retinol binding protein and odor binding proteins (Flower *et al.* 2000). The common structure of the lipocalins has been known for several years (Flower 1996). The lipocalin chain folds in nine-stranded antiparallel β-sheet. A single eight-stranded antiparallel β-sheet forms a continuously hydrogen β-barrel that constitutes the core of the protein, flanked by a three-turn α-helix.

Structure

The purification of β-lactoglobulin was presented fifteen years before its three-dimensional structure was determined (Palmer 1973). The β-lactoglobulin is the major protein component of whey from the milk of ruminants and many other mammals, but it is not present in human mother's milk. With a MW of 18 400 D, it consists of 162 amino acid residues and contains two disulfide bonds and a free thiol group. This protein is usually a dimer under physiological conditions. The most common genetic variants of bovine β-lactoglobulin (*Bos Taurus*) are the A and the B forms, the A variant differing from the B variant by the substitution of Asp64 to Gly and Val118 to Ala (Hambling *et al.* 1992, and cited references). Although about a dozen variants have been observed, mostly because of their different electrophoretic mobilities (Sawyer and Kontopidis 2000, and cited references). The three-dimensional structure was first determined at 6 Å resolution (Green *et al.* 1979); in 1986 Papiz *et al.* showed the remarkable similarity between β-lactoglobulin and retinol-binding protein, suggesting a carrier role for β-lactoglobulin (Papiz *et al.* 1986, Sawyer *et al.* 1998). Presently, the biological function of β-lactoglobulin remains unknown, although having been extensively studied. The structure of the dimer at 1.8 Å resolution displayed in Fig. 9.1 shows the β-barrel (left) and the central cup of calyx (right) (Brownlow *et al.* 1997). The first full NMR assignment of H1, C13 and N15 chemical shift for β-lactoglobulin reveals close resemblance with the X-ray native structure at pH 6.2 in both secondary structure and overall topology (Kuwata *et al.* 1999, Uhrinova *et al.* 1998, 2000). Although the same core structure exists both at low pH using NMR spectroscopy and in crystalline forms, significant differences exist between the NMR and the X-ray structures and insight the role of the flexible loops connecting the β-sheets (Jameson *et al.* 2002).

Binding sites

Binding studies carried out on β-lactoglobulin show that a variety of ligands can to be bound by β-lactoglobulin (Akerstrom *et al.* 2000, Boudaud and Dumont 1996, Charles *et al.* 1996, Guichard and Langourieux 2000, Kontopidis *et al.* 2002, Sawyer *et al.* 1998, Sawyer and Kontopidis 2000). However, the results

Fig. 9.1 Ribbon diagram of the β-lactoglobulin, from PDB-1BEB structure (Brownlow *et al.* 1997).

have been controversial concerning the location of the binding sites. According to a general consensus, the retinoid species bind inside the calyx, by analogy with the behaviour of lipocalins; however, the exact location of the binding site remains uncertain. Crystal structure of complexes with retinol suggests that retinol binds at a surface pocket in a region of high flexibility (Monaco *et al.* 1987), and retinal and retinol have been shown to bind at the same inner site (Cho *et al.* 1994). The X-ray structure of complexes with palmitate (Wu *et al.* 1999) and 12-bromododecanoic acid (Qin *et al.* 1998) reveals that the ligands bind into the central cavity. However, fatty acids and retinoids bind independently and simultaneously to the β-lactoglobulin (Naraya and Berliner 1997); in the same way, retinal and palmitate did not compete for the same site (Wang *et al.* 1999). Based on fluorescence results, fatty acids appear to bind to an external pocket (Frapin *et al.* 1993), whereas NMR study indicates that at neutral pH palmitic acid binds within the calyx site (Ragona *et al.* 2000). Measurement of the interactions between β-lactoglobulin and small hydrophobic ligands was monitored by fluorometry (Dufour and Haertle 1990b; Muresan *et al.* 2001). They revealed that β-ionone binds in the vicinity of a tryptophan residue. More recently, NMR study showed that several amino acids belonging to side chains pointing into the central cavity, are affected by the binding of γ-decalactone, whereas binding of β-ionone affects amino acids located in a groove near the outer surface of the protein (Lübke *et al.* 2002). At last, induced circular dichroism spectra reveal binding of bilirubin at the open end mouth of the β-barrel (Zsila 2003). There is thus evidence suggesting that there is more than one binding site in β-lactoglobulin, the main one being the retinol site and a second site existing in addition to the central calyx, on the outer surface close to the helix (Kontopidis *et al.* 2002; Sawyer *et al.* 1998; Sawyer and Kontopidis 2000).

Stability and denaturation

Despite its stability, β-lactoglobulin exists in various oligomeric states as a function of pH, temperature and concentration.

pH

β-lactoglobulin from ruminant species exists as a monomer below pH 3, and forms a dimer under physiological conditions (Hambling *et al.* 1992). The dimer–monomer equilibrium depends on addition of salts (Joss and Ralston 1996; Sakurai *et al.* 2001) and is determined by a subtle balance between stabilising and destabilising forces: on one hand hydrophobic interactions between the dimer interface favour dimer formation; on the other hand electrostatic repulsions between the positively charged monomers favour the monomer. Structural changes accompanying pH-induced dissociation of the β-lactoglobulin dimer have been investigated by NMR spectroscopy (Fogolari *et al.* 1998; Uhrinova *et al.* 1998). Although the three-dimensional structure of β-lactoglobulin at an acidic pH is very similar to that of a subunit within the dimer at pH 6.2, differences occur in the orientation of two loops and of the flanking three-turn α-helix at the termini. The loop situated at the entry of the hydrophobic cavity is in the 'closed' position at low pH (below 6.2) and in the 'open' position in the pH range 6.2–7.2 (Uhrinova *et al.* 2000). This could explain the better accessibility of the ligands at higher pH values and a better specificity at lower pH values.

Temperature

Heat treatments occurring during food processing should induce structural changes. Despite heat-resistance structural features due to reversible changes (Edwards *et al.* 2002), the flavour binding behaviour was significantly altered by heat treatment at 75 ºC and related to conformational changes and aggregation of β-lactoglobulin (O'Neill and Kinsella 1988). Heat-induced aggregation mechanism has been extensively studied (Hong and Creamer 2002, and cited references). Upon heating, the native dimers progressively dissociate into native monomers and a thiol group becomes solvent accessible. Aggregates are formed via intermolecular thiol-catalysed sulfhydryl group/disulfide bond interchange reactions and non-covalent interactions and to a lesser extent, thiol–thiol oxidation (Hoffman and van Mil 1997, Prabakaran and Damodaran 1997). More recently, it has been shown that non-native dimers and non-native monomers formed during heat-treatment of β-lactoglobulin solutions may play an important role in aggregation at the macromolecular level (Croguennec *et al.* 2003, 2004a; Surroca *et al.* 2002).

Recent studies put forward the role of the medium and the influence of the ionic strength on temperature denaturation (Baussay *et al.* 2004, Labouré *et al.* 2004, Pouzot *et al.* 2004). Indeed, the addition of salt decreases the electrostatic repulsion by anion binding and stabilises the dimer (Sakurai *et al.* 2001).

Kinetic studies display the role of milk concentration, so that thermal denaturation was delayed at higher milk solid concentrations (Anema 2000). At pH 6.6 in various salt conditions, it has been shown that heating induces the formation of at least two molten globule-like states of which stability depends on the nature of exposed thiol residue (Croguennec *et al.* 2004b).

Pressure

Ligand and flavour bindings to β-lactoglobulin have been found to be modified by high pressures (300–900 MPa). Compared to native β-lactoglobulin, the high hydrostatic pressure (HHP)-induced molten globule state of β-lactoglobulin exhibited a significant decrease in affinity and the number of binding sites for retinol. The affinity for palmitic acid, capsaicin, or carvacol ligands is also lowered compared with native β-lactoglobulin, and no more detectable specific binding was found for α-ionone, β-ionone, cinnamaldehyde or vanillin flavours (Yang *et al.* 2003).

Indeed, high hydrostatic pressure treatment resulted in changes in the hydrophobic calyx and surface hydrophobic sites of β-lactoglobulin. It has been demonstrated that pressure reduced emulsifying capacity and foam ability, and enhanced hydrophobicity of β-lactoglobulin (Pittia *et al.* 1996). This behaviour could be related to structural changes. Studies of molecular modifications of β-

lactoglobulin upon exposure to high pressure show only minor irreversible changes (Iametti *et al.* 1997). But despite a similar overall conformation, the architecture of β-lactoglobulin before and after high pressure exposure was supposed to be stabilised by slightly different interactions (Subirade *et al.* 1998). By employing Fourier transform infrared spectroscopy, changes in the secondary structure of β-lactoglobulin variant A were observed at a lower pressure compared with the variant B, indicating a more flexible structure for variant A. Thus β-lactoglobulin A showed a greater proportion of α-helix conformation together with protein aggregation, whereas the changes of β-lactoglobulin B were minor and reversible, without aggregation (Hosseini-nia *et al.* 1999). H^1 NMR study confirmed the difference observed between variants A and B at high pressure (Belloque *et al.* 2000). Until 200 MPa, the core of the protein remains unaltered, but conformational flexibility increases at pressures over 200 MPa, and the core of β-lactoglobulin A seemed to unfold faster than that of β-lactoglobulin B. However, after treatment at 400 MPa, the structure was found to be identical to the native structure after equilibration back to atmospheric pressure for both variants.

More recent studies performed both by intrinsic tryptophan fluorescence and circular dichroism suggested that high hydrostatic pressure at 600 MPa treatment conditions induced a conformational change. Far-ultraviolet circular dichroism (CD) spectra reveal that the secondary structure of β-lactoglobulin converts from native β-sheets to non-native α-helices, whereas the intrinsic tryptophan fluorescence suggests that the aromatic groups become more mobile and more accessible during this transition (Yang *et al.* 2001). This change should be related to moderate pressure (Aouzelleg *et al.* 2004a). However, investigation by Fourier transform Raman spectroscopy suggests the involvement of β-sheet structures in protein aggregation at high pressure as well as an increase in random structures and a decrease in the proportion of the α-helix (Ngarize *et al.* 2004).

In addition to the effect of pH, temperature and pressure, alcohols are found to involve structural changes (Dib *et al.* 1996; Dufour and Haertle 1990a). Ethanol induces a reversible conversion of β-sheets into α-helix, modifying the retinol binding stoechiometry. The same phenomenon of helical structure proportion increase was observed for diols. The modifications in the protein conformation could reduce the number of accessible binding sites on the protein (Fischer and Widder 1997).

9.2.2 Other whey proteins

α-lactalbumin, bovine serum albumin and caseins are studied to a lesser extent for their binding properties towards flavour compounds.

α-lactalbumin

α-lactalbumin was found to bind ketones and aldehydes (Franzen and Kinsella 1974; Jasinski and Kilara 1985) but with a poor flavour binding capacity

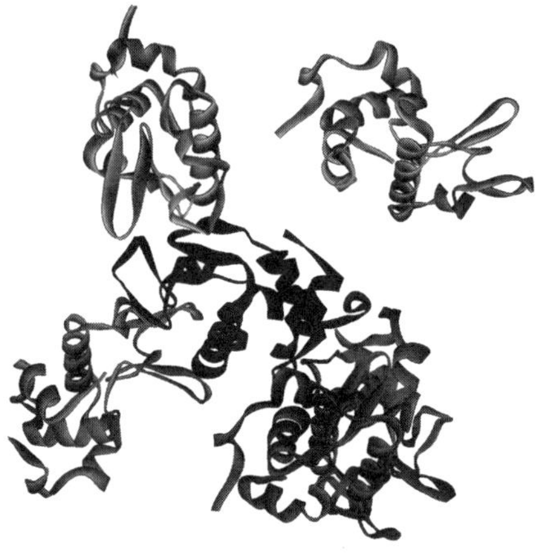

Fig. 9.2 Ribbon diagram of the bovine α-lactalbumin from PDB-1F6S structure (Chrysina *et al.* 2000).

compared with other whey proteins (Jasinski and Kilara 1985). α-lactalbumin is made of six subunits of 122 amino acids, as shown in Fig. 9.2. It has a strong Ca^{2+} binding site, which also binds other cations such as Mg^{2+}, Mn^{2+}, Na^{+}, and K^{+}, and several distinct Zn^{2+} binding sites (Permyakov and Berliner 2000). Effects of temperature and pressure are often carried out together with β-lactoglobulin (Bertrand-Harb *et al.* 2003, Havea *et al.* 2001, Hong and Creamer 2002, Huppertz *et al.* 2004, Karasova *et al.* 1998, Ye *et al.* 2004).

Bovine serum albumin (BSA)

BSA is described as a globular non-glycoprotein with a molecular weight close to 66 430. It is made of 583 amino acid residues and has 17 cystine residues (8 disulphides bridges and 1 free thiol group) (Carter and Ho 1994, Hirayama *et al.* 1990). Fatty acids bind to BSA, probably by hydrophobic interactions, the carbonyl group playing only a minor role in this interaction (Burova *et al.* 1999, and cited references; Morrisett *et al.* 1975, Spector 1975), and polysaturated aldehydes derived from the polyunsaturated acids likely to react with lysine residues to form Shiff base adducts by covalent bonds (Refsgaard *et al.* 2000).

However, it has been found that BSA binds carbonyl compounds with a high affinity (Damodaran and Kinsella 1980b, Jasinski and Kilara 1985) inducing conformational changes of the protein. The chemical reduction of disulfides bridges of BSA decreases its affinity for carbonyl compounds.

The binding affinity is influenced by the chain length, the functional group and the structural state of the protein (Damodaran and Kinsella 1980b). In fact, around 21 binding sites have been reported in BSA, and as many as five to six primary binding sites seem to be involved in carbonyl binding (Jasinski and Kilara 1985). More recent investigation on the release of 2-octanone bound by BSA/pectin complexes reveals that under conditions of weak attractive interaction between BSA and pectin (at pH 6.4) 2-octanone is preferentially bound by BSA, while it is bound by both BSA and pectin under conditions of strong interbiopolymer interaction at pH 4.3, suggesting a competitive binding of BSA and 2-octanone with the polysaccharide matrix (Burova *et al.* 1999). Vanillin interacts with BSA more strongly than with caseinate, by hydrogen bonding and hydrophobic interactions, hydrophobic interactions appearing more important (Chobpattana *et al.* 2002).

The role of small ligands in relation to thermal and pH conditions in denaturation process of BSA has been investigated. In this way, interactions of 2-octanone and vanillin (one aliphatic and one aromatic odorant) have been studied using native, acidic and thermally denatured BSA. The obtained experimental data seem to reflect a reversible plasticisation of the core of the protein by flavour ligands. The plasticised globule state seems to be quite similar to the molten globule state of proteins and the main difference between the native protein molecule and the molten globule state appears to be due to the disruption of the hydrophobic interactions while the plasticised globule state corresponds to the dilution of the hydrophobic core by a lipophilic plasticiser (Burova *et al.* 2003). The role of both hydrophobic and electrostatic interactions was underlined by a study of the binding of 7,8-dihydroxycoumarin (Daphnetin) to BSA (Liu *et al.* 2004).

Caseins

Casein is the major milk protein. Its primary structure is known for most of the genetic variants of α_{s1}, α_{s2}, β and κ-caseins (169 to 209 residues) (Mercier *et al.* 1971, 1972, Brignon *et al.* 1972, 1977, Grosclaude *et al.* 1973, Ribadeau Dumas *et al.* 1972). All caseins exhibit a similar amphiphilic character, the polypeptidic chain folded with the non-polar side chains buried in the interior (Wong *et al.* 1996, and cited references).

The caseins of bovine milk are one class of non-crystallisable proteins, and tertiary structure derived from X-ray crystallography could not be established (Swaisgood 1993). However, sequence-based predictions of secondary structures have been predicted by adjusting to global secondary structure determined by spectral studies. In this way, the secondary structural assignments that had been reconciled with Raman spectroscopic data were used as a starting point to generate a three-dimensional model of α_{s1}-casein (Kumosinski *et al.* 1991a), β-casein (Kumosinski *et al.* 1993a) and κ-casein (Kumosinski *et al.* 1991b, 1993b). The predicted structure of α_{s1}-casein contains a hydrophobic and a hydrophilic domain, which are connected by a segment of α-helix. The β-casein structure refined using minimisation techniques showed a slackly packed asymmetrical structure, with hydrophobic side chains uniformly dispersed over the C terminal and the centre surface of the structure, whereas the N terminal was hydrophilic (Kumosinski *et al.* 1993a). An initial model was first constructed for κ-casein, then refined using energy minimisation techniques and both the initial (Kumosinski *et al.* 1991b) and refined (Kumosinski *et al.* 1993c) models include two sets of anti-parallel β-sheet structure containing predominantly hydrophobic chains.

There are two fundamental functions of caseins: the effective transport of Ca^{2+} and the self-associations that lead to the colloid state (Farrell *et al.* 2002b). The caseins are poor ordered proteins and more flexible than the typical globular proteins (Swaisgood 1993). In this way, caseins present most of the properties of the molten globule state, such as a somewhat compact structure, significant amount of secondary structure, but little tertiary folds, a high degree of hydratation and side chain flexibility (Farrell *et al.* 2002b, Qi *et al.* 2001a). The association of bovine casein with small hydrophobic molecules involves an inclusion mechanism within the hydrophobic interior of micelle-like protein associates.

9.2.3 Proteins of leguminous plants

Proteins of leguminous plants are one of the most promising matters for formulating new forms of food products (Garcia *et al.* 1997a). These proteins are isolated from soy, pea, and beans and may be used as functional additives. Proteins of leguminous plants are related to two major components, vicilin and legumin. Despite different quaternary structures, they seem to be related to a common ancestor because of partial homologies in their primary structure and of a very similar tertiary structure of each subunit (Lawrence *et al.* 1994, Shutov *et al.* 1995). Vicilins have a trimeric structure, common to Canavalin (*Canavalia ensiformis*, monomer 181 amino acid residues (Ko *et al.* 1993, 2000)), Phaseolin (*Phaseolus vulgaris*, trimer 397 amino acid residues (Lawrence *et al.* 1990, 1994) and β-conglycinin (*Glycine max*, trimer 416 amino acid residues), (Maruyama *et al.* 2001). Legumins present a hexameric structure and are found in pea, faba pea, lupin, and especially in soybean.

Soy proteins

Soybean proteins are among the most attractive plant food proteins for humans, as they exhibit a hypocholesterolemic effect and have good nutritional and physicochemical properties (such as gel-forming and emulsifying abilities required for food systems). Soybean proteins are composed of two major components, β-conglycinin 7S (vicilin class) and glycinin 11S (legumin class), which account for about 30% and 40% of the total seed proteins, respectively. These two components are responsible for the nutritional, physicochemical and physiological properties of soybean proteins (Garcia *et al.* 1997b, Maruyama *et al.* 2001, Tumer *et al.* 1981).

The glycinin 11S is more present in plant seeds than the β-conglycinin 7S. The quaternary structure of the glycinin 11S is a hexameric protein related to the legumins. Five major subunits of glycinin have been identified and classified into two groups according to their amino acid sequences (Coates *et al.* 1985, Nielsen *et al.* 1989). Each subunit is composed of an acidic polypeptide (32 kDa) and a basic polypeptide (20 kDa). The determination of crystal structure of soybean 11S legumin showed a hexamer formed by face-to-face stacking of two trimers (Fig. 9.3) (Adachi *et al.* 2001, 2003). The core of the protomer consists of two jelly-roll β-barrels and two extended helix domains (Fig. 9.4).

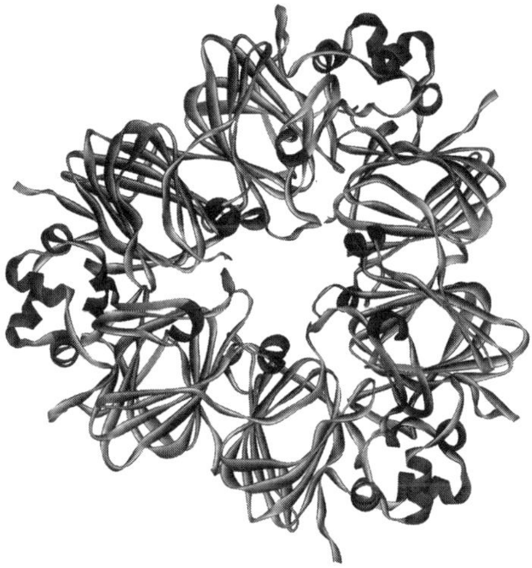

Fig. 9.3 Ribbon diagram of the proglycinin homotrimer from PDB-1FXZ structure (Adachi *et al.* 2001).

The role of the protein structure in binding and release of aroma was investigated in the case of hexyl acetate and revealed that the native legumins molecules possess the higher binding affinity for this compound. Denaturating proteins by heat results in intensive aggregation of protein basic chains and causes an increase in the number of binding sites altogether with a decrease in the value of intrinsic affinity constant. The acid denaturation (pH 3) dramatically alters the native protein structure, producing a loss for hexyl acetate bound to the protein. The quaternary structure that forms a hollow hydrophobic cylinder shows high capacity to bind hydrophobic molecules (Semenova *et al.* 2002a,

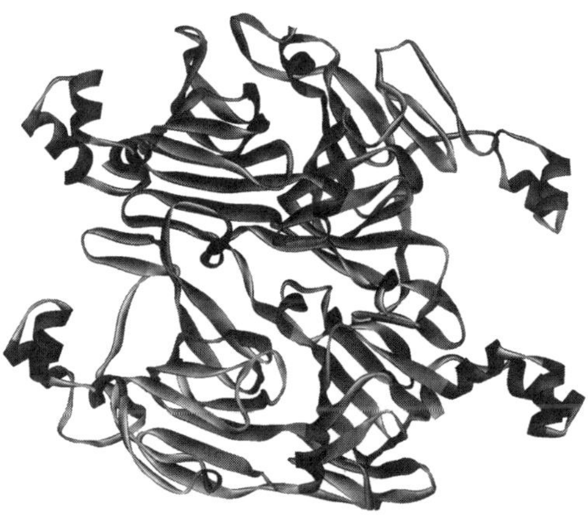

Fig. 9.4 Ribbon diagram of the core protomer model glycinin 11S from PDB-1OD5 structure (Adachi *et al.* 2003).

2002b). Competitive binding of aroma compounds with 11S glycinin suggested that some structures of aroma compounds are more suitable than others for the binding to the protein (Semenova *et al.* 2002c).

9.3 Nature and strength of the interactions

9.3.1 Flavour binding: reversible and irreversible interactions

In the most cases the interactions between protein and aroma compounds are reversible interactions that involve hydrophobic and hydrogen bonding (Guichard 2002, Lubbers *et al.* 1998, McClements 2002). Flavour compounds are able to establish reversible binding as well as covalent irreversible binding. Aldehyde groups can bind covalently the amino groups of proteins via a Shiff base formation. However, vanillin binds to milk proteins through hydrophobic interactions (Chobpattana *et al.* 2002, Grinberg *et al.* 2002, Hansen and Heinis 1991, 1992, McNeill and Schmidt 1993, Mikheeva *et al.* 1998, Ng *et al.* 1989, Relkin and Vermersch 2001) and aliphatic aldehydes (Meynier *et al.* 2004) and volatile disulfides (Adams *et al.* 2001) are connected by covalent irreversible bonds to proteins.

Reversible and weak bonding seems to presume a non-specific hydrophobic nature of interaction. For example, 2-octanone is bound either by the native form and denatured form of bovine serum albumin, whereas the unfolded form of ovalbumin displays significant binding affinity to vanillin (Burova *et al.* 2003; Mikheeva *et al.* 1998; Yang *et al.* 2002, 2003). In this way polar groups such as carbonyl seem to play only a minor role (Burova *et al.* 1999, and cited references; Morrisett *et al.* 1975; Spector 1975). However, some interactions could be related to specific interactions, as in the case of vanillin which does not bind to the denatured form of bovine serum albumin (Burova *et al.*, 2003).

In the case of well structured proteins such as β-lactoglobulin, the role of specific interaction involving polar groups seems to be significant. At least three distinct sites exist on β-lactoglobulin with two main ones being the retinol site (in centre of the β-barrel) and on the outer surface close to the helix. Examination of this region reveals that it is a relatively hydrophobic part of the surface in the vicinity of significant possibilities of hydrogene bond (Sawyer *et al.* 1998). Some studies enhanced the specificity of binding of small ligands to β-lactoglobulin and strongly suggest the implication of polar interactions (Cho *et al.* 1994, Dufour and Haertle 1990b, Guichard and Langourieux 2000, Kontopidis *et al.* 2002, Lübke *et al.* 2002, Narayan *et al.* 1997, Qin *et al.* 1998, Wu *et al.* 1999). In the same way, it has been demonstrated that within one chemical class, affinity for β-lactoglobulin increases with hydrophobic chain length or log P values, except for the terpens (Reiners *et al.* 2000). Indeed, three-dimensional quantitative structure activity relationships (3-D QSAR) molecular modelling studies have shown the existence of two groups of ligands, confirming the presence of at least two binding sites on the β-lactoglobulin and put forward the role of hydrogen bonding (Tromelin and Guichard 2003, 2004).

9.3.2 Binding constants determinations

The determination of binding constants should be carried out according to several physicochemical approaches, taking into account only reversible interaction or both reversible and irreversible binding (Guichard and Langourieux 2000, O'Neill 1996, and cited references; Sostmann *et al.* 1997).

Liquid phase analysis approaches

- Dynamic coupled column liquid chromatography (DCCLC) is based on generating saturated solutions by pumping water through a column packed with glass beads coated with the compound to be measured and was initially introduced for the study of polycyclic aromatic hydrocarbon solubility (May *et al.* 1978). DCCLC was successfully used for the determination of complex formation constants between aroma compounds and β-lactoglobulin (Jouenne and Crouzet 1997a, 2000, Langourieux 1995).
- Affinity chromatography is a technique enabling purification of a biomolecule with respect to biological function or individual chemical structure. Using immobilised β-lactoglobulin it leads to the determination of binding constants of aroma compounds due to a delayed retention time. Moreover, it allows the study of competitive effects with several flavour compounds in solution (Pelletier *et al.* 1998, Reiners *et al.* 2000, Sostmann and Guichard 1998).
- Exclusion chromatography separates molecules on the basis of size. The large molecules move more rapidly through the column, and in this way the molecular weight can be calculated from elution time. The Hummel and Dryer application of this method allows the calculation of binding parameters for reversible interactions between macromolecules and flavour compounds (Dumont 1987, Guichard and Etiévant 1998).
- Equilibrium dialysis technique is a simple effective tool for the study of low affinity interactions between molecules. In a standard equilibrium dialysis assay two chambers are separated by a dialysis membrane. At equilibrium, the concentration of free ligand in solution is the same in both chambers. In the receptor chamber, however, the overall concentration is higher due to the bound ligand component. The concentration of free ligand in the ligand chamber can then be used to determine the binding characteristics (Jasinski and Kilara 1985, Muresan *et al.* 2001, O'Neill and Kinsella 1987).

DCCLC and affinity chromatography allow calculation of global affinity constants for reversible molecular interactions. Stoichiometry of complexes can be found by exclusion chromatography and equilibrium dialysis, and in this way the number of binding sites is determined using Scatchard plot or Hill plot. However, exclusion chromatography takes into account only reversible interactions, whereas equilibrium dialysis does not distinguish between reversible and irreversible interactions.

Headspace analysis

Headspace analysis allows the determination of the partition coefficient which is currently defined as the ratio of the solute concentration in the vapour phase to its liquid phase concentration (Nawar 1966, 1971, Saleeb and Pickup 1978). It is a global measurement, that takes into account both reversible and irreversible interaction. This method is largely employed, but a limitation is the difficulty of detecting compounds with low volatility. However, it can constitute a means of assessing the effect of non-volatile food components on volatile flavour compounds (King and Solms 1983). The static headspace method was proposed to determine Henry's Law constants (Robbins *et al.* 1993), and more recently a mathematical model was proposed (Harrison and Hills 1997) and applied to describe flavour release from aqueous solutions containing flavour-binding polymers (Andriot *et al.* 2000).

9.3.3 Structural changes and binding sites localisation

Structural changes involved on proteins could be investigated by calorimetric approaches and spectroscopic techniques.

Differential scanning calorimetry

Differential scanning calorimetry (DSC) is a calorimetric technique used for measuring the energy necessary to maintain a constant temperature for a studied substance and an inert reference material. It has been widely employed in studying thermally induced structural and phase transitions in biomolecular systems and constitutes a good tool to protein–ligand study (Miles 1994). In this way, conformational changes in food protein in relation to binding to small molecules have been investigated successfully by calorimetric studies such as DSC to account transitions between native and denatured proteins (Aouzelleg *et al.* 2004b, Burova *et al.* 1999, 2003, Chikenji and Kikuchi 2000, Grinberg *et al.* 2002, Stevenson *et al.* 1996).

Fourier transform infrared

Fourier transform infrared (FT-IR) or Raman spectroscopy are one of the most powerful techniques for determining the secondary structure of globular proteins in aqueous solutions (Arrondo *et al.* 1993, Surewicz *et al.* 1993) and are also often employed as investigation tools for monitoring the nature of changes in the conformation of the proteins (Fang and Dalgleish 1997, Farrell *et al.* 2002a, Le Quéré *et al.* 1999, Lefèvre and Subirade 1999, Lübke *et al.* 1999, Nonaka *et al.* 1993, Qi *et al.* 1997, 2001b, Subirade *et al.* 1998, Tian *et al.* 2004). It is a quick approach which allows identification of the types of secondary structure (α-helix or β-sheet) involved in structural changes, but the resolution is too low to determine the precise localisation of these changes.

Circular dichroism spectroscopy

Circular dichroism spectroscopy (CDS) measures differences in the absorption of left-handed polarised light versus right-handed polarised light which arise from structural asymmetry of the molecules observed (for example proteins). The absence of regular structure results in zero CD intensity, while an ordered structure results in a spectrum which may show both positive and negative signal deviations (depending on the observation wavelength). Circular dichroism spectroscopy is an important tool in the structural determinants of proteins and is particularly good for comparing the structures of a protein (Adler *et al.* 1973; Greenfield and Fasman 1969). The real power of CDS is in the analysis of structural changes in a protein submitted to perturbation, or in the comparison of the structure of an engineered protein to the parent protein, particularly changes induced in manufacturing processes or formulation, and studying small conformational changes induced by receptor/ligand complex formation (Clark and Smith 1989, Farrell *et al.* 2001, Zsila 2003).

Fluorescence

Fluorescence is the phenomenon by which absorption of light at a given wavelength by a fluorescent molecule is followed by the emission of light at longer wavelengths. The distribution of wavelength-dependent intensity that causes fluorescence is known as the fluorescence excitation spectrum, and the distribution of wavelength-dependent intensity of emitted energy is known as the fluorescence emission spectrum. Proteins contain three aromatic amino acid residues (tryptophan, tyrosine, and phenylalanine) which may contribute to their intrinsic fluorescence. Tryptophan (as a free acid) demonstrates stronger fluorescence and higher quantum yield than the other two aromatic amino acids (Fasman *et al.* 1966, Teale 1960). The fluorescence of the aromatic residues varies in somewhat unpredictable manner in various proteins. Comparing with the unfolded state of the protein, the quantum yield may be either increased or decreased by the folding (signal quenching could be caused by both NH_2 and COOH). Accordingly, a folded protein can have either greater or less fluorescence than the unfolded form. The intensity of fluorescence is not very informative in itself. The magnitude of intensity, however, can serve as a probe of perturbations of the folded state. The wavelength of the emitted light is a better indication of the environment of the fluorophore. Tryptophan residues that are exposed to water, have maximal fluorescence at a wavelength of about 340–350 nm, whereas totally buried residues emit fluorescence at about 330 nm.

In addition to structural changes also provided by CD and FT-IR, fluorescence could provide an estimation of binding of aroma to food proteins (Damodaran and Kinsella 1980b, 1981b, Dufour and Haertle 1990b, Dufour *et al.* 1992, Frapin *et al.* 1993, Liu *et al.* 2004, Yang *et al.* 2003), but comparison with equilibrium dialysis often showed an overestimation of the binding constant (Guichard 2002, Muresan *et al.* 2001).

Fluorescence, CD and FT-IR, are rapid tools, but for more detailed structural analysis it requires complementary techniques like X-ray crystallography or NMR, which are more complex and time-consuming.

X-ray crystallography

The most common experimental method for obtaining a detailed picture of a protein or a protein–ligand complex is to interpret the diffraction of X-rays from many identical molecules in an ordered array commonly referred to as a crystal (Hodgkin 1950). First, the molecule must be crystallised, and the crystals must be of perfect quality. Protein crystallisation is a critical step, and X-ray crystallographic data fail for any proteins because some are non-crystallisable, for example, caseins (Swaisgood 1993). When X-rays strike a protein crystal they are scattered (diffracted) by individual atoms. Using sophisticated mathematical techniques, an X-ray diffraction pattern can be used to work out the three-dimensional protein structures producing very accurate structural models and even showing how proteins interact with other molecules (Adachi *et al.* 2001, Ko *et al.* 1993, 2000, Lawrence *et al.* 1990, 1994, Monaco *et al.* 1987, Papiz *et al.* 1986, Wu *et al.* 1999).

Nuclear magnetic resonance NMR

Like X-ray crystallography, the main application of nuclear magnetic resonance (NMR) spectroscopy in biology is to determine the structure of proteins (Wüthrich 1994). NMR determines structures of proteins in solution, but is limited to molecules smaller than 30 kD. NMR is the method of choice for small proteins which are not readily crystallised, and yields the relative positions of hydrogen atoms. The results of an NMR analysis are an ensemble of alternative models, in contrast to the unique model obtained by crystallography. In food proteins field, NMR techniques, are used for protein structure determination (Kuwata *et al.* 1999, Uhrinova *et al.* 2000), for binding site localisation (Lübke *et al.* 2002), and for the evaluation of concentration equilibrium constant using nuclear Overhauser effects (Jung *et al.* 2000, 2002, Jung and Ebeler 2003).

In addition to these experimental investigation approaches, computational methods could be applied in proteins interaction studies.

Computational methods: molecular modelling

Molecular modelling should be used in different ways.

In the field of ligand–protein interaction, structural informations on target proteins are used in molecular modelling and binding simulation studies (Colmenarejo *et al.* 2001, Mason *et al.* 2004, Perola and Charifson 2004). However, the three-dimensional structure must be determined either by X-ray crystallography or NMR, and the localisation of the binding site may be known. If lack of information occurs, QSAR (quantitative structure activity relationships) or QSPR (quantitative structure property relationships) ligand approaches are suitable ways to improve the comprehension of interaction between aroma ligands and food proteins (Guth and Fritzler 2004, Marabotti *et al.* 2000, Tromelin and Guichard 2003, 2004).

Molecular modelling can also be used to propose a structural conformation for a protein lacking direct structural data (Kumosinski *et al.* 1991a, 1991b, 1993a, 1993b). In the case of unavailability of the three-dimensional structures of a target protein, homology models can be built if suitable target proteins with high sequence identity are available (Kister *et al.* 2002, Schmitt *et al.* 2002).

9.4 Effect of medium on protein–flavour interactions

Proteins are used in the food industries as substitutes for lipids. But the addition of proteins produces different effects in function of the medium and the matrices. Different sorts of proteins are used: milk proteins (β-lactoglobulin, bovine serum albumin, casein), egg albumin, gelatin and soy proteins.

9.4.1 pH

Some authors studied the interactions between protein in solutions buffered at different pH, and flavour compounds. Interactions of selected flavour compounds with whey protein were investigated (Mills and Solms 1984). They found that the binding of heptanal by whey protein increased between pH 4.66 and pH 6.89, whereas for 2-nonanone, the reverse phenomenon occurred. Damodaran and Kinsella (1980a) showed that the affinity constant between bovine serum albumin and 2-nonanone increased in function of pH between pH 3 and pH 7, and decreased at pH 9.

Jouenne and Crouzet (2000) studied, by headspace analysis and exponential dilution, the effect of pH on the retention of aroma compounds by β-lactoglobulin. Four values of pH (3, 6, 9, 11) and compounds from different chemical classes were studied: 3 methyl ketones (C7–C9), 4 esters (C6–C9), limonene and myrcene. For limonene and myrcene, a 'salting out effect' was noticed for acid pH. The retention of methylketones and esters increased when the pH increased between 3 and 9, whereas at pH 11, the retention decreased dramatically. However, the global variation of retention of terpens was the same as those observed for methyl ketones and ethyl esters. For a constant pH value, the retention of methyl ketones and esters with a maximum for ethyl octanoate increased with the increasing length of the aliphatic chains. These authors interpreted the results obtained by the modification of the structure of the protein. In fact, between pH 3 and pH 9, the flexibility modification of the protein, allowing a better accessibility to the primary or the secondary hydrophobic sites, could explain the increasing retention of these compounds. At pH 11, the decrease of aroma retention was the consequence of the alkaline denaturation of β-lactoglobulin. More recently, van Ruth and Villeneuve (2002) completed the study of Jouenne and Crouzet (2000). Van Ruth and Villeneuve (2002) made some experiments by static headspace analysis to show the impact of the pH (pH 3, 6, 9), at different protein concentrations (0; 0.5; 0.7; 1; 2%) on the air/liquid partition coefficients of 20 compounds varying in functional group and chain length. The compounds tested were: 6 alcohols, 4 esters, 5 ketones, 3 aldehydes, and 2 others. The presence of β-lactoglobulin significantly changes the partition coefficients of 17 on the 20 aroma compounds, except for 3 alcohols (1-propanol, 1-butanol, and 1-nonanol). The increase of pH decreases the partition coefficient, of most of the compounds studied except α-pinene. Van Ruth and Villeneuve concluded that the change of retention of most of the aroma compounds with pH indicated the importance of the conformation of the protein for the exposure of binding sites. The behaviour of α-pinene is opposite with a higher binding affinity at pH 3. The authors supposed that its binding implies a binding site located at the surface rather than in the hydrophobic pocket of the protein.

Moreover, the presence of flavour compounds in mixture with β-lactoglobulin (2%, pH 6) induces an increase of their partition coefficients (van Ruth and Villeneuve 2002). It is the case when propyl acetate, ethyl butyrate, butyl acetate were added to any of the three aldehydes tested. But these effects are only related to the presence of β-lactoglobulin, because no effect of the flavour compounds on each other in water could be detected.

Weel *et al.* (2003) have also studied the effect of pH (3, 5, 6, 7, 9) on the retention of aldehydes (C4, C6, C8) by whey protein (3%), by static headspace analysis. An increasing retention is found with increasing length of aldehydes carbon chain altogether with a higher retention at pH 9.

Thus, all these results showed considerable effects of protein concentration, pH, chemical classes and chain length on the protein–flavour interactions.

9.4.2 Temperature

Beyeler and Solms (1974) found that the binding constants obtained with some flavour compounds for BSA or soy protein did not vary for a temperature ranging between 25 and 45 °C. Moreover, the interactions between 2-nonanone and soy proteins did not change for a temperature between 25 and 45 °C (Damodaran and Kinsella 1981b). Druaux *et al.* (1995) found the same results for the interactions between BSA and γ-decalactone: these interactions are independent of the temperature between 10 and 30 °C. However, the affinity constant of 2-nonanone for soy protein increased considerably between 5 °C (2000 M^{-1}) and 25 or 45 °C (930 M^{-1}) and the number of binding sites decreased (two sites at 5 °C instead of four at 25 °C). These variations of thermodynamic parameters are due to structural changes of the protein: low temperatures induce a precipitation of soy protein (Damodaran and Kinsella 1981b).

In conclusion, the temperature seems to have a real effect on binding properties, but only when the structure of protein is modified. This point is particularly true for heat denaturation.

9.4.3 Thermal denaturation of protein

O'Neill and Kinsella (1988) showed that a heat treatment of β-lactoglobulin at 75 °C during 10 and 20 minutes undergoes a conformational change. This conformational change induces a modification of the interactions between the protein and 2-nonanone: the affinity constant decreases in function of temperature, whereas the number of binding sites increases. The heat treatment induces the aggregation of the protein, which modifies the nature of the binding resulting in a decrease in the association constant and an increase in the number of binding sites.

Conversely, an increase in retention by whey protein when increasing temperature was observed (Hansen and Booker 1996). The heat treatment induces the aggregation of the protein, which modifies the nature of the binding site resulting in a decrease in the association constant and an increase in the number of binding sites.

Moreover, the partial denaturation of soy protein (90 °C, 1h, pH 8, Tris-HCl buffer) increases the affinity constant (1200 M^{-1} instead of 930 M^{-1} for native protein), but the number of binding sites for 2-nonanone in soy protein did not change. This partial denaturation did not alter the oligomeric nature of soy protein, but the quaternary structure of soy protein may undergo some reorganisation of the subunits which may enhance the hydrophobicity of the previously existing sites (Damodaran and Kinsella 1981b).

Andriot *et al.* (1999) studied the interactions between benzaldehyde and β-lactoglobulin dissolved in different media (water solution, or in presence of NaCl or ethanol) in function of heat treatment. In water solution, a heat treatment had no effect on the retention of benzaldehyde whereas in the presence of NaCl or ethanol, retention of benzaldehyde decreased. This could be due to an aggregation of β-lactoglobulin in the presence of NaCl or ethanol.

9.4.4 Salt

The addition of salts modifies the ionic force of the medium, and may modify the interactions between proteins and flavour compounds. Damodaran and Kinsella (1980a) studied the interactions between bovine serum albumin and 2-nonanone in function of an increasing salt concentration. The quantity of 2-nonanone bound to BSA increases with the concentration of sodium sulphate, sodium chlorure, sodium bromure and sodium perchlorate. The efficiency of the anions to increase the fixation capacity follows the Hofmeister series, i.e. $F^- > SO_4^{2-} > Cl^- > Br^- > SCN^- > Cl_3CCO^-$. The anions tend to stabilise or destabilise the protein structure, modifying the conformation of the protein. The stabilising or destabilising effects of anions on proteins has been attributed to their ability to bind the dipolar groups of the macromolecules, and their effect on the structure of water which in turn affect the intrachain hydrophobic interactions and hence the equilibrium structure of protein.

Moreover, Jouenne *et al.* showed that the retention of 2-octanone by β-lactoglobulin increased with increasing sodium chloride concentration (0 to 1 M NaCl) (Jouenne 1997, Jouenne and Crouzet 1997b). For limonene, a salting-out effect was observed at pH 3 for low ionic strength (<0.25 M). When the NaCl concentration increased from this value (0.25 M) to 0.5, and 1 M, retention increased: 3% for 0.5 M and 30% for 1 M. The authors supposed that the decrease of limonene retention between 0 and 0.25 M may be the consequence of NaCl fixation in the proximity of its binding site. The different behaviour of 2-octanone may be explained by the presence of two different binding sites, or by the lower sensibility of 2-octanone to the ionic environment, near the binding site. On the other hand, Andriot *et al.* (1999) found that the amount of benzaldehyde bound by β-lactoglobulin (3%) decreased with the addition of NaCl (50 mM) to the water solution, i.e. the percentage of bound benzaldehyde decreased from 25% in water solution to 18% in NaCl (50 mM).

9.4.5 Co-solvent

The presence of chemical agents such as urea can also influence the interactions. Binding of 2-nonanone to soy proteins decreases when urea concentration decreases (Damodaran and Kinsella 1981a). Urea denaturates the protein, thus inducing an increase of the number of binding sites, but with a lower global affinity for 2-nonanone (Damodaran and Kinsella 1981a). Moreover, dithiothreitol reduces the disulfide bridges of bovine serum albumin, and modifies the binding of 2-nonanone (Damodaran and Kinsella 1980b).

Dufour and Haertle (1990a) studied by circular dichroism and fluorescence the interactions between β-lactoglobulin and retinol. In the presence of 20% ethanol (v/v), one molecule of β-lactoglobulin binds 2 molecules of retinol, whereas in aqueous solution, one molecule of β-lactoglobulin binds 1 molecule of retinol. In these conditions, circular dichroism shows that the native secondary structure of β-lactoglobulin is unchanged.

However, Andriot *et al.* (1999) showed no effect of 13% ethanol on bound benzaldehyde in β-lactoglobulin 3% dispersed in NaCl 50 mM solution.

9.4.6 Foaming

Phillips *et al.* (1990) have investigated the pH effect on foaming properties of whey protein. They found that the maximum foaming stability is obtained at pH 5, near the iso-electric point. Moreover, at pH 5, a heat treatment (80 °C, 10 minutes) induced an increase in the foaming stability of 65%, whereas at pH 4 or pH 7, the heat treatment had no effect on the foaming stability. The addition of low concentration of ethanol increased the foaming stability of β-lactoglobulin at pH 7 (Ahmed and Dickinson (1991) cited by (Marin 1998)). The effect of pressure was also studied on the foaming properties of β-lactoglobulin (Pittia *et al.* 1996). Its structure was modified if pressure was settled between 300 and 900 MPa, thus reducing the emulsifying capacity and foamability compared to native β-lactoglobulin.

Only a few studies were focused on the effects of flavour compounds on foaming properties of β-lactoglobulin. Marin and Relkin (1999) studied the impact of isoamyl acetate addition on foaming properties of β-lactoglobulin, through a conductimetric method and showed an increase of density and stability of the foam formed by β-lactoglobulin. The authors proposed that this enhancement could be explained by the formation of a complex between β-lactoglobulin and isoamyl acetate. A similar effect of benzaldehyde on the foaming properties of β-lactoglobulin was also demonstrated by the same authors (Marin and Relkin 2000). Thus, the addition of flavour compounds interacting with β-lactoglobulin can enhance the foaming properties of β-lactoglobulin.

9.4.7 Impact on protein functionality

A high correlation between flavour binding for several ketones and acetates versus the surface hydrophobicity of proteins was established (Solms and Guggenbuehl 1991).

Seuvre *et al.* (2001) investigated the retention of aroma compounds by β-lactoglobulin in different conditions, such as the hydration of protein and different water activities on the sorption of aroma compounds. When the protein has a low degree of hydration, the absorbed quantities of 2-nonanone and d-linalol were very low. In this condition, for low water content, the sorption was not affected by water activity (between 0.11 and 0.43 water activity values). On the other hand, when the protein is in solution (water activity values closed to 1), the sorption of these two compounds drastically increased. In other words, at low water content, the sorption of the two compounds was in the order of 10 mg per 100 g of dry protein, whereas in solution the sorption reached 400 mg of aroma compound per 100 g of protein. These results illustrate the role of water on interactions between flavour compounds and β-lactoglobulin. The authors interpreted the results by a structural modification of the protein: in fact, at low water content, the protein was more rigid. Moreover, in order to describe the role of β-lactoglobulin on the aroma compound transfer at the air–water interface, Seuvre *et al.* (2001) also determined the sorption kinetic of four flavour compounds in aqueous solutions with and without protein. The presence of protein modified the aroma sorption behaviour: in fact, the quantities of absorbed aroma were higher in the presence of protein.

9.5 Perceptive consequences in food systems: some examples

9.5.1 Impact of interactions on flavour perception

The flavour intensity of benzaldehyde, limonene and citral were determined in the presence of casein and whey protein by quantitative descriptive analysis deviation from reference (Hansen and Heinis 1992). In the presence of whey protein, the benzaldehyde flavour intensity declined, on the other hand, the casein had no effect on benzaldehyde flavour intensity. For limonene, the flavour intensity decreased when the protein concentration (whey protein or casein) increased. For citral, panellists detected no effect on flavour intensity in presence of whey protein or casein. The authors proposed the following hypothesis: the decrease of benzaldehyde and limonene flavour intensity in the presence of whey protein or casein may be due to non-polar interaction for casein and interaction with non-polar binding sites, cysteine-aldehyde condensation, or Schiff base formation with whey protein. The effect of sodium caseinate or whey concentrate on vanillin flavour intensity was also tested using the same method (Hansen and Heinis 1991). The vanillin flavour intensity was moderately lower than the reference for three concentrations of sodium caseinate tested and for two concentrations of whey protein. The authors supposed that this decrease of vanillin flavour intensity was probably due to the cysteine-aldehyde condensation or Schiff base formation. On the other hand, sensory analysis applying a matching test showed that the addition of β-lactoglobulin had no effect on the odour perception of vanillin, but brought about a significant decrease in the flavour perception of eugenol (Reiners *et al.* 2000). These results are interpreted with the different affinity of the two flavour compounds for β-lactoglobulin: the affinity constant of vanillin is lower than that of eugenol. Moreover, addition of β-lactoglobulin (0.5% and 1%) to aqueous solutions of three methyl ketones (2-heptanone, 2-octanone, 2-nonanone) induces a significant decrease in odour intensity (Andriot *et al.* 2000). Recently, the development of a new technique, APCI-MS (atmospheric pressure chemical ionisation-mass spectrometry), allows the study of the protein–flavour interactions in *in vivo* conditions. Thus, the interactions between flavour compounds and milk protein under static (headspace analysis) and dynamic (APCI-MS) conditions were described by Le Guen and Vreeker (2003). First, they studied the interactions between 2-alkenals (C3 to C9), methyl ketones (C3 to C9) and whey protein in *in vitro* conditions by headspace, and after that, in *in vivo* conditions by APCI-MS, to monitor the release of aroma during eating. Headspace analysis indicated that all alkenals were highly retained by milk protein (between 76 and 100% retention). A formation of Schiff base between some amino acids in the milk protein and the carbonyl group of the alkenal is proposed. By APCI-MS measurement, no flavour compound could be detected in the breath in the presence of proteins, confirming by APCI-MS the results obtained by headspace analysis. These results can be explained if the interactions between alkenals and milk protein are irreversible on the time scale of the APCI-MS experiments. Conversely the APCI-MS signal of methyl ketones is not influenced by the presence of milk protein, whereas the headspace analysis shows a great retention of methylketones. The authors supposed that the results could be explained by saliva dilution, and the interactions between flavour and mouth tissue. Weel *et al.* (2003) have also reported some effect of whey protein on the *in vivo* release of aldehydes. They studied the interactions between three aldehydes (C4, C6, C8) and whey proteins at five different pH by headspace measurements and APCI-MS (*in vivo* and *in vitro*). By APCI-MS, a retention of these aldehydes by the protein at pH values of 5–9 was always observed. In contrast to the static headspace conditions, the effect of carbon chain length on the extent of retention was weaker. The aldehyde–protein interactions seem thus to be less significant under dynamic conditions than under static headspace conditions. In the protocol used in this study, the differences observed by the two methods were not due to saliva or mouth coating. In fact, the authors showed that the addition of artificial or human saliva had no effect on the aldehyde–whey protein interactions. Moreover, the effect of mouth coating was studied by comparing headspace concentrations above aldehyde solutions that were rinsed in the mouth with those that were not. The hypothesis of the authors for the *in vivo* aroma release process is that not only will all free aldehydes in the film be released into the exhaled air but also all of the aldehydes will be reversibly bound to the whey protein.

If irreversible interactions occurred, aroma would not be perceived at all during food consumption. Aldehydes tend to react chemically with protein amino groups, resulting in irreversible binding, as Schiff bases (Hansen and Heinis 1991, 1992). A covalent binding was thus observed with aldehydes and arginin present in soy and whey proteins (Plug and Haring 1994). Moreover, sulphur compounds may also interact with protein by irreversible binding. Parker *et al.* (2003) reported a study on the interaction of sulphur containing aroma compounds with proteins in model systems and real food systems. Other irreversible interactions of disulphides were also shown to occur with ovalbumin, β-lactoglobulin and lysosyme.

9.5.2 Fat mimetic

Actually, the demand of consumers for low fat products increases, pressing food industries to develop reduced-fat food. To maintain the properties brought by fat, such as appearance, mouthfeel, texture, juiciness, flavour and storage stability, fat replacers are added. Proteins are usually used as fat mimetics. Lucca and Tepper (1994) reported the functionality of fat replacers in food and described the advantages and limitations of three classes of fat replacers, i.e. proteins-, carbohydrates- and lipid-based ingredients. Protein-based fat mimetics are generally used in frozen dessert, salad dressing, margarine, dairy products and their functional properties are mouthfeel, creaminess and viscosity (Akoh 1998). McClements and Demetriades (1998) wrote a review focused on the role of fat mimetics in food emulsions. They reported the effects of fat replacers on stability, texture, appearance and flavour of food emulsions. They concluded that the initial flavour of an emulsion is altered when the fat content is reduced. But the addition of fat replacers can modify the flavour of emulsions. Ma *et al.* (1997) compared the effects of three types of fat mimetics on rheological characteristics of low-fat Cheddar cheeses, and the relation with the network structure. Low-fat cheeses made with a carbohydrate-based fat mimetics have a network structure more similar to full-fat cheese than the low-fat control or sample made with protein-based fat mimetics. Haque and Aryana (2002) studied the effect of two fat replacers (protein and carbohydrates based) on flavour volatiles of low-fat Cheddar cheese during the first month of ripening at 7 °C. They showed that fat replacers influenced the formation of flavour compounds during initial maturation. Troy *et al.* (1999) showed that blends of the fat replacers can be used to offset the poor quality associated with low-fat beef burgers.

9.6 Conclusion and future trends

This review has pointed out the importance of protein–flavour interactions to understand flavour release and flavour perception.

There is an increasing knowledge on the interactions between flavour compounds and several well characterised food proteins, in simple model systems. However, data are missing on the molecular structures of food interacting with flavour compounds. The binding sites are often not well determined, even for β-lactoglobulin, one of the best characterised and studied milk proteins. Molecular modelling seems to be a useful tool to study molecular interactions, but only few studies were done in this field of research of 'flavour in food'. The first approaches involving structural data only allowed the visualisation of the binding sites. Next step could be a docking approach, but the traditional rigid methods are no more appropriate. New developments using flexible approaches by molecular dynamic simulations allow the prediction of a binding mode of a flexible ligand in a flexible binding pocket. These methods have not been used to our knowledge in the present field of research, but are widely used in pharmacology. Their application to study protein–flavour interactions will help to increase the understanding of flavour release in matrices containing food proteins.

The original approach based on modelling the aroma–protein interaction from the point of view of protein, but using informations only derived from the aroma compound appears to be appropriate to identify binding subsites. This will help in the understanding of the different behaviour or aroma compounds and give some indications on the possible binding sites, together with informations on the specificity of the binding sites. The advantage is that no information is needed on the structure of the protein, which means that this approach can be applied to any protein, if there exists a sufficiently large amount of data on aroma–protein binding constants.

Moreover, the approach based on the structural properties of the ligands could be widen to more complex systems and then QSAR or QSPR models could be established in real food systems, and not only to understand flavour–protein interactions but to try to understand aroma compound volatility in real food systems.

By using *in vivo* dynamic methods (APCI-MS), it was thus possible to point out the type of protein–flavour interactions which may have an impact on flavour perception. The development of these new methods will increase the understanding of the nature of the interactions which can be chemical or physiological.

Proteins are also a source of aroma generation during processing and these additional flavour compounds have also to be taken into account to understand the changes in flavour perception occurring by changing the formulation of protein ingredients.

9.7 Sources of further information

Introduction to theoretical organic chemistry and molecular modelling, William B. Smith (ed.) (1996) VCH Publishers, New York, USA.

This book provides an introduction for chemists with a limited mathematical background. It is written at the level of the advanced undergraduate or first-year graduate student in organic chemistry, whose exposure to theoretical chemistry is relatively recent. It covers molecular modelling computer software, and offers a useful guide to the scope and limitations of each program, along with specific examples of input and output for several of the most popular software.

http://www.accelrys.com/
Accelrys' modelling and simulation products let to create, view, and analyse graphical models of molecules from materials and chemicals, to small drug-like molecules, to DNA and macromolecular proteins.

http://www.tripos.com/
Tripos Discovery Software are dedicated to innovation and discovery in the field of molecular design and informatics.

http://www.jenner.ac.uk/lipocalin/
This webpage is designed to provide up-to-date information about the Lipocalin protein family, whose members are highly diverse and found in a variety of species. Lipocalins are typically small secreted proteins that are characterised by several features, the most prominent being their ability to bind small, hydrophobic molecules such as retinol. Others include their binding to specific cell-surface receptors and their ability to form macro-molecular complexes.

http://www.rcsb.org/pdb/
The PDB is the single worldwide repository for the processing and distribution of 3-D structure data of large molecules of proteins and nucleic acids. New structures are released each Wednesday by 1:00 a.m. Pacific time. Details about the history, function, progress, and future goals of the PDB can be found in the *PDB Annual Reports* and *PDB Newsletters*.

http://www.expasy.org/
The ExPASy (expert protein analysis system) proteomics server of the Swiss Institute of Bioinformatics (SIB) is dedicated to the analysis of protein sequences and structures as well as 2-D PAGE (Disclaimer/References).

http://folding.stanford.edu/results.html
Gives details about some protein folding simulations.

9.8 References

ADACHI M, TAKENAKA Y, GIDAMIS A B, MIKAMI B and UTSUMI S (2001) 'Crystal structure of soybean proglycinin alaB1b homotrimer', *J Mol Biol*, 305 (2), 291–305.

ADACHI M, KANAMORI J, MASUDA T, YAGASAKI K, KITAMURA K, MIKAMI B and UTSUMI S (2003) 'Crystal structure of soybean 11S globulin: Glycinin A3B4 homohexamer', *Proc Natl Acad Sci USA*, 100 (12), 7395–7400.

ADAMS R L, MOTTRAM D S, PARKER J K and BROWN H M (2001) 'Flavor-protein binding: Disulfide interchange reactions between ovalbumin and volatile disulfides', *J*

Agric Food Chem, 49 (9), 4333–4336.

ADLER A J, GREENFIELD N J and FASMAN G D (1973) 'Circular dichroism and optical rotatory dispersion of proteins and polypeptides', *Methods Enzymol*, 27, 675–735.

AHMED M and DICKINSON E (1991) Effects of ethanol on the foaming of food macromolecules. In *Food Polymers, Gels and Colloids*, Dickinson E (ed.), The Royal Society of Chemistry, Cambridge, 503–507.

AKERSTROM B, FLOWER D and SALIER J-S (2000) 'Lipocalins: unity in diversity', *Biochim Biophys Acta*, 1482, 1–8.

AKOH C C (1998) 'Fat replacers', *Food Technol*, 52 (3), 47–53.

ANDRIOT I, MARIN I, FERON G, RELKIN P and GUICHARD E (1999) 'Binding of benzaldehyde by β-lactoglobulin, by static headspace and high performance liquid chromatography in different physico-chemical conditions', *Lait*, 79 (6), 577–586.

ANDRIOT I, HARRISON M, FOURNIER N and GUICHARD E (2000) 'Interactions between methyl ketones and β-lactoglobulin: sensory analysis, headspace analysis, and mathematical modeling', *J Agric Food Chem*, 48 (9), 4246–4251.

ANEMA S G (2000) 'Effect of milk concentration on the irreversible thermal denaturation and disulfide aggregation of beta-lactoglobulin', *J Agric Food Chem*, 48, 4168–4175.

AOUZELLEG A, BULL L A, PRICE N C and KELLY S M (2004a) 'Molecular studies of pressure/temperature-induced structural changes in bovine beta-lactoglobulin', *J Sci Food Agric*, 84 (5), 398–404.

AOUZELLEG A and BULL L A (2004b) 'Differential scanning calorimetry study of pressure/temperature processed beta-lactoglobulin: The effect of dextran sulphate', *Food Res Int*, 37 (10), 933–940.

ARRONDO J L R, MUGA A, CASTRESANA J and GONI F M (1993) 'Quantitative Studies of the Structure of Proteins in Solution by Fourier-Transform Infrared-Spectroscopy', *Prog Biophys Mol Biol*, 59 (1), 23–56.

BATT C A, BRADY J and SAWYER L (1994), 'Design improvements of beta-lactoglobulin', *Trends Food Sci Technol*, 5 (8), 261–265.

BAUSSAY K, LE BON C, NICOLAI T, DURAND D and BUSNEL J P (2004) 'Influence of the ionic strength on the heat-induced aggregation of the globular protein beta-lactoglobulin at pH 7', *Int J Biol Macromol*, 34 (1–2), 21–28.

BELLOQUE J, LOPEZ-FANDINO R and SMITH G M (2000) 'A H-NMR study on the effect of high pressures on beta-lactoglobulin', *J Agric Food Chem*, 48, 3906–3912.

BERTRAND-HARB C, IVANOVA I V, DALGALARRONDO M and HAERTLE T (2003) 'Evolution of beta-lactoglobulin and alpha-lactalbumin content during yoghurt fermentation', *Int Dairy J*, 13 (1), 39–45.

BEYELER M and SOLMS J (1974) 'Interaction of flavor model compounds with soy protein and bovine serum albumin', *Lebensm Wiss Technol*, 7 (4), 217–219.

BOUDAUD N and DUMONT J P (1996) 'Interaction between flavor components and beta-lactoglobulin', *Flavor-Food Interactions*, 633, 90–97.

BRIGNON G, MERCIER J C, RIBADEAU-DUMAS B and DAS B C (1972) 'Primary structure of bovine kappa paracasein', *FEBS Lett*, 27 (2), 301–305.

BRIGNON G, RIBADEAU DUMAS B, MERCIER J C, PELISSIER J P and DAS B C (1977) 'Complete amino acid sequence of bovine alphaS2-casein', *FEBS Lett*, 76 (2), 274–279.

BROWNLOW S, MORAIS CABRAL J H, COOPER R, FLOWER D R, YEWDALL S J, POLIKARPOV I, NORTH A C and SAWYER L (1997) 'Bovine beta-lactoglobulin at 1.8 A resolution – still an enigmatic lipocalin', *Structure*, 5 (4), 481–495.

BUROVA T V, GRINBERG N V, GOLUBEVA I A, MASHKEVICH A Y, GRINBERG V Y and

TOLSTOGUZOV V B (1999) 'Flavour release in model bovine serum albumin/pectin/ 2-octanone systems', *Food Hydrocoll.*, 13 (1), 7–14.

BUROVA T V, GRINBERG N V, GRINBERG V Y and TOLSTOGUZOV V B (2003) 'Binding of odorants to individual proteins and their mixtures. Effects of protein denaturation and association. A plasticized globule state', *Colloid Surface A*, 213 (2–3), 235–244.

CARTER D C and HO J X (1994) 'Structure of serum albumin', *Adv Protein Chem*, 45, 153–203.

CHARLES M, BERNAL B and GUICHARD E (1996) Interactions of beta-lactoglobulin with flavour componds. In *8th Weurman Flavour Research Symposium*; Taylor A J and Mottram D S (eds), Royal Society of Chemistry, Cambridge, 433-436.

CHIKENJI G and KIKUCHI M (2000) 'What is the role of non-native intermediates of beta-lactoglobulin in protein folding?' *Proc Natl Acad Sci USA*, 97 (26), 14273–14277.

CHO Y, BATT C A and SAWYER L (1994) 'Probing the retinol-binding site of bovine beta-lactoglobulin', *J Biol Chem*, 269 (15), 1102–1107.

CHOBPATTANA W, JEON I J, SMITH J S and LOUGHIN T M (2002) 'Mechanisms of interaction between vanillin and milk proteins in model systems', *J Food Sci*, 67 (3), 973–977.

CHRYSINA E D, BREW K and ACHARYA K R (2000) 'Crystal structures of apo- and holo-bovine alpha-lactalbumin at 2.2-A resolution reveal an effect of calcium on inter-lobe interactions', *J. Biol. Chem.*, 275 (47), 37021–37029.

CLARK D C and SMITH L J (1989) 'Influence of alcohol-containing spreading solvents on the secondary structure of proteins: a circular dichroism investigation', *J Agric Food Chem*, 37 (3), 627–633.

COATES J B, MEDEIROS J S, THANH V H and NIELSEN N C (1985) 'Characterization of the subunits of beta-conglycinin', *Arch Biochem Biophys*, 243 (1), 184–194.

COLMENAREJO G, ALVAREZPEDRAGLIO A and LAVANDERA J L (2001) 'Cheminformatic models to predict binding affinities to human serum albumin', *J Med Chem*, 44 (25), 4370–4378.

CROGUENNEC T, BOUHALLAB S, MOLLE D, O'KENNEDY B T and MEHRA R (2003) 'Stable monomerie intermediate with exposed Cys-119 is formed during heat denaturation of beta-lactoglobulin', *Biochem Biophys Res Commun*, 301 (2), 465–471.

CROGUENNEC T, MOLLE D, MEHRA R and BOUHALLAB S (2004a) 'Spectroscopic characterization of heat-induced nonnative beta-lactoglobulin monomers', *Protein Sci*, 13 (5), 1340–1346.

CROGUENNEC T, O'KENNEDY B T and MEHRA R (2004b) 'Heat-induced denaturation/ aggregation of beta-lactoglobulin A and B: kinetics of the first intermediates formed', *Int Dairy J*, 14 (5), 399–409.

DAMODARAN S and KINSELLA J E (1980a) 'Stabilization of proteins by solvents. Effect of pH and anions on the positive copperativivity of 2-nonanone binding to bovine serum albumin', *J Biol Chem*, 255, 8503–8508.

DAMODARAN S and KINSELLA J E (1980b) 'Flavor protein interactions. Binding of carbonyls to bovine serum albumin: thermodynamic and conformational effects', *J Agric Food Chem*, 28 (3), 567–571.

DAMODARAN S and KINSELLA J E (1981a) 'Interaction of carbonyls with soy protein: conformational effect', *J Agric Food Chem*, 29 (6), 1253–1257.

DAMODARAN S and KINSELLA J E (1981b) 'Interaction of carbonyls with soy protein: thermodynamic effects', *J Agric Food Chem*, 29 (6), 1249–1253.

DIB R, CHOBERT J M, DALGALARRONDO M and HAERTLÉ T (1996) 'Secondary structure changes and peptic hydrolysis of beta-lactoglobulin induced by diols', *Biopolymers*, 39 (1), 23–30.

DRUAUX C, LUBBERS S, CHARPENTIER C and VOILLEY A (1995) 'Effects of physico-chemical parameters of a model wine on the binding of gamma-decalactone on bovine serum albumin', *Food Chem*, 53, 203–207.

DUFOUR E and HAERTLÉ T (1990a) 'Alcohol-induced changes of beta-lactoglobulin-retinol-binding stoichiometry', *Protein Eng.*, 4 (2), 185–190.

DUFOUR E and HAERTLÉ T (1990b) 'Binding affinities of beta-ionone and related flavor compounds to beta-lactoglobulin: effects of chemical modifications', *J. Agric. Food Chem.*, 38 (8), 1691–1695.

DUFOUR E, ROGER P and HAERTLÉ T (1992) 'Binding of Benzo(alpha)pyrene, Ellipticine, and Cis-Parinaric Acid to beta-Lactoglobulin: Influence of Protein Modifications', *J Prot Chem*, 11 (6), 645–652.

DUMONT J P (1987) Flavour-protein interactions: a key to aroma persistence. In *Flavour Science and Technology*, Martens M, Dalen G A, Russwurm H, Jr (eds), John Wiley, Chichester, 143–148.

EDWARDS P J B, JAMESON G B, PALMANO K P and CREAMER L K (2002) 'Heat-resistant structural features of bovine beta-lactoglobulin A revealed by NMR H/D exchange observations', *Int Dairy J*, 12 (4), 331–344.

FANG Y and DALGLEISH D G (1997) 'Conformation of beta-lactoglobulin studied by FTIR: effect of pH, temperature, and adsorption to the oil–water interface', *J Colloid Interface Sci*, 196 (2), 292–298.

FARRELL J, H. M., WICKHAM E D, UNRUH J J, QI P X and HOAGLAND P D (2001) 'Secondary structural studies of bovine caseins: temperature dependence of [beta]-casein structure as analyzed by circular dichroism and FTIR spectroscopy and correlation with micellization', *Food Hydrocoll*, 15 (4–6), 341–354.

FARRELL H M, QI P X, WICKHAM E D and UNRUH J J (2002a) 'Secondary structural studies of bovine caseins: Structure and temperature dependence of beta-casein phosphopeptide (1–25) as analyzed by circular dichroism, FTIR spectroscopy, and analytical ultracentrifugation', *J Prot Chem*, 21 (5), 307–321.

FARRELL H M, QI P X, BROWN E M, COOKE P H, TUNICK M H, WICKHAM E D and UNRUH J J (2002b) 'Molten globule structures in milk proteins: Implications for potential new structure–function relationships', *J Dairy Sci*, 85 (3), 459–471.

FASMAN G D, BODENHEIMER E and PESCE A (1966) 'Fluorescence studies on poly-alpha-amino acids: models of protein conformation', *J Biol Chem*, 241 (4), 916–923.

FISCHER N and WIDDER S (1997) 'How proteins influence food flavor', *Food Technol*, 51 (1), 68–70.

FLOWER D (1996) 'The lipocalin protein family: structure and function', *Biochem J*, 318, 1–14.

FLOWER D (2000) 'Beyond the superfamily: the lipocalin receptors', *Biochim Biophys Acta*, 1482, 327–336.

FLOWER D, NORTH A and SANSOM C (2000) 'The lipocalin protein family: structural and sequence overview', *Biochim Biophys Acta*, 1482, 9–24.

FOGOLARI F, RAGONA L, ZETTA L, ROMAGNOLI S, DE KRUIF K G and MOLINARI H (1998) 'Monomeric bovine β-lactoglobulin adopts a β-barrel fold at pH 2', *FEBS Lett*, 436 (2), 149–154.

FRANZEN K L and KINSELLA J E (1974) 'Parameters affecting the binding of volatile flavor compounds in model food systems. I. Proteins', *J Agric Food Chem*, 22 (4), 675–678.

FRAPIN D, DUFOUR E and HAERTLÉ T (1993) 'Probing the fatty acid binding site of beta-lactoglobulins', *J Prot Chem*, 12 (4), 443–449.

GARCIA M C, TORRE M, MARINA M L and LABORDA F (1997a) 'Composition and characterization of soyabean and related products', *Crit Rev Food Sci Nutr*, 37 (4), 361–391.

GARCIA M C, TORRE M, LABORDA F and MARINA M L (1997b) 'Rapid separation of soybean globulins by reversed-phase high-performance liquid chromatography', *J Chromatogr A*, 758 (1), 75–83.

GREEN D W, ASCHAFFENBURG R, CAMERMAN A, COPPOLA J C, DUNNILL P, SIMMONS R M, KOMOROWSKI E S, SAWYER L, TURNER E M C and WOODS K F (1979) 'Structure of bovine [beta]-lactoglobulin at 6A resolution', *J Mol Biol*, 131 (2), 375–397.

GREENFIELD N and FASMAN G D (1969) 'Computed circular dichroism spectra for the evaluation of protein conformation', *Biochem*, 8 (10), 4108–4116.

GRINBERG V Y, GRINBERG N V, MASHKEVICH A Y, BUROVA T V and TOLSTOGUZOV V B (2002) 'Calorimetric study of interaction of ovalbumin with vanillin', *Food Hydrocoll*, 16 (4), 333–343.

GROSCLAUDE F, MAHE M F and RIBADEAU-DUMAS B (1973) 'Primary structure of alpha casein and of bovine beta casein. Correction', *Eur J Biochem*, 40 (1), 323–324.

GUICHARD E (2002) 'Interactions between flavor compounds and food ingredients and their influence on flavor perception', *Food Rev Int*, 18 (1), 49–70.

GUICHARD E and ETIÉVANT P (1998) 'Measurement of interactions between polysaccharides and flavour compounds by exclusion size chromatography: advantages and limits', *Nahr*, 42 (6), 376–379.

GUICHARD E and LANGOURIEUX S (2000) 'Interactions between Beta-lactoglobulin and flavour compounds', *Food Chem*, 71, 301–308.

GUTH H and FRITZLER R (2004) 'Binding studies and computer-aided modelling of macromolecule/odorant interactions', *Chem Biodiv*, 1, 2001–2023.

HAMBLING S G, MCALPINE A S and SAWYER L (1992) β-lactoglobulin. In *Advanced Dairy Chemistry-1*, Fox P F (ed.), Elsevier Applied Science, London.

HANSEN A P and HEINIS J J (1991) 'Decrease of vanillin flavor perception in the presence of casein and whey proteins', *J Dairy Sci*, 74 (9), 2936–2940.

HANSEN A P and HEINIS J J (1992) 'Benzaldehyde, citral, and d-limonene flavor perception in the presence of casein and whey proteins', *J Dairy Sci*, 75 (5), 1211–1215.

HANSEN A P and BOOKER D C (1996) Flavor interactions with caseins and whey proteins. In *Flavor-Food interactions*, McGorrin R J and Leland J V (eds), ACS Symposium Series 633, American Chemical Society, Washington, DC, 75–89.

HAQUE Z U and ARYANA K J (2002) 'Volatiles in lowfat cheddar cheese containing commercial fat replacers', *Food Sci Technol Research*, 8 (2), 188–190.

HARRISON M and HILLS B P (1997) 'Mathematical model of flavor release from liquids containing aroma-binding macromolecules', *J Agric Food Chem*, 45 (5), 1883–1890.

HAVEA P, SINGH H and CREAMER L K (2001) 'Characterization of heat-induced aggregates of beta-lactoglobulin, alpha-lactalbumin and bovine serum albumin in a whey protein concentrate environment', *J Dairy Res*, 68 (3), 483–497.

HENG L, VAN KONINGSVELD G A, GRUPPEN H, VAN BOEKEL M, VINCKEN J P, ROOZEN J P and VORAGEN A G J (2004) 'Protein–flavour interactions in relation to development of novel protein foods', *Trends Food Sci Technol*, 15 (3–4), 217–224.

HIRAYAMA K, AKASHI S, FURUYA M and FUKUHARA K (1990) 'Rapid confirmation and revision of the primary structure of bovine serum albumin by ESIMS and Frit-FAB LC/MS', *Biochem Biophys Res Commun*, 173 (2), 639–46.

HODGKIN D C (1950) 'X-ray analysis and protein structure', *Cold Spring Harb Symp Quant Biol*, 14, 65–78.

HOFFMAN M A M and VAN MIL P (1997) 'Heat-induced aggregation of b-lactoglobulin: role of the free thiol group and disulfide bonds', *J Agric Food Chem*, 45, 2942–2948.

HONG Y H and CREAMER L K (2002) 'Changed protein structures of bovine beta-lactoglobulin B and alpha-lactalbumin as a consequence of heat treatment', *Int Dairy J*, 12 (4), 345–359.

HOSSEINI-NIA T, ISMAIL A A and KUBOW S (1999) 'Pressure-induced conformational changes of beta-lactoglobulin by variable-pressure Fourier transform infrared spectroscopy', *J Agric Food Chem*, 47, 4537–4542.

HUPPERTZ T, FOX P F and KELLY A L (2004) 'High pressure-induced denaturation of alpha-lactalbumin and beta-lactoglobulin in bovine milk and whey: a possible mechanism', *J Dairy Res*, 71 (4), 489–495.

IAMETTI S, TRANSIDICO P, BONOMI F, VECCHIO G, PITTIA P and ROVERE P G DA (1997) 'Molecular modifications of beta-lactoglobulin upon exposure to high pressure', *J Agric Food Chem*, 45, 23–29.

JAMESON G B, ADAMS J J and CREAMER L K (2002) 'Flexibility, functionality and hydrophobicity of bovine beta-lactoglobulin', *Int Dairy J*, 12 (4), 319–329.

JASINSKI E and KILARA A (1985) 'Flavor binding by whey proteins', *Milchwissenschaft*, 40 (10), 596–599.

JOSS L A and RALSTON G B (1996) 'Beta-lactoglobulin B: a proposed standard for the study of reversible self-association reactions in the analytical ultracentrifuge?' *Anal Biochem*, 236 (1), 20–26.

JOUENNE E (1997) Etude des interactions entre la beta-lactoglobuline et les composés d'arôme. Thèse de Doctorat, Université de Montpellier 2, France.

JOUENNE E and CROUZET J (1997a) Measurement of affinity constants by DCCLC. In *Interaction of Food Matrix with Small Ligands Influencing Flavour and Texture*, Schieberle P (ed.), European Communities; Vol. 3, 68–71.

JOUENNE E and CROUZET J (1997b) Influence of sodium chloride and urea on volatile compounds–β-lactoglobulin interactions. In *Interaction of Food Matrix with Small Ligands Influencing Flavour and Texture*, Zacharoff A B (ed.), Office for Official Publications of the European Communities, Vol. 2, 37–43.

JOUENNE E and CROUZET J (2000) 'Effect of pH on retention of aroma compounds by b-lactoglobulin', *J Agric Food Chem*, 48 (4), 1273–1277.

JUNG D-M and EBELER S E (2003) 'Investigation of binding behavior of alpha- and beta-ionones to beta-lactoglobulin at different pH values using a diffusion-based NOE pumping technique', *J Agric Food Chem*, 51, 1988–1993.

JUNG D M, DE ROPP J S and EBELER S E (2000) 'Study of interactions between food phenolics and aromatic flavors using one- and two-dimensional H-1 NMR spectroscopy', *J Agric Food Chem*, 48 (2), 407–412.

JUNG D-A, DE ROPP J S and EBELER S E (2002) 'Application of Pulsed Field Gradient NMR Techniques for Investigating Binding of Flavor Compounds to Macromolecules', *J Agric Food Chem*, 50 (15), 4262–4269.

KARASOVA L, FUKAL L, KODICEK M and RAUCH P (1998) 'Heat Denaturation of alpha-lactalbumin and beta-lactoglobulin', *Chem Pap-Chem Zvesti* 52, 552–552.

KING B M and SOLMS J (1983) Headspace gas-chromatography as a means of assessing the effect of non-volatile food components on volatile flavour compounds. In *Sensory Quality in Foods and Beverages: Definition, Measurement and Control*, Williams A A and Atkin R K (eds), Ellis Horwood, Chichester, 196–202.

KINSELLA J E (1990) 'Flavor perception and binding', *Inform*, 1 (3), 215.

KISTER A E, FINKELSTEIN A V and GELFAND I M (2002) 'Common features in structures and sequences of sandwich-like proteins', *Proc Natl Acad Sci USA*, 99 (22), 14137–14141.

KO T P, NG J D and McPHERSON A (1993) 'The 3-Dimensional Structure of Canavalin from Jack-Bean (Canavalia-Ensiformis)', *Plant Physiol*, 101 (3), 729–744.

KO T P, DAY J and McPHERSON A (2000) 'The refined structure of canavalin from jack bean in two crystal forms at 2.1 and 2.0 A resolution', *Acta Crystallogr D Biol Crystallogr*, 56 (Pt 4), 411–420.

KONTOPIDIS G, HOLT C and SAWYER L (2002) 'The ligand-binding site of bovine beta-lactoglobulin: Evidence for a function?' *J Mol Biol*, 318 (4), 1043–1055.

KUMOSINSKI T F, BROWN E M and FARRELL H M, JR (1991a) 'Three-dimensional molecular modeling of bovine caseins: alpha s1-casein', *J Dairy Sci*, 74 (9), 2889–2895.

KUMOSINSKI T F, BROWN E M and FARRELL H M, JR (1991b) 'Three-dimensional molecular modeling of bovine caseins: kappa-casein', *J Dairy Sci*, 74 (9), 2879–2887.

KUMOSINSKI T F, BROWN E M and FARRELL H M, JR (1993a) 'Three-dimensional molecular modeling of bovine caseins: an energy-minimized beta-casein structure', *J Dairy Sci*, 76 (4), 931–945.

KUMOSINSKI T F, BROWN E M and FARRELL H M, JR (1993b) 'Three-dimensional molecular modeling of bovine caseins: a refined, energy-minimized kappa-casein structure', *J Dairy Sci*, 76 (9), 2507–2520.

KUWATA K, HOSHINO M, FORGE V, ERA S, BATT C A and GOTO Y (1999) 'Solution structure and dynamics of bovine beta-lactoglobulin A', *Protein Sci*, 8 (11), 2541–2545.

LABOURÉ H, CASES E and CAYOT P (2004) 'Heat induced beta-lactoglobulin polymerization: role of the change in medium permittivity', *Food Chem*, 85 (3), 399–406.

LANGOURIEUX S (1995) Aroma compound–matrix interactions by dynamic coupled column liquid chromatography. In *Interaction of Food Matrix with Small Ligands Influencing Flavour and Texture*, 71–75.

LAWRENCE M C, SUZUKI E, VARGHESE J N, DAVIS P C, VAN DONKELAAR A, TULLOCH P A and COLMAN P M (1990) 'The three-dimensional structure of the seed storage protein phaseolin at 3 A resolution', *Embo J*, 9 (1), 9–15.

LAWRENCE M C, IZARD T, BEUCHAT M, BLAGROVE R J and COLMAN P M (1994) 'Structure of phaseolin at 2.2 A resolution. Implications for a common vicilin/legumin structure and the genetic engineering of seed storage proteins', *J Mol Biol*, 238 (5), 748–776.

LE GUEN S and VREEKER R (2003) Interactions between flavour compunds and milk protein under static and dynamic conditions. In *Flavour Research at the Dawn of the Twenty-first Century*, Le Quéré J L and Etiévant P (eds), Lavoisier, Tec & Doc: Paris, France, 182–187.

LE QUÉRÉ J L, LÜBKE M, ANDRIOT I and GUICHARD E (1999) Measurement of interaction between aroma compounds and the protein β-lactoglobulin by spectroscopic methods. In *9th Weurman Flavour Research Symposium*, Schieberle P and Engel K H (eds), Deutsche Forschungsanstalt für Lebensmittelchemie, Garching, Germany, 329–333.

LEFÈVRE T and SUBIRADE M (1999) 'Structural and interaction properties of beta-lactoglobulin as studied by FTIR spectroscopy', *Int J Food Sci Technol*, 34, 419–428.

LIU J Q, TIAN J N, HE W Y, XIE J P, HU Z D and CHEN X G (2004) 'Spectrofluorimetric study of the binding of daphnetin to bovine serum albumin', *J Pharm Biomed Anal*, 35 (3), 671–677.

LUBBERS S, LANDY P and VOILLEY A (1998) 'Retention and release of aroma compounds in foods containing proteins', *Food Technol*, 52 (5), 68–74, 208–214.

LÜBKE M, GUICHARD E and LE QUÉRÉ J L (1999) Infrared spectroscopic study of β-

lactoglobulin interactions with flavour compounds. In *Flavor Release ACS meeting*, Taylor A J and Roberts D (eds), American Chemical Society, Washington, DC, 282–292.

LÜBKE M, GUICHARD E, TROMELIN A and LE QUÉRÉ J-L (2002) 'Nuclear magnetic resonance spectroscopic study of beta-lactoglobulin interactions with two flavor compounds, gamma-decalactone and beta-ionone', *J Agric Food Chem*, 50, 7094–7099.

LUCCA P A and TEPPER B J (1994) 'Fat replacers and the functionality of fat in foods', *Trends Food Sci Technol*, 5 (1), 12–19.

MA L, DRAKE M A, BARBOSA CANOVAS G V and SWANSON B G (1997) 'Rheology of full-fat and low-fat Cheddar cheeses as related to type of fat mimetic', *J Food Sci*, 62 (4), 748–752.

MARABOTTI A, BALESTRERI L, COZZINI P, MOZZARELLI A, KELLOGG G E and ABRAHAM D J (2000) 'HINT predictive analysis of binding between retinol binding protein and hydrophobic ligands', *Bioorg Med Chem Lett*, 10 (18), 2129–2132.

MARIN I (1998) Contribution à l'étude des interactions entre la β-lactoglobuline et certains composés d'arôme. Application à la stabilisation des mousses. Thèse de doctorat, Université Paris XI – Paris Sud, France.

MARIN I and RELKIN P (1999) 'Foaming properties of beta-lactoglobulin: impact of pre-heating and addition of isoamyl acetate', *Int J Food Sci Technol*, 34 (5/6), 517–522.

MARIN I and RELKIN P (2000) 'Interaction properties of beta-lactoglobulin and benz-aldehyde and effect on foaming properties of beta-lactoglobulin', *Food Chem*, 71, 401–406.

MARUYAMA N, ADACHI M, TAKAHASHI K, YAGASAKI K, KOHNO M, TAKENAKA Y, OKUDA E, NAKAGAWA S, MIKAMI B and UTSUMI S (2001) 'Crystal structures of recombinant and native soybean beta-conglycinin beta homotrimers', *Eur J Biochem*, 268 (12), 3595–604.

MASON K, PATEL N M, LEDEL A, MOALLEMI C C and WINTNER E A (2004) 'Mapping protein pockets through their potential small-molecule binding volumes: QSCD applied to biological protein structures', *J Comput-Aided Mol Des*, 18 (1), 55–70.

MAY W E, WASIK S P and FREEMAN D H (1978) 'Determination of the aqueous solubility of polynuclear aromatic hydrocarbons by a coupled column liquid chromatographic technique', *Anal Chem*, 50 (1), 175–179.

McCLEMENTS D J (2002) 'Modulation of globular protein functionality by weakly interacting cosolvents', *Crit Rev Food Sci Nutr*, 42 (5), 417–471.

McCLEMENTS D J and DEMETRIADES K (1998) 'An integrated approach to the development of reduced-fat food emulsions', *Crit Rev Food Sci Nutr*, 38 (6), 511–536.

McNEILL V L and SCHMIDT K A (1993) 'Vanillin interaction with milk protein isolates in sweetened drinks', *J Food Sci*, 58 (5), 1142–1147.

MERCIER J C, GROSCLAUDE F and RIBADEAU-DUMAS B (1971) 'Primary structure of bovine s1 casein. Complete sequence', *Eur J Biochem*, 23 (1), 41–51.

MERCIER J C, URO J, RIBADEAU-DUMAS B and GROSCLAUDE F (1972) 'Primary structure of a kappa B I bovine casein macropeptide', *Eur J Biochem*, 27 (3), 535–547.

MEYNIER A, RAMPON V, DALGALARRONDO M and GENOT C (2004) 'Hexanal and t-2-hexenal form covalent bonds with whey proteins and sodium caseinate in aqueous solution', *Int Dairy J*, 14 (8), 681–690.

MIKHEEVA L M, GRINBERG N V, GRINBERG V Y and TOLSTOGUZOV V B (1998) 'Effect of thermal denaturation on vanillin binding to some food proteins', *Nahr*, 42 (3/4), 185–186.

MILES C A (1994) 'Differential scanning calorimetry (DSC): protein structure probe useful for the study of damaged tendons', *Equine Vet J*, 26 (4), 255–256.

MILLS O E and SOLMS J (1984) 'Interaction of selected flavour compounds with whey proteins', *Lebensm Wiss Technol*, 17 (6), 331–335.

MONACO H L, ZANOTTI G, SPADON P, BOLOGNESI M, SAWYER L and ELIOPOULOS E E (1987) 'Crystal structure of the trigonal form of bovine beta-lactoglobulin and of its complex with retinol at 2.5 Å resolution', *J Mol Biol*, 197, 695–706.

MORRISETT J D, POWNALL H J and GOTTO A M, JR (1975) 'Bovine serum albumin. Study of the fatty acid and steroid binding sites using spin-labeled lipids', *J Biol Chem*, 250 (7), 2487–2494.

MURESAN S, VAN DER BENT A and DE WOLF F A (2001) 'Interaction of beta-lactoglobulin with small hydrophobic ligands as monitored by fluorimetry and equilibrium dialysis: Nonlinear quenching effects related to protein–protein association', *J Agric Food Chem*, 49 (5), 2609–2618.

NARAYAN M and BERLINER L J (1997) 'Fatty acids and retinoids bind independently and simultaneously to beta lactoglobulin', *Biochem*, 36 (7), 1906–1911.

NAWAR W W (1966) 'Some considerations in interpretation of direct headspace gas chromatographic analyses of food volatiles', *Food Technol* 213 (2), 115–117.

NAWAR W W (1971) 'Some variables affecting composition of headspace aroma', *J Agric Food Chem*, 19 (6), 1057–1059.

NG P K W, HOEHN E and BUSHUK W (1989) 'Binding of vanillin by fababean proteins', *J Food Sci*, 54 (1), 105–107.

NGARIZE S, HERMAN H, ADAMS A and HOWELL N (2004) 'Comparison of changes in the secondary structure of unheated, heated, and high-pressure-treated β-lactoglobulin and ovalbumin proteins using Fourier transform Raman spectroscopy and self-deconvolution', *J Agric Food Chem*, 52 (21), 6470–6477.

NIELSEN N C, DICKINSON C D, CHO T J, THANH V H, SCALLON B J, FISCHER R L, SIMS T L, DREWS G N and GOLDBERG R B (1989) 'Characterization of the glycinin gene family in soybean', *Plant Cell*, 1 (3), 313–328.

NONAKA M, LI CHAN E and NAKAI S (1993) 'Raman spectroscopic study of thermally induced gelation of whey proteins', *J Agric Food Chem*, 41 (8), 1176–1181.

O'KEEFE S, WILSON L A, RESURRECCION A P and MURPHY P A (1991) 'Determination of the binding of hexanal to soy glycinin and beta-conglycinin in an aqueous model system using a headspace technique', *J Agric Food Chem*, 39 (6), 1022–1028.

O'NEILL T E (1996) Flavor binding by food proteins: An overview. In *Flavor–Food Interactions*, McGorrin R J and Leland J V (eds), Washington, DC, American Chemical Society, Vol. 633, 59–74.

O'NEILL T and KINSELLA J E (1987) 'Flavor protein interactions: characteristics of 2-nonanone binding to isolated soy protein fractions', *J Food Sci*, 52 (1), 98–101.

O'NEILL T E and KINSELLA J E (1988) 'Effect of heat treatment and modification on conformation and flavor binding by beta-lactoglobulin', *J Food Sci*, 53 (3), 906–909.

PALMER J K (1973) 'Separation of components of aroma concentrates on the basis of functional group and aroma quality', *J Agric Food Chem*, 21 (5), 923–925.

PAPIZ M Z, SAWYER L, ELIOPOULOS E E, NORTH A C T, FINDLAY J B C, SIVAPRASADARAO R, JONES T A, NEWCOMER M F and KRAULIS P J (1986) 'The structure of beta-lactoglobulin and its similarity to plasma retinol-binding protein', *Nature*, 324, 383–385.

PARKER J K, MOTTRAM D S and ADAMS R L (2003) Interactions of sulphur-containing aroma compounds with proteins in both model systems and real food systems. In *Flavour Research at the Dawn of the Twenty-first Century*, Le Quéré J L and Etiévant P (eds), Lavoisier, Tec & Doc: Paris, France, 45–50.

PELLETIER E, SOSTMANN K and GUICHARD E (1998) 'Measurement of interactions between beta-lactoglobulin and flavor compounds (esters, acids, and pyrazines) by affinity and exclusion size chromatography', *J Agric Food Chem*, 46 (4), 1506–1509.

PERMYAKOV E A and BERLINER L J (2000) 'alpha-Lactalbumin: structure and function', *Febs Lett*, 473 (3), 269–274.

PEROLA E and CHARIFSON P S (2004) 'Conformational analysis of drug-like molecules bound to proteins: An extensive study of ligand reorganization upon binding', *J Med Chem*, 47 (10), 2499–2510.

PHILLIPS L G, SCHULMAN W and KINSELLA J E (1990) 'pH and heat treatment effects on foaming of whey protein isolate', *J Food Sci*, 55 (4), 1116–1119.

PITTIA P, WILDE P J, HUSBAND F A and CLARK D C (1996) 'Functional and structural properties of β-lactoglobulin as affected by high pressure treatment', *J Food Sci*, 61, 1123–1128.

PLUG H and HARING P (1994) 'The influence of flavour-ingredient interactions on flavour perception', *Food Qual Prefer*, 5 (1/2), 95–102.

POUZOT M, DURAND D and NICOLAI T (2004) 'Influence of the ionic strength on the structure of heat-set globular protein gels at pH 7 beta-lactoglobulin', *Macromol*, 37 (23), 8703–8708.

PRABAKARAN S and DAMODARAN S (1997) 'Thermal unfolding of beta-lactoglobulin: Characterization of initial unfolding events responsible for heat-induced aggregation', *J Agric Food Chem*, 45 (11), 4303–4308.

QI X L, HOLT C, McNULTY D, CLARKE D T, BROWNLOW S and JONES G R (1997) 'Effect of temperature on the secondary structure of beta-lactoglobulin at pH 6.7, as determined by CD and IR spectroscopy: a test of the molten globule hypothesis', *Biochem J*, 324, 341–346.

QI P X, BROWN E M and FARRELL H M (2001a) ' "New views" on structure-function relationships in milk proteins', *Trends Food Sci Technol*, 12 (9), 339–346.

QI P X, UNRUH J J, WICKHAM E D and FARRELL H M (2001b) 'Structure of bovine beta-casein studied by Fourier-transform infrared spectroscopy and computer modeling', *Biophys J*, 80 (1), 399A–399A.

QIN B Y, CREAMER L K, BAKER E N and JAMESON G B (1998) '12-Bromododecanoic acid binds inside the calyx of bovine [beta]-lactoglobulin', *FEBS Lett*, 438 (3), 272–278.

RAGONA L, FOGOLARI F, ZETTA L, PÉREZ D M, PUYOL P, DE KRUIF K, LÖHR F, RüTERJANS H and MOLINARI H (2000) 'Bovine β-lactoglobulin: Interaction studies with palmitic acid', *Prot Sci*, 9 (7), 1347–1356.

REFSGAARD H H, TSAI L and STADTMAN E R (2000) 'Modifications of proteins by poly-unsaturated fatty acid peroxidation products', *Proc Natl Acad Sci USA*, 97 (2), 611–616.

REINERS J, NICKLAUS S and GUICHARD E (2000) 'Interactions between β-lactoglobulin and flavour compounds of different chemical classes. Impact of the protein on the odour perception of vanillin and eugenol', *Lait*, 80, 347–360.

RELKIN P and VERMERSCH J (2001) Binding properties of vanillin to whey proteins: effects on protein conformational stability and foaming properties. In *Food Colloids*, Dickinson E and Miller R (eds), Royal Society of Chemistry, Cambridge, 282–292.

RIBADEAU DUMAS B, BRIGNON G, GROSCLAUDE F and MERCIER J C (1972) 'Primary structure of bovine beta casein. Complete sequence', *Eur J Biochem*, 25 (3), 505–514.

ROBBINS G A, WANG S and STUART J D (1993) 'Using the static headspace method to determine Henry's law constants', *Anal Chem*, 65, 3113–3118.

SAKURAI K, OOBATAKE M and GOTO Y (2001) 'Salt-dependent monomer-dimer equilibrium of bovine beta-lactoglobulin at pH 3', *Protein Sci*, 10 (11), 2325–2335.

SALEEB F Z and PICKUP J G (1978) Flavor interactions with food ingredients from headspace analysis measurements. In *Flavor of Foods and Beverages*, Charalambous G and Inglett G E (eds), Academic Press, New York, USA, 113–130.

SAWYER L, BROWNLOW S, POLIKARPOV I and WU S Y (1998) 'Beta-lactoglobulin: structural studies, biological clues', *Int Dairy J*, 8, 65–72.

SAWYER L and KONTOPIDIS G (2000) 'The core lipocalin, bovine beta-lactoglobulin', *Bioch Biophys Acta*, 1482, 136–148.

SCHMITT S, KUHN D and KLEBE G (2002) 'A new method to detect related function among proteins independent of sequence and fold homology', *J Mol Biol*, 323 (2), 387–406.

SEMENOVA M G, ANTIPOVA A S, MISHARINA T A and GOLOVNYA R V (2002a) 'Binding of aroma compounds with legumin. I. Binding of hexyl acetate with 11S globulin depending on the protein molecular state in aqueous medium', *Food Hydrocoll*, 16 (6), 557–564.

SEMENOVA M G, ANTIPOVA A S, WASSERMAN L A, MISHARINA T A and GOLOVNYA R V (2002b) 'Binding of aroma compounds with legumin. II. Effect of hexyl acetate on thermodynamic properties of 11S globulin in aqueous medium', *Food Hydrocoll*, 16 (6), 565–571.

SEMENOVA M G, ANTIPOVA A S, BELYAKOVA L E, POLIKARPOV Y N, WASSERMAN L A, MISHARINA T A, TERENINA M B and GOLOVNYA R V (2002c) 'Binding of aroma compounds with legumin. III. Thermodynamics of competitive binding of aroma compounds with 11S globulin depending on the structure of aroma compounds', *Food Hydrocoll*, 16 (6), 573–584.

SEUVRE A M, DIAZ M A E and VOILLEY A (2001) 'Retention of aroma compounds by beta-lactoglobulin in different conditions', *Food Chem*, 77 (4), 421–429.

SHUTOV A D, KAKHOVSKAYA I A, BRAUN H, BAUMLEIN H and MUNTZ K (1995) 'Legumin-like and vicilin-like seed storage proteins: Evidence for a common single-domain ancestral gene', *J Mol Evol*, 41 (6), 1057–1069.

SOLMS J and GUGGENBUEHL B (1991) Physical aspects of flavor applications in food systems. In *Flavour Science and Technology*, Wiley, Chichester, 319–335.

SOSTMANN K, BERNAL B, ANDRIOT I and GUICHARD E (1997) Flavour binding by β-lactoglobulin: different approaches. In *Flavor Perception. Aroma Evaluation*, Kruse H-P and Rothe M (eds), Eigenverlag Universität Postdam: Eisenach, 425–434.

SOSTMANN K and GUICHARD E (1998) 'Immobilized beta-lactoglobulin on a HPLC-column: a rapid way to determine protein-flavour interactions', *Food Chem*, 62, 509–513.

SPECTOR A A (1975) 'Fatty acid binding to plasma albumin', *J Lipid Res*, 16 (3), 165–179.

STEVENSON R J, CHEN X D and MILLS O E (1996) 'Modern analyses and binding studies of flavour volatiles with particular reference to dairy protein products', *Food Res Int*, 29 (3–4), 265–290.

SUBIRADE M, LOUPIL F, ALLAIN A-F and PAQUIN P (1998) 'Effect of dynamic high pressure on the secondary structure of beta-lactoglobulin and on its conformational properties as determined by Fourier transform infrared spectroscopy', *Int Dairy J*, 8 (2), 135–140.

SUREWICZ W K, MANTSCH H H and CHAPMAN D (1993) 'Determination of protein secondary structure by fourier-transform infrared-spectroscopy – a critical assessment', *Biochem*, 32 (2), 389–394.

SURROCA Y, HAVERKAMP J and HECK A J R (2002) 'Towards the understanding of molecular mechanisms in the early stages of heat-induced aggregation of beta-lactoglobulin AB', *J Chromatogr A*, 970 (1-2), 275–285.

SWAISGOOD H E (1993) 'Review and update of casein chemistry', *J Dairy Sci*, 76 (10), 3054–3061.

TEALE F W (1960) 'The ultraviolet fluorescence of proteins in neutral solution', *Biochem J*, 76, 381–388.

TIAN J N, LIU J Q, HE W Y, HU Z D, YAO X J and CHEN X G (2004) 'Probing the binding of scutellarin to human serum albumin by circular dichroism, fluorescence spectroscopy, FTIR, and molecular modeling method', *Biomacromol*, 5 (5), 1956–1961.

TROMELIN A and GUICHARD E (2003) 'Use of catalyst in a 3D-QSAR study of the interactions between flavor compounds and β-lactoglobulin', *J Agric Food Chem*, 51 (7), 1977–1983.

TROMELIN A and GUICHARD E (2004) '2D-and 3D-QSAR models of interaction between flavor compounds and beta-lactoglobulin using catalyst and Cerius2', *QSAR Comb Sci*, 23 (4), 214–233.

TROY D J, DESMOND E M and BUCKLEY D J (1999) 'Eating quality of low-fat beefburgers containing fat-replacing functional blends', *J Sci Food Agric*, 79 (4), 507–516.

TUMER N E, THANH V H and NIELSEN N C (1981) 'Purification and characterization of mRNA from soybean seeds. Identification of glycinin and beta-conglycinin precursors', *J Biol Chem*, 256 (16), 8756–8760.

UHRINOVA S, UHRIN D, DENTON H, SMITH M, SAWYER L and BARLOW P N (1998) 'Complete assignment of 1H, 13C and 15N chemical shifts for bovine beta-lactoglobulin: Secondary structure and topology of the native state is retained in a partially unfolded form', *J Biomol NMR*, 12, 89–107.

UHRINOVA S, SMITH M H, JAMESON G B, UHRIN D, SAWYER L and BARLOW P N (2000) 'Structural Changes Accompanying pH-Induced Dissociation of the β-Lactoglobulin Dimer', *Biochem*, 39 (13), 3565–3574.

VAN RUTH S M and VILLENEUVE E (2002) 'Influence of beta-lactoglobulin, pH and presence of other aroma compounds on the air/liquid partition coefficients of 20 aroma compounds varying in functional group and chain length', *Food Chem*, 79 (2), 157–164.

WANG Q, ALLEN J C and SWAISGOOD H E (1999) 'Binding of lipophilic nutrients to beta-lactoglobulin prepared by bioselective adsorption', *J Dairy Sci*, 82 (2), 257–264.

WEEL K G C, BOELRIJK A E M, BURGER J J, CLAASSEN N E, GRUPPEN H, VORAGEN A G J and SMIT G (2003) 'Effect of whey protein on the in vivo release of aldehydes', *J Agric Food Chem*, 51 (16), 4746–4752.

WONG D W S, CAMIRAND W M and PAVLATH A E (1996) 'Structures and functionalities of milk proteins', *Crit Rev Food Sci Nutr*, 36 (8), 807–844.

WU S Y, PÉREZ M D, PUYOL P and SAWYER L (1999) 'Beta-lactoglobulin binds palmitate within its central cavity', *J Biol Chem*, 274 (1), 170–174.

WÜTHRICH K (1994) 'NMR assignments as a basis for structural characterization of denatured states of globular proteins', *Curr Opin Struc Biol*, 4 (1), 93–99.

YANG J, DUNKER A K, POWERS J R, CLARK S and SWANSON B G (2001) 'Beta-Lactoglobulin molten globule induced by high pressure', *J Agric Food Chem*, 49 (7), 3236–3243.

YANG J, POWERS J R, CLARK S, DUNKER A K and SWANSON B G (2002) 'Hydrophobic probe binding of beta-lactoglobulin in the native and molten globule state induced by high pressure as affected by pH, KIO(3) and N-ethylmaleimide', *J Agric Food Chem*, 50 (18), 5207–5214.

YANG J, POWERS J R, CLARK S, DUNKER A K and SWANSON B G (2003) 'Ligand and flavor binding functional properties of beta-lactoglobulin in the molten globule state induced by high pressure', *J Food Sci*, 68 (2), 444–452.

YE A Q, SINGH H, OLDFIELD D J and ANEMA S (2004) 'Kinetics of heat-induced association of beta-lactoglobulin and alpha-lactalbumin with milk fat globule membrane in whole milk', *Int Dairy J*, 14 (5), 389–398.

ZSILA F (2003) 'A new ligand for an old lipocalin: induced circular dichroism spectra reveal binding of bilirubin to bovine beta-lactoglobulin', *Febs Lett*, 539 (1–3), 85–90.

10

Carbohydrate–flavour interactions

J. Delarue and P. Giampaoli, ENSIA, France

10.1 Introduction

Carbohydrates are present in almost all types of food products. They represent a wide range of molecules that share common elementary units with similar or identical chemical structures. Nevertheless, at least three categories of carbohydrates can be considered depending on their molecular weight: small sugars (i.e. mono- and disaccharides) such as glucose and sucrose, oligosaccharides such as oligofructose, and polysaccharides such as starch or cellulose.

In aqueous solutions, mono- and disaccharides interact with water molecules, which consequently influences the partition behaviour of some aroma compounds. The role of polysaccharides is more diverse as they can considerably change the texture of a food, for example, increasing its viscosity or turning it into a gel. Moreover, polysaccharides can also form inclusion complexes with small molecules such as aroma compounds. For example, cyclic oligosaccharides such as cyclodextrins can form inclusion complexes with volatile compounds and are very good candidates as carrier material for aroma. At lower water contents, carbohydrates of various molecular weights may form a glassy matrix which has excellent retention properties towards aroma compounds. Industry makes advantage of theses properties for flavour encapsulation and for optimal flavour retention during drying processes and other heat treatments.

10.2 Aroma interactions with mono- and disaccharides

Mono- and disaccharides are molecules that elicit a sweet taste and that clearly influence flavour perception as anyone can experience it. Sucrose, for example, is sometimes found to enhance aroma perception whereas some people on the contrary argue that sugar 'kills the flavour' as in wine or in coffee tasting, for example. Thus attention has been paid to the role of these food components on the release of aroma volatiles.

Thanks to their hydroxyl groups, small carbohydrates such as mono- and disaccharides are generally very soluble in water. Hence, sugars in foodstuffs strongly interact with water thus contributing to the physicochemical properties of many food matrices. The departure from Raoult's Law of concentrated solutions of sugars indicates that the hydration number of glucose and fructose approaches respectively 2 and 5 molecules of water per molecule of sugar. Moreover, NMR studies revealed a primary hydration layer comprising an average number of water molecules even higher: 3.7 for glucose and 6.6 for sucrose. Thus mono- and disaccharides could be expected to generate a 'salting out' effect in aqueous systems, meaning that the presence of such non-volatile solutes would increase the concentration of volatiles in the gas phase. On the opposite small sugars may interfere with the diffusion of volatile compounds and thus slow their release down.

10.2.1 Influence of mono- and disaccharides on the static partition of aroma compounds

In the so-called 'salting out' effect, mono- and disaccharides in solution are known to structure water molecules thus decreasing the amount of free water in the food matrix. As a consequence, the concentration of flavour compounds increases in the remaining available free water, which in turn affects the apparent partition equilibrium of the volatile compounds in favour of the gas phase. Several experiments support this idea, although this salting out result is not observed for any flavour compound. Moreover, the release of some volatiles was shown to decrease in the presence of mono- and disaccharides.

Nawar (1966, 1971) observed, for example, a salting out effect when sucrose or glucose were added in water solutions containing highly volatile compounds such as dimethylketone but conversely a decreased release for aroma compounds of higher molecular weight such as heptanone and heptanal. It is not clear, however, whether these results were statistically significant or not.

Hansson *et al.* (2001) found that sucrose and inverted sugar increased the release of five out of six flavour compounds tested (isopentyl acetate, cis-3-hexenyl acetate, ethyl hexanoate, L-menthone and linalool) for sugar concentrations over 60% (w/w) in a pectin/sugar aqueous solution. Noticeably, the behaviour of limonene was not influenced by the addition of sugars in contrast with the previous volatiles which are more polar. The effect was even more pronounced for inverted sugar (i.e. glucose + fructose) than for sucrose at the same mass concentration, which suggests that inverted sugar (whose molar concentration is actually higher) structures more water molecules and results in a salting out effect.

Given these different behaviours, some authors investigated whether the headspace concentration of volatile compounds after equilibration above solutions containing small sugars could be related to their chemical or physicochemical characteristics. The general trend is that polar compounds show an enhanced volatility whereas the headspace concentration of less-polar compounds tends to decrease with sucrose or glucose added. A more detailed analysis of experimental data shows that a more refined model is needed.

Nahon *et al.* (1998) also observed contrasted aroma behaviours in the release of volatiles in a model orange drink containing varying sucrose concentration (0–60% w/v). They noted that volatile compounds with low GC retention times (1-ethoxy-1-methoxyethane; ethyl acetate; 1,1-diethoxyethane; 2-pentanone; methyl butanoate and 1-penten-3-one) were significantly affected by higher sucrose concentrations. On the other hand, increasing sucrose concentration resulted in a lower release of two compounds exhibiting higher retention times (ethyl hexanoate and octanal) and did not change the release of compounds with intermediate retention times.

Friel *et al.* (2000) tried to predict headspace concentrations from physicochemical characteristics of volatiles. Amongst them, (Log P)^2, LUMO energy (energy related to dispersion energies of polar solutes in solution) and a connectivity index (connectivity indices are topological indices, which describe the way in which the atoms of a molecule are bonded together) were found to be better predictors. The model obtained was based on experimental data showing the effect of sucrose aqueous concentration ranging from 0 to 65% (w/v) on the headspace concentration of 40 volatiles above the solution. As previously, some volatiles were displaced from the solutions containing high sucrose level (e.g. linalool) whilst others appeared to increase their solubility in the solution (e.g. ethyl decanoate). Eventually, the presence of sucrose did not change the headspace concentration of particular compounds (e.g. acetaldehyde). These results are summarised in Table 10.1 with other data obtained by Voilley *et al.* (1977).

It should be noted that these results were observed for important additions of sucrose (65%), which is not representative of a food formulation, except maybe for diet versions of a regular 'full-sugar product'. For example, in a more complex but also more realistic system (fat-free yoghurt), Decourcelle *et al.* (2004) found no effect of fructose (1.4 to 4.0% w/w) nor of fructo-oligosaccharides (1.4 to 4.0% w/w) on the release of 15 compounds of a strawberry aroma assessed by headspace solid-phase microextraction (SPME).

10.2.2 Dynamic aspects of aroma retention and release

Investigations realised at equilibrium are of prime importance to understand partition of volatile compounds between the food matrix and the gas phase surrounding it. Indeed, they provide useful indications of volatility changes due to carbohydrates (compared to water). However, dynamic phenomena such as diffusion of aroma compounds in the matrix must also be considered when studying either flavour release in-mouth or food processes such as drying or encapsulating for which aroma losses might be critical.

Table 10.1 Ratio of headspace concentration in sugar solutions relative to water of 52 volatile compounds (Friel *et al.* 2000)

Volatile compound	Ratio	Volatile compound	Ratio
Ethyl decanoate	0.54	3-ethyl, 2-methyl pyrazine	1.34
Menthofuran	0.66	Pyrazine	1.37
α-damascenone	0.67	Ethyl pentaoate	1.42
Ethyl methyl furan	0.71	2-ibutyl 3-meo pyrazine	1.45
Dimethyl sulphide	0.72	Ethyl hexanoate	1.50
2-isopropyl phenol	0.77	Acetone	1.50
Benzaldehyde	0.83	Pyrrole	1.54
(E)-2-hexenal	0.88	Diacetyl	1.64
Methyl salicylate	0.89	Octanol*	1.71
2,3-diethyl pyrazine	0.91	Acetone*	1.72
Anethole	0.96	Ethanol	1.80
2,6-dimethyl cyclohexanone	0.98	Methylfuran	1.81
Acetadehyde	1.05	Cyclohexanone	1.84
Furan	1.11	Isoamyl acetate	1.89
Heptyl acetate	1.13	Diethyl succinate	1.92
2,5-dimethylpyrazine	1.14	Valeronitrile	1.93
Eugenol	1.16	Ethyl acetate	1.98
Octan-2-one	1.23	Ethyl butyrate	1.99
Hexan-2-one	1.24	Hexanol	2.05
Pentan-3-one	1.25	Hexyl acetate	2.07
Isoamylbutyrate	1.28	Butanone	2.14
Methyl acetate	1.29	1,4-cineole	2.18
Isobutyl thiazole	1.31	Furfuryl acetate	2.21
Acetylthiophene	1.32	Terpinolene	2.22
Menthone	1.33	Linalool	2.46
Guaiacol	1.34	Caryophyllene	2.65

* Data from Voilley *et al.* (1977)

Understanding the contribution of mono- and disaccharides to in-mouth aroma release and flavour perception implies to study aroma release under dynamic conditions. Some authors carried out such investigations *in vitro* using model mouth systems while others performed *in vivo* measurements using an API–MS interface.

Using a flavour release vessel simulating mouth conditions, Roberts *et al.* (1996) measured the dynamic release of seven aroma compounds from aqueous solutions thickened with either sucrose, guar gum or carboxymethylcellulose (CMC). Their results show a decreased release of the most volatile compounds (α-pinene, ethyl-2-methylbutyrate and 1,8-cineole) when the solution was thickened with sucrose as compared with guar gum and CMC at comparable viscosity level. The release was further decreased when the sucrose concentration increased from 52 to 69% w/w. This reduced volatility, however, could not be explained by the increase of viscosity, since the solutions containing guar gum and CMC were adjusted to the same viscosity. More likely, this effect could result from steric hindrance. The reduced water activity at this range of sucrose

Table 10.2 Diffusivity of water and volatiles during drying at 60 °C (Voilley and Le Meste 1985)

Substrates	D.E.	Water content (g/100 g d.w.)	Diffusivity ($\times 10^{11}$ m^2s^{-1})					
			Water	Acetone	Ethyl acetate	Propan-2-ol	Diacetyl	n-hexanol
Glucose	100	50	4.7	8.0	9.8	4.8	–	–
Maltose	50	50	4.6	2.1	3.7	0.2	0.8	0.7
Maltotriose	33	50	4.5	1.7	2.3	–	–	0.6
Glucose syrup	62.5	50	4.1	2.9	4.3	–	1.4	–
–	31	50	2.6	1.3	1.7	0.6	0.7	–
–	20	50	1.6	0.1	0.2	0.01	0.1	0.07

concentrations affects aroma release, although no clear explanation can be given to explain this phenomenon. Eventually, the authors suggest that sucrose–sucrose interactions may occur, thus forming hydrophobic regions entrapping non-polar aroma compounds in a sort of inclusion complex (see description of the formation of such complexes with amylose and cyclodextrins later in this chapter).

At lower sucrose concentration (2.5–10% w/w), Lethuaut *et al.* (2004) found no influence on the *in vivo* release of different volatiles in model custards. Using a similar system with model strawberry yogurts, Mei *et al.* (2004) found that the release of ethyl butanoate and of (Z)-hex-3-enol were significantly suppressed by sucrose (0–12% w/w) and by 55 DE high-fructose corn syrup (0–10% w/w).

Another concern for the food and aroma industries is the retention of aroma compounds in low-water content matrices, as used in flavour encapsulation or, more generally in drying and baking processes. The food matrices that undergo these treatments often contain a large amount of carbohydrates, among which are polysaccharides as well as mono- and disaccharides. Polysaccharides play a major role in the retention of aroma compounds under such conditions, however, small sugars may also have a significant influence on the retention.

Because these drying and baking processes are dynamic, the release (or the retention) of aroma compounds must be studied taking into account dynamic phenomena. Voilley and Le Meste (1985) measured the diffusivity of five aroma compounds in different sugar solutions (Table 10.2). Their results from drying experiments indicate that diffusivity decreased when molecular weight of sugars (substrate) increased and that this effect could vary depending on the nature of the volatile compound (highest diffusivity for ethyl acetate and lowest diffusivity for the alcohols). The latter observation indicates that hydrogen or hydrophobic interactions with carbohydrates are involved in the observed behaviour. These effects are also observed by direct measurement in more diluted solutions but to a lesser extent than in drying experiments. Overall, this decrease in aroma diffusivity can be seen as a result of the increase in viscosity due to water losses (or increased sugar concentration).

To summarise, it can be seen that the influence of mono- and disaccharides on flavour release is rather limited. Experimental results are contrasting, especially when concentration changes remain within realistic proportions regarding food formulation. This indicates that differences in flavour perception when sugars are added to a food or a drink are more likely due to perceptual interactions rather than to differences in aroma release behaviour. These considerations are in agreement with Davidson *et al.* (1999) who showed, using *in vivo* release monitoring together with sensory time intensity measurement, that the change in flavour perception of mint chewing gums while chewing was more dependent on sucrose perception kinetics than on menthone release.

10.2.3 Influence of the matrix physical state

Carbohydrate matrices undergo various phase transitions during drying or during baking. These phenomena result primarily from the properties of the polysaccharides (see later in this chapter), but mono- and disaccharides also play a role. Levi and Karel (1995) measured the retention of n-propanol from a sucrose matrix and from the sucrose:raffinose system (raffinose is a tri-saccharide, made of a galactose, a glucose and a fructose unit). These two systems allowed them to compare matrices with different structures (crystallised and amorphous) and under different physical states (glassy and rubbery). They observed a concomitant sucrose crystallisation and propanol release. The sucrose:raffinose system which is less prone to crystallisation shows a rapid initial release of propanol above the glass transition temperature (T_g) attributed to the collapse of the matrix, followed by a slower diffusional release.

In more complex systems also made of polysaccharides, the small sugars may act as plasticisers on carbohydrate polymers. They thus lower the matrix glass transition temperature. As a result, the presence of mono- and disaccharides tends to favour the collapse of carbohydrate matrices used for flavour encapsulation. On the other hand, mono- and disaccharides still slow the diffusion of aroma compounds through the matrix. Consequently, in such matrices low molecular weight carbohydrates plug the molecular level holes between the long entangled polymer chains and thus slow the release of volatile compounds after the initial loss due to collapse (Levine *et al.* 1992).

10.3 Structure of polysaccharides

10.3.1 Chemical structure of starch

Starch is a natural polymer which is synthesised by higher plants as a reserve of carbohydrates. Main sources of starch are cereals (maize, wheat, rice, etc.), tubers (potato, manioc) or leguminous plants (pea). As a food ingredient, starch can be used as such or it can be modified, depending on its use and on the functional properties that are sought (thickener, stabiliser, carrier, etc.).

Speaking of chemistry, starch is a polymer of D-glucose, with two main polymeric forms which are amylose and amylopectin. Amylose is a linear polymer made of (1–4) linked α-D-glucopyranosyl units with some molecules branched by (1–6) α linkages. The degree of polymerisation (DP) of native amylose corresponds to approximately 600 to 6000 units. Amylopectin is a highly branched molecule also made of α-D-glucopyranosyl units linked up by 1–4 bonds. However, in the case of amylopectin, 5–6% of 1–6 bonds allows junctions between the linear parts of the polymer. Amylopectin thus forms clusters of either short (DP 15–20) or long chains (DP 45–55) as well as internal chains of higher DP (>60) (Buléon *et al.* 1998).

10.3.2 Physical structures of starch

Amylose and amylopectin give starch a semi-crystalline structure that is typical of the starch granule in its native form. The crystalline structure represents between 15 and 45% of the starch granule as shown by X-ray diffraction. Two types of structure can be observed: the A type structure which is typical of cereal starches and the B type structure which is typical of starches from tubers (Buléon *et al.* 1998). In the presence of particular molecules, starch forms inclusion complexes in which amylose only is involved. These inclusion complexes have been first reported from starch capacity to fix iodine (Rutschmann and Solms 1990d, Nuessli *et al.* 1997) or small molecules like isopropanol, butanol, glycerol, naphtol, etc. These complexes give rise to amylose structures commonly referred to as V-type structures. These structures can be distinguished depending on the molecules that are entrapped. Six categories have thus been determined: V iodine, V isopropanol, V butanol, V naphthol, V glycerol and Vh. To date, only four of these structures are well known: V6I, V6II, V6III and V8. The associated number designates the number of glucose residues per turn of amylose helix involved in the complex (i.e. 6 or 8 residues) and the Roman numeral designates the inter-helix spacing (Rappenecker and Zugenmaier 1981, Buléon and Delage 1990, Helbert and Chanzy 1994).

10.3.2 Modified starches

When heated in the presence of water, native starch undergoes important physical changes. At temperatures above 60 °C and with excess of water, starch granules swell irreversibly until the native granular structure vanishes; starch loses its crystalline state and amylose chains are released. This phenomenon is referred to as starch gelatinisation.

Pre-gelatinised starches can be obtained by heating with appropriate control of humidity. In practice, aqueous starch suspensions containing 30–40% of dry matter are dried on tumble. Carbohydrate hydroxyl groups of starch can be substituted or can react with several functional groups. The main reactions involved are esterification and etherification. Main grafted substituents are acetate, hydroxypropyl and n-octenylsuccinate functions.

Starch cross-linking results in the bonding of different starch chains between them through phosphate or adipate groups.

Full or partial hydrolysis of starch can also be obtained chemically or by enzymatic action (mainly α-amylase). Depending on the degree of hydrolysis, products of varying molecular weights are obtained. These molecules are sorted according to their dextrose equivalent (DE) which is defined as the percentage of reducing power (free reducing endings) relatively to glucose. Three categories of products from hydrolysis can thus by distinguished: maltodextrins (DE < 20), glucose syrups ($100 > \text{DE} > 20$) and monosaccharides (DE $= 100$). Cyclodextrins (CD) are used as encapsulating agents for aroma compounds. They are derived enzymatically from starch. There are numerous enzyme options for cyclodextrin production. One way of production and purification has been described by Lee and Kim (1991) and consisted in simultaneous hydrolysis of starch and synthesis of cyclodextrins by cyclodextrin glycosyltransferase (CGTase). These compounds are cyclic polymers of α-D-glucopyranosyl units linked up by 1–4 bonds. The cycle can be made of 6 (αCD), 7 (βCD) or 8 (γCD) glucose units, and the resulting three-dimensional structures exhibit a hydrophobic central cavity and a hydrophilic external surface. This conformation results from the preferential positioning of hydroxyl groups outward, which gives cyclodextrins the ability to trap small hydrophobic molecules into their cavity.

10.4 Interaction mechanisms with aroma compounds

Several phenomena are involved in the interactions between aroma compounds and food matrices (Taylor 1998, Druaux and Voilley 1997, Hau *et al.* 1996):

- Diffusion of molecules in the matrix or through a porous system.
- Binding of aroma compounds on the matrix by the way of molecular interactions (adsorption, inclusion, trapping, etc.).
- Partition of aroma compounds between the different phases of the food system.

In polysaccharidic matrices diffusion phenomena and molecular interactions are most frequently considered as predominant.

10.4.1 Physicochemical phenomena involved in retention of aroma

Diffusion

Several phenomena can reduce the diffusion of flavour molecules in a food matrix and consequently their release in the gas phase. These phenomena are viscosity changes (due, for example, to the presence of hydrocolloids), trapping phenomena like caging in by gel effect or trapping in micro-regions. Diffusion is, as expected, dependent on the steric and chemical properties of the aroma compounds. Diffusion in the matrix is therefore easier when the size, the molecular weight and the steric hindrance of the flavour compound decrease (Goubet *et al.* 1998, Reineccius 1988).

The presence of water makes diffusion easier for flavour molecules that are water soluble, that is to say the more polar molecules: diacetyl and 3-methylbutanal in extruded starch (Hau *et al.* 1998) hexanol on corn starch (Boutboul *et al.* 2000) and losses during freeze drying from maltodextrin (Voilley and Rifai 1982).

Caging in by gel effect

Polysaccharides such as hydrocolloids (gums, carboxymethylcellulose, alginates, pectin, etc.) are well known to form gels in the presence of water. These gels are made of networks of macromolecular chains that structure water and entrap free water and flavour molecules. This is a physical interaction phenomenon. Most authors (Hollowood *et al.* 2002, Rega *et al.* 2002, Roberts *et al.* 1996, Guichard *et al.* 1991, Baines and Morris 1987) observed a decrease of flavour intensity induced by the presence of hydrocolloids. As an example,

Guichard (1996) observed such a clear decrease of headspace concentration and of the perceived flavour intensity by adding pectin to strawberry jam.

Trapping in micro-regions
This phenomenon occurs when encapsulating a flavour by a drying process. During the dehydration of an aqueous solution of a mixture of a solution containing polysaccharides and flavour compounds, hydrogen bonds between polysaccharides and water molecules are gradually replaced by hydrogen bonds between polysaccharide chains when water molecules are removed. As a result, the polymeric chains form amorphous micro-regions that entrap the volatile compounds, thus dramatically reducing their diffusion (Goubet *et al.* 1998).

Molecular interactions
At the molecular level, polysaccharides may interact in various ways with aroma compounds. Molecular interactions include hydrogen bonding, London interactions, hydrophobic interactions or molecular inclusion. Molecular inclusion has been the most widely studied phenomenon. Hydrogen bonds result from electrostatic interactions between electronegative atoms such as oxygen or nitrogen (such atoms are said to be acceptors of hydrogen bonds) and hydrogen atoms (donors) involved in covalent bonding. These bonds which are of weak energy (25 to 40 kJ.mol^{-1}) are observed between hydroxyl or carboxyl functions of polysaccharides and polar functions of aroma compounds (alcohols, aldehydes, ketones) (Maier 1975; Le Thanh *et al.* 1992; Boutboul *et al.* 2000). These authors observed higher retention of primary alcohols on native corn starch or maltodextrin.

London forces also called Van der Waals forces are involved between temporarily induced dipoles in non-polar molecules. Hydrophobic interactions result from a hydrophobic effect rather than from an interactive binding. It is related to polar molecules or part of polar molecules that have a weak affinity for water. Such solutes tend to gather and to associate by Van der Waals bonds (induced dipole moments). Meanwhile, surrounding water molecules reorganise and associate to each other by hydrogen bonds. However, these forces have never been shown in flavour-carbohydrate interaction studies.

Molecular inclusion
Inclusion complexes formed by aroma compounds in the amylose helical structure of starch (Osman-Ismail and Solms 1973) or in β-cyclodextrin have been observed (Lindner *et al.* 1981). Molecular inclusion mainly involves hydrophobic interactions, although hydrogen bonds and London forces also play a role.

10.4.2 Chemical properties of aroma compounds influencing polysaccharide/aroma molecular interactions

Volatility, chemical function and chain length are important parameters in flavour polysaccharide interactions.

In some cases such as flavour encapsulation or food drying processes, the loss of aroma is directly related to the diffusion properties of the aroma compounds in the matrix and to interactions they may develop with polysaccharides. As a result, the loss of flavour compounds increases with their volatility which mainly depends on the molecular size and the polarity of the molecule.

Meanwhile the functional groups play an important role in the interaction with the matrix. Alcohols are generally considered as best candidates to interact with polysaccharidic matrices. Next are aldehydes and ketones (Goubet *et al.* 1998, Maier 1975, Boutboul *et al.* 2000) since like alcohols, these flavour compounds may link to the matrix by the way of hydrogen bonds.

Accordingly, for flavour molecules of equal carbon numbers, the interactions with polysaccharides increase with the polarity of the flavour molecule. For example, 1-octanol is more retained on a maltodextrin matrix than octanone (Bangs and Reineccius 1981).

Finally, for a given homologous series, sorption increases with the chain length. For example, ethyl butyrate is less retained than ethyl hexanoate on arabic gum (Rosenberg *et al.* 1990). Similarly during the freeze-drying of starch hydrolysates, retention of esters increases in the following order: ethyl propanoate, ethyl butanoate, ethyl pentanoate (Voilley 1995).

10.5 Example of retention in amylaceous matrices and their derivatives

10.5.1 β-cyclodextrin – aroma complexes

β-cyclodextrins are widely used for the encapsulation of aroma since they may entrap within their hydrophobic cavity aroma molecules generally considered also as hydrophobic. This cavity can host various molecules with molecular weight ranging between 80 and 250 g.mol^{-1}. Bulkier molecules can be trapped if they have a lateral chain that may enter the cavity. The stability of these complexes depends on several factors: the size and the shape of the guest molecule, its polarity and its hydrophobicity (Blyshak *et al.* 1989). Most studies on β-cyclodextrin complexes have been carried out in aqueous solutions (Pagington 1986). Nevertheless some seldom studies on complexes in dry conditions showed similar interactions (Sun *et al.* 1999). The hydrophobic character of the cavity allows the hosting of non-polar molecules or non-polar parts of molecules. The interactions increase with the size of the molecule or with the size of the non-polar group.

The guest molecules that give rise to the most stable complexes contain a non-polar part of appropriate size that allows it to fit in the β-cyclodextrin cavity and a polar group that stays outside and is stabilised by hydrogen bonds with the hydroxyl functions of the β-cyclodextrin. Sun *et al.* (1999) thus observed strong interactions between β-cyclodextrin and n-butanol, benzaldehyde and cyclopentanone. These compounds indeed have a linear carbon chain or a phenyl group (non-polar) as well as an alcohol, an aldehyde or a ketone function (polar).

In order to obtain the maximum retention, it is necessary to mix both ingredients in aqueous solution (Pagington 1986) which is logical since water enhances hydrophobic interactions between aroma compounds and the cyclodextrin internal cavity.

10.5.2 Starch and derivatives – aroma interaction

Several factors were shown to modify the retention of aroma compounds on a starch matrix (Whorton 1995, Goubet *et al.* 1998). In this section, we will examine retention phenomena for matrices of varying levels of hydration, or varying concentrations of starches.

Influence of the matrix physical state

Polysaccharidic matrices are composed either of amorphous and crystalline zones or of amorphous material solely. Sorption of aroma compounds in the matrix occurs in amorphous regions, whereas crystalline zones are inaccessible because they are made of polymeric chains organised in tight structures. For matrices of comparable chemical structure, the more crystalline is the material, the least the retention of aroma compounds is. This is the reason why the matrices chosen to encapsulate flavours are mainly amorphous materials (e.g. maltodextrins, glucose syrups, modified starches, β-cyclodextrins).

Weakly hydrated polysaccharidic materials are in a glassy state at room temperature. However, such materials exhibit a glass transition. Indeed when the material is heated above its glass transition temperature it changes into a rubbery state. Then the free volume of the polymer increases as a result of greater molecular motion and of disentanglement of the macromolecular chains. This allows aroma compounds to diffuse more readily in the material (Whorton 1995, Delarue and Giampaoli 2000).

The presence of water lowers the glass transition temperature of polysaccharides. Above a given water content, glass transition occurs at room temperature. Water induces swelling of the matrix and acts as a plasticiser for the material. Water molecules penetrate between the chains, making them more mobile and thus facilitating the penetration of other molecules like aroma compounds (Roos and Karel 1991, Whorton and Reineccius 1995).

Hau *et al.* (1998) observed that retention of diacetyl on crushed starch extrudates was more important when the samples were in a rubbery state (27 to 43% of humidity) than when they were in a glassy state (19% of humidity).

In the case of encapsulated compounds, the change to the rubbery state facilitates the diffusion of compounds from the bulk of the matrix towards the surface, which results in a loss of aroma by evaporation in air (Whorton 1995). This is why aroma encapsulation necessitates a stable material in an amorphous glassy state that allows the trapping of the aroma molecules and reduction of their mobility.

Voilley and Rifai (1982) measured the loss of aroma from freeze-dried maltodextrins flavoured with 25% of aroma during storage under various relative humidity conditions. The authors observed a good retention for a relative humidity lying between 0 and 50%. Between 50 and 75% of relative humidity, retention dropped dramatically and stabilised to a plateau between 75 and 100%. The authors notice a breakdown of the matrix structure between 30 and 75% of relative humidity. They ascribed this phenomenon to the glass transition induced by the plasticisation of the material by water. Indeed the matrix retains aroma compounds well, whereas in the rubbery state aroma compounds diffuse from the bulk to the outside of the matrix.

The glass transition is a critical situation for which other phenomena may occur such as crystallisation of amorphous zones or breakdown of the matrix.

The change from the glassy state to the rubbery state, either induced by heat or because of plasticisation by water, favours the relaxation of polymer chains, which can consequently facilitate crystallisation under suitable conditions. Aroma and water molecules are then expelled from the crystalline regions towards the amorphous parts of the matrix and eventually towards the surface of the material (Roos and Karel 1991b, Whorton 1995). Total crystallisation of the carrier thus leads to a complete release of the compounds that were initially encapsulated.

Sorption of aroma compounds increases with the porosity and with the specific surface of the blank matrix. Thus microporous materials may act as an absorbent for aroma (Zeller *et al.* 1999). Boutboul *et al.* (2002) studied the retention of aroma compounds on different starch matrices by inverse gas chromatography. They showed that the specific area was the main factor which explains the differences of retention on the different matrices.

Amylose complexes – aroma compounds in hydrated conditions

Starch is able to form inclusion complexes with a variety of ligands (iodine, fatty acids, emulsifiers, flavour substances). It is likely that these binding interactions mainly involve amylose since amylopectin exhibits a weak capacity to form complexes (Godshall and Solms 1992). In these complexes, the guest molecule is entrapped within the amylose helix thanks to hydrophobic interactions. These complexes are of high stability. An everyday example of such a phenomenon is the flavour loss when bread stales. Indeed this is due to the trapping of volatile compounds in amylose helices (Reinneccius 1992).

The study of starch/aroma complexes has been conducted mostly by the Osman-Ismail, Rutschmann and Solms team between 1972 and 1990 and has been reviewed in detailed (Godshall and Solms 1992). Since then, investigations continued with Heinemann *et al.* (2001, 2003). It should be noted that all these experiments have been done with gelified potato starch dispersed in water at a low concentrations (1–2%). Complexes of starch have been observed with several aroma compounds: menthone, decanal, 1-naphthol, limonene, carvone, geraniol, fenchone, camphor and thymol (Rutschmann and Solms 1990a–f, Nuessli *et al.* 1997). The bonding parameters of these complexes are reported in Table 10.3.

Ligands with a cyclic structure such as limonene and menthone are trapped in helices constituted of seven glycosyl residues per helical turn and bicyclic

Table 10.3 Bonding parameters of aroma compounds with potato starch (Rutschmann and Solms 1990a,b)

Ligand	Value of dissociation constant ($mol.L^{-1}$)	Bmax	Hill coefficient
Limonene	2.74×10^{-3}	5.40	2.05
Menthone	4.92×10^{-4}	43.94	1.43
Decanal	3.82×10^{-4}	13.78	2.10
1-naphthol	1.97×10^{-4}	62.35	2.22

compounds that are bulkier such as 1-naphthol are trapped in helices exhibiting an eight residues per turn pattern (Rustchmann and Solms 1990c). Different conclusions can be drawn from these studies:

- The stability of the complex decreases when the steric hindrance of the ligand increases. Linear chain ligands thus allow a better stability. Furthermore, this stability increases with the carbon chain length.
- Depending on the nature of the ligand, two binding mechanisms have been proposed (Rutschmann and Solms 1990a): on one hand the inclusion of polar ligands inside the amylose helix by the way of hydrophobic interactions, and on the other hand adsorption of polar ligands on the external part of the helix or onto another amylose or amylopectin chain by the way of hydrogen bonding with starch hydroxyl groups. Ring and Whittam (1991) also suggested the existence of dipole–dipole Van der Waals interactions. It is likely that the inclusion of the volatile compound adjust so that its dipole moment is anti-parallel to that of the amylose helix (resulting from the sum of the dipole moments of its glucose units).
- The best complexing agents are molecules that carry both a lipophilic part and a hydrophilic part (like decanal), the carbon chain being inclused in the amylose helix and the polar group being stabilised out of the helix.

Hau *et al.* (1998) used static headspace analysis to measure the sorption of three aroma compounds on extruded wheat starch samples containing 19–43% of water. They observed that the retention of polar compounds (diacetyl and 3-methylbutanal) increased when the water content of starch increased, whereas the retention of heptane which is a non-polar compound did not change. The authors explained this phenomenon by the water solubility of the volatile compounds increasing in that order: heptane $\ll$ 3-methylbutanal $<$ diacetyl. Most polar compounds solubilise in the free water portion of the material, which allows them to penetrate the matrix by diffusion.

Voilley and Rifai (1982) studied the retention of aroma compounds encapsulated in a freeze-dried maltodextrin matrix under varying relative humidity conditions. For a relative humidity equilibrium corresponding to the onset of solvating water, there is an important loss of volatile compounds due to the increase of diffusivity. Then retention decreases in the following order: n-hexanol, benzaldehyde, diacetyl, ethyl acetate, n-propanol, and acetone. The authors note that the retention decreased when the water solubility increased. In another study, Voilley (1995) measured the diffusivity of n-hexanol, diacetyl and acetone in a maltodextrin-based matrix as a function of its water content. Indeed the diffusivity coefficients increased with the water content and with the water solubility of these compounds, that is to say in the following order: n-hexanol < diacetyl < acetone.

Nuessli *et al.* (1995) observed that complexation of amylose by high concentration aroma compounds (decanal and fenchone) leads to the gelification of starch aqueous solutions. Gel formation would be due to aggregation of insoluble amylose-ligand complexes. Measurements of the viscoelastic properties of the mixture together with the binding of iodine on starch have revealed gel formation when all amylose inclusion sites are saturated by aroma compounds.

10.6 Conclusion

The retention of aroma compounds in a polysaccharidic matrix is mainly due to two factors:

1. Diffusion of aroma compounds in the matrix, which increases when the temperature of the matrix or its water content increases and when the size or the molecular weight of the volatile compound decreases. Furthermore, in the presence of water, diffusion increases with water solubility of aroma compounds.
2. Physicochemical interactions between the aroma compounds and the matrix. The higher the affinity of a compound for the polysaccharide, the more interactions will be created and the more it will be retained. Thus polar volatile molecules with a long carbon chain will be preferentially retained in low-moisture systems. High-moisture systems undergo fewer constraints and some particular physical conformations may form such as amylose helices, for example. In this case, interactions are closer to what is observed with proteins. Interactions such as hydrophobic interactions take place and are strengthened by the presence of water molecules and by exclusion phenomena that inhibit the polysaccharide hydrophilic sites. Figure 10.1 summarises the various effect according to the degree of hydrolysis of polysaccharides and the relative humidity of the matrix.

Today, polysaccharides are extensively used in the development of new products for aroma encapsulation. For example, many people focused on the controlled release of aroma compounds in confectionery products (mints, chewing gums). However, the understanding of carbohydrate–flavour interactions is not only relevant for the flavour industry. It also has important applications for cereal-based products that undergo thermal processes such as baking (cookies, crackers, cakes), extrusion cooking (breakfast cereals, snacks, dry pet food, etc.) or other drying processes (spray drying, tumble drying, etc.).

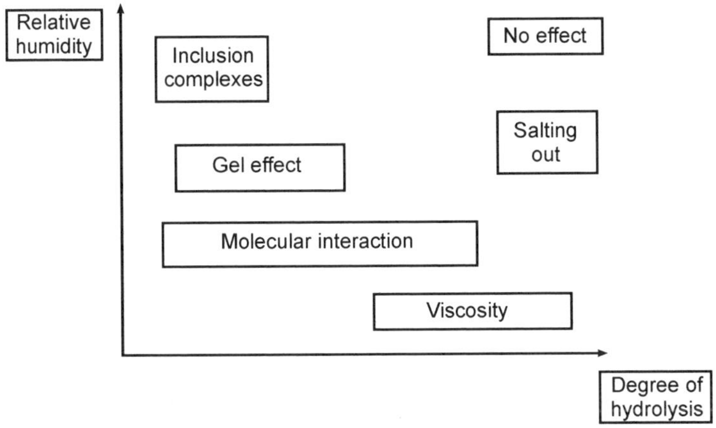

Fig. 10.1 Predominant mechanisms of interactions between aroma compounds and polysaccharides depending on their degree of hydrolysis and on the relative humidity of the matrix.

Similarly, the role of carbohydrates in aroma retention in coffee drying gave rise to an impressive piece of research in the industry and offered perspectives for the launch of instant coffees with controlled aroma characteristics and foaming properties. Besides, the use of carbohydrates such as cyclodextrins or amylose in order to trap aroma compounds during thermal or drying processes is also an interesting perspective.

The flavour stability during food storage is also sought by the use of carbohydrates. As revealed by numerous patents, industrial researchers investigate the retention properties of multi-component matrices that combine, for example, monosaccharides, dextrins, starches or other hydrocolloids, possibly together with other kinds of materials such as lipid membranes, etc. In this perspective, many studies remain to be conducted in order to improve the amount of the aroma that can be retained in polysaccharide matrices.

The retention of flavour compounds using covalent bonding is a rather uncertain perspective. However, it is possible to use heat treatment or enzymatic hydrolysis in order to release glycosidically bound volatile compounds that are naturally present as precursors in some fruits. It is therefore possible to make advantage of this to modify the flavour of fruit purees, fruit juices and wine (Crouzet *et al.* 1996). Interestingly, it can be noted that enzymatic hydrolysis or lipolysis is used for the controlled release of fragrance compounds in fabric softeners.

Besides, some non-food applications can result from the study of flavour–carbohydrate interactions. For example, cyclodextrins, which are now made available at low cost, are used in many applications in the flavour industry as well as in drug formulation and in cosmetics. Notably, they can be used as an efficient odour eliminator material for home care air fresheners.

10.7 References

ARVISENET G., LE BAIL P., VOILLEY A. and CAYOT N. (2002) Influence of physicochemical interactions between amylose an aroma compounds on the retention of aroma in food like matrices. *Journal of Agricultural and Food Chemistry*, 50, 7088–7093.

BAINES Z.V. and MORRIS E.R. (1987) Flavour/taste perception in thickened systems: the effect of guar gum above and below c*. *Food Hydrocolloids*, 1 (3), 197–205.

BANGS W.E. and REINECCIUS G.A. (1981) Influence of dryer infeed matrices on the retention of volatile flavor compounds during spray drying. *Journal of Food Science*, 47, 254–259.

BLYSHAK L.A., DODSON K.Y., PATONAY G., WARNER I. M. and MAY W. E. (1989) Determination of cyclodextrin formation constants using dynamic coupled-column liquid chromatography. *Analytical Chemistry*, 61(9), 955–960.

BOUTBOUL A., GIAMPAOLI P., FEINGENBAUM A. and DUCRUET V. (2000) Use of inverse gas chromatography with humidity control of the carrier gas to characterize aroma-starch interactions. *Food Chemistry*, 71, 387–392.

BOUTBOUL A., GIAMPAOLI P., FEINGENBAUM A. and DUCRUET V. (2002) Influence of the nature and treatment of starch on aroma retention. *Carbohydrate Polymers*, 47, 73–82.

BOUTBOUL A., LENFANT F., GIAMPAOLI P., FEIGENBAUM A. and DUCRUET V. (2002) Use of IGC to determine thermodynamic parameters of aroma starch interactions. *Journal of Chromatography A*, 969, 9–16.

BULÉON A., COLONNA P., PLANCHOT V. and BALL S. (1998) Starch granules: structure and biosynthesis. *International Journal of Biological Macromolecules*, 23, 85–112.

BULÉON A. and DELAGE M. M. (1990) Single crystals of V amylose complexed with isopropanol and acetone. *International Journal of Biological Macromolecules*, 12 (2), 25–33.

CROUZET J., CHASSAGNE D. and SAKHO M. (1996) Contribution of glysidically bound volatile compounds to processed food aroma. In: *Contribution of Low- and Non-Volatile Materials to the Flavor of Foods.* Pickengagen W., Ho C.T. and Spanier A.M. (eds), Allured Publishing Corporation: Carol Stream, IL, 207–216.

DAVIDSON, J.M., LINFORTH, R.S.T., HOLLOWOOD, T.A. and TAYLOR, A.J. (1999) Effect of sucrose on the perceived flavour intensity of chewing gum. *Journal of Agricultural and Food Chemistry* 47, 4330–4340.

DECOURCELLE N., LUBBERS S., VALLET N., RONDEAU P. and GUICHARD E. (2004) Effect of thickeners and sweeteners on the release of blended aroma compounds in fat-free stirred yoghurt during shear conditions. *International Dairy Journal*, 14, 783–789.

DELARUE J. and GIAMPAOLI P. (2000) Study of interaction phenomena between aroma compounds and carbohydrate matrices by inverse gas chromatography. *Journal of Agricultural and Food Chemistry*, 48, 2372–2375.

DRUAUX C. and VOILLEY A. (1997) Effect of food composition and microstructures on volatile flavour release. *Trends in Food Science and Technology*, 8 (11), 364–368.

ETZEL M. R. (1993) Thermodynamic mechanisms for the retention of aroma compounds during drying of foods. *AIChE Symposium Series*, 89 (297), 32–38.

FRIEL E. N., LINFORTH R. S. T. and TAYLOR A. J. (2000) An empirical model to predict the headspace concentration of volatile compounds above solutions containing sucrose. *Food Chemistry,* 71, 309–317.

GODSHALL M. A. and SOLMS I. (1992) Flavor and sweetener interactions with starch. *Journal of Food Technology*, 46, 140–145.

GOUBET I., DAHOUT C., SEMON E., GUICHARD E., LE QUERE J.L. and VOILLEY A. (2001) Competitive binding of aroma compounds by beta-cyclodextrin. *Journal of Agricultural and Food Chemistry*, 49, 5916–5922.

GOUBET I., LE QUERE J.-L. and VOILLEY A. J. (1998) Retention of aroma compounds by carbohydrates: influence of their physicochemical characteristics and of their physical state. A review. *Journal of Agricultural and Food Chemistry*, 46, 1981–1990.

GUICHARD E. (1996) Interactions between pectins and flavor compounds in strawberry jam. In *Food–flavor Interactions*, McGorin R.J. and Leland J.V. (eds), American Chemical Society: Washington, DC, 118–142.

GUICHARD E., ISSANCHOU S., DESCOURVIERES A. and ETIEVANT P. (1991) Pectin concentration, molecular weight and degree of esterification: influence on volatile composition and sensory characteristics of strawberry jam. *Journal of Food Science*, 56, 1621–1627.

HANSSON A., ANDERSSON J. and LEUFVÉN A. (2001) The effect of sugars and pectin on flavour release from a soft drink-related model system. *Food Chemistry,* 72, 363–368.

HAU M.Y.M., GRAY D.A. and TAYLOR A.J. (1996) Binding of volatiles to starch. In *Food–flavor Interactions.* McGorin, R.J. and Leland, J.V. (eds), American Chemical Society: Washington, DC, 109–117.

HAU M.Y.M., GRAY D.A. and TAYLOR A.J. (1998) Binding of volatiles to extruded starch at low water contents. *Flavour and Fragrance Journal*, 13, 77–84.

HEINEMANN C., CONDE-PETIT B., NUESSLI J. and ESCHER F. (2001) Evidence of starch inclusion complexation with lactones. *Journal of Agricultural and Food Chemistry*, 49, 1370–1376.

HEINEMANN C., ESCHER F. and CONDE-PETIT B. (2003) Structural features of starchlactone inclusion complexes in aqueous potato starch dispersions: The role of amylose and amylopectin. *Carbohydrate Polymers*, 51, 159–168.

HELBERT W. and CHANZY H. (1994) Single crystals of V amylose complexed with n-butanol or n-pentanol: structure features and properties. *International Journal of Biological Macromolecules*, 16 (4), 207–213.

HOLLOWOOD T.A., LINFORTH R.S.T. and TAYLOR A.J. (2002) The effect of viscosity on the perception of flavour. *Chemical Senses*, 27, 583–591.

JOUQUAND C., DUCRUET V. and GIAMPAOLI P. (2004) Partition coefficients of aroma compound in polysaccharide solutions by the phase ratio variation method. *Food Chemistry*, 85 (3), 467–474.

LEE Y.D. and KIM H.S. (1991) Enzymatic production of cyclodextrins from unliquefied corn starch in an attrition bioreactor. *Biotechnology and Bioengineering*, 37 (9), 795–801.

LE THANH M., THIBEAUDEAU P., THIBAUT M.A. and VOILLEY A. (1992) Interactions between volatile and non-volatile compounds in the presence of water. *Food Chemistry*, 43, 129–135.

LETHUAUT L., WEEL K.G.C., BOERLIJK A.E.M. and BROSSARD C.D. (2004). Flavor perception and aroma release from model dairy desserts. *Journal of Agricultural and Food Chemistry*, 52, 3478–3485.

LEVI G. and KAREL M. (1995) The effect of phase transitions on release of n-propanol entrapped in carbohydrate glasses. *Journal of Food Engineering*, 24, 1–13.

LEVINE H., SLADE L., VAN LENGERICH B. and PICKUP J.G. (1992) *Double-encapsulated compositions containing volatile and/or labile components, and processes for prepara-*

tion and use thereof. United States Patent No. 5087461: Nabisco Brands Inc.

LINDNER K., SZENTE L. and SZEJTLI J. (1981) Food flavoring with b-cyclodextrin complexed flavour substances. *Acta alimentaria*, 10, 175–186.

MAIER H.G. (1975) Binding of volatile aroma substances to nutrients and foodstuffs. In: H. Maarse and P. J. Groenen, *Proceedings of the International Symposium on Aroma Research held at the Central Institute for Nutrition and Food Research TNO*, Zeist, the Netherlands, 26–29 May 1975. Centre for Agricultural Publishing and Documentation, Wageningen, 143–157

MEI J.B., REINECCIUS G.A., KNIGHTON W.B. and GRIMSRUD E.P. (2004) Influence of strawberry yogurt composition on aroma release. *Journal of Agricultural and Food Chemistry*, 52, 6267–6270.

NAHON D.F., NAVARRO Y KOREN P.A., ROOZEN J.P. and POSTHUMUS M.A. (1998) Flavor release from mixtures of sodium cyclamate, sucrose, and an orange aroma. *Journal of Agricultural and Food Chemistry*, 46, 4963–4968.

NAWAR W.W. (1966) Some considerations in interpretation of direct headspace gas chromatographic analysis of food volatiles. *Food Technology*, 19, 115–117.

NAWAR W.W. (1971) Some variables affecting composition of headspace aroma. *Journal of Agricultural and Food Chemistry*, 19, 1057–1059.

NUESSLI J., SIGG B., CONDE-PETIT B. and ESCHER F. (1997) Characterization of amylose-flavor complexes by DSC and X-ray diffraction. *Food Hydrocolloids*, 11, 27–34.

OSMAN-ISMAIL, F. and SOLMS J. (1973) The formation of inclusion compounds of starches with flavor substances. *Lebensmittel Wissenschaft und Technologie*, 6, 4, 147–150.

PAGINGTON, J.S. (1986) Beta-cyclodextrin and its uses in the flavour industry. In G.G. Birch and M.G. Lindley (eds) *Developments in Food Flavours*. London: Elsevier Applied Science Publishers, 131–150.

RAPPENECKER G. and ZUGENMAIER P. (1981) Detailed refinement of crystal structure of Vh-amylose. *Carbohydrate Polymers*, 89, 11–19.

REGA B., GUICHARD E. and VOILLEY A. (2002) Flavour release from pectin gels: effects of texture, molecular interactions and aroma compounds diffusion. *Sciences des Aliments*, 22, 235–248.

REINECCIUS G.A. and RISCH S.J. (1986) The encapsulation of artificial flavors by b-cyclodextrin. *Perfumer and Flavorist*, 11, 1–6.

REINECCIUS G.A. (1988) The spray drying of food flavors. In: *Flavor Encapsulation*. S.J. Risch and G.A. Reineccius (eds), American Chemical Society: Washington, D.C., 55–66.

RIFAI M.K. and VOILLEY A. (1991) Collapse and aroma retention during storage of freeze-dried products. *Journal of Food Science and Technology*, 28, 135–137.

RING S.G. and WHITTAM M.A. (1991) Linear dextrins – solid forms and aqueous solution behavior. In: *Biotechnology of Amylodextrin Oligosaccharides*, R.B. Friedman (ed.), *ACS Symposium Series 458*, American Chemical Society: Washington, DC, 273–293.

ROBERTS D.D., ELMORE J.S., LANGLEY K.R. and BAKKER J. (1996) Effect of sucrose, guar gum, and carboxymethylcellulose on the release of volatile flavor compounds under dynamic conditions. *Journal of Agricultural and Food Chemistry*, 44, 1321–1326.

ROOS Y. and KAREL M. (1991) Plasticizing effect of water on thermal behavior and crystallization of amorphous food models. *Journal of Food Science*, 56, 38–43.

ROSENBERG M., KOPELMAN I.J. and TALMON Y. (1990) Factors affecting retention in spray-drying microencapsulation of volatile materials. *Journal of Agricultural and Food*

Chemistry, 36, 1288–1294.

RUTSCHMANN M.A. and SOLMS J. (1989) Formation of inclusion complexes of starch with different organic compounds. I. Method of evaluation of binding profiles with menthone as an example. *Lebensmittel Wissenschaft und Technologie*, 22, 40–244.

RUTSCHMANN M.A. and SOLMS J. (1990a) Formation of inclusion complexes of starch with different organic compounds. II. Study of ligand binding in binary model systems with decanal, 1-naphthol, monostearate and monopalmitate. *Lebensmittel Wissenschaft und Technologie*, 23, 70–79.

RUTSCHMANN M.A. and SOLMS J. (1990b) Formation of inclusion complexes of starch with different organic compounds. III. Study of ligand binding in binary model systems with (-)limonene. *Lebensmittel Wissenschaft und Technologie*, 23, 80–83.

RUTSCHMANN M.A. and SOLMS J. (1990c) Formation of inclusion complexes of starch with different organic compounds. IV. Ligand binding and variability in helical conformations of V amylose complexes. *Lebensmittel Wissenschaft und Technologie*, 23, 84–87.

RUTSCHMANN M.A. and SOLMS J. (1990d) Formation of inclusion complexes of starch with different organic compounds. V. Characterization of complexes with amperometric iodine titration, as compared with direct quantitative analysis. *Lebensmittel Wissenschaft und Technologie*, 23, 88–93.

RUTSCHMANN M.A. and SOLMS J. (1990e) Formation of inclusion complexes of starch in ternary model systems with decanal, menthone, and 1-naphthol. *Lebensmittel Wissenschaft und Technologie*, 23, 457–464.

RUTSCHMANN M.A. and SOLMS J. (1990f) The formation of ternary inclusion complexes of starch with menthone and monostearate – a possible food model system. *Lebensmittel Wissenschaft und Technologie*, 23, 451–456.

RUTSCHMANN M.A. and SOLMS J. (1991) Inclusion complexes of potato starch – a binding model with synergism and antagonism. *Lebensmittel Wissenschaft und Technologie*, 24, 473–475.

SUN D., CHEN J., LU W. and ZHENG X. (1999) Heats of adsorption of some organic compounds on β-cyclodextrin determined by gassolid chromatography. *Journal of Chromatography A*, 864, 293–298.

TAYLOR A.J. (1998) Physical chemistry of flavour. *International Journal of Food Science and Technology*, 33, 53–62.

VOILLEY A. (1995) Flavor encapsulation: influence of encapsulation media on aroma retention during drying. In: *Encapsulation and Controlled Release of Food Ingredients*. Risch S.J. and Reinnecius G.A. (eds), American Chemical Society: Washington, DC, 169–179.

VOILLEY A. and LE MESTE M. (1985) Aroma diffusion: the influence of water activity and of molecular weight of the other solutes. In D. Simatos and J.-L. Multon (eds), *Properties of Water in Foods*. Dordrecht, Netherlands: Martinus Nijhoff.

VOILLEY A. and RIFAI M. (1982) Fixation de substances d'arôme sur des supports alimentaires et activité de l'eau. *Sciences des Aliments*, hors série II, 53–60.

VOILLEY A., SIMATOS D. and LONCIN M. (1977) Gas phase concentrations of volatiles in equilibrium with a liquid aqueous phase. *Lebensmittel Wissenschaft und Technologie*, 10, 45–49.

WHORTON C. (1995) Factors influencing volatile release from encapsulation matrixes. In: *Encapsulation and Controlled Release of Food Ingredients*. Risch S.J. and Reinnecius G.A. (eds), American Chemical Society: Washington, DC, 134–142.

WHORTON B.C. and REINECCIUS G.A. (1995) Evaluation of mechanisms associated with the release of encapsulated flavor materials from maltodextrin matrices. In: *Encapsulation and Controlled Release of Food Ingredients*. Risch S.J. and Reinnecius G.A. (eds), American Chemical Society: Washington, DC, 143–160.

ZELLER B.L., SALEEB F.Z. and LUDESCHER R.D. (1999) Trends in development of porous carbohydrate food ingredients for use in flavor encapsulation. *Trends in Food Science and Technology*, 9, 389–394.

ZHAO J.G., MADSON M.A. and WHISTLER R.L. (1996) Cavities in porous corn starch granules provide a large storage space. *Cereal Chemistry*, 73, 379–380.

11

Modelling aroma interactions in food matrices

K. B. de Roos, Givaudan, Nederland BV, The Netherlands

11.1 Introduction

To be perceptible, aroma compounds must be released into the air and transferred to the olfactory epithelium in the nose. Since the composition of the released aroma varies with the composition and texture of the product (Nelson and Hoff 1968, Buttery *et al.* 1969), a flavouring can generate totally different odour sensations when applied to different foods. To solve this problem flavour companies and universities have spent much time on studying aroma–matrix interactions. These studies have resulted in mathematical models that can be applied in computer programs to predict the aroma release from different product matrices.

Computer modelling has become an important tool in flavour creation (De Roos and Wolswinkel, 1994, Sonnenberg *et al.* 2002). It is not only a powerful tool to shorten time to market but also for improving flavour performance and increasing cost effectiveness. Much of the time consuming trial and error that used to be part of flavour creation is being eliminated in this way. In this chapter the various models developed for predicting the aroma release will be discussed.

Two different types of aroma–matrix interactions can be distinguished:

- Physical interactions, determined by:
  - The air–product partition coefficient of the aroma compound. This coefficient is a measure for the volatility of the aroma compound in the product medium.
  - The mass transfer coefficient. This coefficient determines the kinetics of the aroma release.
- Chemical interactions. In contrast with physical interactions, these interactions result in a change of the aroma molecules.

In this review, we will restrict ourselves to physical interactions. First, the mathematical models describing the thermodynamic equilibria (section 11.2) and the kinetics of the aroma release (section 11.3) will be discussed. Section 11.4 deals with the various mechanistic models that describe the aroma retention and release: the *in vitro* aroma release is addressed in section 11.4.1, the aroma retention during food processing in section 11.4.2, the *in vivo* release from liquid and solid foods in section 11.4.3 and the aroma persistence after swallowing in section 11.4.4. Empirical modelling of the aroma–matrix interactions is discussed in section 11.5. The final sections discuss applications and future trends and provide sources of further information.

11.2 Phase partitioning: aroma release under equilibrium conditions

11.2.1 Factors affecting phase partitioning

An aroma compound, allowed to equilibrate between the product and air phase, distributes over the two phases according to the air–product partition coefficient P_{ap} which is defined as:

$$P_{ap} = C_a/C_p \tag{11.1}$$

where C is the concentration (kg/m^3) and the subscripts a and p refer to the air and product phase, respectively.

A product is in equilibrium with the air if there is no effective mass transport between the phases. This means that the amount going from one phase to another is identical to that going in the opposite direction (de Roos 2000). At equilibrium, an aroma compound is homogeneously distributed in each of the phases.

Since the aroma release takes place in a closed system, there is conservation of mass:

$$M_{total} = M_a + M_p = C_a V_a + C_p V_p \tag{11.2}$$

where M is the mass of the aroma compound (kg) and V the phase volume (m^3). Combination of equations 11.1 and 11.2 yields then the following relationship for the amount released under equilibrium conditions:

$$M_a = M_{total} \frac{P_{ap}}{P_{ap} + V_p/V_a} \tag{11.3}$$

This equation is the equation for a batch extraction process, where M_a represents the amount extracted by the air phase. If initially all aroma was in the product phase, then $M_{total} = M_p{}^0 = C_p{}^0 V_p$ from which follows that:

$$C_a = C_p{}^0 \frac{P_{ap}}{P_{ap} V_a/V_p + 1} \tag{11.4}$$

The air–product partition coefficient P_{ap} of an aroma compound is a measure for its volatility in the product. The volatility of aroma compounds in products is influenced by many factors, among which the following will be discussed in more detail:

- Phase partitioning in emulsions.
- Aroma binding by dissolved (macro)molecules.
- Acid–base equilibria.
- Solvent properties of mixed phases.
- Physical state.
- Temperature.

11.2.2 Modelling of phase partitioning

Phase partitioning in emulsions

The volatility of aroma compounds in emulsions can be calculated from the volume fractions f_o and f_w of lipid (oil) and water phase and their volatility in these phases:

$$P_{ap} = C_a/C_p = \frac{C_a}{f_o C_o + f_w C_w} = \frac{1}{f_o P_{oa} + f_w P_{wa}} = \frac{1}{f_o/P_{ao} + f_w/P_{aw}} \qquad 11.5$$

where P_{aw} and P_{ao} are the air–water and air–oil partition coefficients. For most aroma compounds, P_{aw} is much higher than P_{ao}. This difference in volatility reflects the preference of aroma compounds for water and oil and is therefore a measure for their hydrophobicity/lipophilicity. The quotient P_{aw}/P_{ao} is equivalent to the oil–water partition coefficient P_{ow}:

$$P_{aw}/P_{ao} = (C_a/C_w)/(C_a/C_o) = C_o/C_w = P_{ow} \qquad 11.6$$

Traditionally octanol–water partition coefficients have been used as a measure for the hydrophobicity of aroma compounds, traditionally octanol–water partition coefficients have been used. However, since triglycerides constitute the major lipid fraction of most foods, it makes more sense to use triglyceride–water (oil–water) partition coefficients. Octanol–water partition coefficients satisfactorily predict the partitioning of aroma compounds with polar amphiphilic lipids, such as phospholipids (Chen *et al.* 1994) and mono- and diglycerides, but for predicting the partitioning with neutral lipids they are less suitable (Rabe *et al.* 2003a, 2004).

With partly water-soluble lipids one has to distinguish between two types of interaction: interactions with dissolved lipids and interactions with aggregates such as oils droplets and micelles. Only the second type of interaction is correctly described by equation 11.5. The first type of interactions can be described as a (1:1) binding process between a flavour compound F and a dissolved lipid molecule L (Hussam *et al.* 1995):

$$L + F \leftrightarrow LF \qquad 11.7$$

of which the binding constant K is given by:

$$K = C_w^{FL}/(C_w^F C_w^L) \qquad 11.8$$

C_w^F and C_w^L are the molar concentrations (mole/m^3) of the unbound aroma and lipid molecules in aqueous solution, and C_w^{FL} the molar concentration of the flavour–lipid complex. Since the total concentration C_w of the flavour compound in the aqueous phase is given by $C_w = C_w^{FL} + C_w^F$, it follows from equation 11.8 that the concentration of free flavour molecules in water is given by:

$$C_w^F = C_w/(1 + KC_w^L) \qquad 11.9$$

So, the dissolved lipids reduce the concentration of free flavour molecules in aqueous solution by a factor of $(1 + KC_w^L)$. Since it is the concentration of the free molecules that exerts a vapour pressure, the concentrations in the air are reduced by the same factor. It follows then from equation 11.5 that aroma compound concentrations in the air at equilibrium are given by:

$$C_a = \frac{C_p}{f_o/P_{ao} + f_w(1 + KC_w^L)/P_{aw}} \qquad 11.10$$

The interactions of aroma compounds with dissolved lipid molecules are weak compared with the interactions with the separated lipid phase and can often be neglected if they occur at the same time (Hussam *et al.* 1995). The lipid droplet size in emulsions does not significantly affect the phase partitioning (Linforth *et al.* 2002, Carey *et al.* 2002).

Aroma binding by dissolved (macro)molecules

The interactions of aroma compounds with dissolved molecules other than lipids are also adequately described by equation 11.10. Among the interactions that have been studied are those with caffeine (King and Solms 1982, Tunaley *et al.* 1985), phenols (King and Solms 1982) maltodextrin (Semenova *et al.* 2000), starch (Escher *et al.* 2000, Kant *et al.* 2003), cyclodextrin (Kant *et al.* 2004) and protein (O'Neill 1996, Lübke *et al.* 2000, Andriot *et al.* 2000, Le Guen and Vreeker 2003).

Complications occur with heterogeneous macromolecules such as proteins, which have several binding places with a different affinity to aroma molecules. Proteins interact with aroma compounds through hydrogen bonding, hydrophobic interactions, ionic interactions and/or interactions with amino and thiol groups. For example, aldehydes can bind to proteins through hydrophobic interactions or interactions with amino and thiol groups (Kallen 1971). Their volatility in an aqueous protein solution is then given by:

$$P_{ap} = P_{aw}/\{1 + (K_{P1} + K_{P2} + \ldots K_{Pn})C_w^P\} \qquad 11.11$$

where P (upper case) refers to protein and subscripts 1 through n to the different protein–aldehyde interactions. The interactions with amino groups are strongly pH dependent (Lübke *et al.* 2000, Weel *et al.* 2003). At low pH the amino group is protonated and does then not interact with aldehydes. In that case, only the hydrophobic interactions determine the concentrations of free aldehyde molecules in aqueous solution.

In practice, it is difficult to work with relationship 11.11 because C_w^P is often unknown due to interactions of the proteins with molecules other than aroma ones. Therefore, it is a fortunate circumstance that the *in vivo* aroma release is hardly affected by proteins (Le Guen and Vreeker 2003) and certainly if lipids are also present in the product (Roberts and Pollien 2000, Seuvre *et al.* 2000).

Acid–base equilibria

This is a special case of aroma binding where the ionic form is the bonded form of the aroma compound. So, the volatility of a nitrogen compound in aqueous solution is then (de Roos and Sarelse 1996):

$$P_{ap} = P_{aw}/(1 + H^+/K_a) \qquad 11.12$$

where K_a is the ionisation constant. The volatility of acids is given by:

$$P_{ap} = P_{aw}/(1 + K_a/H^+) \qquad 11.13$$

The effects of aroma binding are additive. For example, the headspace concentration of a nitrogen compound over milk is given by:

$$C_a = \frac{C_p}{f_o/P_{ao} + f_w[1 + K_P C_w^P + H^+/K_a]/P_{aw}} \qquad 11.14$$

Solvent properties of mixed phases

The solubility of an aroma compound in a product influences also its volatility. It is for that reason that alcohol (Conner *et al.* 1998, Boelrijk *et al.* 2003), salt (Rabe *et al.* 2003b) and sugar (Nahon *et al.* 2000, Rabe *et al.* 2003c) affect the aroma volatility in water. Laub *et al.* (1980) have proposed the following empirical relationship for calculating the retention P_{ma} $(= 1/P_{am})$ in homogeneous mixtures of solvents:

$$P_{ma} = f_1 P_{1a} + f_2 P_{2a} + \ldots + f_n P_{na} \qquad 11.15$$

where f_n is the volume fraction of compound.

Dwarakanath and Pope (1998) used a different relationship to describe the partitioning between an aqueous solution and a mixture of *n* organic compounds:

$$\log P_{mw} = f_1 \log P_{1w} + f_2 \log P_{2w} + \ldots + f_n \log P_{nw} \qquad 11.16$$

Since $P_{mw} = P_{aw}/P_{am}$, it follows then that the mixture–air partition coefficient P_{am} is given by:

$$\log P_{ma} = f_1 \log P_{1a} + f_2 \log P_{2a} + \ldots + f_n \log P_{na} \qquad 11.17$$

To explain the effects of sodium chloride (Rabe *et al.* 2003b) and sugar (Rabe *et al.* 2003c) on the aroma solubility in water quantitatively, it must be assumed that the solubility in the hydration shell around the solutes is low to negligible. So, the effective fraction of salt in solution is much higher than the added volume fraction f_s of dry salt! This phenomenon is known as 'salting out'. The effect is that the aroma compound concentrations in the free water and the air in equilibrium with it increase by a higher factor than expected on the basis of the added volume of salt.

Physical state
If the physical state changes from liquid to crystalline, the partitioning of the aroma compounds between air and product is also changing. For example, during ice formation, the crystal lattice excludes dissolved molecules (Thijssen 1974) thus increasing their concentrations in the remaining liquid phase and the gas phase that is in equilibrium with it. Recent studies with solid fats suggest that fat crystallisation has a similar effect (Roudnitzky *et al.* 2003, Roberts *et al.* 2003a). Rabe *et al.* (2003a,d), on the other hand, did not find any effect of the physical state of triglycerides on the aroma release. The different effects of fat solidification on phase partitioning might be due to the degree of crystallinity of the fat phase. Amorphous and less rigid polymorphous crystalline areas can more easily incorporate solutes than the rigid crystalline structures.

The fact that rigid structures exclude solutes might also explain why the shell of structured water around ionic and polar solutes hardly participates in phase partitioning. Only small hydrophilic molecules seem to be able to penetrate into the rigid hydrophilic structure of the hydration shell (Rabe *et al.* 2003b).

Temperature
If the enthalpy of vaporisation ΔH_v is constant, the aroma volatility at T K can be calculated with the aid of the following form of the Clausius–Clapeyron equation (Nelson and Hoff 1968):

$$\log\left(\frac{P_{an}^{T}}{P_{ap}^{303}}\right) = \frac{\Delta H_v}{4.576}\left(\frac{1}{303} - \frac{1}{T}\right) \qquad 11.18$$

where P_{an}^{T} and P_{ap}^{303} are the partial vapour pressures at T and 303 K, respectively.

Since in practice ΔH_v varies with temperature, the predictions are only satisfactory over narrow temperature intervals. The usefulness of the equation is further limited by the fact that ΔH_v depends on the nature of the aroma compound and the composition of the product medium. Therefore, attempts to predict air–product partition coefficients as a function of temperature have remained relatively scarce (Muntz and Roberts 1987). The available data show that the temperature affects the volatility of aroma compounds in water and lipids to a different degree (Hall and Anderson 1983) and consequently also the partitioning between lipids and water (Kertes and King 1987, Kinkel *et al.* 1981, Leo *et al.* 1971). More fundamental studies on the effect of temperature on the phase partitioning have been done in the domain of gas–liquid chromatography (e.g. Curvers *et al.* 1985a,b).

11.3 Mass transfer: aroma release under non-equilibrium conditions

11.3.1 Factors affecting aroma transfer

When the headspace concentrations over an aqueous solution are diluted by air sweeping across its surface, mass transport from water to air takes place in an effort to restore the phase equilibria. This results in concentration gradients in

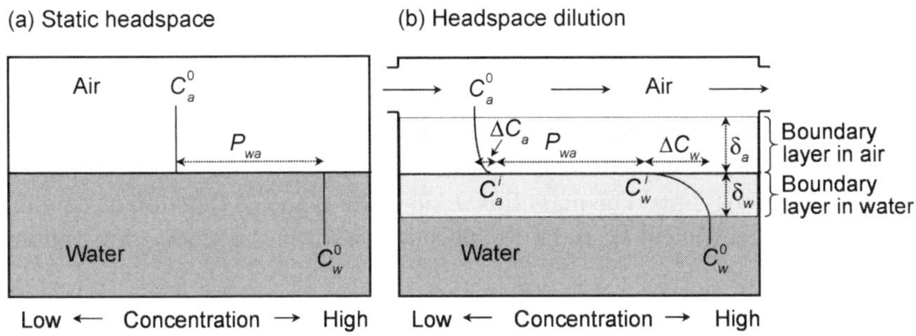

Fig. 11.1 Effect of headspace dilution on the concentration gradients of aroma compounds in water and air. P_{wa} is the water–air partition coefficient (reciprocal of the volatility).

the water and gas phases as depicted in Fig. 11.1b. The degree of non-equilibrium, represented by the concentration gradients ΔC_w and ΔC_a, is the driving force for mass transport. If R is the resistance to that transport, the mass flux J in either phase is given by (Darling *et al.* 1986):

$$J_w = \Delta C_w / R_w = k_w (C_w^i - C_w^0) \quad 11.19$$

$$J_a = \Delta C_a / R_a = k_a (C_a^0 - C_a^i) \quad 11.20$$

where J is the mass flux per unit time and area (kg/m^{2}s), k the mass transfer coefficient (m/s) and C^0 and C^i are the aroma compound concentrations (kg/m^3) in the bulk phase and at the interface, respectively. The subscripts a and p refer to the air and product phases. Since it is the aroma compound concentrations at the product surface that determine the maximum concentrations in the air, it is clear that under non-equilibrium conditions the maximum headspace concentrations predicted by equations 11.1 and 11.5 will never be achieved.

The mass flux through the boundary layers δ_a and δ_w must be the same at any distance from the interface, because no concentrations build up in these layers. So, if the resistance to mass transfer in the gas phase is small and can be neglected ($C_w^i = C_a^i / P_{aw} = C_a^0 / P_{aw}$), the following relationships must exist:

$$J = k_a (C_a^i - C_a^0) = k_w (C_w^0 - C_w^i) = k_o (C_w^0 - C_a^0 / P_{aw}) \quad 11.21$$

where k_o is the overall mass transfer coefficient.

When the mass transfer coefficients in air and water are comparable, i.e., differ by a factor of less than 10, the resistances in both phases have to be taken into account. In that case, the following relationship for the overall mass transfer coefficient k_o should be used (Marin *et al.* 1999, 2000):

$$\frac{1}{k_o} = \frac{1}{k_a} + \frac{P_{aw}}{k_w} \quad 11.22$$

If $k_a = k_w / P_{aw}$, the resistances to mass transfer in the two phases are identical. If $P_{aw} < k_w / k_a$, the release is thermodynamically (P_{aw}) controlled and if $P_{aw} > k_w / k_a$, the release is strongly kinetically (k_w / k_a) controlled.

The concentration gradients ΔC_w generated at the water surface increase with increase of the aroma volatility (P_{aw}), the mass transport rates in air (k_a) and the resistance to mass transfer in water ($1/k_w$). So, the higher the value of k_a/k_w, the more difficult it will be to replenish depleted concentrations at the water surface. At very high flow rates over the water surface and/or very low diffusion rates in the aqueous phase, depletion of volatile compounds at the water surface is exhaustive and $C_p^i \rightarrow 0$. The mass flux J_w in water is then a function of only the mass transfer coefficient k_w and the concentration C_w in the product bulk phase:

$$J_w = -k_w C_w^0 \qquad 11.23$$

The value of the mass transport coefficient k depends on the composition of the product and the diffusion mechanisms that are involved.

11.3.2 Effect of diffusion mechanism on mass transport

In a stagnant phase the only mechanism of mass transport is the molecular or static diffusion, which is caused by the random movement of the molecules. The rate of molecular diffusion is determined by the diffusion coefficient D. According to the Stokes–Einstein equation, D decreases with increase of the radius r of the molecule:

$$D = kT/6\eta\pi r \qquad 11.24$$

where k = Boltzman's constant (J/K), T = temperature (K) and η = dynamic viscosity (Pa s). Typical molecular diffusivities in air are 10^{-5} m^2/s, in water 10^{-9} m^2/s and in vegetable oils 10^{-10} m^2/s. Since the radii r of the aroma molecules do not vary much, the differences between the diffusion coefficients of aroma compounds in pure liquids are small. So, under static conditions the release from such systems is similar for all aroma compounds.

In emulsions the diffusion coefficients D_e are varying more widely. This because of the different diffusion coefficients in the lipid and aqueous phase and the unequal distribution of the aroma compounds over these phases. The diffusion coefficient in emulsions is given by (Overbosch *et al.* 1991):

$$D_e = \frac{f_w D_w + f_o D_o P_{ow}}{f_w + f_o P_{ow}} \qquad 11.25$$

Since the diffusion constants in lipids are lower than those in aqueous solutions, the static diffusion in emulsions will decrease with the lipophilicity of the aroma compounds. Binding to (macro)molecules or particles can have similar effects on the diffusion rates. In contrast with the thermodynamically controlled release, which is related to the concentrations of free molecules in aqueous solution, the kinetically controlled release is more closely related to the total concentrations in the product.

Thickening agents do not have a major effect on the molecular diffusion as long as the mobility of the water molecules is not substantially affected (Darling *et al.* 1986, Bylaite *et al.* 2003, Malone *et al.* 2000). The diffusion coefficients in hydrocolloid solutions are given by (Waggoner *et al.* 1993):

$$D = D_w(1 - \Phi_h)^2/(1 + \Phi_h)^2 \tag{11.26}$$

where D and D_w are the diffusion coefficients (m^2/s) in the thickened system and pure water, respectively, and Φ_h is the volume fraction of the hydrocolloid. For hydrocolloid volume fractions of up to 1% this means that the diffusion coefficients in water decrease by not more than 4%.

In dynamic systems the most important diffusion mechanism is the eddy or convective diffusion. Eddy diffusion transports elements or eddies of the fluid or gas from one location to another, carrying with them the dissolved aroma molecules. The rate of eddy diffusion is completely independent of aroma compound type. In contrast with molecular diffusion, eddy diffusion can become strongly inhibited by thickening agents already at low concentrations (Darling *et al.* 1986, De Roos and Wolswinkel 1994, Roberts *et al.* 1996). If gelation takes place, the eddy diffusion can increase again when the gel breaks down into particles and allows their transport by eddy diffusion (Darling *et al.* 1986).

The negative effect of viscosity on the eddy diffusion can be nullified by putting more kinetic energy into the system. Through effective stirring the concentrations gradients at the surface of viscous solutions can be almost completely eliminated. In that case, we will see no effect of viscosity on the aroma release (Rabe *et al.* 2003c).

The eddy diffusion of the aroma compounds is not much affected by binding to lipids or macromolecules because the bonded aroma molecules are usually transported at the same rate as the free molecules. If the partitioning in the product phase would be instantaneous, the eddy diffusion-controlled aroma release would be simply related to the aroma compound concentrations in the product. However, the decrease of the aroma intensity of emulsions with increasing oil droplet size shows that under kinetically controlled conditions this is not the case (Linforth *et al.* 2002, Carey *et al.* 2003, Weel, 2004).

Whether static or eddy diffusion (or both) is affecting the aroma release can be concluded from the change of the release over time. At constant forced convection, eddy diffusion results in a steady state release during which the relative surface concentrations C_w^i/C_w remain constant (Fig. 11.2b). The result is

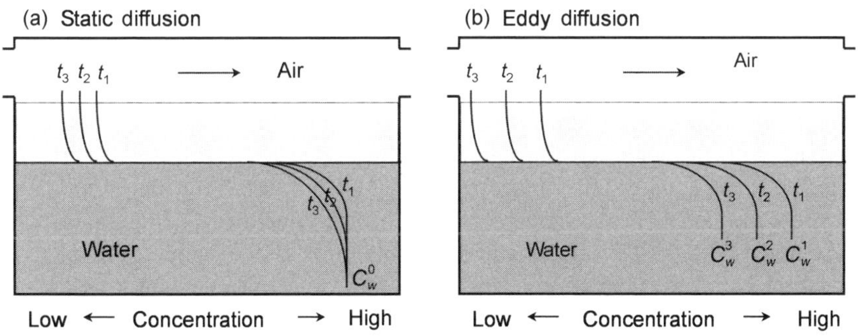

Fig. 11.2 Change of the concentration gradients of an aroma compound with time during its release from (a) a static and (b) a dynamic aqueous solution.

that the concentrations C_w and C_w^i as well as the absolute release rates decrease logarithmically with time (de Roos 2000).

Under static conditions, the concentration C_w^i at the product surface decreases more sharply because the thickness δ_w of the boundary layer is increasing with time (Fig. 11.2a). Therefore, the flavour release rates initially decrease sharply and then level off because the concentrations in the bulk phase remain constant (Darling *et al.* 1986). A plot of the logarithm of the release rate versus time gives a concave curve for the static diffusion versus a straight line for the eddy diffusion (De Roos 2000). In general, eddy diffusion is much more effective in reducing the concentration gradients at the product surface than static diffusion. The diffusion mechanism affects the release models through its effect on the mass transfer coefficient.

11.3.3 Modelling of the mass transfer

Release from static systems

Some time after the one-dimensional diffusion from static solutions has started, the aroma concentrations at the surface are fairly constant. Darling *et al.* (1986) derived that under these conditions the mass transfer coefficient k_p is given by $\sqrt{(D/\pi t)}$. So, the mass flux from a static solution at time t is then as follows:

$$J_p = (\sqrt{D/\pi t})(C_p^i - C_p^0) \qquad 11.27$$

There is good agreement with experimental data but only at relatively long times when the surface concentrations no longer rapidly decrease.

Another model used to describe the release from static systems is the stagnant film model (Whitman 1923). The model assumes that the boundary layers at the interface are stagnant and of constant thickness and that mass transport through these layers is by static diffusion (Hills and Harrison 1995, Marin *et al.* 2000). In that case, $k = D/\delta$. So, the mass flux is then given by:

$$J_p = (D_p/\delta_p)(C_p^i - C_p^0) \qquad 11.28$$

Release from dynamic systems

In dynamic systems the boundary layers are often not stagnant. An interfacial mass transfer theory such as the penetration theory (Higbie 1935) can then better be used to describe the release. According to this theory, mass transfer between the phases takes place when a volume element from the bulk phase comes into contact with the interface for a finite time interval t_e. During this fixed time contact, mass transfer takes place by molecular diffusion (Darling *et al.* 1986, Overbosch *et al.* 1991, Harrison *et al.* 1997). Subsequently, the volume element is remixed with the bulk phase and the whole process is repeated. According to equation 11.27, the momentary mass flux at time t between the volume elements is given by:

$$J_p = \sqrt{D/\pi t}(C_p^i - C_p^t) \qquad 11.29$$

where C_p^t is the aroma compound concentration in the bulk phase at time t. From this follows then that the average flux over a finite time interval t_e is given by:

$$J_p = \frac{1}{t_e}\int_0^{t_e} \sqrt{D/\pi t}(C_p^i - C_p^t)dt = 2\sqrt{D/\pi t_e}(C_p^i - C_p^t) \qquad 11.30$$

Note that according to this theory the flux is greatest for the shortest contact times (i.e. at turbulence).

An important extension of the penetration theory is the surface renewal theory (Danckwerts 1951). The difference between both theories is that the fixed contact time t_e is replaced by the average of a wide variety of different contact times with different degrees of penetration. The average flux over a time period t is then:

$$J_p = \sqrt{Ds}(C_p^i - C_p^t) \qquad 11.31$$

where s is the fractional surface renewal rate.

Because t_e and s cannot be measured independently, k_p has either to be measured directly from experimental data or to be calculated with the aid of relationships for the convective mass transport in tubes or stirred vessels. The latter approach is discussed under section 11.4.1 (Release from dynamic phases).

Another model, which considers the aroma release as a sequence of extractions of the product phase with air, requires the determination of the volume fraction of air that is in equilibrium with a fixed volume fraction of product during a short contact time t_e. This is equivalent to determining the quotient k_a/k_p of the mass transfer coefficients in air and product. This model, known as the non-equilibrium partition model, will also be discussed under section 11.4.1 (Release from dynamic phases).

11.4 Mechanistic modelling of aroma release

The mechanistic models vary with the conditions of the release. Therefore, the discussion of these models has been subdivided in separate sections that cover the following situations: release in model systems (11.4.1), aroma retention during heat processing (11.4.2), *in vivo* aroma release from liquid and solid foods (11.4.3) and aroma persistence after swallowing (11.4.4).

11.4.1 Modelling of *in vitro* aroma release

Release from static phases

It is the aroma released under static conditions that is in general determining the odour of a product prior to consumption. Models describing this release often assume that the resistance in the product phase controls the release and that resistance in the gas phase can be neglected (Darling *et al.* 1986). In that case $C_p^i = C_a^0 P_{ap}$ and the flux is then:

$$J_p = \sqrt{D_p/\pi t}(C_a^0/P_{ap} - C_p^0) \qquad 11.32$$

C_a^0 initially sharply decreases with time. So, after some time $C_a^0/P_{aw} \ll C_w^0$ and can also be neglected. With these assumptions, indeed a good fit with theory was obtained for the long time release.

Overbosch *et al.* (1991) made a further step by relating the concentration gradient ΔC_p to the diffusion coefficients in air and product:

$$J_p = C_p\sqrt{D_p/\pi t}\ \frac{P_{ap}\sqrt{D_a/D_p}}{1 + P_{ap}\sqrt{D_a/D_p}} = C_p\sqrt{D_p/\pi t}\ \frac{P_{ap}}{P_{ap} + \sqrt{D_p/D_a}} \qquad 11.33$$

The total amount M_t released per surface area A as a function of time is then given by:

$$M_a^t = 2C_pA\sqrt{D_pt/\pi}\ \frac{P_{ap}}{P_{ap} + \sqrt{D_p/D_a}} \qquad 11.34$$

Equation 11.34 resembles equation 11.3, which describes the release at equilibrium. The quotient $\sqrt{(D_p/D_a)}$ in equation 11.34 is equivalent to the quotient V_p/V_a in equation 11.3. In equation 11.34 the volumes of the boundary layers that have been in con-tact with each other during time t are given by $(A\sqrt{D_pt})/(A\sqrt{D_at}) = \sqrt{(D_p/D_a)}$. The term $2C_pA\sqrt{(D_pt/\pi)}$ in equation 11.34 is equivalent to M_{total} in equation 11.3 and represents the mass of the aroma compound in the boundary layer that has been subjected to extraction with air during a time period t.

Equation 11.34 confirms the conclusion in section 11.3.1 that the concentrations gradients in the product phase are a function of the volatility P_{ap} and the mass transport (diffusion) coefficients in water and air. If $P_{ap} \gg \sqrt{(D_p/D_a)}$, the quotient $P_{ap}/[P_{ap} + \sqrt{(D_p/D_a)}]$ approaches the value of 1. The release is then very similar for all compounds and said to be strongly diffusion ($\sqrt{(D_p/D_a)}$) controlled. At low values of P_{ap}, the quotient varies with the value of P_{ap} and the release is then thermodynamically (P_{ap}) controlled.

Marin *et al.* (1999, 2000) have used the stagnant film model to describe the release from a stagnant aqueous phase. Since the mass transfer coefficients in water were found to be higher than expected on the basis of the diffusion coefficients (due to the turbulence created by the airflow over the water surface), an apparent mass transfer coefficient was defined, which is equivalent to the effective diffusion coefficient D divided by an equivalent limiting layer δ_p. The long-term pseudo-steady-state release at time t is then given by the following simple relationship:

$$C_a^t/C_a^0 = \frac{1}{1 + v_a/k_oA} \qquad 11.35$$

where C_a^t/C_a^0 is the concentration in the gas phase at time t relative to the equilibrium headspace concentration at $t = 0$, and v_a is the airflow rate (m/s). Since k_o is a function of the air–water partition coefficient (see equation 11.22), the relative concentration in the headspace is a simple function of the partition coefficient and the gas flow rate. Under the conditions used, P_{aw} controlled the release at $P_{aw} < 10^{-3}$, whereas at $P_{aw} > 10^{-3}$ the resistance to mass transfer in water controlled the release.

Release from dynamic phases

McNulty and Karel (1973a) were among the first to develop a flavour release model. The model describes flavour release in the mouth from o/w emulsions assuming that: (a) flavour compounds are transferred from oil to water when the phase equilibria are disturbed by dilution with saliva and (b) only the aqueous flavour concentrations stimulate perception. Although the model has also been used to explain differences in odour perception, it deals specifically with the perception of taste in the mouth (McNulty 1987). The kinetic component of the aroma release is not incorporated in the model. In view of the low significance of the model for the in vivo aroma release it will not be discussed here in further detail.

Harrison *et al.* (1997) and Harrison and Hills (1997b) used the penetration theory of mass transfer to model the aroma release from stirred liquids in closed systems. In this type of 'model mouth' systems, the aroma concentrations in the gas phase can build up till equilibrium has been achieved. It was assumed that the exchange of aroma molecules between the bound and unbound state is very fast compared to the transport of flavour across the water–gas interface which means that the latter is the rate-determining step. Based on these assumptions the following equation was derived for the release from stirred solutions with surface area A:

$$C^t = C_p^0 \frac{P_{ap}}{P_{ap}V_a/V_p + 1}\left(1 - \exp\left\{-\left[\frac{P_{ap} + V_p/V_a}{P_{ap}}\right]\frac{k_p A}{V_p}t\right\}\right) \qquad 11.36$$

The expression in brackets is the time dependent factor for the release in a closed system. At $t = \infty$ the equilibrium headspace concentration is achieved:

$$C_a^\infty = C_p^0 \frac{P_{ap}}{P_{ap}V_a/V_p + 1} \qquad 11.37$$

This equation is again equivalent to equation 11.3 (substitute C by M/V). The model predicts higher release rates with increase of stirring rate and decrease of viscosity, which is in agreement with the experiment (Bakker *et al.* 1998).

Most other release models apply to open systems, which better imitate the situation in the mouth and throat when air is passing through. When it is assumed that the air becomes then saturated with aroma, the following relationship for the amount M_a^t of an aroma compound released at time t can be derived (Overbosch *et al.* 1991):

$$\frac{M_a^t}{M_p^0} = 1 - \exp\left[-v_a P_{aw}\frac{A}{V_p}t\right] \qquad 11.38$$

where v_a is the gas flow rate. Since the outflowing air is assumed to be in equilibrium with the product, the amount released is here independent of the mass transport coefficient.

A different approach was used by Banavara *et al.* (2002) and Rabe *et al.* (2002, 2004) for modelling the aroma release from stirred solutions. The investigators assumed that the mass transport in the liquid phase is the rate-determining step ($k_o = k_p$) and that k_p can be calculated with the following relationship for the convective mass transport in stirred vessels:

$$k = 0.026\ \mathrm{Re}^{0.8}\ Sc^{1/3} \frac{P}{H_{\ln}} \tag{11.39}$$

where: Re $=$ Reynolds number $= \eta/\rho D$
Sc $=$ Schmidt's number $= d^2 n\rho/\eta$
P $= D(\delta^2/V_f)(p/S)^{0.5}$
$H_{\ln}$ $=$ measure for average distance to the surface of the liquid;

where d is the diameter of the stirrer (m), ρ the density of the liquid phase (kg/m^3), n the number of stirrer rotations per second (s^{-1}), δ the effective film thickness of air flowing over the liquid–gas interface (m), V_f the volumetric flow rate of air (m^3/s), p the partial pressure of the aroma (Pa) and S the solubility of the flavour molecule in water (kg/m^3).

Under the conditions used, the release was strongly thermodynamically (P_{ap}) controlled (Banavara *et al.* 2002). This indicates that stirring was effective in reducing, and even in eliminating, the aroma concentration gradients near the surface of the fluids. Therefore, the release in emulsions was simply related to the aqueous aroma concentrations (Rabe *et al.* 2004) and the release from sugar solutions not significantly affected by differences in solution viscosity (Rabe *et al.* 2003c).

De Roos and Wolswinkel (1994) developed the non-equilibrium partition model for describing the aroma release from dynamic systems. The model is premised on the following assumptions:

- The aroma of the product is extracted consecutively with infinitesimal volumes V_a^* of air.
- During each successive extraction, equilibrium is achieved at the air-product interface between small volume fractions V_p^* of product and V_a^* of air.
- After each extraction the initial aroma concentrations at the surface of the product are restored by eddy diffusion before the next extraction takes place.

The retained fraction after the first extraction is then given by:

$$M_p^1/M_p^0 = \frac{V_p^*}{V_p}\left[\frac{P_{pa}}{P_{pa} + V_a^*/V_p^*}\right] + (1 - V_p^*/V_p) \tag{11.40}$$

where $1 - V_p^*/V_p$ stands for the non-extracted part of the product. Please note the similarity with equation 11.34 ($M_p^0 V_p^*/V_p = 2C_w A\sqrt{(D_w t/\pi)}$).

After n consecutive extractions and remixing steps the fraction released is given by:

$$M_a^n/M_p^0 = 1 - \left\{\frac{V_p^*}{V_p}\left[\frac{P_{pa}}{P_{pa} + V_a^*/V_p^*}\right] + (1 - V_p^*/V_p)\right\}^n \tag{11.41}$$

The number of extractions n is related to time. The quotient V_a^*/V_p^* is equivalent to the quotient of the mass transfer coefficients k_a/k_p; k_a and k_p determine the thickness of the boundary layers that are in contact with each other during each short time contact t_e, so $V_a^*/V_p^* = (Ak_a t_e)/(Ak_p t_e) = k_a/k_p$. The value of V_a^*/V_p^* is determined directly by fitting with experimental data (de Roos and Graf 1995). The value of n is linked to that of V_p^*/V_p and always very large. Since in dynamic systems the diffusion rates are very similar, the same values of V_a^*/V_p^* and V_p^*/V_p can be used for all aroma compounds. The model predicts lower release rates with increase of viscosity, in agreement with experiment (De Roos and Wolswinkel 1994).

In principle, equation 11.41 applies to a completely eddy diffused controlled release. If part of the mass transport in the product takes place by molecular diffusion, this will be reflected by an increase of V_a^*/V_p^* due to the lower rate of the molecular diffusion (lower value of V_p^*). Equation 11.41 will then also have to take into account the effect of flavour binding on the molecular diffusion. Nevertheless, the unmodified equation predicts the release from static systems already quite satisfactorily (De Roos and Graf 1995).

11.4.2 Modelling of aroma retention during heat processing and storage

Most studies on aroma retention during heat processing have been done on liquid aqueous systems (Steinke *et al.* 1989, Stanford and McGorrin 1994, Lindstrom and Parliment 1994). The studies demonstrate that the retention increases with decreasing volatility of the aroma compounds in the product base. Another factor that affects the release is the water loss during heating (de Roos and Graf 1995). The steam developed stimulates mass transfer in the product and serves as a vehicle to transport aroma compounds across the product–air boundary.

As expected, aroma loss from stagnant systems is strongly diffusion controlled. This is the reason why aroma loss during spray drying shows better correlation with dynamic than with static headspace concentrations (Bangs and Reineccius 1990). With increase of water loss and product viscosity, the relative importance of the molecular diffusion in the mass transport is growing as demonstrated in studies by Kerkhof and Thijssen (1975), Bruin and Luyben (1980), Bruin (1992), Coumans *et al.* (1994) and Goubet *et al.* (1998). These studies, most of which were done on concentrated solutions of sugars or maltodextrin, have resulted in two theories on the retention of volatile components in low moisture systems: the selective diffusion theory of Thijssen and Rulkens (Rulkens 1973), and the micro-region entrapment theory of Flink and Karel (1970).

The selective diffusion theory explains the high aroma retention in viscous hydrophilic systems on the basis of the low diffusivity of aroma compounds (D_A) in these systems in comparison with the diffusivity of water (D_W). The rapid decrease of the quotient D_A/D_W with decrease of moisture content is the reason of the high aroma retention at high solids content. The micro-region entrapment theory assumes that during moisture loss micro-regions are formed

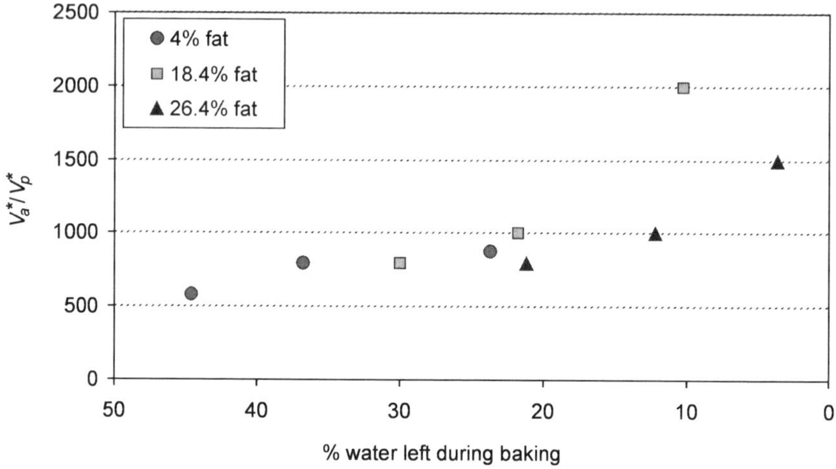

Fig. 11.3 Effect of water content on the resistance to mass transfer (expressed as V_a^*/V_p^*) during microwave baking of cakes of different fat content (according to de Roos and Graf 1995). Bake times: 1, 2 and 3 minutes.

in the food matrix in which aroma molecules are trapped. Both theories apply to aroma retention in hydrophilic matrices but do not provide useful relationships for predicting aroma retention in biphasic systems.

The diffusion of aroma compounds in a solid hydrophilic medium is strongly hindered, if possible at all. So, it can happen that at very low moisture levels, the aroma compounds generated during Maillard reactions become entrapped in the hydrophilic phase and can no longer partition into lipid phase and into the air. These aroma compounds can then only be released during consumption when the hydrophilic phase is hydrated or dissolved. The orthonasal aroma of solid products has to come from compounds already present in the lipid phase, which is a poor barrier for aroma compounds (De Roos 2003).

De Roos and Graf (1995) have used the non-equilibrium partition model to relate aroma loss during baking to the aroma volatility in the dough. Cakes of different fat content were baked while keeping the total amount of water and fat constant. Fitting with experimental data indicated that the resistance to mass transfer in the dough (expressed as V_a^*/V_p^*) increases with increasing water loss (Fig. 11.3). The resistance in high-fat cakes is relatively low suggesting that a high fat content facilitates mass transport. This might be attributed to the larger surface area and continuity of the fat phase, both of which facilitate the aroma transport.

11.4.3 Modelling of *in vivo* aroma release

Mechanisms of aroma transport from product to nose

When modelling the *in vivo* aroma release, it makes sense to distinguish between the initial burst of aroma release and the aroma persistence (see Fig. 11.4):

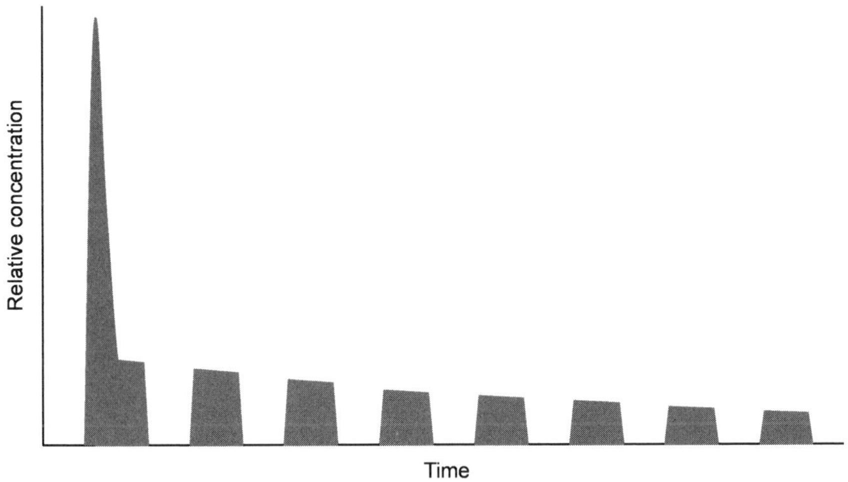

Fig. 11.4 Schematic picture of the breath-by-breath concentrations of an aroma compound in the exhaled air.

- The high initial burst of volatile release that immediately follows swallowing has been associated with the transfer of air from the mouth; here the product has had the chance to build up relatively high headspace concentrations (Normand *et al.* 2004, Linforth *et al.* 2003). Another explanation for the initial burst of aroma release is that the act of swallowing has caused a transient increase of the interfacial mass transfer due to a higher turbulence (Wright *et al.* 2003). Since the concentrations of volatiles in the breath immediately after swallowing are similar to those observed in the breath during chewing, the same processes are assumed to be involved (Linforth *et al.* 2003). Data analysis suggested that with each chew air is pumped out of the mouth into the throat, where it joins the exhaled breath (Hodgson *et al.* 2003).
- The release during the second and next exhalation after swallowing is assumed to take place from a thin film remaining on the surface of the pharynx after swallowing (Normand *et al.* 2004, Linforth *et al.* 2003, Wright *et al.* 2003). During exhalation, the aroma compounds present in this film are exposed to relatively large airflows, which results in a strongly kinetically controlled release. This explains why the aroma concentrations in the later exhalations are lower than those in the first exhalation and why this difference is most pronounced for the most volatile compounds.

The whole pattern of aroma release during the successive exhalations strongly resembles the release from static liquids (Marin *et al.* 1999, 2000). According to equation 11.35, the low concentrations released during the second and next exhalations have to be attributed to higher air flow rates.

To obtain the same aroma impact in a different product, each aroma compound should have the same maximum intensity in the new as in the original product. Since the aroma concentrations in the first breath are responsible for the maximum intensity of the aroma, it is these concentrations that have to be the same.

Resistance to mass transfer during consumption

Comparison of the release under equilibrium conditions with that during drinking indicates that the latter release is strongly diffusion controlled (Linforth *et al.* 2002) and for the most volatile compounds even completely diffusion controlled according to equation 11.23 (Fig. 11.5). Analysis of the published data indicates that the release is completely diffusion controlled at $P_{aw} > 0.01$ ($\approx$ volatility of ethyl acetate in water). At $P_{aw} < 0.01$ the release becomes more thermodynamically controlled. This means that during drinking the mass transport in air is about 100 times faster than that in water ($V_a^*/V_w^* = k_a/k_w \approx 100$). The observed trend in the relative release rates was independent of whether the concentrations were measured in the breath from the nose or in the breath from the mouth. This suggests that in both cases the rate-determining step is the transport across the water–gas interface in the mouth.

The high resistance to mass transfer during consumption is also evident from the effects of alcohol, lipids and proteins on the aroma release. The large effects of alcohol (Boelrijk *et al.* 2003) on the aroma volatility are reflected by only minor effects on the release during consumption, typical for a kinetically controlled release. The major effects of lipids (De Roos and Wolswinkel 1994, Roberts *et al.* 2003b, Doyen *et al.* 2001, Miettinen *et al.* 2003, Weel 2004),

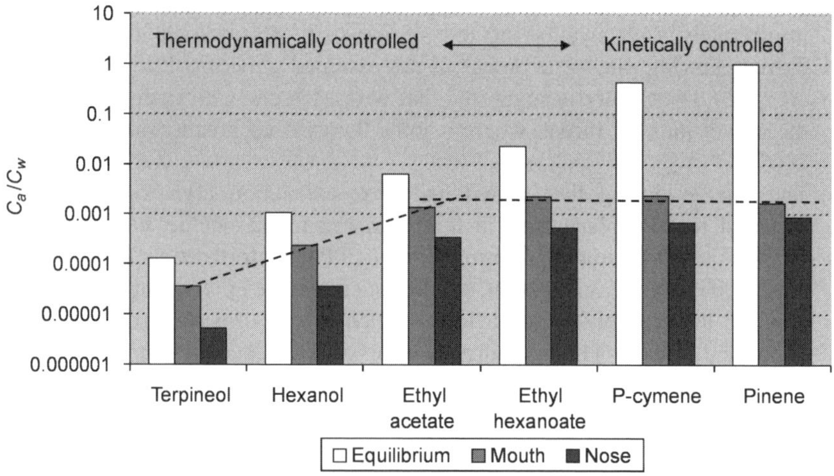

Fig. 11.5 Aroma concentrations released from water under equilibrium conditions compared with those released into the breath from mouth and nose. The latter concentrations were calculated from the air–water partition coefficients and data published by Linforth *et al.* (2000).

proteins (Le Guen 2003, Weel *et al.* 2003), starch (Kant *et al.* 2003) and cyclodextrin (Kant *et al.* 2004) on the aroma volatility are also levelled down to only minor effects on the *in vivo* aroma release.

The only minor effects of aroma binding on the dynamic release demonstrates that the exchange of aroma molecules between the bound and unbound state is fast compared to the transport of aroma across the water–gas interface (see also section 11.3.2). However, there are exceptions. In contrast with the fast hydrophobic aroma–protein interactions (Le Guen and Vreeker 2003), the amino-carbonyl interactions of aldehydes with proteins seems to be clearly slower than the liquid–air interfacial mass transfer. This explains the relatively strong effect of proteins on the *in vivo* release of aldehydes. However, at low pH the interactions of aldehydes with proteins are weak or absent (Lübke *et al.* 2000, Zhou and Decker 1999). Only the hydrophobic interactions between proteins and aldehydes are then influencing the *in vivo* release (Weel *et al.* 2003).

The effects of viscosity (Hollowood *et al.* 2002, Kora *et al.* 2004, Malone *et al.* 2003, Linforth and Taylor 2000) and temperature (Boelrijk *et al.* 2003) on the *in vitro* release are also levelled down under *in vivo* conditions. This implies that differences in product viscosity and temperature can often be neglected when predicting the *in vivo* aroma release.

As already mentioned at the beginning of this section, there are strong indications that the persistent release is more mass transfer controlled than the initial release. This has most likely to be attributed to the higher airflow rates to which the product is exposed during the later exhalations (see equation 11.35). With the limited data available it is difficult to make an accurate estimate of the factor by which the quotient V_a^*/V_w^* will exceed that during the first exhalation. However, the very similar concentrations of ethanol (Log $P_{aw} = -3.7$) in the first and later breaths after its consumption in aqueous solution (Linforth *et al.* 2003) indicate that the persistent release of ethanol is not mass transfer controlled in contrast with that of the more volatile compounds (Linforth and Taylor 2000, Normand *et al.* 2004). This would mean that during the persistent release $200 < V_a^*/V_w^* < 10^{3.7}$.

Modelling of the release from liquid products

The mathematical models that describe the aroma release from liquid products are adaptations of models developed for model systems and assume that aroma-rich air in the mouth is transferred directly into the nose. Harrison and Hills (1997a) and Harrison (1998) have adapted the mathematical models for the release in closed systems for the effects of breathing and saliva dilution. The release equations are an extension of equation 11.36 and are taking into account the interactions with lipids and macromolecules as well as the effect of viscosity. The models assume that the *in vivo* release is simply related to the air–product partition coefficient, i.e. related to the aroma concentrations in the aqueous phase rather than to the total aroma compound concentrations in the product.

Modelling of the release from solid products

Aroma compounds in solid products are normally first released into the saliva before they are transferred to the mouth's headspace and the nose. When modelling the release from solid products, it is important to distinguish between the following release mechanisms:

- *Product dissolution*. Characteristic for this release mechanism is that all compounds are released at the same rate and to the same extent (i.e. 100%). So, the release rate is a simple function of the rate of product dissolution.
- *Product extraction*. In this case the release from product to saliva is incomplete. The aroma compounds are released at different rates determined by the product–saliva partition coefficients.

During the consumption of solid foods, both release mechanisms are often playing a role. When the aroma release is through product dissolution, the rate of dissolution and the saliva flow rate determine the concentrations in the saliva. The latter concentrations are just a factor x lower than those in the product. For the prediction of the aroma release, one can then use the equations for liquid products taking into account that the product has been diluted by a factor x.

The aroma release by dissolution of a solid product can be described by the stagnant-layer theory of interfacial mass transfer (Hills and Harrison 1995, Harrison and Hills 1996, Wright *et al.* 2003a,b, Harrison *et al.* 1998). Assuming that the resistance in the product phase controls the release, the mass transfer from product to saliva is given by:

$$\frac{dM}{dt} = J \cdot A = \frac{D}{\delta} A(C_p - C_w^t / P_{aw}) \tag{11.42}$$

where A is the surface area of the product–saliva interface and C_p and C_w are the aroma concentrations in the product and the saliva bulk phase. The release from gelatine gels is related to the melting temperature of the gel, which in turn is related to the decreasing sucrose concentrations at the surface of the gel (Harrison and Hills 1996).

Since the rate of product dissolution is linearly related to the surface area of the product, release equation 11.42 has been extended to incorporate functions that account for changes in product sphericity and fragmentation during chewing (Harrison *et al.* 1998, Wright *et al.* 2003a). Although all models neglect the resistance to mass transfer over the saliva–air interface, there is good agreement with *in vitro* and *in vivo* release experiments.

Moore *et al.* (2000) studied the release from gels containing dispersed fat particles. Using the release model of Harrison and Hills (1997a) good agreement with sensory time intensity data was obtained provided that the effects of aroma removal by breathing and swallowing were incorporated in the model. Fat particle size was found to have a negligible effect on the time–intensity profile of the release indicating that the fat–water partitioning in the gelled emulsion particles is fast compared to the diffusion from the particles. Variation of gel firmness at same solids content was also found to have a negligible effect, which is in agreement with equation 11.26 for the static diffusion in thickened systems.

Lian (2000) has proposed a more general model for the aroma release from solids taking into account the resistance to mass transfer in both solid and fluid phase:

$$\frac{1}{k_o} = \frac{1}{k_w} + \frac{1}{k_p P_{pw}} \tag{11.43}$$

For the release from gelled emulsion spheres with radius r into a continuous sink of water ($C_w = 0$), the following relationship for the amount released at time t is then obtained:

$$M_w^t / M_p^0 = 1 - \exp\left[-\frac{3}{r} t \frac{k_w}{P_{pw} + k_w/k_p}\right] \tag{11.44}$$

This equation degenerates to the Crank equation if the diffusion in the particle is rate limiting and to the Sherwood equation when the mass transfer in the surrounding liquid phase becomes rate limiting.

In practice, the *in vivo* aroma release from gelled emulsion particles is also influenced by the mass transfer across the liquid–air interface. At a certain particle size, the rate-limiting step is switching from the mass transfer across the particle–liquid interface to that across the liquid–air interface (Malone and Appelqvist 2003). A simple relationship has been derived for the critical size above which the release of aroma from the dispersion of gelled emulsion particles is affected by the size of the particles (Lian *et al.* 2004).

That in general the resistance to mass transfer over the saliva–air interface cannot be neglected became also clear during attempts to apply the non-equilibrium partition model to the prediction of the aroma release from chewing gum (de Roos 2003). A good fit with experimental data could only be obtained if the resistance to mass transfer across the saliva–air interface was taken into account.

11.4.4 Modelling of aroma persistence after swallowing

Models describing the persistent aroma release assume that the mass transport in the thin coating of the throat is the rate-determining step. In the model described by Wright *et al.* (2003) the basic release equation is extended to cover the release that begins with the second breath after swallowing:

$$\frac{dM_a^t}{dt} = Ak_f(C_f - C_a/P_{af}) - (p + q \sin \omega t)C_a^t \tag{11.45}$$

where the last term is the loss of aroma from the airway during periodic breathing with a frequency ω and amplitude q. The subscript f refers to the film covering the throat and the term p accounts for the constant irreversible loss from the air volume as a result of absorption by the body. The model provided good fits with sensory time intensity curves for aqueous solutions.

Normand *et al.* (2004) developed a multiple-extraction model for predicting the aroma persistence. This model assumes that the aroma release is the result of a series of equilibrium batch extractions:

$$M_a^n/M_w^0 = \frac{P_{aw}}{P_{aw} + V_w/V_a}\left(\frac{V_w/V_a}{P_{aw} + V_w/V_a}\right)^{n-1} \quad 11.46$$

A good fit with experimental data is only obtained if apparent air–water partition coefficients are being used. Since these are calculated from the diffusion-controlled release into the breath, they are much more equal than the real air–water partition coefficients. In this respect, this model differs from the non-equilibrium partition model, which uses real air–water partition coefficients and attributes the uniformity of the release to mass transfer effects (see section 11.4.1).

Equation 11.46 describes the release that begins with the second breath after swallowing. For prediction of the more long-term release, aroma absorption into a layer of mucosa underneath the layer of saliva must be assumed to be involved. The model correctly predicts the trends observed experimentally.

11.5 Empirical modelling of aroma release

Empirical models use statistical methods to relate experimental data to physico-chemical parameters. With the aid of the quantitative structure property relationships (QSPR) approach, relationships can be derived without completely understanding the processes that are involved. This approach is of particular interest for such complex processes as the aroma release during consumption. For good results the selection of relevant parameters is important (Linforth *et al.* 2000, Linforth 2002). In QSPR, all of these parameters can usually be estimated provided sufficient data are available to describe the variations in these parameters. A disadvantage of the method is that for high accuracy a large set of experimental data is required.

QSPR has been used to produce relationships for predicting air–water partition coefficients (Katritzky *et al.* 1998), time-intensity profiles of aroma release including maximum intensity and persistence (Linforth *et al.* 2000, Linforth and Taylor 2000), the dependence of the *in vivo* release on the aroma volatility (Linforth *et al.* 2002), and the effect of lipids on the aroma release (Carey *et al.* 2002). An example of the type of relationships obtained is the following relationship for the aroma persistence P_A in the breath after consumption of an aqueous solution (Linforth and Taylor 2000):

$$P_A = 114 - 24LogP - 16O - 23Log\rho_L - 8.0C_{ARB} - 14Logp^2 - 2.6Log\rho_L^2 + 3.3LogP^3 \quad 11.47$$

where P is the estimated octanol–water partition coefficient, ρ_L the estimated vapour pressure of the aroma compound, O the ether linkage count and C_{ARB} the carbonyl group count.

The QSPR approach has also been used for reformulating aroma compositions for same impact in different products (Sonnenberg *et al.* 2002). The process starts with comparing the headspace concentrations of a set of aroma compounds in the original and new product. The differences are then related to various physical parameters, such as log *P* and boiling point, using the QSPR approach. The best regression equation resulting from this process is used for calculating the concentrations in the new product.

11.6 Applications

One of the most useful applications of the release models is the prediction of the aroma release during consumption and the aroma retention during processing. Computer programs have been developed that allow flavourists to optimise aroma compositions for high performance at low costs through:

- Optimising flavour release by:
  - Rebalancing aroma compositions to compensate for the differences in aroma release from different products (De Roos and Wolswinkel 1994, Sonnenberg *et al.* 2002).
  - Improving flavour effectiveness by selecting the most effective aroma chemicals for a particular application, e.g., long-lasting aroma compounds for chewing gums (De Roos and Wolswinkel 1994).
- Creation of heat-stable flavourings (De Roos and Mansencal 2003) by:
  - Selecting aroma compounds that combine high retention during heating with high release during consumption.
  - Rebalancing of aroma compositions to compensate for anticipated losses during heating and storage.

Other useful future applications in the flavour area could be:

- Compensation for anticipated loss of volatiles during spray drying.
- Calculation of costs per odour unit by linking the *in vivo* release with odour thresholds. This allows elimination of less effective aroma chemicals from aroma compositions.
- Optimisation of the time–intensity profiles of the aroma release (Malone *et al.* 2000).

Examples of useful applications in flavour analysis are:

- Prediction of gas and liquid chromatographic retention times (Curvers *et al.* 1985a,b).
- Selection of the optimum aroma separation or recovery method.
- Estimation of equilibration times during headspace analysis.

11.7 Future trends

11.7.1 Model systems

To simulate the *in vivo* aroma release more closely, future model system studies will have to be carried out under more kinetically controlled conditions. The replacement of the 'artificial mouth' by an 'artificial throat' will result in better simulation of the *in vivo* aroma release (Weel 2004).

11.7.2 Aroma retention

Prediction of aroma retention during heating will show that encapsulation or *in situ* generation will often be required to achieve effective aromatisation, certainly if the aroma compounds are very volatile (dimethyl sulphide, Strecker aldehydes) or chemically unstable (2-acetylpyrroline and thiols). Future heat-stable flavourings might therefore be expected to contain flavour precursors and encapsulated aroma compounds (De Roos 2003).

11.7.3 *In vivo* aroma release

Models will be further improved by taking into account that the release in mouth and throat is strongly kinetically controlled and therefore much less affected by binding to lipids and (macro)molecules than the volatility.

11.7.4 Relating aroma release to perception

The perception of aroma molecules is not only related to their release but also to their absorption in the nasal cavity (Keyhani *et al.* 1997). Moreover, taste and texture are influencing aroma perception (Baek *et al.* 1999, Taylor *et al.* 2003, Weel *et al.* 2002). Therefore, future research may be expected to focus on better understanding the relationship between aroma release and perception.

11.7.5 Relating aroma release to consumer preference

Most interesting would be to know how the aroma release patterns influence consumer preference. For example, is aroma persistence always positive or only in certain applications? Because of the high practical significance of answers to this kind of questions, this subject deserves more attention in future.

11.8 Sources of further information

The theory of aroma–matrix interactions is taken from the fields of physical chemistry and chemical engineering. For the basic physical chemistry of these interactions the reader is referred to the article of Taylor (1998). For a more in-depth study of mass transfer phenomena the following books are recommended:

CRANK J (1975) *The Mathematics of Diffusion*, 2nd edn, Oxford University Press, New York.
COULSON J M and RICHARDSON J F (1993) *Chemical Engineering*, Pergamon Press, Oxford, UK.
CUSSLER E L (1997) *Mass Transfer in Fluid Systems*, 2nd edn, Cambridge University Press, UK.
FOURNIER R L (1998) *Basic Transport Phenomena in Biomedical Engineering (Chemical Engineering)*, CRC Press, Boca Raton, FL, USA.
SARAVACOS G D and MAROULIS Z B (2001) *Transport Properties of Foods*, Marcel Dekker, New York.
SHYY W, OUYANG H, BLOSCH E, THAKUR S S and LIU J (1997) *Computational Techniques for Complex Transport Phenomena*, Cambridge University Press, UK.

11.9 References

ANDRIOT I, HARRISON M, FOURNIER N and GUICHARD E (2000) 'Interactions between methylketones and β-lactoglobulin: sensory analysis, headspace analysis and mathematical modelling', *J Agric Food Chem*, 48, 4246–4251.
BAEK I, LINFORTH R S T, BLAKE A and TAYLOR A J (1999) Sensory perception is related to the rate of change of volatile concentration in-nose during eating of model gels, *Chem Senses*, 24, 155–160.
BAKKER J, BOUDAUD N and HARRISON M (1998) 'Dynamic release of diacetyl from liquid gelatine in the headspace', *J Agric Food Chem*, 46, 2714–2720.
BANAVARA D S, RABE S, KRINGS U and BERGER R G (2002) 'Modelling of flavour release from water', *J Agric Food Chem*, 50, 6448–6452.
BANGS W E and REINECCIUS G A (1990) 'Prediction and flavour retention during spray drying: an empirical approach', *J Food Sci*, 55, 1683–1685.
BOELRIJK A, BASTEN W, BURGERING M, GRUPPEN H, VORAGEN F and SMIT G (2003) 'The effect of co-solvent on the release of key flavours in alcoholic beverages; comparing in vivo with artificial mouth-MS nose measurements', in Le Quéré J L and Etiévant P X, *Flavour Research at the Dawn of the Twentieth-first Century*, Paris, Editions Tec & Doc, 204–207.
BRUIN S (1992) 'Flavor retention in dehydration processes', in Singh R P and Wirakartakusumah M A, *Advances in Food Engineering*, Boca Raton, FL, USA, CRC Press, 15–32.
BRUIN S and LUYBEN K C A M (1980) 'Drying of food materials: a review on recent developments', in Muyumdar A S, *Advances in Drying*, Washington, Hemishere Publ Corp, Chapter 6.
BUTTERY R G, LING L C and GUADAGNI D G (1969) Food volatiles: volatilities of aldehydes, ketones and esters in dilute water solution', *J Agric Food Chem*, 17, 385–489.
BYLAITE E, MEYER A S and ADLER-NILSEN J (2003) 'Changes in macroscopic viscosity do not affect the release of aroma aldehydes from pectinaceous food model system of low sucrose content', *J Agric Food Chem*, 51, 8020–8026.
CAREY M, ASQUITH T, LINFORTH R S T and TAYLOR A J (2002) 'Modelling the partition of volatile compounds from a cloud emulsion', *J Agric Food Chem*, 50, 1985–1990.
CAREY M, LINFORTH R and TAYLOR A (2003) 'Factors affecting dynamic flavour release from emulsions', in Le Quéré J L and Etiévant P X, *Flavour Research at the Dawn of the Twentieth-first Century*, Paris, Editions Tec & Doc, 212–215.

CHEN N, ZHANG Y, TERABE S and NAKAGAWA T (1994) 'Effect of physico-chemical properties and molecular structure on the micelle–water partition coefficient in micellar electrokinetic chromatography', *J Chromatogr*, 678, 327–323.

CONNER J M, BIRKMYRE L, PATERSON A and PIGGOTT J R (1998) 'Headspace concentration of ethyl esters at different alcoholic strengths', *J Sci Food Agric*, 77, 121–126.

COUMANS W J, KERKHOF P J A M and BRUIN S (1994) 'Theoretical and practical aspects of aroma retention in spray drying and freeze drying', *Drying Technology*, 12 (1&2), 99–149.

CURVERS J, RIJKS J and CRAMERS C (1985a) 'Temperature programmed retention indices: calculation from isothermal data. Part I: Theory', *J High Resolut Chromatogr Chromatogr Commun*, 8, 607–610.

CURVERS J, RIJKS J and CRAMERS C (1985b) 'Temperature programmed retention indices: Calculation from isothermal data. Part II: Results with nonpolar columns', *J High Resolut Chromatogr Chromatogr Commun*, 8, 611–617.

DANCKWERTS P V (1951) 'Significance of liquid–film coefficients in gas absorption', *Ind Eng Chem*, 43, 1460–1467.

DARLING D F, WILLIAMS D and YENDLE P (1986) 'Physico-chemical interactions involved in aroma transport processes from solution' in Birch G G and Lindley M G, *Interactions of Food Components*, London, Elsevier Applied Science, 165–188.

DE ROOS K B (2000) 'Physicochemical models of flavour release from foods', in Roberts D D and Taylor A J, *Flavour Release*, Washington, DC, American Chemical Society, 126–141.

DE ROOS K B (2003) 'Effect of texture and microstructure on flavour retention and release', *Intern Dairy J*, 13, 593–605.

DE ROOS K B and GRAF E (1995) 'Non-equilibrium partition model for predicting flavor retention in microwave and convection heated foods', *J Agric Food Chem*, 43, 2204–2211.

DE ROOS K B and MANSENCAL R (2003) 'Poor performance of flavourings in heat processed products: A problem of poor flavour retention or poor flavour release?', in Le Quéré J L and Etiévant P X, *Flavour Research at the Dawn of the Twentieth-first Century*, Paris, Editions Tec & Doc, 27–32.

DE ROOS K B and SARELSE J A (1996) 'Volatile acids and nitrogen compounds in prawn powder', in Taylor A J and Mottram D S, *Flavour Science. Recent Developments*, Cambridge, The Royal Society of Chemistry, 13–18.

DE ROOS K B and WOLSWINKEL C (1994) 'Non-equilibrium partition model for predicting flavour release in the mouth', in Maarse H and Van der Heij H G, *Trends in Flavour Research*, Amsterdam, Elsevier Science B.V., 15–32.

DOYEN K, CAREY M, LINFORTH R S T, MARIN M and TAYLOR A J (2001) 'Volatile release from an emulsion: headspace and in mouth studies', *J Agric. Food Chem*, 49, 804–810.

DWARAKANATH V and POPE G A (1998) 'New approach for estimating alcohol partition coefficients between nonaqueous phase liquids and water', *Environ Sci Technol*, 32, 1662–1666.

ESCHER F E, NUESSLI J and CONDE-PETIT B (2000) 'Interactions of flavor compounds with starch in food processing' in Roberts D D and Taylor A J, *Flavour Release*, Washington, DC, American Chemical Society, 230–245.

FLINK J M and KAREL M (1970) 'Retention of organic volatiles in freeze-dried solution of carbohydrates', *J Agric Food Chem*, 18, 259–297.

GOUBET I, LE QUÉRÉ J-L and VOILLEY A J (1998) 'Retention of aroma compounds by carbohydrates; Influence of their physicochemical characteristics and their physical state. A review', *J Agric Food Chem*, 46, 1981–1990.

HALL G and ANDERSON J (1983) 'Volatile fat oxidation products II. Influence of temperature on volatility of saturated, mono- and di-unstaturated aldehydes in liquid media', *Lebensm Wiss u Technol*, 16, 362–366.

HARRISON M (1998) 'Effect of breathing and saliva flow on flavour release from liquid foods, *J Agric Food Chem*, 46, 2727–2735.

HARRISON M, CAMPBELL S and HILLS B P (1998) 'Computer simulation of flavour release from solid foods in the mouth', *J Agric Food Chem*, 46, 2736–2743

HARRISON M and HILLS B P (1996) 'A mathematical model to describe flavour release from gelatine gels' *Int J Food Sci Technol*, 31, 167–176.

HARRISON M and HILLS B P (1997a) 'Effects of air flow on flavour release from liquid emulsions in the mouth' *Int J Food Sci Technol*, 32, 1–9.

HARRISON M and HILLS B P (1997b) 'A mathematical model of flavour release from liquids containing aroma binding macromolecules', *J Agric Food Chem*, 45, 1883–1890.

HARRISON M, HILLS B P, BAKKER J and CLOTHIER T (1997) 'Mathematical models of flavour release from liquid emulsions', *J Food Sci*, 62, 653–658, 664.

HIGBIE R (1935) 'The rate of absorption of a pure gas into a still liquid during short periods of exposure', *Trans Am Inst Chem Eng*, 25, 36–60.

HILLS B P and HARRISON M (1995) 'Two-film theory of flavour release from solids', *Intern J Food Sci Technol*, 30, 425–436.

HODGSON M, LINFORTH R S T and TAYLOR A J (2003) 'Simultaneous real-time measurements of mastication, swallowing, nasal airflow, and aroma release', *J Agric Food Chem*, 51, 5052–5057.

HOLLOWOOD T A, LINFORTH R S T and TAYLOR A J (2002) 'The effect of viscosity on the perception of flavour', *Chem Senses*, 27, 583–591.

HUSSAM A, BASU S C, HIXON M and OLUME Z (1995) 'General method for the study of solute–surfactant association equilibria of volatile solutes by headspace gas chromatography', *Anal Chem*, 67, 1459–1464.

KALLEN R G (1971) 'Equilibriums for the reaction of cysteine and derivatives with formaldehyde and protons', *J Am Chem Soc*, 93, 6227–6235.

KANT A, LINFORTH R S T, HORT J and TAYLOR A J (2004) 'Effect of β-cyclodextrin on aroma release and flavour perception', *J Agric Food Chem*, 52, 2028–2035.

KANT A, LINFORTH R S T and TAYLOR A J (2003) 'Flavour retention and release from starch solutions', in Le Quéré J L and Etiévant P X, *Flavour Research at the Dawn of the Twentieth-first Century*, Paris, Editions Tec & Doc, 103–106.

KATRITZKY A R, WANG Y L, SILD S, TAMM T and KARELSON M (1998) 'QSPR studies on vapor pressure, aqueous solubility, and prediction of water air partition coefficients', *J Chem Inform Compu Sci*, 38, 720–725.

KERKHOF, P J A M and THIJSSEN H A C (1975) 'Quantitative study on the effects of process variables on aroma retention during drying of foods', in Maarse H and Groenen P J, *Aroma Res., Proceedings of A.I.Ch.E. meeting on Dehydration and Concentration of Foods*, The Netherlands, Cent. Agric. Publ. Doc. Wageningen, 167–192.

KERTES A S and KING C J (1987) 'Extraction chemistry of low molecular weight aliphatic alcohols', *Chem Rev*, 87, 687–710.

KEYHANI K, SCHERER P W and MOZELL M M (1997) A numerical model of nasal odorant transport for the analysis of human olfaction', *J Theor Biol*, 186, 279–301.

KING B M and SOLMS J (1982) 'Interactions of volatile flavour compounds with propyl gallate and other phenols as compared with caffeine', *J Agric Food Chem*, 30, 838–840.

KINKEL J F M, TOMLINSON E and SMIT P (1981) 'Thermodynamics and extrathermodynamics of organic solute liquid-liquid distribution between water and 2,2,4-trimethylpentane', *Int J Pharm*, 9, 121–136.

KORA E P, SOUCHON I, LATRILLE E, MARTIN N and MARIN M (2004) 'Composition rather than viscosity modifies the aroma compound retention of flavoured stirred yogurt', *J Agric Food Chem*, 52, 3048–3056.

LAUB R J, PURNELL J H and SUMMERS D M (1980) 'Infinite dilution liquid–gas partition coefficients of some volatile solutes with binary solvents exhibiting upper consolute temperatures', *J Chem Soc Faraday I*, 76, 362–367.

LE GUEN S and VREEKER R (2003) 'Interactions between flavour compounds and milk proteins under static and dynamic conditions', in Le Quéré J L and Etiévant P X, *Flavour Research at the Dawn of the Twentieth-first Century*, Paris, Editions Tec & Doc, 182–187.

LEO A, HANSCH C and ELKINS D (1971) 'Partition coefficients and their uses', *Chem Rev*, 71, 525–616.

LIAN G (2000) Modelling flavor release from oil containing gel particles', in Roberts D D and Taylor A J, *Flavour Release*, Washington, DC, American Chemical Society, 201–211.

LIAN G, MALONE M E, HOMAN J E and NORTON I T (2004) 'A mathematical model of volatile release in mouth from the dispersion of gelled emulsion particles', *J Controlled Release*, 98, 139–155.

LINDSTROM T R and PARLIMENT T H (1994) 'Microwave volatilization of aroma compounds', in Parliment T H, Morello M J and McGorrin R J, *Thermally Generated Flavors. Maillard, Microwave and Extrusion Processes*, Washington, DC, American Chemical Society, 405–413.

LINFORTH R (2002) 'Modelling flavour release', in Taylor A J, *Food Flavour Technology*, Sheffield UK, Academic Press, 185–209.

LINFORTH R and TAYLOR A J (2000) 'Persistence of volatile compounds in the breath after their consumption in aqueous solutions', *J Agric Food Chem*, 48, 5419–5423.

LINFORTH R S T, FRIEL E N and TAYLOR A J (2000) 'Modelling aroma release from foods using physicochemical parameters', in Roberts D D and Taylor A J, *Flavour Release*, Washington, DC, American Chemical Society, 166–178.

LINFORTH R, HODGSON M and TAYLOR A (2003) 'Studies of retronasal flavour delivery', in Le Quéré J L and Etiévant P X, *Flavour Research at the Dawn of the Twentieth-first Century*, Paris, Editions Tec & Doc, 143–147.

LINFORTH R, MARTIN F, CAREY M, DAVIDSON J and TAYLOR A J (2002) 'Retronasal transport of aroma compounds', *J Agric Food Chem* 50, 1111–1117.

LÜBKE M, GUICHARD E and LE QUÉRÉ J-L (2000) 'Infrared spectroscopic study of β-lactoglobulin interactions with flavor compounds', in Roberts D D and Taylor A J, *Flavour Release*, Washington, DC, American Chemical Society, 282–292.

MALONE M E and APPELQVIST I A M (2003) 'Gelled emulsion particles for controlled release of lipophilic volatiles during eating', *J Controlled Release*, 90, 227–241.

MALONE M E, APPELQVIST I A M, GOFF T C, HOMAN J E and WILKINS J P G (2000) 'A novel approach to the selective control of lipophilic flavour release in low fat products', in Roberts D D and Taylor A J, *Flavour Release*, Washington DC, American Chemical Society, 212–227.

MALONE M E, APPELQVIST I A M and NORTON I T (2003) 'Oral behaviour of food hydrocolloids and emulsions. Part 2. Taste and aroma release', *Food Hydrocolloids*, 17, 775–784.

MARIN M, BAEK I and TAYLOR A J (1999) 'Volatile release from aqueous solutions under dynamic headspace dilution conditions', *J Agric Food Chem*, 47, 4750–4755.

MARIN M, BAEK I and TAYLOR A J (2000) 'Flavor release as a unit operation: a mass transfer approach based on a dynamic headspace dilution method', in Roberts D D and Taylor A J, *Flavour Release*, Washington, DC, American Chemical Society, 153–165.

McNULTY P B (1987) 'Flavour release- elusive and dynamic', in Blanshard J M V and Lillford P, *Food Structure and Behaviour*, London, Academic Press, 245–248.

McNULTY P B and KAREL M (1973) 'Factors affecting flavour release and uptake in O/W emulsions. I. Release and uptake models. *J Food Technol*, 8, 309–318.

MIETTINEN S-M, HYVÖNEN L and TUORILA H (2003) 'Timing of intensity perception of a polar vs nonpolar aroma compound in the presence of added vegetable fat in milk', *J Agric Food Chem*, 51, 5437–5443.

MOORE I P T, DODDS T M, TURNBULL R P and CRAWFORD R A (2000) 'Flavor release from composite dairy gels: a comparison between model predictions and time-intensity experimental studies', in Roberts D D and Taylor A J, *Flavour Release*, Washington, DC, American Chemical Society, 281–294.

MUNTZ C and ROBERTS P V (1987) 'Air-water phase equilibria of volatile organic solutes', *Research and Technology*, 62–69.

NAHON D F, HARRISON M and ROOZEN J P (2000) 'Modelling flavor release from aqueous sucrose solutions, using mass transfer and partition coefficients', *J Agric Food Chem*, 48, 1278–1284.

NELSON P E and HOFF J E (1968) 'Food volatiles: Gas chromatographic determination of partition coefficients in water–lipid systems', *J Food Sci*, 33, 479–482.

NORMAND V, AVISON S and PARKER A (2004) 'Modelling the kinetics of flavour release during drinking', *Chem Senses*, 29, 235–245.

O'NEILL T E (1996) 'Flavor binding by food proteins: an overview', in McGorrin R J and Leland J V, *Flavor-Food Interactions*, Washington, DC, American Chemical Society, 59–74.

OVERBOSCH P, AFTEROF W G M and HARING P G M (1991) 'Flavour release in the mouth', *Food Rev Int*, 7, 137–184.

RABE S, KRINGS U, BANAVARA D S and BERGER R G (2002) 'Computerized apparatus for measuring dynamic flavour release from liquid food matrices', *J Agric Food Chem*, 50, 6440–6447.

RABE S, KRINGS U, ZORN H and BERGER R G (2003a) 'Lipid molarity affects liquid/liquid aroma partitioning and its dynamic release from oil/water emulsions', *Lipids*, 38, 1075–1084.

RABE S, KRINGS U and BERGER R G (2003b) 'Initial dynamic release from sodium chloride solutions', *Eur Food Res Technol*, 218, 32–39.

RABE S, KRINGS U and BERGER R G (2003c) 'Dynamic flavour release from sucrose solutions', *J Agric Food Chem*, 51, 5058–5066.

RABE S, KRINGS U and BERGER R G (2003d) 'Influence of oil-in-water emulsion characteristics on initial dynamic flavour release', *J Sci Food Agric*, 83, 1124–1133.

RABE S, KRINGS U and BERGER R G (2004) 'Dynamic flavour release from miglyol/water emulsions: modelling and validation', *Food Chem*, 84, 117–125.

ROBERTS D D, ELMORE J S, LANGLEY K R and BAKKER J (1996) 'Effects of sucrose, guar gum, and carboxymethylcellulose on the release of volatile flavour compounds under dynamic conditions', *J Agric Food Chem*, 44, 1321–1326.

ROBERTS D D and POLLIEN P (2000) 'Relative influence of milk components on flavor compound volatility', in Roberts D D and Taylor A J, *Flavour Release*, Washington, DC, American Chemical Society, 321–332.

ROBERTS D D, POLLIEN P and WATZKE B (2003a) 'Experimental and modelling studies showing the effect of lipid type on flavor release from milk-based liquid emulsions', *J Agric Food Chem*, 51, 189–195.

ROBERTS D D, POLLIEN P, ANTILLE N, LINDINGER C and YERETZIAN C B (2003b) 'Comparison of nosespace, headspace and sensory intensity ratings for the evaluation of flavour absorption by fat', *J Agric Food Chem*, 51, 3636–3642.

ROUDNITZKY N, IRL H, ROUDAUT G and GUICHARD E (2003) 'Influence of fat nature on flavour release', in Le Quéré J L and Etiévant P X, *Flavour Research at the Dawn of the Twentieth-first Century*, Paris, Editions Tec & Doc, 136–139.

RULKENS W H (1973) *Retention of volatile trace components in drying aqueous carbohydrate solutions*, Ph.D. thesis, The Netherlands, Technical University of Eindhoven.

SEMENOVA M G, ANTIPOVA A S, BELYAKOVA L E, POLIKARPOV Y N, MISHARINA T A, TERENINA M B and GOLOVNYA R V (2000) 'Influence of maltodextrins with different dextrose equivalent on the interaction between hexyl acetate and legumin in an aqueous medium', in Roberts D D and Taylor A J, *Flavour Release*, Washington, DC, American Chemical Society, 260–273.

SEUVRE A M, ESPINOSA-DIAZ and VOILLEY A (2000) 'Influence of the food matrix structure on the retention of aroma compounds', *J Agric Food Chem*, 48, 4296–4300.

SONNENBERG S, FINKE A, KLAMT A, LOHRENZ J, BÜRGER T and MATTHIESEN S (2002) 'Selection method for aromatic substances', Patent No WO 02/051263.

STANFORD M A and McGORRIN R J (1994) 'Flavor volatilization in microwave food model systems', in Parliment T H, Morello M J and McGorrin R J, *Thermally Generated Flavors. Maillard, Microwave and Extrusion Processes*, Washington, DC, American Chemical Society, 414–436.

STEINKE J A, FRICK C M, GALLAGHER J A and STRASSBURGER K J (1989) 'Influence of microwave heating on flavor', in Parliment T H, McGorrin R J and Ho C-T, *Thermal Generation of Aromas*, Washington, DC, American Chemical Society, 519–525.

TAYLOR A J (1998) 'Physical chemistry of flavour', *Int J Food Sci Technol*, 33, 53–62.

TAYLOR A, HOLLOWOOD T, DAVIDSON J, COOK D and LINFORTH R (2003) 'Relating taste and mouthfeel stimuli to overall flavour perception', in Le Quéré J L and Etiévant P X, *Flavour Research at the Dawn of the Twentieth-first Century*, Paris, Editions Tec & Doc, 194–199.

THIJSSEN H A C (1974) 'Freeze-concentration', in Spicer A, *Adv Preconc Dehydr Foods Symp*, New York, Wiley, 115–149.

TUNALEY A, FRANKLIN J G, GRIFFITHS N M and KING B M (1985) 'The effect of caffeine on the odor intensity of some flavour volatiles', *Lebensm-Wiss u Technol*, 18, 238–241.

VOILLEY A J (1995) 'Flavour encapsulation. Influence of encapsulation media on aroma retention during drying', in Risch S J and Reineccius G A, *Encapsulation and Controlled Release of Food Ingredients*, Washington, DC, American Chemical Society, 169–179.

WAGGONER R A, BLUM F D and MACELROY J M D (1993) 'Dependence of the solvent diffusion coefficient on concentration in polymer solutions', *Macromolecules*, 26(25), 6841–6848.

WEEL K G C (2004) *Release and Perception of Aroma Compounds During Consumption*, Thesis Wageningen University, The Netherlands, ISBN 90-8504-014-0.

WEEL K G C, BOELRIJKA E M, ALTING A C, VAN MIL P J J M, BURGER J J, GRUPPEN H, VORAGEN A G J and SMIT G (2002) 'Flavour release and perception of flavoured protein gels: perception is determined by texture rather than by release, *J Agric Food Chem*, 50, 5149–5155.

WEEL K G C, BOELRIJKA E M, BURGER J J, CLAASSEN N E, GRUPPEN H, VORAGEN A G J and SMIT G (2003) 'Effect of whey protein on the vivo release of aldehydes, *J Agric Food Chem*, 51, 4746–4752.

WHITMAN W G (1923) 'Preliminary experimental confirmation of the two-film theory of gas absorption', *Chem Metall Eng*, 29, 146–148.

WRIGHT K M and HILLS P H (2003a) 'Modelling flavour release from a chewed bolus in the mouth: Part I. Mastication', *Intern J Food Sci Technol*, 38, 351–360.

WRIGHT K M and HILLS P H (2003b) 'Modelling flavour release from a chewed bolus in the mouth: Part II. The release kinetics', *Intern J Food Sci Technol*, 38, 361–368.

WRIGHT K M, HILLS P H, HOLLOWOOD T A, LINFORTH R S T and TAYLOR A J (2003) 'Persistence effects in flavour release from liquids in the mouth', *Intern J Food Sci Technol*, 38, 343–350.

ZHOU S and DECKER E A (1999) 'Ability of amino acids, dipeptides, polyamines and sulfhydryls to quench hexanal, a saturated aldehyde lipid oxidation product', *J Agric Food Chem*, 47, 1932–1936.

12

Flavour release from liquid food products

A. E. M. Boelrijk, G. Smit and K. G. C. Weel, NIZO Food Research, The Netherlands and J. J. Burger, Quest International Nederland BV, The Netherlands

12.1 Introduction

The sensory impression of a food product is an important factor in consumers' appreciation. Much research effort is therefore dedicated worldwide to the sensory properties of food products. In this respect, the term 'flavour' is much used. Generally, this term refers to the combined perception of aroma, taste, and texture at the time of food consumption (Taylor 1996).

In recent years, mass-spectrometric techniques were developed to measure the release of aroma compounds in a person's breath, while he or she eats a product (Taylor and Linforth 1996, Lindinger *et al.* 1998). The research on *in vivo* aroma release was boosted by the high sensitivity and time resolution of these new techniques, compared with earlier attempts. Many research groups started to work in this suddenly emerging discipline. Now it had become possible to study the link between the aroma release profile of a product and its aroma perception.

Ever since, many *in vivo* aroma release experiments have been done with solid and semi-solid products like gels, vegetables, dairy products and chewing gums (Brauss *et al.* 1998, Brauss *et al.* 1999, Davidson *et al.* 1999, Linforth *et al.* 1999, Baek *et al.* 1999, Weel *et al.* 2002, Taylor *et al.* 2001). However, only a few studies have focused on liquid systems (Linforth and Taylor 2000, Doyen *et al.* 2001, Linforth *et al.* 2002).

See Fig. 12.1 for a schematic overview of a beverage and its main ingredients.

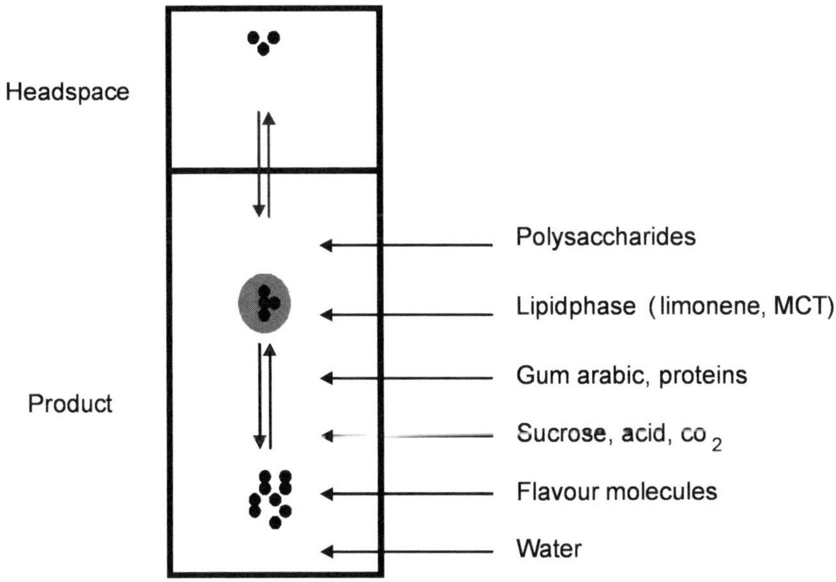

Fig. 12.1 Schematic overview of a beverage.

Non-alcoholic beverages are an important group of liquid food products. If the composition of a soft drink is changed, the aroma release profile might change as well due to changes in volatility of the aroma compounds, which consequently could affect the overall perception. There has been much industrial and scientific interest in the replacement of bulk sweeteners by intense sweeteners in order to reduce caloric intake and cariogenic activity (O'Brien Nabors and Gelardi 1991, Nawar 1971, Voilley *et al.* 1977, Land and Reynolds 1981, Le Thanh *et al.* 1992, Nahon *et al.* 1998a, Friel *et al.* 2000, Deibler and Acree 1999, Hansson *et al.* 2001a,b).

In earlier studies possible interactions between bulk or intense sweeteners and aroma have been evaluated using static headspace analysis. Effects like pH and type and concentration of sweetener were reported to have different effects on a number of aroma compounds and interactive effects between them appeared to be important as well (Nawar 1971, Voilley *et al.* 1977, Land and Reynolds 1981, Le Thanh *et al.* 1992, Nahon *et al.* 1998a, Friel *et al.* 2000, Deibler and Acree 1999, Hansson *et al.* 2001a,b). However, these results did not lead to a general conclusion about the role of interactions between sweeteners (conventional and intense) for release. The same holds for interactions between other ingredients and aroma compounds and the effects these interactions have on aroma release from beverages.

The aim of this chapter is, therefore, to summarise the facts with respect to *in vivo* aroma release from beverages and to discuss the phenomena that determine the extent of aroma release in these products.

12.2 *In vivo* aroma release measurements during drinking of beverages

12.2.1 Intra- and interpersonal variation in aroma release and perception

Solid foods reside in the mouth for a relatively long time (30 to 60 s). This period is sufficiently long to obtain an accurate release pattern during *in vivo* APCI-MS or PTR analysis. Liquids, however, are usually kept in the mouth for a much shorter period of time: people tend to swallow liquids quickly after having taken the product into their mouth, and aroma release takes place during this short time interval. Small irregularities in drinking patterns can, therefore, have a large effect on the amount of aroma released. These irregularities can originate from swallowing, depth of breath, jaw and tongue movements, saliva flow, interactions with mouth mucosa and so on.

It was reported that for liquids the highest aroma release signal is generally found in the first expiration after swallowing (Linforth and Taylor 2000). No gas is transferred from the oral cavity to the nasal cavity as long as no opening of the barrier formed by the tongue and the soft palate occurs either by swallowing or by vigorous tongue and mouth movements (Buettner and Schieberle 2000a). The existence of such an anatomical barrier has been studied by the use of videofluoroscopy and real-time magnetic resonance imaging during the swallowing process (Buettner *et al.* 2001). The studies done on aroma release of liquid systems used an approach of swallowing followed by exhalation. The volatile concentration of the first peak in the release signal after swallowing, the so-called 'swallow breath', is taken as a measure for aroma release (Linforth and Taylor 2000, Doyen *et al.* 2001, Linforth *et al.* 2002, Weel *et al.* 2003b).

When *in vivo* aroma release measurements are performed, intra- and interpersonal variation have to be properly dealt with. The sources of intra- and interpersonal variation in the absolute amount of aroma released between people are many: differences in mouth, nose and throat geometry, and chewing, swallowing, and breathing behaviour and saliva production (Brown *et al.* 1996, Pionnier *et al.* 2004, Wright *et al.* 2003). *In vivo* MS-Nose or PTR measurements can be used to study these effects accurately.

An example of how interpersonal variation in aroma release can influence aroma perception is given by the following experiment (Fig. 12.2), conducted in our laboratory. Simultaneous time–intensity (TI) and *in vivo* aroma release measurements could distinguish three groups in a panel assessing dairy desserts by simultaneous time–intensity (TI) and *in vivo* aroma release measurements. They were instructed to swallow after 20 seconds. One group had their highest release during the initial chewing. A second group had initial release, but they peaked at swallowing. Finally, one panellist had no release at all before swallowing, and the aroma released only upon swallowing

The TI profiles of these three groups closely matched their release curves. This illustrates how opening of the tongue–velum border during chewing can vary between persons and can result in differences in aroma release and

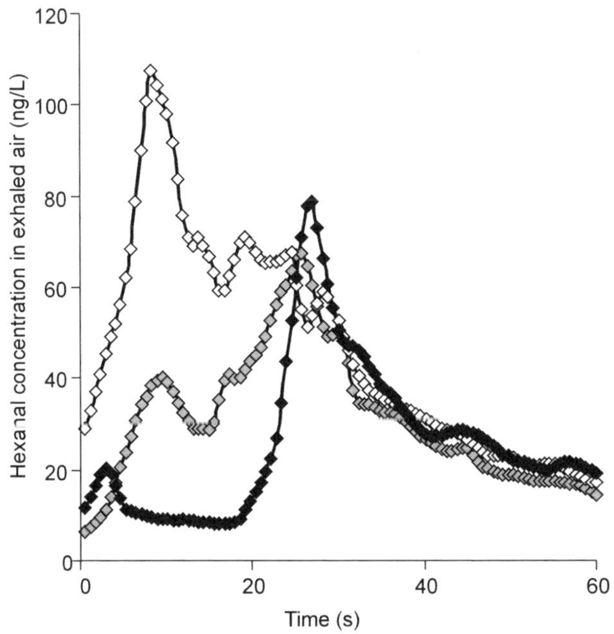

Fig. 12.2 Hexanal release curves (ng/l air) of dairy desserts, representing three typical profiles. Each curve is the average of six replicates of a single assessor.

perception. It also illustrates the importance of considering individual release data, instead of simply averaging across people (Lethuaut *et al.* 2004).

When *in vivo* aroma release of liquids is measured, reduction of experimental variation is of critical importance for obtaining meaningful results. One irregular breath does not change this release profile for solids dramatically, but it will have a large effect on the single release peak after swallowing a liquid sample. Reproducible release measurements of liquids are not possible without control of swallowing, breathing and mouth movements. For this purpose, a protocol was developed in our laboratory accounting for these factors by which the relative standard deviation of trained panellists could be reduced to 10–15%.

The drinking situation of the protocol resembles natural drinking as much as possible, which consists of the uptake of a sample followed by the swallowing of it. Figure 12.3 shows an example of an *in vivo* release curve of citral from an aqueous solution, and acetone as breath indicator, obtained with the use of the protocol. During the first exhalation, no release of citral takes place. This observation agrees with the existence of an anatomical barrier between the buccal and nasal cavities as was described before (Buettner *et al.* 2001). Then the sample is 'chewed' and swallowed and subsequent exhalation results in a high release of citral. The area of this peak is taken as a value for the aroma release. When the data in Fig. 12.3 are inspected in more detail, it is observed that the peak maxima of the exhalation peaks of citral are split into three. This is due to the chewing movements that are made once per second and thus three

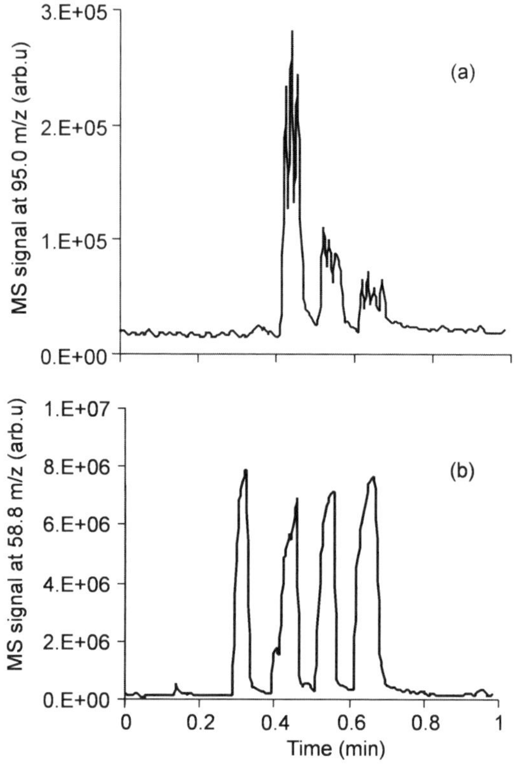

Fig. 12.3 Example of raw data of a single *in vivo* measurement of release of citral (a) and acetone (b). (Copyright ACS.)

times per exhalation. Because acetone releases from the lungs, and is not in contact with the mouth its peaks do not display this splitting.

Persistence of aroma compounds in the breath was defined as the ratio between the peak heights in the release signal of the first and second exhalations after swallowing (Linforth and Taylor 2000). To get an impression of the persistence of the aroma compounds used in the current study, the ratio of the peak area values of the first and second peak of all measurements of all panellists was averaged for three aroma compounds. Persistence values of 58%, 64%, and 38% were found for citral, geranyl acetate and nonanal, respectively. The persistence value found for nonanal is surprisingly high, in comparison with the results of Linforth and Taylor (Linforth and Taylor 2000), who reported a persistence value of 6% for both hexanal and decanal. Despite the on-average relatively high release in the second breath after swallowing for the compounds used, we have chosen to focus on the release in the first peak after swallowing, because this peak still represents the highest release and because it is generated at an important time interval for aroma perception during drinking.

12.2.2 Training of the panellists

It requires only a few hours of practice, spread out over a few days, to train panellists in the use of this protocol. During the training sessions, the panellists received direct feedback about their produced aroma release curves on the PC screen. This was found a valuable assistance in the learning process. The relative standard deviation of the values produced by a fully trained panellist is between 10 and 20%. This level did not decrease upon additional training, and was considered acceptable for the studies under consideration. An average untrained volunteer will produce data with a relative standard deviation of 20 to 30%. This value can increase further if no strict protocol is used.

12.3 Physiological factors important for *in vivo* aroma release of beverages

Various physiological factors can influence the release of aroma compounds under *in vivo* conditions, and the effect of these factors can differ between people. Foods can undergo temperature changes once introduced in the mouth. A higher temperature leads to a stronger partitioning of volatiles in the air phase at equilibrium conditions. This has been demonstrated previously for a range of aroma compounds (Roberts and Acree 1995, Brown *et al.* 1996, Deibler and Acree 1999). In the case of solid products, chewing efficiency can influence aroma release. Different chewing styles and large person-to-person variation was observed while chewing gels (Wright *et al.* 2003, Brown *et al.* 1996). Pionnier and co-workers were able to partly relate interpersonal differences in aroma release under *in vivo* conditions to differences in respiratory and masticatory parameters (Pionnier *et al.* 2004). Since liquid products experience only very limited oral processing, these issues appear to be less important for beverages.

12.3.1 Aroma–saliva interactions

The interactions between aroma compounds and saliva or mucous membranes have been investigated in recent years by several groups (Buettner and Schieberle 2000b, Friel and Taylor 2001, Van Ruth *et al.* 2001, Linforth *et al.* 2002). The flow rate and composition of saliva can vary greatly within and among people, caused by the degree of hydration, body position, exposure to light, the smell of food, smoking, previous stimulation, and climatological circumstances (Dawes 1981). Possible effects of saliva on aroma release are dilution, interactions between aroma compounds and saliva constituents, and enzymatic activity. Aldehydes and other aroma compounds can interact with salivary proteins, especially mucin, as shown by static headspace measurements (Friel and Taylor 2001, Van Ruth *et al.* 2001) and by the SOOM-technique (spit-off odourant measurements) (Buettner and Schieberle 2000b). Losses in the oral cavity were observed for aldehydes and esters during 1 minute of contact time (Buettner and Schieberle 2000b). In addition, esters, thiols and aldehydes were found to be subject to enzymatic conversion upon contact with saliva within a period of 10 minutes (Hussein *et al.* 1983, Buettner 2002a,b).

In another study, Linforth and co-workers (Linforth *et al.* 2002) investigated the effect of factors such as absorption to epithelia of mouth, nose and pharynx, and dilution by saliva, on aroma release. It was demonstrated that the mass transfer from aqueous solution in the mouth into exhaled air was a major factor affecting the actual released amount of aroma.

12.3.2 Swallowing and formation of a liquid film in the throat

For liquid products, it has been shown that swallowing determines the *in vivo* aroma release rather than the preceding oral processing. This concept was first mentioned by De Roos and Wolswinkel in 1994 and again later, in 1996, when Land introduced the 'swallow-breath' principle, which is 5–15 ml of air that is pumped into the nose retronasally, as a result of swallowing. This plug of air has been in close masticatory contact with the food or drink in the mouth (De Roos and Wolswinkel 1994, Land 1996). This plug of air is therefore important for aroma perception. It has become clear that the highest *in vivo* aroma release signal for liquids is generally found in the first exhalation after swallowing (Linforth and Taylor 2000). It was shown that no air is transferred from the oral cavity to the nasal cavity as long as no opening of the barrier, formed by the tongue and the soft palate, occurs by swallowing or by vigorous tongue and mouth movements (Buettner and Schieberle 2000a, Buettner *et al.* 2001). In a recent study, Hogson and co-workers combined synchronised measurements of mastication, swallowing, breath flow and aroma release and demonstrated that an average chew pumps a volume of 26 ml of air from the oral cavity into the throat. This is especially relevant for the situation when the food is not swallowed immediately. Nasal airflow and associated aroma release were not detected during swallowing, but airflow and release were obvious directly after the swallowing event. The volume of the retronasal pathway was calculated to be 48 or 72 ml (depending on the way of calculation) (Hogson *et al.* 2003).

After swallowing, the majority of the sample disappears into the oesophagus (Fig. 12.4), but a thin layer of the liquid sample remains on the surface of the pharynx. Buettner and co-workers visualised the formation of such a coating by video fluoroscopy, when a volunteer swallowed viscous oral contrast medium (Buettner *et al.* 2002). During the exhalation following a swallow, a steep gradient in aroma concentration exists between the thin liquid layer on the surface of the pharynx and the exhaled air that passes over this surface. This exhaled air flows into the nasal cavity and along the olfactory epitlium where the aroma compounds are perceived.

12.4 Development of the artificial throat

If one accepts the importance of swallowing and the formation of a liquid film in the throat for aroma release, it seems logical to incorporate these facts in a

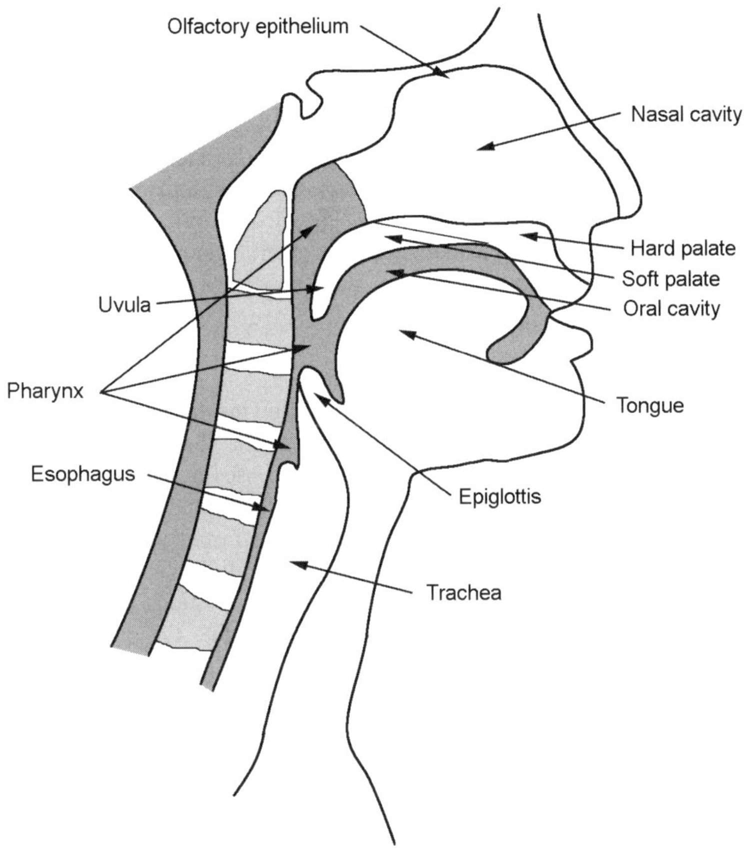

Fig. 12.4 The human throat. Adapted from 'Interactive atlas of human anatomy, version 2.0'. Gastrointestinal system edition, Frank H. Netter, MD, Novartis.

device if one aims to make a model system able to simulate *in vivo* aroma release.

All model systems that were published so far had been designed to mimic the aroma release in the mouth and are based on the shared principle of a certain amount of foodstuff (usually in liquid form) containing aroma compounds and other ingredients of interest (carbohydrates, proteins, lipids) that is put into a vessel and stirred or shaken in different ways (and heated to 37 °C, in most cases) (Lee 1986, Van Ruth *et al.* 1994, 1995, 2000, Roberts and Acree 1995, Naßl *et al.* 1995, Elmore and Langley 1996, Bakker *et al.* 1998, Springett *et al.* 1999, Rabe *et al.* 2002, Deibler *et al.* 2001, Roberts *et al.* 1996, Van Ruth and Roozen 2000, Banavara *et al.* 2002). Air is sampled from the headspace or nitrogen is purged through the liquid phase. The volatile compounds present in the stream of gas released from the model system are analysed in-line or batch-wise by a direct mass spectrometry (MS) technique or by trapping the compounds on absorbing materials followed by gas chromatography–mass spectrometry (GC–MS) (Lee 1986, Van Ruth *et al.* 1994, Roberts and Acree 1995, Naßl *et al.* 1995, Elmore and Langley 1996, Bakker *et al.* 1998, Springett *et al.* 1999, Rabe *et al.* 2002). All kinds of apparatus-related and product-related parameters have been studied using these model systems, which has increased insight in the process of volatile release in general (Deibler *et al.* 2001, Roberts *et al.* 1996, Van Ruth *et al.* 1995, 2000, Van Ruth and Roozen, 2000, Banavara *et al.* 2002). Since there was (to our knowledge) no model system that incorporated the swallowing act and the subsequent formation of a liquid film in the throat, our laboratories attempted to develop such a device.

12.4.1 Description of the artificial throat

As mentioned above, during the exhalation following the swallowing, a steep gradient in aroma concentration exists between the thin liquid layer on the surface of the pharynx and the exhaled air that passes over this surface. It has been suggested that the major part of the aroma compounds present in this thin liquid layer coating the throat will release almost instantaneously during this exhalation, due to the large surface area: volume ratio (Weel *et al.* 2003a). A model system that aims to simulate the dynamic conditions of *in vivo* aroma release of liquids should therefore be able to simulate this process. Thus, a model system is needed in which a thin layer of liquid is exposed to a relatively large flow rate, in order to approach the dynamic *in vivo* release conditions. An advantage of a model system with high correlation to the *in vivo* situation is the fact that parameters influencing the release can be easily changed and studied to increase understanding of *in vivo* aroma release. Moreover, the experimental variation of an artificial device is usually smaller than *in vivo* measurements. In addition, an artificial throat can be used to measure the aroma release of samples that are too dangerous or too repulsive to be analysed *in vivo*.

The developed artificial throat consists of a vertical glass tube, in which a thin liquid layer is formed, when a liquid sample flows down (Weel *et al.* 2004b, Davidson 2005). See Fig. 12.5 for a schematic overview of the artificial throat configuration. The MS–Nose sampling capillary samples air from the top end of the tube. An essential part of the system is a 3 mm thick tube of Viton rubber in the middle of the glass tubing that can be closed and opened by a clamp. Above this rubber section several liquids can be added simultaneously or subsequently from syringes through capillaries, ending in the glass tubing. A water mantle surrounds the glass tubing and is coupled to a water bath, equipped with a thermostat (set to 37 °C). An air inlet, which is pointing upwards, is located below the water mantle. The liquid can leave the system from the down end of the glass tubing.

At the start of the experiment, the clamp is closed and 4 ml of liquid is loaded above the clamp. When using human or artificial saliva, 2 ml saliva or water is poured in first, followed by 2 ml aroma solution. After a contact time the clamp is opened. The liquid pours down along the glass tubing. A thin liquid layer remains on the surface. Ten seconds after opening of the clamp, a stream of air

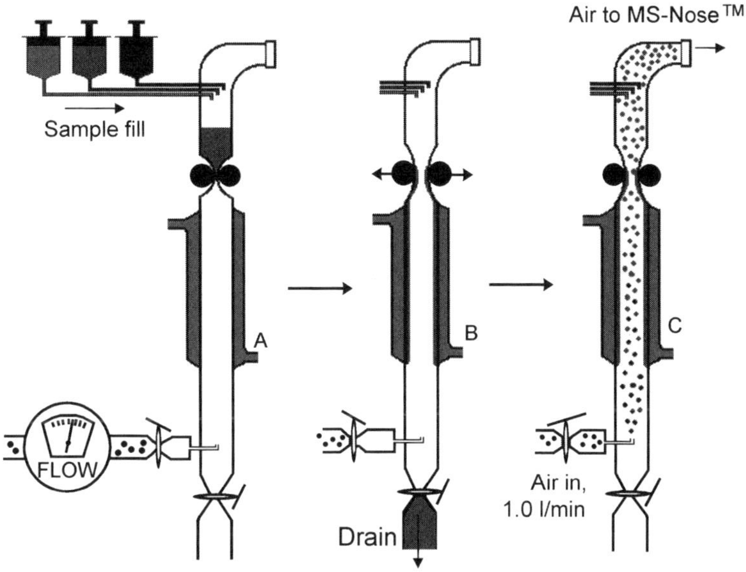

Fig. 12.5 Schematic overview of the artificial throat configuration.

(1.0 l/min) enters the tube and flows upwards, where it can freely flow out of the system, while a small part of the air is sampled by the MS–Nose.

12.4.2 Aroma release from the artificial throat

The design of the artificial throat was inspired on the hypothesis that the majority of aroma release from liquids *in vivo* originates from the layer that remains at the inner surface of the human throat after swallowing (Weel *et al.* 2003a). The results obtained with the equipment are in line with this hypothesis (Weel *et al.* 2004b). With respect to the sample volume, the amount of liquid that remains on the surface is only slightly dependent on the amount of liquid passing through, as long as the total amount is much larger than the amount constituting the film at the inner surface. When the aroma concentration is higher, there is proportionally more aroma compound available for release as expected. At equilibrium conditions a higher temperature leads to a stronger partitioning of volatiles into the air phase. This has been demonstrated previously for a range of aroma compounds (Roberts and Acree 1995, Deibler and Acree 1999). However, the sample temperature does not influence the total aroma release, neither *in vivo* nor in the artificial throat, because all aroma present in the thin film will rapidly release, once the exhalation or airflow starts. A decrease in measured release with increasing flow rate is also expected, because only a part of the air exhaled either by the artificial throat or the panellists, is sampled. The aroma concentration is determined in the part of the air sampled. When the flow rate is higher, the aroma compounds are more diluted, resulting in a lower concentration. The total released amount of aroma does not change within the range of flow rates studied, because all aroma compounds present will release. The decrease in peak width at higher flow rate indicates that the reservoir of aroma molecules present in the thin liquid film (coating the inner surface of both the human throat and the artificial throat) is exhausted faster.

12.4.3 Importance of throat surface

Gravity driven clearance of the liquid layer on the glass surface is the major factor responsible for the decrease in release in the artificial throat at longer time intervals between opening of the clamp and application of the airflow. *In vivo*, there are additional processes involved, explaining the stronger loss of aroma compounds there. These are fluid excretion of the mucosa and partitioning of the aroma compounds into the mucus (Buettner *et al.* 2002, Weel *et al.* 2003a). It was shown previously that the extent of the interaction between aroma compounds and mucosa depends heavily on the contact time (Weel *et al.* 2003a). Buettner and others (Buettner *et al.* 2002) reported a decrease of 30–40% upon 10 minute rinsing of pyrazine solutions in the mouth. The measurements presented in Table 12.1 show the relative change in release amounts of aroma (%) from the artificial throat or *in vivo* with 10 s and 30 s between swallowing and exhalation, compared to no time in between. In general, the decrease in release upon longer residence times in the (artificial) throat is larger in the human throats compared to the glass artificial throat. This is expected as the inner surface of the human throat is not as inert as the glass surface of the artificial throat.

12.4.4 Effect of sample viscosity on aroma release from the artificial throat

The effect of viscosity was studied with and without addition of saliva. When the viscosity of the samples increases, a thicker layer remains at the inner surface of the tube. This thicker layer will contain proportionally more aroma

Table 12.1 Relative change in released amounts of aroma (%) from the artificial throat or *in vivo* with 10 s and 30 s between swallowing and exhalation, compared to no time between swallowing and exhalation

	Artificial throat		Panellist 1		Panellist 2	
	10 s	30 s	10 s	30 s	10 s	30 s
Diacetyl	−17%	−25%	−26%	−45%	−35%	−64%
Ethylbutyrate	−17%	−26%	−38%	−56%	−20%	−58%
Butanal	−19%	−15%	−68%	−78%	−57%	−77%
Hexanal	−25%	−26%	−66%	−73%	−54%	−76%
Nonanal	−36%	−43%	−63%	−78%	−56%	−88%

compounds, and consequently the total released amount will be higher. In contrast to the artificial throat results, *in vivo* measurements showed no effect of viscosity on the amount of aroma released (Cook *et al.* 2003a,b, Weel *et al.* 2004b). This can be caused by differences in swallowing mechanism and by dilution with saliva. Swallowing in the artificial throat is driven by gravity, while under *in vivo* conditions the liquid is forced downwards by pharyngeal peristalsis. It seems reasonable to assume that the film thickness formed *in vivo* might therefore be different (i.e. more viscous solutions may not form thicker layers due to shear thinning behaviour). Future versions of the artificial throat should take peristalsis effects into account.

12.5 Effect of ingredients on aroma release from beverages

12.5.1 Effect of bulk and intense sweeteners

Carbohydrates can change the volatility of aroma compounds, but the extent of this effect is carbohydrate and aroma compound dependent. Examples are known in which the volatility of aroma compounds in water is lowered by addition of polysaccharides, as a result of non-specific molecular interactions (Godshall 1997). On the other hand, mono- and disaccharides have been shown to cause a salting-out effect, by lowering the amount of free water (Voilley *et al.* 1977). The competing effects of intermolecular attractions and salting-out in each specific case make it difficult to formulate a general theory of carbohydrate-aroma interactions.

The food and beverage industry has a large interest in the effects of replacement of bulk sweeteners with intense sweeteners on aroma perception (O'Brien Nabors and Gelardi 1991). So far, the cause of the reported sensory changes in aroma perception due to this replacement is not clear (Larson-Powers and Pangborn 1978, Wiseman and McDaniel 1991, Von Sydow *et al.* 1974, Nahon *et al.* 1996, 1998b). The study of Nahon and others (1996), for example, clearly showed for a blackcurrant soft drink that changing the type of sweetener (from sucrose to equisweet mixtures of intense sweeteners) changes the sensory aroma attributes from a blackcurrant to a strawberry character. A possible cause for the reported sensory changes could be a change in aroma release upon bulk sweetener replacement.

Therefore, we have used the protocol in section 12.2 for *in vivo* aroma release measurements of beverages to test the effect of sweetener type (sucrose or intense sweetener) on aroma release in a lemon-lime type beverage. The effect of various types of sweeteners in presence or absence of citric acid on aroma release was studied. The sweetener type and citric acid did not significantly influence the aroma release, in either the static or the *in vivo* measurement method (Weel *et al.* 2003b).

Since the sweetener concentrations used are the ones that are normally applied in beverages (5–15%), and the protocol used to assess *in vivo* aroma release is a sensitive and accurate *in vivo* method, it was concluded that substitution of sucrose by intense sweeteners has no profound effect on *in vivo* aroma *release*. However, that is not true for the effect that the substitution of sucrose by intense sweeteners has on aroma *perception*. How can these seemingly contradictory conclusions be understood? Since differences in aroma release cannot account for the differences in aroma perception, these perceptual differences have so far been explained as psychophysical effects, i.e. that the aroma perception changes, because of differences in textural perception between solutions of sucrose and intense sweeteners, or because of cognitive interactions between the sensory perceptions of, for instance, a blackcurrant or lemon-lime aroma and bulk or intense sweetness.

Earlier studies have described an effect of sucrose on aroma release for high sucrose concentrations (up to 65% w/v) (Voilley *et al.* 1977, Nawar 1971, Nahon *et al.* 1998a, Hansson *et al.* 2001a, Friel *et al.* 2000, Le Thanh *et al.* 1992). In order to compare these studies with our work, static headspace measurements of aroma compounds were performed with solutions with a range of sucrose concentrations, in which the concentration of aroma compound was corrected ('volume fraction corrected') or not corrected ('total volume') for the volume fraction of sucrose. The results of these experiments are shown in Fig. 12.6. The data points in this figure indicate that the static headspace concentrations of the 'total volume' samples show a clear increase in static headspace concentration of all three compounds upon addition of high concentrations of sucrose. This is because the same amount of aroma compounds is put in a reduced volume of water; i.e. the mole fraction of the aroma compounds is increased (Voilley *et al.* 1977). A higher concentration of aroma compounds in the water phase results in a higher headspace concentration if the partition coefficient is not changed, according to Henry's law. Once the aroma concentration was made proportional to the molar volume of water, in the case of the 'volume fraction corrected' samples, there is no longer a significant effect of sucrose on headspace concentration of citral and geranyl acetate. However, nonanal seems to interact with sucrose, because the headspace concentration decreases for the 'total volume' samples. These results illustrate how the direction of the effect (increase or decrease) is determined by the correction for added sucrose and by the type of aroma compound. Most authors do not correct their data for sucrose volume fraction, and results should, therefore, be compared carefully (Voilley *et al.* 1977, Nawar 1971, Nahon *et al.* 1998a, Hansson *et al.* 2001a, Friel *et al.* 2000, Le Thanh *et al.* 1992).

The effect of starch compared with oligosaccharides is different as it is a polymer well known for its aroma binding properties. Under specific conditions starch molecules can form helical structures. The inside of these helical structures form hydrophobic regions which can entrap especially more hydrophobic volatiles forming inclusion complexes (Solms 1986). A similar effect appears for cyclodextrin, which is composed of six to eight glucopyranose rings and is able to include lipophilic aroma compounds into its hydrophobic core (Qi and Hedges 1995, Kant *et al.* 2004).

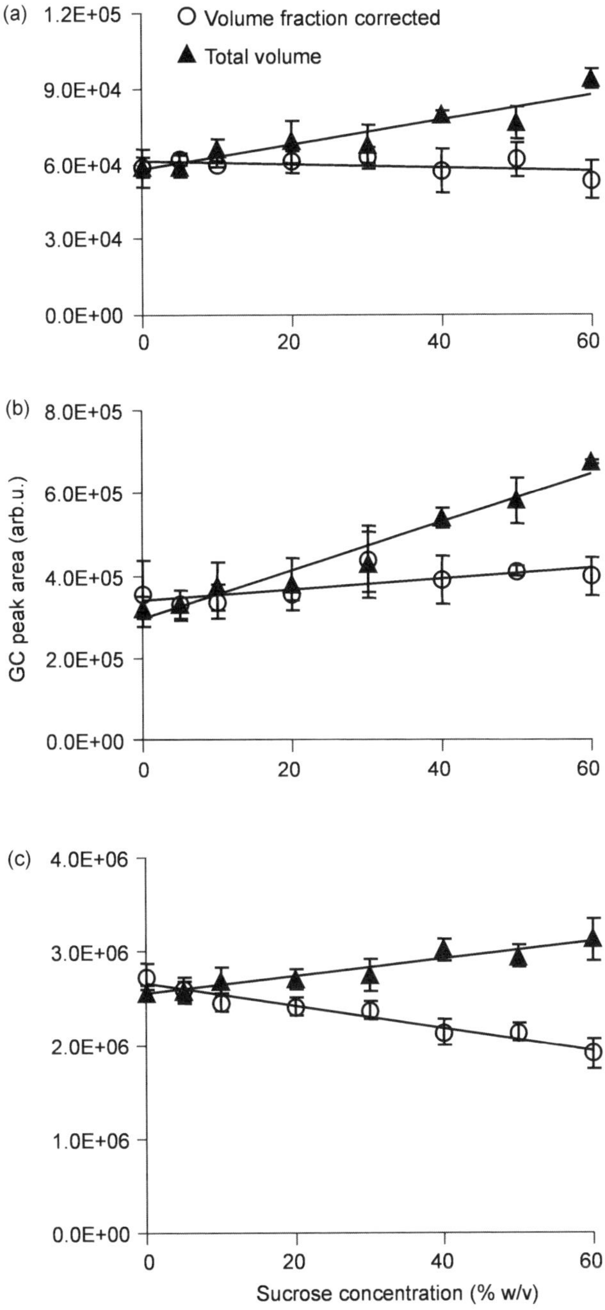

Fig. 12.6 Averaged HS–GC peak area (arbitrary units) of citral (a), geranyl acetate (b) and nonanal (c), above solutions of different sucrose concentrations. Added amount of the aroma compounds, was either corrected (○) or not (▲) for the volume occupied by sucrose. Error bars represent standard deviations. (Copyright ACS.)

12.5.2 The effect of proteins

Proteins and aroma compounds can interact via relatively weak reversible physical adsorption through Van der Waals or hydrophobic interactions, and strongly via covalent or ionic chemical binding (Fischer and Widder 1997). The subject of interactions between aroma compounds and proteins, and parameters affecting these interactions, has been reviewed (Guichard 2002). The main factors are the physicochemical properties of the aroma compound, temperature, ionic conditions, presence of ethanol, and the conformation of the food protein.

Much research effort has been invested in the interactions between aroma compounds and β-lactoglobulin. This has revealed an increase of the binding constants with increasing carbon chain length for the binding of small molecules as aldehydes, ketones, esters and alcohols (Pelletier *et al.* 1998, Reiners *et al.* 2000, Mills and Solms 1984, Guichard and Langourieux 2000, O'Neill and Kinsella 1987, Andriot *et al.* 2000), strongly suggesting hydrophobic interactions, which have a reversible character (Mills and Solms 1984). β-lactoglobulin is reported to have two separate binding sites for hydrophobic ligands (Cho *et al.* 1994, Narayan and Berliner 1997, Wu *et al.* 1999). While retinol (Cho *et al.* 1994) and fatty acids (Wu *et al.* 1999, Ragona *et al.* 2000) are reported to bind in the central cavity of β-lactoglobulin, information about the exact binding side of other ligands is sometimes contradictory (Dufour and Haertlé 1990, Sostmann and Guichard 1998, Lübke *et al.* 1999, Pelletier *et al.* 1998, Narayan and Berliner 1998). A recent study of the binding sites of two aroma molecules, γ-decalactone and β-ionone, using nuclear magnetic resonance spectroscopy, demonstrated binding of the former compound into the central cavity, and binding to a groove near the outer surface of the protein of the latter (Lübke *et al.* 2002).

The studies mentioned above, describing the interactions between β-lactoglobulin and various ligands, give no information about the behaviour of these interactions in the dynamic situation in the mouth during the consumption of a food. In an attempt to understand the dynamic conditions of food consumption, several mathematical models were developed (De Roos and Wolswinkel 1994, Harrison and Hills 1997b). The model developed by Harrison and Hills (1997b) predicts dynamic volatile release from solutions containing aroma binding macromolecules. The model is based on first order kinetics to describe the reversible binding between the aroma compound and the polymer and on the penetration theory of interfacial mass transfer to describe the aroma release across the liquid–gas interface. The latter has been shown to be the rate-limiting step for aroma release in most situations (Harrison and Hills 1997b). Andriot and co-workers (Andriot *et al.* 2000) followed the initial release of volatiles from protein solutions by static headspace measurements after different times of equilibration (15–2700 s). These experimental data were fitted to the model developed by Harrison and Hills (1997b) and, in general, there was good agreement between experiment and theory (Andriot *et al.* 2000). However, the model was not validated by in-mouth measurements. In addition, possible interactions between aroma compounds and saliva, and the effect of mucous membranes of the oral, nasal and pharyngeal cavities on aroma release were not taken into account.

Aldehyde-protein interactions and their effect on aldehyde release have been investigated previously for static headspace, mouth model, and *in vivo* measurements (Friel and Taylor 2001, Van Ruth *et al.* 2001, Weel *et al.* 2003a). Considerable reversible interactions were found during static headspace measurements (binding up to 90%). The aim of a study done in our lab was to reveal the relevance for aroma release of interactions between food proteins and aroma compounds under in-mouth conditions for liquid products. This was done by measuring and comparing the interactions between aldehydes and whey proteins under static and dynamic headspace conditions as well as under in-mouth conditions during consumption (Weel *et al.* 2003a). Retention of aldehydes by whey proteins in solutions buffered at a range of pH values was studied under all three conditions. Static headspace measurements showed a clear increase in retention in the presence of whey proteins for aldehydes with longer carbon chains and for buffer solutions with higher pH values. For *in vivo* aldehyde release measurements, these effects were much less pronounced. Experiments showed that the presence of saliva or the binding of aldehydes to the surface of the oral cavity were not responsible for this effect (Weel *et al.* 2003a). Based on the videofluorescopy measurements that show that a thin film of solution remains in the pharynx after swallowing (Buettner *et al.* 2002), it was hypothesized that the results obtained with the *in vivo* measurements described above can be explained by the fact that the exhalation after swallowing will act as an efficient extraction step and will release both the free aldehydes present in this thin film. Figure 12.7 gives an indication of the interaction between a homologous range of aldehydes and whey protein in the static headspace (HS) conditions, in the mouth model, in the artificial throat and under *in vivo* conditions. From the results it is clear that the *in vivo* results deviate from the results obtained with static headspace and the mouth model. The artificial throat results seem more similar to the *in vivo* results (Weel *et al.* 2004b).

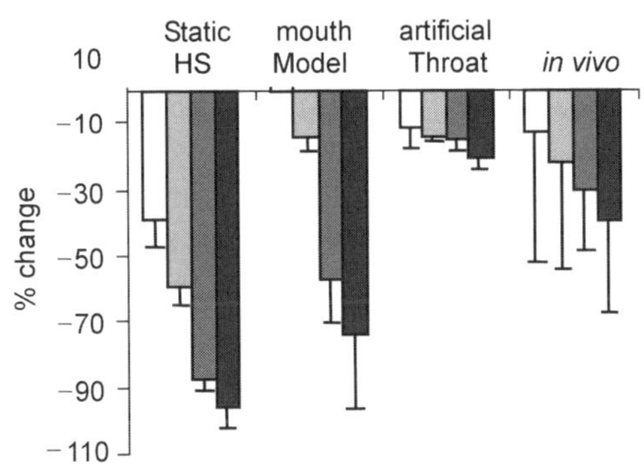

Fig. 12.7 Relative change (%) in release of butanal (white bars), hexanal (light grey bars), octanal (dark grey bars), and nonanal (black bars) due to presence of 3% whey protein for different systems (graph contains data from Weel *et al.* (2003a)). Error bars represent standard deviations. (Copyright ACS.)

12.5.3 The effect of lipids

Lipids have a large influence on aroma release as compared to hydrophilic food ingredients as proteins and polysaccharides (De Roos 1997). Even at low concentrations, the presence of a lipophilic phase in the system can have a large influence, particularly for strongly hydrophobic aroma compounds. Addition of only 0.5% miglyol (a triglyceride of caprylic and capric acids) to water results in a stronger decrease in volatility on 2-nonanone than addition of 3% β-lactoglobulin (Seuvre *et al.* 2000). Knowledge of the impact of lipids on aroma release is valuable for the formulation of increasingly popular low-fat products (De Roos 1997, Overbosch *et al.* 1991). Reduction of the fat content results in a higher release and perception and less lingering of aroma compounds, depending on their hydrophobicity (Brauss *et al.* 1999). Predicting release and perception of aroma compounds from food products that contain an oil phase has been the goal of several mathematical (McNulty 1987, De Roos and Wolswinkel 1994, Harrison and Hills 1997a, Harrison *et al.* 1997) and empirical (Carey *et al.* 2002) models. These studies demonstrated that the oil content of an emulsion and the hydrophobicity of the aroma compounds are key factors for predicting the release. Several other studies have also indicated this, using either analytical (Malone *et al.* 2003, Haahr *et al.* 2000, Doyen *et al.* 2001, Carey *et al.* 2002, Van Ruth *et al.* 2002) or sensorial methods (Miettinen *et al.* 2003), or using both (Roberts *et al.* 2003a). Some authors (De Roos and Wolswinkel 1994, Doyen *et al.* 2001, Malone *et al.* 2003, Roberts *et al.* 2003a) have compared the effect of oil content in liquid emulsions on *in vivo* aroma release or on sensory aroma perception with static headspace measurements. These studies have shown that the effect of oil content on *in vivo* aroma release or perception is smaller than expected from equilibrium headspace studies. In many food systems, including soft drinks, lipids are present as oil droplets dispersed in an aqueous phase. This is therefore the most prevalent form of lipids in model food systems for studies of aroma release. The oil fraction of the emulsion and the hydrophobicity of the aroma compounds are key factors predicting the release (Malone *et al.* 2003, Haahr *et al.* 2000, Doyen *et al.* 2001, Carey *et al.* 2002, Van Ruth *et al.* 2002, Miettinen *et al.* 2003, Roberts *et al.* 2003a, Weel *et al.* 2004a). The type of lipid can also have an effect on aroma release. Increase in solid fat content was found to increase the release (Roberts *et al.* 2003b, De Roos 1997). In our laboratory, the effects of oil content and oil droplet size distribution on release of a set of ester compounds have been studied under *in vivo*, static headspace, and artificial throat conditions (Weel *et al.* 2004a). The effect of oil content on orthonasal and retronasal perceived intensity of ethyl hexanoate was studied using a seven-person panel. With increasing oil content and with a higher hydrophobicity of the aroma compound, a stronger decrease in aroma release was found. This effect was stronger under static headspace conditions than under *in vivo* and artificial throat conditions, and the sensory intensity of ethyl hexanoate was perceived stronger orthonasally than retronasally. The lowest effective oil content was determined for all systems (Table 12.2). Of the compounds tested, droplet size distribution only influenced the in vivo release of geranyl acetate.

Table 12.2 Lowest oil contents from emulsions that differ significantly from 0% oil emulsions, for various analytical techniques. All tested by single sided t-tests ($\alpha = 0.05$), except panel results (single-sided Fisher LSD test, to take the panellist effect into account, adapted from Weel *et al.* 2004a)

	Ethyl butanoate	Ethyl hexanoate	Geranyl acetate
Panellist 1	(>5%)[1]	0.5%	0.1%
Panellist 2	(>5%)[1]	1%	0.05%
HS-GC	0.5%	0.01%	0.01%
AT (1 min)	0.5%	0.2%	0.05%
Panel orthonasal	Nm	0.2%	Nm
Panel retronasal	Nm	1%	Nm
Panel fattiness	Nm	0.5%	Nm

[1] A possible significant effect at higher oil content
[2] Not measured

The artificial throat results correlated well with *in vivo* release, giving support to the assumption that a thin layer of liquid remaining in the throat after swallowing determines aroma release.

The aroma intensity was assessed both orthonasally (sniffing of a freshly opened bottle), and retronasally (swallowing the sample, followed by exhalation). The effect of oil content was stronger under static headspace conditions than under *in vivo* conditions, and stronger for orthonasal perceived intensity than for retronasal perceived intensity. Based on the results it could be calculated that the relatively large breath flow along a hypothesised thin emulsion layer in the throat after swallowing does not only transport ester compounds from the water phase, but also those from the oil phase (through the water phase). This would explain the smaller effect of oil content on *in vivo* aroma release (Weel *et al.* 2004a).

12.6 Conclusion

In the late 1990s, new techniques were developed that combined a high sensitivity and a high time resolution to allow the measurement of the release of aroma compounds in the breath of persons during consumption. Each individual breath is registered. This created the possibility of investigating the effect of food properties on *in vivo* aroma release and to correlate these results directly to the sensory perception of the aroma and the product as a whole.

In vivo aroma release measurements of solid food products give release profiles, over a typical time period of 10 to 20 exhalations. These profiles are characterised by an initial increase to a certain maximum, usually associated with swallowing, followed by a decrease. It is not possible to record such a profile for liquid products. When a liquid is placed in the mouth, it is swallowed directly, leaving a few seconds only to measure the *in vivo* aroma release. Full

Table 12.3 Effect of product parameters and saliva on aroma release for different analysis methods, based on Weel *et al.* (2003, 2004a, 2004b)

	HSGC	Mouth models	*In vivo* release	Artificial throat
Carbohydrates, i.e. saccharose	Effect on release	Not analysed	No effect on release	Not analysed
Proteins, i.e. whey protein	Strong aroma retention	Strong aroma retention	Weak aroma retention	Weak aroma retention
Lipids, i.e. MCT-oil	Strong aroma retention	Not analysed	Weak aroma retention	Weak aroma retention
Saliva	Dilution, aroma binding and enzymatic conversions	Dilution	Minimal dilution	Minimal dilution

control of breathing, swallowing, and mouth movements are necessary for a reproducible *in vivo* aroma release measurement. Using this protocol, the effect of different ingredients was evaluated. Table 12.3 summarises the results obtained.

What is generally found is that the effects of ingredients on aroma release are much larger under static headspace conditions than under dynamic, *in vivo* conditions. Especially in the case of liquid products, aroma release occurs when exhaling air just after swallowing. A thin liquid layer remains in the human throat after swallowing, and subsequent exhalation transports the aroma compounds present in that layer to the nose for detection. A strong aroma concentration gradient exists between this thin layer and the large stream passing it. Under these dynamic conditions, factors that determine partitioning under static headspace conditions and in mouth models are less relevant. These factors are for instance reversible binding to macromolecules or partitioning into hydrophobic phases. These effects are stronger under static headspace conditions than under *in vivo* conditions. Moreover, static headspace studies reported large effects of saliva on aroma release, but *in vivo* aroma release studies of liquids with MS–Nose showed that saliva plays only a minor role in aroma release. This concept allowed development of the artificial throat, whose release profiles correlate better with *in vivo* results. The device can be used to increase understanding of *in vivo* aroma release. Aroma release of beverages is determined by the principle of total exhaustion of aroma compounds present in a thin liquid layer, both *in vivo* and in the artificial throat.

All these results together clearly show that static headspace analysis and mouth model measurements are not always relevant enough for *in vivo* aroma release and, therefore, should be considered carefully. Only by using methodology like the MS–Nose or PTR–MS for *in vivo* aroma release, it is possible to reveal the effects of compositional product parameters and physiological parameters under actual *in vivo* conditions. In case of liquid samples, the thin layer in the throat has to be taken into account, based on a better understanding of the behaviour of aroma molecules in beverages during consumption, it will be possible to develop and innovate liquid drinks in a more directed manner with respect to (added) aroma components. This is particularly of interest for the development of healthier drinks, which should still taste as good as the reference products.

12.7 Acknowledgements

The work presented in this chapter has been funded by Imperial Chemical Industries PLC (project SRF 2015).

12.8 References

ANDRIOT I, HARRISON M, FOURNIER N and GUICHARD E (2000) Interactions between methyl ketones and beta-lactoglobulin: sensory analysis, headspace analysis, and mathematical modeling, *J. Agric. Food Chem.*, 48 (9), 4246–4251.

BAEK I, LINFORTH R S T, BLAKE A and TAYLOR A J (1999) Sensory perception is related to the rate of change of volatile concentration in-nose during eating of model gels, *Chem. Senses*, 24 (2), 155–160.

BAKKER J, BOUDAUD N and HARRISON M (1998) Dynamic release of diacetyl from liquid gelatin in the headspace, *J. Agric. Food Chem.*, 46 (7), 2714–2720.

BANAVARA D S, RABE S, KRINGS U and BERGER R G (2002) Modeling dynamic flavor release from water, *J. Agric. Food Chem.*, 50 (22), 6448–6452.

BRAUSS M S, LINFORTH R S T, CAYEUX I, HARVEY B and TAYLOR A J (1999) Altering the fat content affects flavor release in a model yogurt system, *J. Agric. Food Chem.*, 47 (5), 2055–2059.

BRAUSS M S, LINFORTH R S T and TAYLOR A J (1998) Effect of variety, time of eating, and fruit-to-fruit variation on volatile release during eating of tomato fruits (*Lycopersicon esculentum*), *J. Agric. Food Chem.*, 46 (6), 2287–2292.

BROWN W E, DAUCHEL C and WAKELING I (1996) Influence of chewing efficiency on texture and flavour perceptions of food, *J. Texture Stud.*, 27, 433–450.

BUETTNER A (2002a) Influence of human saliva on odorant concentrations. 2. Aldehydes, alcohols, 3-alkyl-2-methoxypyrazines, methoxyphenols, and 3-hydroxy-4,5-dimethyl-2(5H)-furanone, *J. Agric. Food Chem.*, 50 (24), 7105–7110.

BUETTNER A (2002b) Influence of human salivary enzymes on odorant concentration changes occurring *in vivo*. I. Esters and thiols, *J. Agric. Food Chem.*, 50 (11), 3283–3289.

BUETTNER A, BEER A, HANNIG C and SETTLES M (2001) Observation of the swallowing process by application of videofluoroscopy and real-time magnetic resonance imaging – consequences for retronasal aroma stimulation, *Chem. Senses*, 26 (9), 1211–1219.

BUETTNER A, BEER A, HANNIG C, SETTLES M and SCHIEBERLE P (2002) Physiological and analytical studies on flavor perception dynamics as induced by the eating and swallowing process, *Food Qual. Pref.*, 13, 497–504.

BUETTNER A and SCHIEBERLE P (2000a) Exhaled odorant measurement (EXOM) – a new approach to quantify the degree of in-mouth release of food aroma compounds, *Lebensm. -Wiss. Technol.*, 33 (8), 553–559.

BUETTNER A and SCHIEBERLE P (2000b) Influence of mastication on the concentrations of aroma volatiles – some aspects of flavour release and flavour perception, *Food Chem.*, 71 (3), 347–354.

CAREY M E, ASQUITH T, LINFORTH R S T and TAYLOR A J (2002) Modeling the partition of volatile aroma compounds from a cloud emulsion, *J. Agric. Food Chem.*, 50 (7), 1985–1990.

CHO Y, BATT C A and SAWYER L (1994) Probing the retinol-binding site of bovine beta-lactoglobulin, *J. Biol. Chem.*, 269 (15), 11102–11107.

COOK D J, HOLLOWOOD T A, LINFORTH R S T and TAYLOR A J (2003a) Oral shear stress predicts flavour perception in viscous solutions, *Chem. Senses*, 28, 11–23.

COOK D J, LINFORTH R S T and TAYLOR A J (2003b) Effects of hydrocolloid thickeners on the perception of savory flavors, *J. Agric. Food Chem.*, 51, 3067–3072.

DAVIDSON J M, LINFORTH R S T, HOLLOWOOD T A and TAYLOR A J (1999) Effect of sucrose on the perceived flavor intensity of chewing gum, *J. Agric. Food Chem.*, 47 (10), 4336–4340.

DAVIDSON P (2005) EP1494027A1; Artificial throat *European Patent Application.*

DAWES C (1981) Factors influencing protein secretion in human saliva, *Front. Oral Physiol.*, 3, 125–137.

DE ROOS K B (1997) How lipids influence food flavor, *Food Technol.*, 51 (1), 60–62.

DE ROOS K B and WOLSWINKEL K (1994) Non-equilibrium partition model for predicting flavour release in the mouth, in Maarse H and van der Heij D G, *Trends in Flavour Research*, Amsterdam, The Netherlands, Elsevier Science, 15–32.

DEIBLER K and ACREE T (1999) 'Effects of beverage bas conditions on flavor release, in Roberts D D and Taylor A J, *Flavor Release*, Washington, DC, American Chemical Society, 333–340.

DEIBLER K D, LAVIN E H, LINFORTH R S T, TAYLOR A J and ACREE T E (2001) Verification of a mouth simulator by in vivo measurements, *J. Agric. Food Chem.*, 49 (3), 1388–1393.

DOYEN K, CAREY M, LINFORTH R S T, MARIN M and TAYLOR A J (2001) Volatile release from an emulsion: headspace and in-mouth studies, *J. Agric. Food Chem.*, 49 (2), 804–810.

DUFOUR E and HAERTLÉ T (1990) Binding affinities of beta-ionone and related flavor compounds to beta-lactoglobulin: effects of chemical modifications, *J. Agric. Food Chem.*, 38 (8), 1691–1695.

ELMORE J S and LANGLEY K R (1996) Novel vessel for the measurement of dynamic flavor release in real time from liquid foods, *J. Agric. Food Chem.*, 44, 3560–3563.

FISCHER N and WIDDER S (1997) How proteins influence food flavor, *Food Technol.*, 51 (1), 68–70.

FRIEL E N, LINFORTH R S T and TAYLOR A J (2000) An empirical model to predict the headspace concentration of volatile compounds above solutions containing sucrose, *Food Chem.*, 71 (3), 309–317.

FRIEL E N and TAYLOR A J (2001) Effect of salivary components on volatile partitioning from solutions, *J. Agric. Food Chem.*, 49 (8), 3898–3905.

GODSHALL M A (1997) How carbohydrates influence food flavor, *Food Technol.*, 51 (1) (1), 63–67.

GUICHARD E (2002) Interactions between flavor compounds and food ingredients and their influence on flavor perception, *Food Rev. Int.*, 18 (1), 49–70.

GUICHARD E and LANGOURIEUX S (2000) Interactions between beta-lactoglobulin and flavour compounds, *Food Chem.*, 71 (3), 301–308.

HAAHR A M, BREDIE W L P, STAHNKE L H, JENSEN B and REFSGAARD H H F (2000) Flavour release of aldehydes and diacetyl in oil/water systems, *Food Chem.*, 71, 355–362.

HANSSON A, ANDERSSON J and LEUFVEN A (2001a) The effect of sugars and pectin on flavour release from a soft drink-related model system, *Food Chem.*, 72 (3), 363–368.

HANSSON A, ANDERSSON J, LEUFVEN A and PEHRSON K (2001b) Effect of changes in pH on the release of flavour compounds from a soft drink-related model system, *Food Chem.*, 74 (4), 429–435.

HARRISON M and HILLS B P (1997a) Effects of air flow-rate on flavour release from liquid emulsions in the mouth, *Int. J. Food Sci. Technol.*, 32 (1), 1–9.

HARRISON M and HILLS B P (1997b) Mathematical model of flavor release from liquids containing aroma binding macromolecules, *J. Agric. Food Chem.*, 45 (5), 1883–1890.

HARRISON M, HILLS B P, BAKKER J and CLOTHIER T (1997) Mathematical models of flavor release from liquid emulsions, *J. Food Sci.*, 62 (4), 653–658.

HOGSON M, LINFORTH R S T and TAYLOR A J (2003) Simultaneous real-time measurements of mastication, swallowing, nasal airflow, and aroma release, *J. Agric. Food Chem.*, 51, 5052–5057.

HUSSEIN M M, KACHIKIAN R and PIDEL A R (1983) Analysis for flavor residuals in the mouth by gas chromatography, *J. Food Sci.*, 48 (6), 1884–1885.

KANT A, LINFORTH R S, HORT J and TAYLOR A J (2004) Effect of beta-cyclodextrin on aroma release and flavor perception, *J. Agric. Food Chem.*, 52 (7), 2028–2035.

LAND D G (1996) In McGorrin R J and Leland J, *Flavor–food interactions*, Washington, DC, ACS, 2–11.

LAND D G and REYNOLDS J (1981) In Schreier P, *Flavour '81*, Berlin, Germany, Walter de Gruyter, 701–705.

LARSON-POWERS N and PANGBORN R (1978) Paired comparison and time-intensity measurements of the sensory properties of beverages and gelatins containing sucrose or synthetic sweeteners, *J. Food Sci.*, 43, 41–46.

LE THANH M, THIBEAUDEAU P and THIBAUT M A (1992) Interactions between volatile and non-volatile compounds in the presence of water, *Food Chem.*, 43, 129–135.

LEE W E (1986) A suggested instrumental technique for studying dynamic flavor release from food products, *J. Food Sci.*, 51 (1), 249–250.

LETHUAUT L, WEEL K G, BOELRIJK A E and BROSSARD C D (2004) Flavor perception and aroma release from model dairy desserts, *J. Agric. Food Chem.*, 52 (11), 3478–3485.

LINDINGER W, HANSEL A and JORDAN A (1998) On-line monitoring of volatile organic compounds at pptv levels by means of proton-transfer-reaction mass spectrometry (PTR–MS) – Medical applications, food control and environmental research, *Int. J. Mass Spectrom.*, 173 (3), 191–241.

LINFORTH R, MARTIN F, CAREY M, DAVIDSON J and TAYLOR A J (2002) Retronasal transport of aroma compounds, *J. Agric. Food Chem.*, 50 (5), 1111–1117.

LINFORTH R and TAYLOR A J (2000) Persistence of volatile compounds in the breath after their consumption in aqueous solutions, *J. Agric. Food Chem.*, 48 (11), 5419–5423.

LINFORTH R S T, BAEK I and TAYLOR A J (1999) Simultaneous instrumental and sensory analysis of volatile release from gelatine and pectin/gelatine gels, *Food Chem.*, 65 (1), 77–83.

LÜBKE M, GUICHARD E and LE QUÉRÉ J-L (1999) Infrared spectroscopy of beta-lactoglobulin interactions with flavour compounds, in Roberts D D and Taylor A J, *Flavor Release*, Washington, DC, American Chemical Society, 282–292.

LÜBKE M, GUICHARD E, TROMELIN A and LE QUÉRÉ J-L (2002) Nuclear magnetic resonance spectroscopic study of beta-lactoglobulin interactions with two flavor compounds, gamma-decalactone and beta-ionone, *J. Agric. Food Chem.*, 50 (24), 7094–7099.

MALONE M E, APPELQVIST I A M and NORTON I T (2003) Oral behaviour of food hydrocolloids and emulsions. Part 2. Taste and aroma release, *Food Hydrocoll.*, 17, 775–784.

McNULTY P B (1987) Flavour release – elusive and dynamic, in Blanshard J M V and Lillford P, *Food Structure and Behaviour*, London, UK, Academic Press, 245–258.

MIETTINEN S M, HYVÖNEN L and TUORILA H (2003) Timing of intensity perception of a polar vs nonpolar aroma compound in the presence of added vegetable fat in milk, *J. Agric. Food Chem.*, 51, 5437–5443.

MILLS O E and SOLMS J (1984) Interaction of selected flavour compounds with whey proteins, *Lebensm. -Wiss. Technol.*, 17, 331–335.

NAHON D F, NAVARRO Y KOREN P A, ROOZEN J P and POSTHUMUS M A (1998a) Flavor release from mixtures of sodium cyclamate, sucrose, and an orange aroma, *J. Agric. Food Chem.*, 46 (12), 4963–4968.

NAHON D F, ROOZEN J P and DE GRAAF C (1996) Sweetness flavour interactions in soft drinks, *Food Chem.*, 56 (3), 283–289.

NAHON D F, ROOZEN J P and DE GRAAF C (1998b) Sensory evaluation of mixtures of sodium cyclamate, sucrose, and an orange aroma, *J. Agric. Food Chem.*, 46 (9), 3426–3430.

NARAYAN M and BERLINER L J (1997) Fatty acids and retinoids bind independently and simultaneously to beta-lactoglobulin, *Biochemistry*, 36 (7), 1906–1911.

NARAYAN M and BERLINER L J (1998) Mapping fatty acid binding to beta-lactoglobulin: ligand binding is restricted by modification of Cys 121, *Prot. Sci.*, 7 (1), 150–157.

NAßL K, KROPF F and KLOSTERMEYER H (1995) A method to mimic and to study the release of flavour compounds from chewed food, *Z. Lebensm. Unters. F.*, 201, 62–68.

NAWAR W W (1971) Some variables affecting composition of headspace aroma, *J. Agric. Food Chem.*, 19, 1057–1059.

O'BRIEN NABORS L and GELARDI R C (1991) In O'Brien Nabors L and Gelardi R C, *Alternative Sweeteners*, New York, NY, Marcel Dekker, 1–10.

O'NEILL T E and KINSELLA J E (1987) Binding of alkanone flavors to beta-lactoglobulin: effects of conformational and chemical modification, *J. Agric. Food Chem.*, 35, 770–774.

OVERBOSCH P, AGTEROF W G M and HARING P G M (1991) Flavor telease in the mouth, *Food Rev. Int.*, 7 (2), 137–184.

PELLETIER E, SOSTMANN K and GUICHARD E (1998) Measurement of interactions between beta-lactoglobulin and flavor compounds (esters, acids, and pyrazines) by affinity and exclusion size chromatography, *J. Agric. Food Chem.*, 46 (4), 1506–1509.

PIONNIER E, CHABANET C, MIOCHE L, LE QUÉRÉ J-L and SALLES C (2004) 1. *In vivo* aroma release during eating of a model cheese: relationships with oral parameters, *J. Agric. Food Chem.*, 52, 557–564.

QI Z H and HEDGES A R (1995) Use of cyclodextrins for flavours, in Ho C T, Tan C T and Tong C H, *Flavor technology*, Washington DC, American Chemical Society, 231–243.

RABE S, KRINGS U, BANAVARA D S and BERGER R G (2002) Computerized apparatus for measuring dynamic flavor release from liquid food matrices, *J. Agric. Food Chem.*, 50, 6440–6447.

RAGONA L, FOGOLARI F, ZETTA L, PEREZ D M, PUYOL P, DE KRUIF K, LOHR F, RUTERJANS H and MOLINARI H (2000) Bovine beta-lactoglobulin: Interaction studies with palmitic acid, *Prot. Sci.*, 9 (7), 1347–1356.

REINERS J, NICKLAUS S and GUICHARD E (2000) Interactions between beta-lactoglobulin and flavour compounds of different chemical classes. Impact of the protein on the odour perception of vanillin and eugenol, *Lait*, 80 (3), 347–360.

ROBERTS D D and ACREE T E (1995) Simulation of retronasal aroma using a modified headspace technique: investigating the effects of saliva, temperature, shearing, and oil on flavor release, *J. Agric. Food Chem.*, 43, 2179–2186.

ROBERTS D D, ELMORE J S, LANGLEY K R and BAKKER J (1996) Effect of sucrose, guar gum, and carboxymethylcellulose on the release of volatile flavour compounds under dynamic conditions, *J. Agric. Food Chem.*, 44, 1321–1326.

ROBERTS D D, POLLIEN P, ANTILLE N, LINDINGER C and YERETZIAN C (2003a) Comparison of nosespace, headspace, and sensory intensity ratings for the evaluation of flavor absorption by fat, *J. Agric. Food Chem.*, 51, 3636–3642.

ROBERTS D D, POLLIEN P and WATZKE B (2003b) Experimental and modeling studies showing the effect of lipid type and level on flavor release from milk-based liquid emulsions, *J. Agric. Food Chem.*, 51, 189–195.

SEUVRE A M, DIAZ M A E and VOILLEY A (2000) Influence of the food matrix structure on the retention of aroma compounds, *J. Agric. Food Chem.*, 48 (9), 4296–4300.

SOLMS J (1986) Interactions of non-volatile substances in foods, in Birch G G and Lindley M G, *Interactions of food components*, London, UK, Elsevier Applied Science, 189–210.

SOSTMANN K and GUICHARD E (1998) Immobilized beta-lactoglobulin on a HPLC-column: a rapid way to determine protein–flavour interactions, *Food Chem.*, 62 (4), 509–513.

SPRINGETT M B, ROZIER V and BAKKER J (1999) Use of fiber interface direct mass spectrometry for the determination of volatile flavor release from model food systems, *J. Agric. Food Chem.*, 47, 1125–1131.

TAYLOR A J (1996) Volatile flavor release from foods during eating, *Crit. Rev. Food Sci. Nutr.*, 36 (8), 765–784.

TAYLOR A J, BESNARD S, PUAUD M and LINFORTH R S T (2001) *In vivo* measurement of flavour release from mixed phase gels, *Biomol. Eng.*, 17, 143–150.

TAYLOR A J and LINFORTH R S T (1996) Flavour release in the mouth, *Trends Food Sci. Tech.*, 7 (12), 444–448.

VAN RUTH S M, DE VRIES G, GEARY M and GIANNOULI P (2002) Influence of composition and structure of oil-in-water emulsions on retention of aroma compounds, *J. Sci. Food Agric.*, 82 (9), 1028–1035.

VAN RUTH S M, GROSSMANN I, GEARY M and DELAHUNTY C M (2001) Interactions between artificial saliva and 20 aroma compounds in water and oil model systems, *J. Agric. Food Chem.*, 49 (5), 2409–2413.

VAN RUTH S M, O'CONNOR C H and DELAHUNTY C M (2000) Relationships between temporal release of aroma compounds in a model mouth system and their physico-chemical characteristics, *Food Chem.*, 71 (3), 393–399.

VAN RUTH S M and ROOZEN J P (2000) Influence of mastication and saliva on aroma release in a model mouth system, *Food Chem.*, 71 (3), 339–345.

VAN RUTH S M, ROOZEN J P and COZIJNSEN J L (1994) In Maarse H and van der Heij D G, *Trends in Flavour Research*, Amsterdam, The Netherlands, Elsevier Science, 59–64.

VAN RUTH S M, ROOZEN J P and COZIJNSEN J L (1995) Changes in flavour release from rehydrated diced bell peppers (*Capsicum annuum*) by artificial saliva components in three mouth model systems, *J. Sci. Food Agric.*, 67, 189–196.

VOILLEY A, SIMATOS D and LONCIN M (1977) Gas phase concentration of volatiles in equilibrium with a liquid aqueous phase, *Lebensm. -Wiss. Technol.*, 10, 45–49.

VON SYDOW E, MOSKOWITZ H, JACOBS H and MEISELMAN H (1974) Odor-taste interactions in fruit juices, *Lebensm. -Wiss. Technol.*, 7, 18–24.

WEEL K G, BOELRIJK A E, BURGER J J, JACOBS M A, GRUPPEN H, VORAGEN A G and SMIT G (2004a) Effect of emulsion properties on release of esters under static headspace, *in vivo*, and artificial throat conditions in relation to sensory intensity, *J. Agric. Food Chem.*, 52, (21), 6572–6577.

WEEL K G, BOELRIJK A E, BURGER J J, VERSCHUEREN M, GRUPPEN H, VORAGEN A G and SMIT G (2004b) New device to simulate swallowing and *in vivo* aroma release in the throat from liquid and semiliquid food systems, *J. Agric. Food Chem.*, 52 (21), 6564–6571.

WEEL K G C, BOELRIJK A E M, ALTING A C, VAN MIL P J J M, BURGER J J, GRUPPEN H, VORAGEN A G J and SMIT G (2002) Flavor release and perception of flavored whey protein gels: perception is determined by texture rather than by release, *J. Agric. Food Chem.*, 50 (18), 5149–5155.

WEEL K G C, BOELRIJK A E M, BURGER J J, CLAASSEN N E, GRUPPEN H, VORAGEN A G J and SMIT G (2003a) Effect of whey protein on the *in vivo* release of aldehydes, *J. Agric. Food Chem.*, 51, 4746–4752.

WEEL K G C, BOELRIJK A E M, BURGER J J, GRUPPEN H, VORAGEN A G J and SMIT G (2003b) A protocol for measurement of *in vivo* aroma release from beverages, *J. Food Sci.*, 68, 1123–1128.

WISEMAN J J and MCDANIEL M R (1991) Modification of fruit flavours by aspartame and sucrose, *J. Food Sci.*, 56, 1668–1670.

WRIGHT K M, SPRUNT J, SMITH A C and HILLS B P (2003) Modelling flavor release from a chewed bolus in the mouth: Part I. Mastication, *Int. J. Food Sci. Technol.*, 38, 351–360.

WU S Y, PEREZ M D, PUYOL P and SAWYER L (1999) Beta-lactoglobulin binds palmitate within its central cavity, *J. Biol. Chem.*, 274 (1), 170–174.

Part III

Influences on flavour perception

13

The process of flavour release

R. Linforth and A. Taylor, Nottingham University, UK

13.1 Introduction

Flavour release is a sequential process. It can start prior to ingesting the foodstuff, as aromas are sampled by sniffing (orthonasally), then continue, as the food is processed in mouth. Finishing only when the last of the aroma molecules disappear from the breath. The most important aspect of flavour release is that aroma molecules leave the bolus and arrive at the olfactory epithelium in the nose where they can be sensed. Hence this chapter is focused on flavour delivery *in vivo*.

Flavour release will depend on the nature of the foodstuff itself and the way in which it is consumed. The state of the foodstuff can change substantially as it is broken down by chewing, and mixed with saliva. The action of chewing and mouth movement not only breaks down the food and mixes it with saliva, but also transfers aroma molecules into the pharynx; exhalation will transfer them from the pharynx, through to the nose. The other main mouth action, swallowing, also delivers aroma molecules in the gas phase to the throat. The main function of swallowing is, however, to move the food through the pharynx, on its route to the stomach. This leaves behind a residue of the foodstuff in the pharynx, which then acts as a reservoir for further flavour release, with aroma molecules partitioning directly into exhaled breath before delivery to the nose.

In addition to the influence of the food matrix and human physiology on flavour release, the physical properties of the aroma compounds themselves have a substantial impact. This is dependent on their partitioning from the bolus, through media such as saliva, on into the breath.

13.2 Influence of the foodstuff on flavour release

The nature of the foodstuff and the changes that happen to it during eating will affect the availability of aroma molecules for delivery to the olfactory epithelium. Foodstuffs are, however, extremely diverse, varying in factors such as their texture, state of hydration, polarity and homogeneity, all of which will affect flavour release.

13.2.1 Orthonasal delivery

One of the first points in the eating process where an aroma stimulus can be delivered is orthonasally (inhalation) prior to ingestion. Orthonasal delivery of aroma molecules is highly dependent on the state of the foodstuff (bolus) and the way in which the food is introduced into the mouth. For instance, the foodstuff may be cold and of low internal molecular mobility (e.g. ice cream) reducing the tendency of aroma molecules to move into the gas phase. Alternatively, the matrix may offer minimal resistance to aroma movement to the product–air interface and aroma volatilisation (e.g. beverages). These differences are further enhanced by the eating process itself.

Differences in orthonasal aroma delivery can be highlighted by considering the fundamental differences between the aroma release of wine and chewing gum. With wine, all of the aroma molecules are dissolved in a matrix with minimal viscosity such that convective and diffusive forces can replenish any aroma molecules lost from the surface as they partition into the gas phase. The glass, the wine is held in, forms a chamber to present the gas phase aroma to the nose, which (given that we typically inhale prior to ingesting and swallowing liquid (Preiksaitis and Mills 1996)) is ideal for orthonasal sampling of aroma. In contrast, chewing gum is a foodstuff in which the flavour molecules have low mobility (based on diffusion rather than convection) and the main phase of release occurs in mouth. Here, it is the action of chewing that results in the replenishment of aroma molecules at the surface of the bolus, allowing partitioning into saliva and the gas phase. Prior to ingestion the limited mobility of molecules in the matrix and the mode of ingestion itself combine to minimise the orthonasal aroma delivery route.

13.2.2 Sample ingestion

The introduction of the food itself into the mouth is in itself a regulated process. The sample size depends on the individual's physiology (dentition, etc.), and their initial assessment of the properties of the foodstuff. Once in the mouth, there may be transfer of aroma compounds from the foodstuff into the breath retronasally (exhalation), resulting in aroma delivery to the olfactory epithelium. It is, however, important to remember that the primary objective of consumption is to process the ingested food and convert the food into a bolus suitable for swallowing.

13.2.3 The effect of matrix structure on flavour release

During consumption, aroma delivery from the foodstuff to the gas phase in the mouth is typically via an aqueous boundary layer. The process where aroma transfers from the food to the boundary layer of saliva and then through to the gas phase was termed the two film theory (Hills and Harrison 1995). The key feature of the model was the description of the stagnant aqueous and gas phase layers around the food, which can restrict the transfer of aroma from the food to the saliva and from the saliva to the breath. However, for aroma compounds to leave the food and enter the saliva phase they must be sufficiently soluble in water. A similar principle exists in drug delivery, where there is a recognised optimum in terms of hydrophobicity – hydrophilicity for the transfer of compounds across membranes or aqueous domains (Dearden and James 1998).

Studies of aroma delivery from gel systems, showed that ρ-cymene and limonene had the greatest release in the initial 10 s of consumption (Taylor *et al.* 2001, Harvey *et al.* 2000), a period when the gel would be least coated with saliva. Other less hydrophobic compounds reached their maxima much later (at around 30 s), when the gels were disintegrating, due primarily to the effects of chewing or melting. At this point the in-breath concentration of the more hydrophobic compounds had decreased, despite the presence of a substantial remainder of the gel in mouth. These differences in flavour release behaviour, show that strongly hydrophobic compounds may behave differently from more polar molecules and preferentially release from the bolus itself. Therefore when investigating the way in which matrices are processed during eating and the impact this has on flavour release, we need to remember that different flavour molecules will show a different behaviour in the same matrix.

Gels are ideal systems for the study of flavour release. Not only are many gels made from food grade material, but, gels have many of the features typical of foodstuffs. They have a moderate to high water content (semi-solids) which can carry dissolved flavour molecules. They can be fractured and sheared by the teeth, resulting in particle size reduction and dispersion in the saliva phase. Essentially, they are broadly representative of a physical state intermediate between that of solutions and low moisture foods. In addition, they are easily manipulated and can be produced with a range of physical properties and volatile content.

Substantial differences in flavour release might be expected between gelling agent types based on their different bulk properties. Gels that can melt in mouth (e.g. gelatine) would be fractured, before finally being dispersed and dissolving in saliva, at a rate dependent on their composition (Harrison and Hills 1996). Whereas brittle gels, with melting points much higher than body temperature (e.g. agarose) would simply fracture during chewing and form a dispersion of particles with an increased surface area.

However, the effect of the gelling agent on flavour release does not appear to be that great, with agarose and gelatine gels having broadly similar flavour release profiles (Taylor *et al.* 2001). This implies that a dispersion of fractured gel particles has similar release properties to a melted gel. *In vivo* the dispersion and interfacial areas of the two gels may, in fact, be fairly similar. Saliva is designed to coat and lubricate foods rather than to act as a medium for their dissolution (which in many cases would require substantially greater amounts of saliva). It can take several seconds for viscous solutions to mix with saliva in mouth (Lucas *et al.* 2004). Before this the biphasic system has a distinct interfacial area, between the solution and the saliva. Consequently, the rate of dispersion and melting of a gelatine gel may restrict the potential for volatile release, as it may still remain localised in distinct regions, comparable to the dispersion of the fractured brittle gel. In this instance it is not the matrix that is the rate limiting step controlling flavour delivery, but, the subsequent processes after flavour has entered the saliva phase.

One of the other factors that can be changed in gel formulation is the concentration of the gelling agent, this would be expected to affect both the properties of the gel and the way that it was consumed (mastication–matrix interactions). Weaker gels should be easier to break down and would be expected to release more flavour compared with stronger gels. Equally a weaker gel might be easier to process in the mouth, resulting in a weaker chewing action and an earlier onset of swallowing.

Studies of gels with different concentrations of gelling agents show that there were no major differences in the maximum intensity of aroma release as the gel strength was varied (Baek *et al.* 1999, Lethuaut *et al.* 2004). Differences in the temporal dimension of flavour release were, however, observed, essentially dependent on how we eat foods with different structural properties. Weaker gels gave a faster rate of flavour release, with an earlier maximum in flavour intensity and an earlier fall in breath volatile concentration (Baek *et al.* 1999). The impact of texture on flavour release could, however, be minimised by the use of an enforced eating protocol (Lethuaut *et al.* 2004) which effectively eliminated differences in release associated with the way in which we have learnt to process/eat foods.

13.2.4 Influence of the physical properties of flavour molecules on release

One of the major factors that affects flavour release is the physical properties of the aroma molecules themselves. The release of a range of aroma compounds from a single gel formulation varied by 10 000-fold in their maximum intensity of release (Linforth *et al.* 2000). This is far greater than the effects of the gel type or its concentration combined. The key physico-chemcial factors accounting for the differences in release properties were associated with the hydrophobicity of the compound and its volatility (vapour pressure). The compounds exhibiting the greatest intensity of release were hydrophobic rather than hydrophilic and more volatile rather than less. These properties will not result in major changes to the diffusion of the compounds, nor their convective movements in bulk components of the gel saliva phase. They are essentially related to the differences in aqueous solubilities and the product–air partition coefficients of the compounds.

In addition to the impact of the volatiles on the intensity of delivery, the temporal dimension was also altered despite the fact that the gels were identical in formulation and consequently consumed in a similar way. These differences were also related to the hydrophobicity and volatility of the compounds (Linforth *et al.* 2000). The more water soluble and lower vapour pressure compounds were the most persistent, these compounds would be the most likely to absorb to the nasal epithelia (Keyhani *et al.* 1997) and hence also the least likely to be removed by exhaled air. Overall, the studies on gels demonstrate that textural attributes may impact on flavour release, but, that much larger differences exist as a result of the physical chemistry of the flavour molecules themselves.

13.2.5 Flavour release from solutions, slurries and low moisture foods

With volatiles in highly hydrated systems with less structure than gels (i.e. the range from solutions to slurries) the volatiles are effectively already in solution. They do not have to leave a solid matrix before they can enter the aqueous phase on route to the gas phase. These systems may mix with saliva resulting in dilution, however, typically the volume of the sample is much greater than the volume of saliva in mouth (1–2 ml) such that dilution effects are minimal over a short period of time (Linforth *et al.* 2002). The fact that the food system has minimal structure results in its ability to spread out rapidly during eating to form a large area for release. However, the lack of structure also impacts on mouth movements (Jack and Gibbon 1995), and as a bolus it is easily and readily swallowed reducing the time available for flavour release to take place. Hence, for these systems there is typically a combined release phase associated with ingestion and swallowing (which are often one connected action), followed by persistence. Further delivery of flavour during this phase (after the cessation of chewing and swallowing), only occurring in the event of mouth movements, which are associated with mouth clearance activity or involuntary swallowing actions as saliva accumulates in-mouth.

The other extreme situation is the dry or low moisture foodstuff. Here the objective of the eating process is to lubricate and hydrate the bolus to allow it to be swallowed with minimal potential for choking (Mioche *et al.* 2003). The process typically involves the fracture of the foodstuff in the mouth as a result of chewing, producing a large dry surface that can readily absorb saliva. At the start of eating the food may quickly absorb all in-mouth saliva. Further saliva is released which then coats and hydrates the bolus although often only to the point that it now adheres to itself to form a cohesive bolus. During this process volatiles may be liberated from the structure as a result of fracture (as in the case of gels), but, there is the potential for flavour release from the surface of the food matrix as it is progressively hydrated. Hydration however, does not stimulate the release of all compounds equally. In a model biscuit system, it was observed that hydration had a greater impact on methyl butanal than anethole, hence the effect can be considered compound dependent (Brauss *et al.* 1999). The dry foodstuff potentially offers the least opportunity for flavour release due to the nature of its consumption. These foods tend to result in a sticky bolus (compared with brittle gels), such that the bolus is more localised rather than spread around the oral cavity. A key feature of such foods, however, is their residence time in mouth (with associated mouth movement) as hydration and lubrication take place. This allows more time for the release process to take place, in addition during swallowing the food is more likely to adhere to the back of the tongue and in the pharynx (Buettner *et al.* 2001) prolonging flavour delivery.

13.3 Losses of flavour molecules after leaving the bolus

Once the flavour molecules have left the bolus and entered saliva, they may be vulnerable to enzymic degradation or absorption into the epithelium in-mouth. This has been studied by taking solutions of aroma compounds into the mouth and then spitting them out after a fixed time interval. This showed that the losses of compounds in-mouth was not the same for all compounds, with the longer chain length aldehydes and esters showing the greatest losses (Buettner and Schieberle 2000a, 2000b, 2000c). Losses of up to 50% were observed over a period of 1 minute. However, shortening the duration of the experiment to 5 seconds reduced losses of all but the most hydrophobic compounds studied, to about 10%, demonstrating that mastication time is an important factor. In addition to the effects of a compound's polarity on in-mouth losses, the concentration also appeared to be important with higher concentrations showing less loss. This showed that the systems resulting in the loss can be saturated by the amount of volatile present, implying a limited number of absorptive sites or enzymic activity.

A further factor in the loss of volatiles in mouth was the matrix effect. Studies in which aqueous samples were compared with fresh orange juice, showed that losses of some compounds (particularly the longer chain aldehydes) were reduced by the presence of other orange juice components (Buettner and Schieberle 2000a). This is likely to have been due to the interaction of these compounds with the orange juice minimising the availability of these compounds for absorptive losses or enzymic degradation.

Investigation of the mode of loss showed that for many compounds the losses in mouth were simply due to absorptive processes (Buettner 2004). Other compounds showed much higher losses than expected on the basis of absorption alone. These compounds were degraded by enzymes present in saliva in a series of *in vitro* experiments. Some compounds showed only small enzyme dependent losses (20% decrease over 10 minutes) whereas others such as furfurylthiol showed much greater loss (90% over 15 minutes). Hence, there can be substantial flavour losses, particularly for compounds dispersed in saliva in mouth, with the extent of such losses depending on the time taken to consume the foodstuff and the flavour molecule itself.

13.4 The transfer of volatiles to the gas phase *in vivo*

13.4.1 The environment

All of the main phases of flavour delivery involve dynamic rather than static conditions. Orthonasal sampling will depend on the gas phase around the product, which in the open air will most certainly be dynamic. The only instances where the consumer experiences a near static headspace orthonasally, is on the opening of a food container. However, this situation very rapidly changes to a dynamic one, hence the aroma concentration decays rapidly. The mouth is a separate compartment to one side of the upper respiratory tract and hence thought of as a region where volatiles can reach equilibration. This may be so if the velum (the seal between the mouth and the throat) remains closed. However, the seal of the velum can be broken by mouth movements (Buettner *et al.* 2001).

When this happens, air is drawn into the mouth as the jaw falls; similarly air is displaced into the pharynx as the jaw rises. The mouth has a maximum capacity of around 80 ml, but, probably contains a gas phase closer to 40 ml during eating. The volume of air transferred in each chewing action was measured as 13 ml (Hodgson *et al.* 2003), which given that we chew approximately 100 times a minute, means that 1300 ml of air is pumped in and out of a 40 ml chamber every minute. Hence the mouth can be a very dynamic environment for gas movement, which may disturb the equilibrium state. The upper airway is the third major region where volatiles can partition into the breath, either from residues of the food in the pharynx after swallowing or from volatiles partitioning in and out of the nasal mucosa. Typical gas flow rates through the upper airway during exhalation and inhalation are in the region of 160 ml/s (Hodgson *et al.* 2003), hence this too is a dynamic situation where equilibration may not be achieved. High gas flow rates will create two opposing effects, they will sweep away the volatiles disturbing the equilibrium whilst simultaneously decreasing the thickness of the gas phase boundary layer enhancing volatile movement into the breath.

13.4.2 Mass transfer under dynamic gas flow conditions

The key transition *in vivo* is the transfer of volatiles from the aqueous continuum to the gas phase. This is the main point influenced by dynamic air movements. Many studies have observed the headspace volatile concentration under dynamic conditions, but few studies have referenced them back to the static equilibrium state. By using the static equilibrium state as a reference point, the behaviour of a compound is expressed relative to the maximum concentration it may achieve under a given set of conditions; rather than comparing different compounds with each other. Marin and co-workers (1999) found that the equilibrium gas phase above an aqueous solution could be readily disturbed by a gas flow of 70 ml/min through a 25 ml headspace. Relative to the volume of the gas phase, these flow rates were much lower than those observed *in vivo*.

The depletion of a compound in the gas phase did not proceed at the same rate for all compounds. The headspace concentration of compounds with low air–water partition coefficients ($K_{aw} = 10^{-5}$), were much more stable than that of compounds with higher air/water partition coefficients (10^{-2}). This was caused by differences in the rate of depletion of the compounds from the interface between the water and air phases. This was expressed mathematically (Marin *et al.* 1999) as the ratio of the partition coefficient to the mass transfer coefficient in the aqueous phase (k_1): effectively the proportion of a compound that has to cross the interface to maintain the equilibrium vs. the rate of delivery to the interface (expressing diffusion and convective forces).

The bulk of the solution was effectively unavailable as a reservoir to maintain the equilibrium, this was dependent on a small volume of the total sample at the interface. Consequently, the situation is comparable to trying to equilibrate a relatively large headspace from a small amount of sample. The compounds with low partition coefficients have to move a small proportion of the molecules from the sample to the gas phase to reach equilibrium, a process that does not substantially deplete the sample volatile content. In contrast, a compound with a K_{aw} of 10^{-2} has to move a much greater proportion of the molecules across the interface to reach equilibrium. The larger the headspace–sample ratio the greater the depletion of the compound concentration in the sample. These differences are increased under dynamic conditions, where the volume of the headspace can be thought of as expanding with time.

Compared with the Marin headspace system, *in vivo* there is more potential for bulk convective movements (e.g. during chewing) and the time scales are much shorter, reducing the time available for depletion at the interface. However, the gas flow rates are much higher, and a shorter time scale may, in fact, reduce the time available for any sample movement to influence surface volatile concentrations.

13.4.3 Differences between flavour molecules *in vivo*

In vivo studies of the headspace and breath volatile concentration for aqueous solutions, showed that compounds with high K_{aw} were present at low concentrations in breath relative to headspace (Linforth *et al.* 2002). Compounds with lower partition coefficients were observed at concentrations much closer to those observed in the headspace. This strongly suggests that the same phenomenon occurs *in vivo*, surface depletion of volatiles at the interface affecting delivery. What made this finding even more surprising was the fact that the *in vivo* analysis was based on swallowing experiments, where the sample would have passed through the pharynx during a phase when inhalation and exhalation ceased (Hodgson *et al.* 2003). This implies that the in mouth/pharynx gas phase volume in contact with the sample, was much greater than the volume of sample available for partitioning, even under these low breath flow conditions.

It is important to remember that these results are relative to the static equilibrium state for each compound. On an absolute scale a compound with a high K_{aw} coefficient, will still be present at a higher concentration in the breath than one with a lower one. The absolute breath volatile concentrations are, however, more similar than expected from static headspace studies, as the low partition coefficient compound enters the gas phase efficiently, whereas the compound with the high partition coefficient does not.

Studies comparing flavour release from viscous solutions and water, showed that there were no significant differences in breath volatile concentration between them when they were swallowed (Hollowood *et al.* 2002, Cook *et al.* 2003). This implies that the limited surface available for volatile transfer was similar in both cases despite the differences in viscosity. The small decrease in flavour release at the highest viscosities may have been due to the ability of individuals to easily swallow these samples, which in turn affected sample interface behaviour.

The influence of the partition coefficient on flavour release under dynamic condition has further implications for systems where aroma–matrix interactions take place. Lipids, proteins and β-cyclodextrin are all capable of reversible interactions (as opposed to the formation of covalent bonds) with aroma compounds, effectively altering the partition coefficient. When these interactions are reversible the flavour molecules in the system behave more like compounds with low air/water partition coefficients than if they were dissolved in water. Under dynamic conditions when aroma losses from the aqueous phase occur, molecules are liberated from the matrix they are interacting with, restoring the volatile concentration in solution. Hence the depletion of the volatile concentration at the interface is minimised. This in turn can stabilise the volatile concentration in the gas phase and hence flavour release.

Consequently the differences in volatile concentration observed under static equilibrium conditions may be substantially different *in vivo* under dynamic conditions. Under static equilibrium conditions the totally aqueous system will have a much higher headspace volatile concentration than that containing aroma binding components. However, under dynamic conditions the totally aqueous system delivers volatile inefficiently compared with that with the lower partition coefficient hence the two systems become more similar. Reversible interactions between flavour molecules and the food matrix have been observed for lipids (Doyen *et al.* 2001, Roberts *et al.* 2003), β-cyclodextrin (Kant *et al.* 2004) and protein (Le Guen and Vreeker 2004, Weel *et al.* 2003) all of which have shown better release for the aroma binding systems *in vivo* than expected on the basis of their static headspace behaviour. This, of course, relies on the fact that the dissociation of the aroma matrix complex is fast enough to allow release during consumption (Harrison *et al.* 1997), otherwise the effect would not be observed.

13.5 Aroma delivery to the upper airway and nose

13.5.1 Absorption in the lungs and nose

Once the volatiles have entered the gas phase they have one of two basic options, exhalation or absorption (temporary or permanently) to surfaces in the upper airway. Compounds present in the gas phase of the pharynx are directly in the path of exhaled or inhaled air. Upon inhalation they are absorbed by the large surface area of the lungs, an organ specifically designed for the exchange of molecules in and out of the gas phase. This can be observed during real-time breath analysis, if compounds are in the laboratory atmosphere they are detected in inhaled air at higher concentrations than in exhaled air. This results in a series of troughs in the baseline volatile signal as 'cleaner' air is exhaled. These absorptive losses are likely to be effectively irreversible as the volatiles move into the relatively large volume of the blood system.

The nose can be thought of as the second major absorptive surface in the airway. Here the surfaces are not primarily designed for absorption, this is merely one of the consequences of their role in humidifying inhaled air. The partitioning of aroma compounds from the breath stream into the nasal epithelium can remove up to 75% of the molecules from the gas phase in a single pass (Keyhani *et al.* 1997), dependent on the physicochemical properties of the compound. The absorbed compounds are typically available for subsequent repartitioning into the gas phase, hence the continued detection of dimethyl pyrazine in the breath for over 1 minute after a single inhalation of headspace (Linforth *et al.* 2002). The gradual deposition and desorption of volatiles from the upper airway (the 'wash in–wash out' principle (Medinsky *et al.* 1993)) will occur throughout eating, this will affect the in-nose volatile concentration, dependent on the duration of the eating process.

13.5.2 Flavour molecule transport in the upper airway

One of the other main features of the upper airway is its ability to transfer flavour molecules from pharynx to nose with minimal mixing. This can be clearly observed when solutions of flavour compounds are swallowed (Linforth and Taylor 2000, Linforth *et al.* 2002). The main pulse of aroma in the breath is a sharp peak at the start of the exhalation following swallowing. As in most chromatography, the lack of tailing and peak symmetry suggest virtually no dead volumes in the upper airway. Because of this it is possible to study the delivery of flavour molecules from the mouth to the exhaled air stream, with the potential to observe each movement of aroma as a peak in breath volatile concentration.

13.5.3 Flavour delivery from the mouth

Flavour molecules present in the gas phase in mouth can be transferred to the pharynx only if the seal between the velum and the back of the tongue opens. Real-time breath analyses using the method developed by Soeting and Heidema (1988) showed that there was considerable variation in the pattern of aroma release between individuals. Some individuals would produce profiles with regular peaks corresponding to volatiles present in each exhalation, whereas others were associated with less regular flavour delivery (so clear was the difference that the former group were termed 'regular breathers'). The samples used in these experiments were, however, flavoured oil, which may not have been very palatable and may have influenced the panellists eating pattern and hence aroma delivery. Analyses of real time flavour release from mint flavoured sweets by Atmospheric Pressure Chemical Ionisation (APCI; Linforth *et al.* 1996) showed regular delivery of flavour to the breath throughout the eating time course. In this case the mints were chewed, and swallowing would have occurred, both events may have resulted in flavour delivery. Chewing and swallowing patterns were not recorded, but, the frequency of peaks in the breath profile was consistent with flavour delivery as a result of chewing actions, which are far more frequent than swallowing events (Wright *et al.* 2003).

Studies using videofluoroscopy have demonstrated that during both the swallowing and chewing process the seal between the velum and the back of the tongue can open connecting the gas phase of the mouth to that in the pharynx (Buettner *et al.* 2001). It was not possible to simultaneously measure the breath volatile content, but, parallel breath analyses for helium or flavour molecules showed their movement from the mouth to the nose during vigorous mouth movements or swallowing. Earlier studies by Land (1996) had showed a small amount of breath (5 to 15 ml) exiting in the nostril during swallowing actions which would also be consistent with volatile delivery. However, in these preliminary experiments a bubble flow meter was used on its own and it was not possible to analyse the real-time breath volatile content, or record the swallowing process restricting interpretation.

More recent work combining electromyography (EMG), breath flow (using a nasal turbine), real-time breath volatile analysis and swallowing (using a laryngograph) analysis clearly demonstrates that both chewing and swallowing events are associated with flavour delivery (Hodgson *et al.* 2003). To investigate chewing, panellists were instructed to chew four times during each exhalation; four peaks of volatile laden air were observed in the breath. The EMG response was slightly offset from the breath volatile peaks, these offsets were later attributed to the time of transmission of the volatile laden air through the upper airway. The breath flow rate analysis, showed that the appearance of volatile laden air was associated with pulses in the breath air flow rate as air was forced into or drawn out of the upper airway by the mouth. These pulses were not observed during normal breathing, when the panellists were not chewing.

Swallowing analysis (Fig. 13.1; Hodgson *et al.* 2003), showed that following inhalation there was a small amount of air exhaled just before the swallow itself. Then, air flow ceased (swallowing apnoea) during the main phase of the swallow, as the bolus was delivered to the oesophagus. The final phase was the exhalation after swallowing, the swallow breath, which delivered the main air flow from the lungs and carried volatile laden air through the nose. The flavour molecules did not arrive at the end of the nostril immediately, but, there was a slight delay due to the dead volume of the upper airway (Damm *et al.* 2002, Hodgson *et al.* 2003).

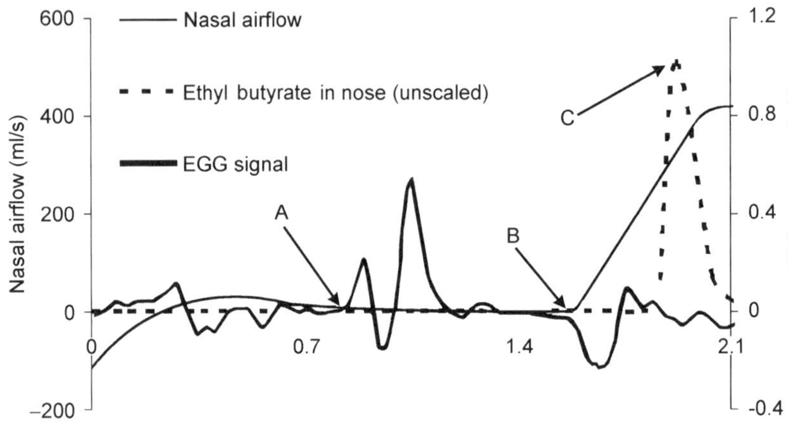

Fig. 13.1 Temporal profile of an individual swallowing a solution of ethyl butyrate; showing, nasal air flow rate, flavour release profile (ethyl butyrate) and swallowing phase (determined by electroglottography (EGG)). Breath flow ceases followed by a double peak in the EGG signal at the start of swallowing (A). After swallowing, exhalation begins (positive air flow) marked by a trough in the EGG signal (B). Shortly after this a pulse of volatile was detected in the exhaled breath (C). Reprinted with permission from Hodgson *et al.* Copyright (2003) American Chemical Society.

The duration of the swallowing apnoea has mainly been studied in the exploration of swallowing disorders rather than flavour delivery. These confirm the lack of air flow through the upper airway during swallowing and the fact that the apnoea can increase when the bolus is more difficult to swallow, i.e. a cookie compared to a thin liquid (Preiksaitis and Mills 1996).

The small air flow observed immediately before the swallowing action itself, may have been the gas flow that Land measured, since the only other exhaled air typically observed is the full exhalation following swallowing, which is much greater in volume than 15 ml. This initial small airflow was probably associated with breath from the lungs, since changes in the shape of the upper airway as the soft pallet rises are unlikely to further reduce the limited dimensions of the upper airway (Damm *et al.* 2002). This air would be unlikely to contain a significant amount of flavour compounds, since it actually precedes the main pharyngeal phase of swallowing (Hodgson *et al.* 2003).

13.5.4 Relative importance of chewing and swallowing actions

Chewing and swallowing are the two main processes moving volatiles from the mouth through into the upper airway. The importance of each to the overall process will depend on the way in which the mouth processes food, and hence the food itself. Buettner and co-workers have considered the swallowing action to play the greater role in flavour delivery (2001, 2002). This is based on their videofluoroscopy studies in which the velum opening during swallowing was more distinct than that associated with chewing. Additional evidence came from the more regular appearance of helium (held in the mouth) in the breath during swallowing compared with regular mouth movements, and the observation that, the major phase of volatile delivery to the breath when a solution was swallowed was in the first breath following swallowing. With little volatile found in the breath before or after this exhalation.

Real-time breath analyses have shown the appearance of peaks associated with swallowing actions for some panellists, but, not others (Haahr *et al.* 2003). However, in this instance, the swallowing actions were scheduled for specific times which may have changed delivery patterns relative to less conscious actions. In addition, the APCI system did not resolve one breath from another making the release profile difficult to interpret.

Other real-time breath analyses without fixed eating protocols have shown eating events with regular peaks (corresponding to chewing), but, no larger, less frequent peaks which would correspond with swallowing actions (Grab and Gfeller 2000, Harvey *et al.* 2000, Linforth *et al.* 2004). High temporal resolution breath-by-breath analysis monitoring the intensity of release during chewing and a swallow with the same bolus, showed that the two events were equal in magnitude (Linforth *et al.* 2004). This may explain the lack of large peaks in release related to swallowing events. It is important to remember that in this case the events are conscious actions, which may alter release, and variations in human physiology combined with the frequency of chewing and swallowing events will also affect their significance.

Clear examples showing high dependencies of flavour release on chewing or swallowing, are chewing gum and boiled sweets. Chewing gum is normally chewed at approximately 100 chews a minute (Wright *et al.* 2003b), with most chewing events during exhalation resulting in volatile laden air entering the upper airway (Linforth *et al.* 2004). Swallowing events occurred less frequently, they did not however appear to produce higher concentrations of volatile in the breath. The boiled sweet was sucked rather than chewed and the weak mouth movements resulted in few instances of regular (i.e. during each breath) air movements into the upper airway. Here, the main flavour delivery events were associated with the less frequent swallowing actions, with gaps of approximately 30 s in between.

In this case of the boiled sweets, the lower air flow through the mouth may allow more time for volatile equilibration. Consequently the peaks may have been more intense even though less frequent. However, this may be offset by reduced mass transfer in mouth due to the lower mouth movements. The two effects are closely linked and experimentation to separate and evaluate these two factors have yet to be performed.

The persistence of volatiles in mouth and their subsequent transmission to the pharynx after the main eating event can lead to spikes or pulse of release late in the eating time course. These stand out in this region of the profile where the breath volatile concentration is lower; they can also appear in averaged curves since they occur at roughly the same point in the eating time course (Linforth *et al.* 1996). Consequently in studies of persistence panellists may be asked to refrain from swallowing actions after the main bolus has been swallowed to avoid disruption of the release curve (Hodgson *et al.* 2004).

However, what is apparent (and often overlooked) from the many real-time analyses performed, is that the profile of peaks added to the time course by chewing or swallowing are often merely spikes on top of a gradually changing basal breath flavour content (Linforth *et al.* 2004), which is caused by the absorption and desorption of volatiles from the upper airway (Medinsky *et al.* 1993, Keyhani *et al.* 1997).

13.5.5 Breath flavour profiles after chewing or swallowing

The profile of the breath volatile concentration following a swallowing or chewing event is also different. Chewing transfers a gas phase pulse of aroma into the pharynx which can pass through the upper airway as a distinct peak if its physico-chemical characteristics minimise absorption to the nasal epithelia (Linforth *et al.* 2002, Hodgson *et al.* 2003). If substantial absorption occurs then a residue of aroma is deposited which gradually washes-out of the upper airway on subsequent exhalations (Medinsky *et al.* 1993). Swallowing transfers not only gas phase volatiles, but, also the bolus which leaves a residue in the throat for further flavour delivery. This typically results in a profile where there is a peak near the start of the exhalation, followed by a distinct shoulder (Fig. 13.2).

The difference between the peak and the shoulder (or second exhalation, which is typically just a little lower) has been used as a measure of persistence (Linforth and Taylor 2000, Buffo *et al.* 2005). However, the two phases of the

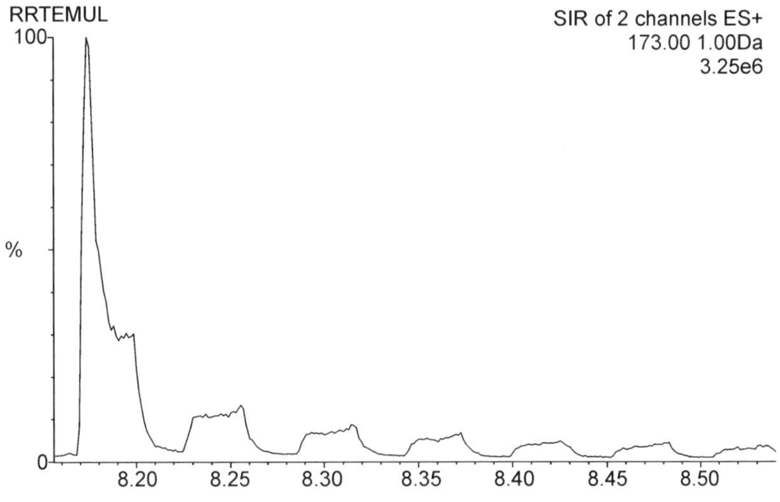

Fig. 13.2 Ethyl octanoate release profile in the breath after swallowing 15 ml of an aqueous solution. A high initial breath concentration falls over the course of the first exhalation to a plateaux, which declines in successive breaths. Each breath being separated by a return to baseline as lab air was sampled.

exhalation are dependent on different processes operating under different conditions and should be considered separately.

The peak at the start of the exhalation is caused by gas transfer from the mouth and the partitioning of volatiles into the gas phase in the pharynx as the bolus passes through. For compounds with efficient mass transfer (low K_{aw}) the breath volatile content was close to that of the static equilibrium. The concentration only decreased during exhalation due to absorption of volatiles to the mucosa in the nose and dilution on route (Linforth *et al.* 2002). For the low K_{aw} compounds the intensity of first peak is normally very reproducible, presumably due to the system achieving near maximum delivery. The shoulder associated with low K_{aw} compounds is greater than that observed for high K_{aw} compounds reflecting efficient (relative to the static equilibrium) delivery of volatile from the in throat residue.

Compounds with high K_{aw} values are typically more variable in their intensity (Robert *et al.* 2004). This is likely to arise from their inefficient delivery to the gas phase. With efficiencies of delivery of often less than 1% of those of lower K_{aw} compounds, the chances for worse or better delivery (and hence variation) are much greater. This will be dependent on factors such as surface movements of the bolus during consumption. Following the peak in the profile, the breath volatile concentration falls to a greater extent than for the low K_{aw} compounds reflecting the poor mass transfer of these compounds under the dynamic conditions of tidal breath flow.

The food matrix, via its influence on the partition coefficient of aroma compounds can alter the ratio of the first peak to the shoulder. Ethyl octanoate consumed in an emulsion rather than water has a much greater shoulder relative to the initial peak (Linforth *et al.* 2004). Furthermore the concentration of ethyl octanoate in the first peak is much closer to the static equilibrium headspace concentration when consuming ethyl octanoate in the emulsion compared to water (Linforth *et al.* 2002). These differences basically reflect the changes in mass transfer as the lipophilic molecules partition into the lipid phase decreasing the partition coefficient.

Clearly the way in which a foodstuff is consumed can have an impact on flavour delivery as well as release. The use or not of an eating protocol is an important consideration for both analytical and sensory investigations of flavour. An eating protocol may help standardise the way in which individuals eat, it may, however, eliminate differences in release and the pattern of release that occur naturally as we interact with our food.

13.6 Persistence of flavour release

Persistence is the phase of flavour release often also termed aftertaste in a perceptual context. It relates to the continued release of aroma molecules to the nasal cavity after the bolus has been consumed. Attempts to analyse persistence have included experiments where solutions were consumed and the decline in breath volatile content measured (Linforth and Taylor 2000, Buffo *et al.* 2005). Persistence was defined as the volatile concentration in the second exhalation after swallowing the bolus, relative to the volatile content of the first. This, however, is comparing two very different events, one associated with flavour release as the bolus passes through the pharynx at a time of zero air flow. The second, the breath volatile content during normal tidal breath flow. Such a comparison relates to other processes; persistence will therefore be defined as the changes in breath volatile concentration as a result of the tidal flow of breath after the cessation of the main chewing and swallowing events.

Studies using videofluoroscopy (Buettner *et al.* 2001) have shown that during swallowing part of the bolus can become detached from the bulk of the material and be deposited on the back of the tongue or in the pharynx. Consequently, in addition to food residue in mouth, a trace of food will also be present in the tongue–pharynx region directly exposed to the upper airway. Flavour released from food residues in the pharynx can directly enter the gas stream and pass through to the nose, where as further release of flavour from in-mouth food residues, will depend on swallowing actions or gas exchange between the pharynx and mouth due to mouth movements (e.g. tongue movements to clear residual bolus from the teeth).

The decline of volatiles after eating or drinking has been shown to depend not only on the effect of the matrix such as lipid content (Linforth and Taylor 2006), but also the volatile compounds themselves (Ingham *et al.* 1995, Linforth *et al.* 2000, 2004). Hence models should either include parameters related to the molecules themselves, or generate parameters that describe these differences.

The decline in the breath volatile content during this phase of the eating process has been modelled using different approaches. One feature that the models share is the ability to model the volatile decay curves, excluding the volatile content of the swallow breath. This data point did not readily fit the models, again suggesting that the release mechanisms should be considered separately.

Wright and co-workers (2003) developed a mechanistic model to describe the process. This involved an exponential decay function related to the mucosal area/volume ratio and mass transfer in the saliva phase; components related to loss of volatile from the system (irreversible binding or losses to lungs/atmosphere); air flow and a sinusoidal wave function. The latter function was a key component since not only were the peaks in the breath volatile content (exhalations) modelled, but also the troughs in between (inhalation). The model resulted in theoretical release curves that very closely matched the *in vivo* release curves. However, in order to get the curves to fit it was necessary to use breath flow rates as high as 15 000 ml/s which are excessively high. This may indicate difficulties in fitting the model to the data, possibly suggesting the need to consider additional factors, or, problems with the fitting process itself.

Other models have approached the decay process as two separate phases, an early phase where the volatile concentration declines rapidly and a second phase where the decay is much less (Normand *et al.* 2004). The two phases represented, the release of flavours mainly from saliva early in the time course, and subsequently a three phase system with air, saliva and a mucosal layer, which were slower in release. From the model it was possible to calculate *in vivo* partition coefficients for the flavour molecules. The partition coefficients were found to be rather similar *in vivo* for compounds which are very different *in vitro*. This is likely to arise from the depletion of the higher K_{aw} compounds from the air–saliva interface such that further release relies on mass transfer in the liquid phase (diffusion and convection). Consequently they have similar release properties to low K_{aw} compounds which are less dependent on mass transfer in the liquid phase, where loss through partitioning is not as great. Hence, as found elsewhere, the low K_{aw} and high K_{aw} compounds are much closer in their release properties under dynamic conditions.

The simplest of the models was that of Hodgson and co-workers (2004). This simply involved fitting a power law function of the form $C = C_1 t^{-p}$ to the data, where C is the concentration at any point in time (t), C_1 is the concentration of volatile in the breath 1 minute after consumption and p is the decay constant. This curve fitted the real time decay curves for volatiles in the breath after swallowing. The model does not include parameters that account for differences between compounds, but, was used to generate values for C_1 and p to allow their comparison. C_1 (which is the breath volatile concentration at 1 minute) was found to be lower for compounds with lower K_{aw} values, as might be expected and inversely correlated with p. Therefore compounds with low K_{aw} values had the lowest decay constants and the lowest breath concentrations, whereas compounds with high K_{aw} values decayed much faster and had correspondingly higher breath volatile contents. The air flow rate is an important factor in flavour release with higher flow rates potentially removing greater amounts of flavour compounds (Harrison and Hills 1997, Marin *et al.* 2000). The power law function was fitted to curves from individuals breathing at different (but, constant) rates. This did not change the shape of the curve (as defined by p), which was dependent more on the compound, but, C_1 which decreased from 0.7 at a tidal breath flow of 120 ml/s to 0.2 at 250 ml/s. Reflecting the loss of volatile from the system reducing the amounts available to partition into the breath.

13.7 Future trends

Before the development of instrumentation to allow the real-time measurement of the volatile content of the breath, the only measures of flavour release that could be performed were using sensory analysis. Sensory data is also dependent on the multi-modal nature of perception and it was often impossible to determine whether it was an interaction of the senses, or physical chemistry that was the cause of perceived differences.

Now that we can actually measure the temporal dimension of the signal delivered to the olfactory epithelium we have been able to understand how aromas actually get out of foods and are transported through the upper airway.

The sensory impact of the actual stimulus can be investigated, plus with coupling to fMRI, we have the capacity to measure the entire process stimulus, neuronal activity and perception (from fork to thought).

It is important to remember that similar aroma release profiles do not always equate to a similar perceptual experience, taste, texture and colour will be major factors affecting our perception of flavour. Although the actual flavour stimulus is important, it is only part of the picture.

13.8 Sources of further information

In addition to articles published in journals, the series of books published following the triennial European flavour conference, the Weurman Symposium (see Buettner (2004) for a reference to the latest edition), are regular, important, snapshots encapsulating our current knowledge of flavour and its behaviour. The American Chemical Society symposium series books, published by the Agriculture and Food Division are also excellent sources of material. These are typically focused on specific themes in flavour such as the role of lipids or flavour release itself (see, for example, Grab and Gfeller 2000).

13.9 References

BAEK I, LINFORTH R S T, BLAKE A and TAYLOR A J (1999) 'Sensory perception is related to the rate of change of volatile concentration in-nose during eating of model gels', *Chem Senses*, 24, 155–160.

BRAUSS M S, BALDERS B, LINFORTH R S T, AVISON S and TAYLOR A J (1999) 'Fat content, baking time, hydration and temperature affect flavour release from biscuits in model-mouth and real systems', *Flav Frag J*, 14, 351–357.

BUETTNER A (2004) 'Prolonged retronasal aroma perception – a phenomenon influenced by physiological factors and food matrix composition', in Le Quéré J L and Étiévant P X, *Flavour Research at the Dawn of the Twenty-first Century*, Paris, Intercept, 188–193.

BUETTNER A, BEER A, HANNIG C, SETTLES M and SCHIEBERLE P (2002) 'Physiological and analytical studies on flavor perception dynamics as induced by the eating and swallowing process', *Food Qual Pref*, 13, 497–504.

BUETTNER A, BEER A, HANNIG C and SETTLES M (2001) 'Observation of the swallowing process by application of videofluoroscopy and real-time magnetic resonance imaging – consequences for retronasal aroma stimulation', *Chem Senses*, 26, 1211–1219.

BUETTNER A and SCHIEBERLE P (2000a) 'Changes in the concentrations of key fruit odourants induced by mastication', in Roberts D D and Taylor A J, *Flavor Release*, Washington, American Chemical Society, 87–98.

BUETTNER A and SCHIEBERLE P (2000b) 'Influence of mastication on the concentrations of aroma volatiles – some aspects of flavour release and flavour perception', *Food Chem*, 71, 347–354.

BUETTNER A and SCHIEBERLE P (2000c) 'Exhaled odorant measurement (EXOM) – A new approach to quantify the degree of in-mouth release of food aroma compounds', *Lebensm.-Wiss Technol – Food Sci Technol*, 33, 553–559.

BUFFO R A, RAPP J A, KRICK T and REINECCIUS G A (2005) 'Persistence of aroma compounds in human breath after consuming an aqueous model mixture', *J Agric Food Chem.*, 89(1), 103–108.

COOK D J, HOLLOWOOD T A, LINFORTH R S T and TAYLOR A J (2003) 'Oral shear stress predicts flavour perception in viscous solutions', *Chem Senses*, 28, 11–23.

DAMM M, VENT J, SCHMIDT M, THEISSEN P, ECKEL, H E, LOTSCH J and HUMMEL T (2002) 'Intranasal volume and olfactory function', *Chem Senses*, 27, 831–839.

DEARDEN J C and JAMES K C (1998) 'Quantitative structure–activity relationships and drug design'. In *Introduction to the Principles of Drug Design and Action*, Smith H J ed., Harwood Academic Publishers, Australia, 167–207.

DOYEN K, CAREY M, LINFORTH R S T, MARIN M and TAYLOR A J (2001) 'Volatile release from an emulsion: headspace and in-mouth studies', *J Agric Food Chem, 49*, 804–810.

GRAB W and GFELLER H (2000) 'Flavorspace: technology for the measurement of fast dynamic changes of flavor release during eating', in Roberts D D and Taylor A J, *Flavor Release*, Washington, American Chemical Society, 33–43.

HAAHR A M, MADSEN H, SMEDSGAARD J, BREDIE W L P, STAHNKE L H and REFSGAARD H H F (2003) 'Flavor release measurement by atmospheric pressure chemical ionization ion trap mass spectrometry, construction of interface and mathematical modeling of release profiles', *Anal Chem, 75*, 655–662.

HARRISON M and HILLS B P A (1996) 'Mathematical model to describe flavour release from gelatine gels', *Int J Food Sci Technol*, 31, 167–176.

HARRISON M and HILLS B P A (1997) 'Effects of air flow-rate on flavour release from liquid emulsions in the mouth', *Int J Food Sci Technol*, 32, 1–9.

HARRISON M, HILLS B P, BAKKER J and CLOTHIER T (1997) 'Mathematical models of flavor release from liquid emulsions', *J Food Sci, 62*, 653–658 and 644.

HARVEY B A, BRAUSS M S, LINFORTH R S T and TAYLOR A J (2000) 'Real-time flavor from foods during eating', in Roberts D D and Taylor A J, *Flavor Release*, Washington, American Chemical Society, 22–32.

HILLS B P and HARRISON M (1995) 'Two-film theory of flavour release from solids', *Int J Food Sci Technol*, 30, 425–436.

HODGSON M, LINFORTH R S T and TAYLOR A J (2003) 'Simultaneous real-time measurements of mastication, swallowing, nasal airflow and aroma release', *J Agric Food Chem*, 51, 5052–5057.

HODGSON M, PARKER A, LINFORTH R S T and TAYLOR A J (2004) '*In vivo* studies on the long term persistence of volatiles in the breath', *Flav Frag J.*, 19, 470–475.

HOLLOWOOD T A, LINFORTH R S T and TAYLOR A J (2002) 'The effect of viscosity on the perception of flavour', *Chem Senses*, 27, 583–591.

INGHAM E E, LINFORTH R S T and TAYLOR A J (1995) 'The effect of eating on the rate of aroma release from mint-flavoured sweets', *Lebensm.-Wiss Technol*, 28, 105–110.

JACK F R and F GIBBON (1995) 'Electropalatography in the study of tongue movement during eating and swallowing (a novel procedure for measuring texture-related behaviour)', *Int J Food Sci Technol*, 30, 415–423.

KANT A, LINFORTH R S T, HORTE J and TAYLOR A J (2004) 'Effect of beta-cyclodextrin on aroma release and flavor perception', *J Agric Food Chem*, 52, 2028–2035.

KEYHANI K, SCHERER P W and MOZELL M M A (1997) 'Numerical model of nasal odorant transport for the analysis of human olfaction', *J Theor Biol*, 186, 279–301.

LAND D G (1996) 'Perspectives on the effects of interactions on flavor perception: an overview', in McGorrin R L and Leland J, *Flavor–Food Interactions*, Washington, American Chemical Society, 2–11.

LE GUEN S and VREEKER R (2004) 'Interactions between flavour compounds and milk proteins under static and dynamic conditions', in Le Quéré J L and Étiévant P X, *Flavour Research at the Dawn of the Twenty-first Century*, Paris, Intercept, 182–187.

LETHUAUT L, WEEL K G C, BOELRIJK A E M and BROSSARD C D (2004) 'Flavor perception and aroma release from model dairy desserts', *J Agric Food Chem*, 52, 3478–3485.

LINFORTH R S T, BAEK I and TAYLOR A J (1999) 'Simultaneous instrumental and sensory analysis of volatile release from gelatine and pectin/gelatine gels', *Food Chem*, 65, 77–83.

LINFORTH R S T, FRIEL E N and TAYLOR A J (2000) 'Modeling aroma release from foods using physicochemical parameters', in Roberts D D and Taylor A J, *Flavor Release*, Washington, American Chemical Society, 166–178.

LINFORTH R, HODGSON M and TAYLOR A (2004) 'Studies of retronasal flavour delivery', in Le Quéré J L and Étiévant P X, *Flavour Research at the Dawn of the Twenty-first Century*, Paris, Intercept, 143–147.

LINFORTH R S T, INGHAM K E and TAYLOR A J (1996) 'Time course profiling of volatile release from foods during the eating process', in Taylor A J and Mottram D S, *Flavour Science Recent Developments*, Cambridge, The Royal Society of Chemistry, 361–368.

LINFORTH R, MARTIN F, CAREY M, DAVIDSON J and TAYLOR A J (2002) 'Retronasal transport of aroma compounds', *J Agric Food Chem*, 50, 1111–1117.

LINFORTH R and TAYLOR A J (2000) 'Persistence of volatile compounds in the breath after their consumption in aqueous solutions', *J Agric Food Chem*, 48, 5419–5423.

LINFORTH R S T and TAYLOR A J (2006) 'Release of flavor from emulsions under dynamic sampling conditions', in Weenan, H, *Chemistry, Flavor and Texture of Lipid Containing Foods*, Washington, American Chemical Society, 159–170.

LUCAS P W, PRINZ J F, AGRAWAL K R and BRUCE I C (2004) 'Food texture and its effect on ingestion, mastication and swallowing', *J Texture Studies*, 35, 661–665.

MARIN M, BAEK I and TAYLOR A J (1999) 'Volatile release from aqueous solutions under dynamic headspace dilution conditions', *J Agric Food Chem*, 47, 4750–4755.

MARIN M, BAEK I and TAYLOR A J (2000) 'Flavor release as a unit operation: a mass transfer approach based on a dynamic headspace dilution method', in Roberts D D and Taylor A J, *Flavor Release*, Washington, American Chemical Society, 153–165.

MEDINSKY M. A, KIMBELL J S, MORRIS J B, GERDE P and OVERTON J H (1993) 'Advances in biologically based models for respiratory tract uptake of inhaled volatiles', *Fundam Appl Toxicol*, 20, 265–272.

MIOCHE L, BOURDIOL P and MONIER S (2003) 'Chewing behaviour and bolus formation during mastication of meat with different textures', *Arch Oral Biol*, 48, 193–200.

NORMAND V, AVISON S and PARKER A (2004) 'Modeling the kinetics of flavour release during drinking', *Chem Senses*, 29, 235–245.

PREIKSAITIS H G and MILLS C A (1996) 'Coordination of breathing and swallowing: effects of bolus consistency and presentation in normal adults', *J Appl Physiol*, 81, 1707–1714.

ROBERTS D D, POLLIEN P, ANTILLE N, LINDINGER C and YERETZIAN C (2003) 'Comparison of Nosespace, headspace, and sensory intensity ratings for the evaluation of flavor absorption by fat, *J Agric Food Chem*, 51, 3636–3642.

ROBERTS D D, POLLIEN P, YERETZIAN C and LINDINGER C (2004) 'Nosespace analysis with Proton-Transfer-Reaction mass spectrometry: Intra- and interpersonal varaition', in Deibler K D and Delwiche J, *Handbook of Flavor Characterization*, New York, Marcel Dekker, 151–162.

SOETING W J and HEIDEMA J A (1988) 'Mass spectrometric method for measuring flavor concentration time profiles in human breath', *Chem Senses*, 13, 607–617.

TAYLOR A J, BESNARD S, PUAUD M and LINFORTH R S T (2001) '*In vivo* measurement of flavour release from mixed phase gels', *Biomol Eng*, 17, 143–150.

WEEL K G C, BOELRIJK A E M, BRUGER J J, CLAASSEN N E, GRUPPEN H, VORAGEN A G J and SMIT G (2003) 'Effect of whey protein on the in vivo release of aldehydes', *J Agric Food Chem*, 51, 4746–4752.

WRIGHT K M, HILLS B P, HOLLOWOOD T A, LINFORTH R S T and TAYLOR A J (2003a) 'Persistence effects in flavour release from liquids in the mouth', *Int J Food Sci Technol*, 38, 343–350.

WRIGHT K M, SPRUNT J, SMITH A C and HILLS B P (2003b) 'Modeling flavour release from a chewed bolus in the mouth: Part 1. Mastication', *Int J Food Sci Technol*, 38: 351–360.

14

Genetic influences on taste

J. Prescott, James Cook University, Australia

14.1 Introduction

As consumers, our sensory and hedonic experiences of foods and beverages are most often of an overall flavour percept that depends upon the integration of multiple, distinct sensory qualities – in particular, olfactory and taste qualities (Prescott 2004). In contrast to the rich, perhaps limitless, inventory of odours that contribute to flavours, however, the sense of taste may consist of no more than a handful of distinct qualities. Nevertheless, given the ubiquity of one or more of these qualities in any flavour, it can be argued that taste forms the substrate for flavours in general.

In recent years, our understanding of taste mechanisms has grown substantially. Distinct transduction mechanisms have been described not just for the traditional taste qualities of sweet, sour, salty and bitter (Glendinning, Chaudhari and Kinnamon 2000), but also for amino acids in general (Nelson, Chandrashekar *et al.* 2002), and glutamate in particular (Chaudhari, Landin and Roper 2000). Two primary forms of taste transduction have been described: G-protein coupled receptors are implicated in sweetness, bitterness, and *umami* (glutamate taste) transduction, while saltiness and sourness use ionic mechanisms (Glendinning, Chaudhari and Kinnamon 2000; Montmayeur and Matsunami 2002). For some qualities, in particular sweetness and bitterness, there are multiple classes of molecular structures that can produce these qualities. This has produced speculation that there may be multiple types of sweetness and bitterness, perhaps reflected in different receptor types. By contrast, there is less variation for saltiness or sourness, which are determined primarily by activation of channels in the taste cell membrane sensitive to Na^+ or H^+ ions, respectively, although there may be multiple intracellular mechanisms involved (Glendinning, Chaudhari and Kinnamon 2000).

Despite potential variations in transduction mechanisms, especially for sweetness and bitterness, the predominant view of taste is still largely of a limited set of basic qualities. While this list may yet be incomplete – for example, there is some evidence that perception of fat may rely on a distinct oral transduction mechanism (Gilbertson 1998; Laugerette *et al.* 2005) – a limited set of basic tastes and associated transduction mechanisms nevertheless strongly implies adaptive significance for these qualities. This is consistent with the recognition that sensitivities to sweetness (as a signal for *calories*), bitterness (*toxins*) and saltiness (*sodium*) are likely to have important roles in regulating intake of nutrients and avoidance of toxins (Hladik, Pasquet and Simmen 2002; Scott 1992). Such interpretations are supported by data showing high positive correlations between the LD_{50} (a measure of toxicity, such that a low LD_{50} indicates high toxicity) and palatability of taste compounds in rats (Scott and Mark 1987). Most taste qualities – sweetness, sourness, bitterness, and *umami* – can also be shown to elicit stereotyped hedonic responses (facial expressions, suckling responses) very shortly after birth in humans (Steiner, Glaser *et al.* 2001). Together with the apparent preservation of characteristic hedonic responses to taste qualities *per se* in adulthood and across cultures differing substantially in diet (Prescott 1998), this has led to the hypothesis that fixed patterns of taste hedonics are innate and similarly serve underlying adaptive purposes.

14.2 Individual differences in taste perceptions

Nevertheless, there are individual differences in hedonic responses to taste, in particular when those tastes are experienced in a food/beverage context. One such example is the distinction between taste likers and dislikers – individuals who show increasing or decreasing hedonic functions, respectively, with increases in sweetener or salt concentrations (Pangborn 1970). Moreover, the perception of tastes also shows individual differences that may modify hedonic responses to tastes, and ultimately this will influence how the overall flavour featuring that taste is perceived and accepted.

With the exception of some bitter compounds (see below), there are limited data on how such variation is distributed within populations, or on the origins of individual differences that might be *specific* to that taste. In the case of both saltiness and sourness, thresholds are thought to be normally, or close to normally, distributed (Blakeslee and Salmon 1935). Sensitivity to sweetness, as measured by thresholds, appears to be unimodally distributed, with the research to date showing unremarkable variation across individuals for either sucrose or saccharin (Blakeslee and Salmon 1935). The distribution of thresholds to monosodium glutamate (MSG), the prototypical *umami* tastant, spans a 500-fold range of concentrations (Lugaz, Pillias and Faurion 2002). Studying MSG concentrations that were perceived to be iso-intense with 29 mM NaCl, Lugaz, Pillias and Faurion (2002) found population distributions that deviated from normality. For both thresholds and supra-threshold ratings, there were a minority of so-called *MSG hypo-tasters* who perceived only small differences between the qualities associated with NaCl and MSG, as well as a smaller percentage of subjects who appeared to be completely insensitive to the glutamate ion, since they were unable to distinguish iso-molar MSG and NaCl solutions.

Sensitivity to bitterness shows most variation in humans, with early reports of thresholds across several bitter compounds showing only modest agreement (Blakeslee and Salmon 1935). Studies of relationships between bitter compounds suggest a high degree of complexity in the mechanisms of bitterness transduction. Yokomukai, Cowart and Beauchamp (1993) found distinct individual differences in ratings of the bitterness of quinine sulfate and urea relative to each other. Although across a large group, the concentration steps of these compounds were rated equally, around 30% of subjects found quinine more intense than urea, with the same percentage showing the opposite pattern. Similar distinct patterns of individual differences for some other bitter compounds were also found when the subjects were separated in their responses to quinine and urea. Delwiche, Buletic and Breslin (2001a) demonstrated significant correlations between intensity ratings of a number of bitter compounds representing different chemical classes, including amines, salts, amino acids, alkaloids, phenols, and the thiourea compound, 6-n-propylthiouracil (PROP; also see below). However, cluster analysis based on bitterness ratings revealed that the compounds fell into four clusters, with one cluster containing only PROP. Their interpretation was that such clusters may represent distinct transduction mechanisms, and most probably different receptor types.

Recent studies of taste genetics have, in fact, revealed the existence of multiple sweetness and bitterness receptor genes. Demonstration of the existence of multiple taste receptors suggests three ways in which taste responses could vary: (1) there may exist multiple types of quality (e.g., multiple types of sweetness or bitterness), one or more of which may be reflected in a given compound; (2) the different receptors may be expressed to differing degrees in different individuals; and (3) for any given receptor type, there may be genetic polymorphisms (that is, variations in the amino acid sequences in the receptor protein). In the case of sweetness, genes for three receptor proteins – *TAS1R1, 2, 3* – have been identified in mice and humans, although little other variation in the structure of these proteins has been reported (Li, Staszewski *et al.* 2002; Liao and Schultz 2003; Kim, Breslin *et al.* 2004). One primary mechanism for sweetness transduction involves a dimer consisting of TAS1R2/TAS1R3. Another dimer involving these same receptor proteins, in this case TAS1R1/TAS1R3, is receptive to *umami* tastes (Li, Staszewski *et al.* 2002). Lugaz, Pillias and Faurion (2002) speculated that the origin of the variations they observed in MSG sensitivity may be polygenic, however, there appears to be no firm evidence of this to date. In the case of bitterness, the existence of a very wide range of chemical structures that are bitter to humans has for some time suggested the possibility that there may be many different forms of receptor, and this has largely been confirmed. To date, 24 potentially functional bitter receptor genes (the *TAS2R* family of genes) have been reported (Shi, Zhang *et al.* 2003). Polymorphisms have also been identified in some *TAS2R* genes and they may be independent contributors to variations in bitterness (Kim, Breslin *et al.* 2004).

14.3 PTC/PROP bitterness

The existence of substantial individual variation in sensitivity to the bitterness of the thiourea compounds, phenylthiocarbamide (PTC) and 6-n-propylthiouracil (PROP), has been known for over 70 years. However, interest in responses to PTC/PROP bitterness has, more recently, grown considerably as it has become apparent that variations in responses to these compounds may be associated with differences in food and flavour perceptions and preferences, in general. Details of the discovery of PTC phenotypes by Fox (1931) and the subsequent investigation of its genetic basis (Blakeslee and Salmon 1931; Snyder 1931) have been covered in recent years in several reviews by Bartoshuk and colleagues (Bartoshuk, Duffy *et al.* 1996; Fast, Duffy and Bartoshuk 2002) and by Guo and Reed (2000), and will not be described in detail here. In summary, it is well accepted that sensitivity to these compounds is bimodally distributed (Reed, Bartoshuk *et al.* 1995), with 25–30% of Caucasian populations being highly insensitive. This percentage has been shown, however, to vary substantially over different populations, although some proportion of PTC/PROP non-tasters (NTs) has been found in almost all population groups studied (Guo and Reed 2000; Bell and Song 2004).

PTC/PROP sensitivity was originally measured in terms of detection thresholds (Harris and Kalmus 1949). Later studies using supra-threshold intensity ratings have shown that a PTC/PROP concentration considered moderately to strongly bitter by the majority of the population is perceived as either weakly, or not at all, bitter by NTs (Hall, Bartoshuk *et al.* 1975; Bartoshuk 2000). Those who find the compound bitter – tasters – also show considerable variation in both threshold and ratings of intensity. Relatively recently in the study of these compounds, Bartoshuk, Fast *et al.* (1992) identified a sub-group of tasters (estimated as 20–25% of the population) that rate PROP as profoundly bitter. Such *super-tasters* (STs) may have been previously overlooked, due initially to the widespread use of detection (threshold) criteria for taster group classification, and subsequently by the use of particular rating scales (e.g., 9-point category scales) that mitigated against discriminating this group from the *medium-tasters* (MTs) (Bartoshuk, Fast *et al.* 2004a). Although women are more sensitive to PROP than are men and STs are more likely to be female, the inheritance pattern does not appear to be sex-linked (Bartoshuk, Duffy and Miller 1994; Guo and Reed 2000; Reed, Bartoshuk *et al.* 1995).

It was initially thought that PTC/PROP tasting genotypes displayed a simple Mendelian inheritance pattern, with tasting (*T* – dominant) and non-tasting (*t* – recessive) alleles (Guo and Reed 2000). It was proposed that the most sensitive tasters – STs – were those with two dominant alleles (Bartoshuk 1993).

However, since some children of insensitive parents are apparently sensitive to PTC, this explanation cannot be complete and the possibility that it may be a polygenic trait has been considered (Reed 2004). The picture that has most recently emerged is that polymorphisms in the bitter taste receptor gene TAS2R38 on chromosome 7 are associated with specific high, low and intermediate PTC/PROP sensitivity phenotypes (Kim, Breslin *et al.* 2004; Kim, Jorgenson *et al.* 2003; Bufe, Breslin *et al.* 2005).

14.3.1 PTC/PROP tasting and sensitivity to other sensory qualities

Much of the recent interest in PTC/PROP tasting, at least among taste scientists, has been related to findings that sensitivity to the bitterness of these compounds appears to act as an index of a broader sensitivity to taste in general. While PROP ratings have been reported to be poorly correlated with those of other bitter compounds (Delwiche, Buletic and Breslin 2001a), tasters have nevertheless been reported as rating the bitterness of urea, sucrose octa-acetate and denatonium benzoate (Mela 1989), sodium and potassium benzoate, potassium chloride (Bartoshuk *et al.* 1988), quinine (Leach and Noble 1986) and caffeine (Hall, Bartoshuk *et al.* 1975) as more intense than do NTs. Research using other taste qualities has also reported higher taste intensities for PROP tasters than for NTs for the sweetness of sucrose and saccharin (Bartoshuk 1979; Gent and Bartoshuk 1983), the saltiness of NaCl (Bartoshuk, Duffy *et al.* 1998) and the sourness of citric acid (Prutkin, Fast *et al.* 1999). In some studies (e.g., Bartoshuk, Fast *et al.* 1992; Prescott, Ripandelli and Wakeling 2001), STs are also found to be significantly more sensitive to other tastes than are MTs.

PROP taster group differences have also been demonstrated in more complex model systems consisting of mixtures of two or more taste qualities (Prescott, Ripandelli and Wakeling 2001; Yiee, Duffy and Bartoshuk 2003). In the Prescott, Ripandelli and Wakeling (2001) studies (see Fig. 14.1), STs gave higher ratings than NTs and/or MTs to the saltiness and bitterness of NaCl/QHCl mixtures, the bitterness of sucrose/QHCl mixtures, the sweetness of sucrose/citric acid mixtures, and the saltiness and sourness of NaCl/citric acid mixtures. Similar effects were evident in ratings of the overall intensity of these mixtures, particularly at the higher concentrations of the individual tastants. One of the interesting findings in these studies was that the ability of one taste to suppress another (in this case, bitter suppressing sweetness) was dependent upon PROP taster groups, in that suppression was only evident for STs and to a lesser extent MTs. Since such complex taste mixture interactions will be a strong influence on the balance of tastes with a food matrix, and as a consequence on the overall food and beverage flavour, this finding raises the possibility that PROP status may be a strong influence on flavour intensities via modulation of mixture suppression.

Strong relationships between PROP tasting and the perception of oral somatosensory (tactile, irritant) qualities have also been reported (see Prescott, Bartoshuk and Prutkin (2004a) for a review). In studies using model systems, oral irritation produced by capsaicin, the primary irritant in chillies (Karrer,

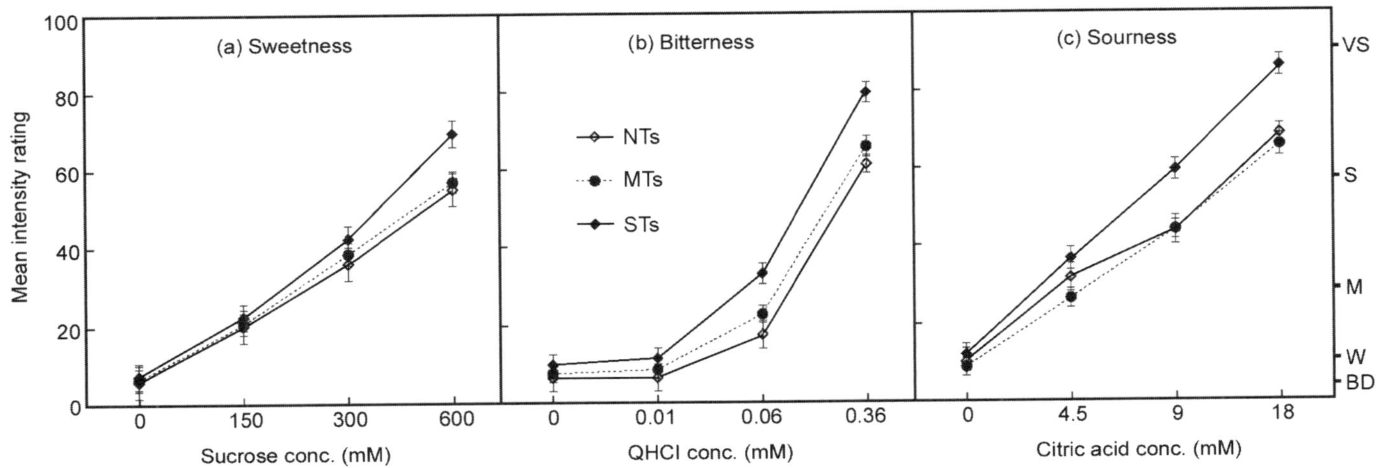

Fig. 14.1 Mean (± sem) ratings of (a) sweetness, (b) bitterness and (c) sourness as a function of increasing tastant concentrations, in PROP taster groups. Note that in each case the variation in tastant occurred in the context of a mixture with another tastant, namely (a) citric acid; (b) sucrose; and (c) sodium chloride. Also shown on the right-hand edge of (c) are the label abbreviations from the Labelled Magnitude Scale: VS – Very strong; S – strong; M – Moderate; W – Weak; BD – Barely detectable. Data adapted from Prescott, J., Ripandelli, N. and Wakeling, I. (2001) Intensity of tastes in binary mixtures in PROP non-tasters, medium-tasters and super-tasters. *Chemical Senses*, 26, 993–1003, by permission of Oxford University Press.

Bartoshuk *et al.* 1992; Prescott and Swain-Campbell 2000), and by other irritants such as cinnamaldehyde (Prescott and Swain-Campbell 2000) and ethanol (Prescott and Swain-Campbell 2000; Bartoshuk, Conner *et al.* 1993; Duffy and Peterson 2000) has consistently been shown to be rated as more intense by tasters than by NTs. The perception of oral tactile sensations, whether produced mechanically (Essick, Chopra *et al.* 2003) or by increasing viscosity or fat content (Prescott, Bartoshuk and Prutkin 2004a) also shows a positive relationship with ratings of PROP intensity.

14.3.2 Sensory properties within foods

There are a relatively large number of findings of differences between PTC/PROP taster groups in their hedonic response to foods and their sensory properties, and particularly to foods dominated by particular tastes. Clearly, the relatively fixed pattern of hedonic responses to basic tastes do not entirely determine our response to those tastes within food and beverage flavours. Hence, while bitterness *per se* appears to be universally disliked (Prescott 1998), beer and coffee are two of the most consumed beverages worldwide. Such preferences will also reflect the effects of various types of associative learning and culture that are independent of genetic factors (Rozin 1982). Nevertheless, PROP/PTC tasters have more food aversions than non-tasters, especially to bitter-tasting foods (Drewnowski and Rock 1995; Fischer, Griffin *et al.* 1961; Glanville and Kaplan 1965). A range of bitter foods, including Brussels sprouts, cabbage, broccoli and spinach (Kaminski, Henderson and Drewnowski 2000; Turnbull and Matisoo-Smith 2002; Drewnowski, Henderson *et al.* 1999), caffeinated coffee (Drewnowski, Henderson *et al.* 1999), grapefruit juice (Drewnowki, Henderson and Shore 1997; Glanville and Kaplan 1965) have been reported as more bitter and/or less preferred by tasters than by NTs. Differences between tasters and NTs have also been found with foods that are sour such as lemon juice, vinegar, and sauerkraut (Glanville and Kaplan 1965) or 'sharp', for example, cheese (Anliker, Bartoshuk *et al.* 1991). Sweet likers are reported as more likely to be NTs, while tasters, both children and adults, were more likely than NTs to show a dislike for sweet foods (Looy and Weingarten 1992). The implications of such findings for food choice in general have recently begun to receive attention, with some data suggesting a role for taste variations in diet-related disease states (see Tepper (2004) and Duffy, Lucchina and Bartoshuk (2004) for overviews and data suggestive of these links; Drewnowski (2004) and Mattes (2004) for, respectively, data unsupportive of, and a sceptical approach to, such links).

It is assumed that differences in hedonic responses to foods and their sensory characteristics between taster groups are based on the existence of differences between PROP taster groups in the perception of tastes and/or somatosensory qualities within foods. However, most research to date has been based on model systems of tastes, chemesthetic and tactile qualities, and there have been relatively few studies addressing PROP taster group differences in the context of real foods or beverages. Pickering, Simunkova and DiBattista (2004) have recently compared PROP taster groups in their ratings of acidity, bitterness and astringency in three wines selected on the basis that they differed in the latter two qualities. They found that, overall, ratings of all three sensory properties were significantly lower in NTs than in MTs and STs. In addition, NTs were found to be least sensitive to variations in bitterness between the different wines. In the only study of the relationship between PROP tasting and sensitivity to systematic variations of qualities within foods, Prescott, Soo *et al.* (2004b) asked taster groups to discriminate variations in bitterness, sweetness or sourness within two foods (yoghurt; cream cheese) and a beverage (orange juice). Particularly in the case of bitterness and sourness, tasters and especially STs, were able to discriminate smaller variations in tastant concentration than NTs. In a second experiment, PROP taster groups rated the sweetness, sourness and oral irritation of carbonated fruit drinks that varied in citric acid and CO_2 concentrations. Ratings of sourness and irritation (see Fig. 14.2) were highest for STs and lowest for NTs. There were no group differences for sweetness ratings, which is consistent with a previous failure to find a relationship between PROP ratings and the sweetness of foods over a range of sucrose concentrations (Drewnowski, Henderson and Barratt-Fornell 1998). These data provide a basis for reported differences of PROP groups in their hedonic responses to foods and

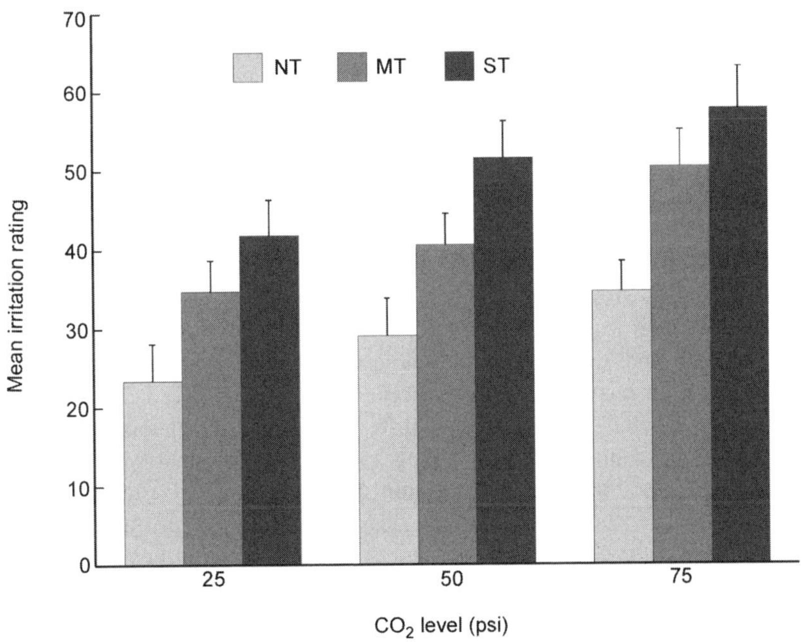

Fig. 14.2 Mean (± sem) ratings of irritation produced by increasing concentrations of CO_2 in a fruit-flavoured beverage in PROP taster groups. At each carbonation level, STs report stronger irritation than do MTs, whose ratings are in turn higher than those of NTs.

suggest, for example, that preferences for, and consumption patterns of, carbonated drinks might differ between taster groups.

In addition to such responses to irritation in beverages, PROP tasting appears to index responses to somatosensory qualities within foods more generally, as it does within model systems. A number of studies have investigated responses to variations in fat content, presumably mediated by response to fat's tactile properties. Tepper and Nurse (1997) found that supra-threshold ratings by both STs and MTs discriminated between 10% and 40% fat content in salad dressings, while those of NTs did not. Duffy, Bartoshuk *et al.* (1996) reported that STs gave higher creaminess ratings than NTs to high fat (>11%) dairy products when evaluating a series ranging from 0.5% to 54% fat. Similarly, STs were found to have superior discriminative ability when comparing milk samples with differing fat contents (Prescott *et al.* 2004a). A recent study in which PROP taster groups profiled a range of dairy foods that varied in fat content from 1.5 to 36% found evidence that STs are more sensitive to textural properties generally in such foods than are NTs (Kirkmeyer and Tepper 2004). This seems a well-established association, despite two studies showing no relationship between PROP tasting and fat perception (Drewnowski, Henderson and Barratt-Fornell 1998; Yackinous and Guinard 2001), both of which may be criticised on methodological grounds (see Bartoshuk, Fast *et al.* (2004a) for a detailed consideration of methodological problems in the measurement of PROP status). A number of different sensory properties contribute to creaminess perception, in particular viscosity and the density of fat globules (Richardson and Booth 1993). The perception of both qualities have been found to vary as a function of PROP tasting (Prutkin 2002; Prescott, Bartoshuk and Prutkin 2004a).

14.4 The role of taste anatomy

Given the multiple receptor genes for bitter and sweet, and the distinct transduction mechanisms for different oral somatosensory modalities (Finger and Simon 2000), how are the relationships between PROP tasting and the perception of taste and somatosensory qualities mediated? Although sensitivity to PROP has been associated with a specific genetic locus (Reed 2004), it is also strongly associated with individual differences in taste anatomy. Miller and Reedy (1990) demonstrated in young adults that the density of fungiform papillae (FP; the structures that contain the collections of taste cells as taste *buds*) on the anterior tongue can vary across individuals by as much as 16-fold. Within these subjects, two distinct and non-overlapping groups varying in FP density, and taste buds and taste pores within papillae, could be identified. The mean ratings of the intensity of a range of concentrations of sucrose, NaCl and PROP (but not citric acid and quinine) were shown to be higher in subjects with greater densities of FP taste pores. Bartoshuk, Duffy and Miller (1994) reported similar significant associations between PROP intensity ratings and density of both FP and taste pores, while a recent study reported a correlation of 0.86

(Spearman's ρ) between FP density and PROP ratings (Essick, Chopra *et al.* 2003). Consistent with this, PROP thresholds were significantly negatively correlated with both papillae and pores, while dividing subjects into NTs, MTs and STs showed a greater than five-fold increase in taste pore density between NTs and STs. Similar relationships have also been reported in children (Hutchinson, Shahbake *et al.* 2002). Figure 14.3 shows these dramatic differences in FP densities on the anterior tongues of two subjects. To date, there are no data on the genetic basis or heritability of such variations in taste structures.

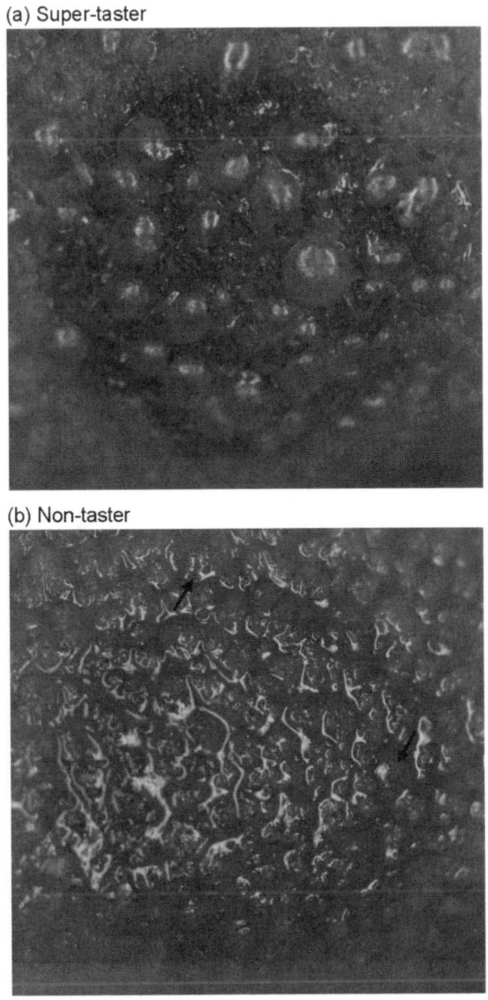

Fig. 14.3 Sections of the anterior tongue surface stained with blue food dye in (a) a PROP super-taster, and (b) a non-taster. The fungiform papillae (FP) stain poorly and so are contrasted against the blue background. Many of these are easily apparent for the ST. In the NT, only two FP are evident (black arrows) for the same tongue area. Images courtesy of Maryam Shahbake, University of Western Sydney.

The salience of such anatomical variation to understanding PROP tasting and its correlates lies in the fact that taste intensities reflect spatial summation – that is, they are at least partly a function of the number of FP and taste pores stimulated (Miller and Bartoshuk 1991, Smith 1971). Thus, the associations between PROP tasting and FP density may be sufficient to explain the relationships seen between PROP tasting and suprathreshold measures of the intensity of other tastes.

Variations in FP densities are also likely to underlie relationships between PROP and oral somatosensory qualities (Prescott, Bartoshuk and Prutkin 2004a). Fungiform papillae are heavily innervated not only by taste afferents within the *chorda tympani* (VIIth) nerve, but also by *trigeminal* (Vth) nerve fibres mediating chemesthetic (oral irritation) and tactile qualities (Farbman and Hellekant 1978). It has been estimated that 75% of the innervation of FP is by trigeminal fibres, which both enter the taste bud and also occupy the surrounding epithelium, terminating at the apex of the papilla (Farbman and Hellekant 1978; Whitehead, Beeman and Kinsella 1985). Hence, greater FP density will mean higher densities of fibres mediating sensations of oral irritation produced by capsaicin or alcohol, as well as qualities such as viscosity. As a consequence, the link between sensitivity to PROP and creaminess in foods (Tepper and Nurse 1997; Duffy, Lucchina and Bartoshuk 2004) may also be determined by individual differences in FP density.

14.4.1 Tactile components of tastes

As well as mediating such tactile and chemesthetic qualities in flavour, there is increasing evidence that the activation of somatosensory fibres, in particular those sensitive to temperature and touch, may also contribute to the experience of taste qualities. Compounds associated with basic taste qualities, such as NaCl (salty), citric acid (sour) and quinine (bitter), have been shown to produce irritation at moderate to high intensities (Gilmore and Green 1993; Green and Hayes 2003). Various categories of somatosensory stimuli are also capable of inducing taste sensations. Cardello (1981) noted that around 25% of individual FP responded to tactile stimulation with a taste quality, in particular sourness. More recently, taste qualities have been elicited by heated and cooled probes placed on areas containing FP, and circumvallate papillae on the rear of the tongue (IXth – *glossopharyngeal n.*) (Cruz and Green 2000), and by the application of the prototypical 'pure' irritant, capsaicin, to circumvallate papillae (Green and Hayes 2003). Further evidence points to the necessity of concurrent tactile stimulation as crucial to perceptions of taste localisation within the mouth (Pfaffman and Bartoshuk 1989; Todrank and Bartoshuk 1991).

One implication of these data is that differences in taste intensities seen as a function of PROP group differences may reflect, to greater or lesser degrees, somatosensory activation by these 'taste' stimuli. Certainly, the magnitude of PROP group differences for the taste of other compounds is generally much lower that that for somatosensory stimuli such as capsaicin irritation. This is not surprising since, as noted earlier, FP receive considerably more innervation by somatosensory than by taste fibres (Farbman and Hellekant 1978). Given the close anatomical relationship between trigeminal fibres and taste cells (Whitehead, Beeman and Kinsella 1985), it becomes highly likely that our taste perceptions are really effectively taste/somatosensory experiences (Green 2003), and that this will be reflected in differences between PROP groups.

14.4.2 Supertasters and taste anatomy

The magnitude of the differences in FP densities between taster groups led Bartoshuk (1993) to suggest that they might be a better indicator of genetic differences than ratings of PROP taste. Recent advances in understanding PROP tasting genetics have suggested that STs are indeed more than just inheritors of two dominant PROP alleles. Bartoshuk, Davisson *et al.* (2004b) divided subjects for whom PROP thresholds and suprathreshold ratings had been obtained on the basis of their PROP genotype. As might be expected, those with two dominant PROP gene alleles (TT) had the lowest thresholds and gave highest ratings for PROP, *tt* subjects had highest thresholds and lowest ratings, with *Tt* subjects giving intermediate values. However, the differences between these groups in suprathreshold ratings were not of the same magnitude as seen when groups are defined using PROP ratings. Specifically, when groups were defined using genotype, the suprathreshold ratings of the *TT* group were only somewhat higher than those of the *Tt* group, in contrast to the often very much higher ratings of bitterness shown by STs rather than MTs. Thus, divisions based on PROP ratings appear to reflect more than just differences in PROP genotypes.

It is possible that this discrepancy may be due to the involvement of other bitter genes. However, since the variation in intensity for PROP is much greater than for any other taste, it has been suggested that STs are those tasters who are homozygous for the dominant allele (*TT*) *and* who have a large number of fungiform papillae (Bartoshuk, Fast *et al.* 2004a). This provides a plausible explanation for the sensitivity of the relationship between PROP tasting and the intensity of other tastes to the type of scale used, which has been seen in some studies (Schifferstein and Frijters 1991; Smagghe and Louis-Sylvestre 1998). If the scale does not easily distinguish STs from other tasters (as category scales may fail to do due to ceiling effects), then the influence of the anatomical differences that underlie the association between PROP and other taste intensities will be minimised.

In effect, then, and as noted earlier, the relationships seen between PROP and other taste intensities are based on fungiform papillae numbers, rather than PROP sensitivity *per se*, and this will be most evident when STs are distinguished from other tasters. A way of assessing this, albeit indirectly, is by comparing responses to other tastes within subjects who are tasters – that is, comparisons between MTs and STs. Prescott, Ripandelli and Wakeling (2001), in their study of PROP tasting and binary taste mixtures, provide data that seems to support the idea that associations between PROP and other taste intensities are related primarily to non-genetic factors. Figure 14.1 shows ratings of intensity for one of the tastes in each of three different binary mixtures – sweet/bitter; salty/bitter; salty/sour. In each case, differences are only evident between STs and both MTs and NTs, rather than between these latter groups. Since taste intensities are a function of spatial summation, it is likely that FP density accounts for the greater part of these differences.

In a more direct study of the relationship between FP density and PROP genetics, Delwiche, Buletic and Breslin (2001b) found no relationship for NTs between FP counts and PROP intensity, whereas a relationship was found for this group with the intensity of quinine. This suggests strongly that PROP non-tasting is primarily accounted for by purely genetic factors. By contrast, PROP tasters showed positive correlations between FP counts, and quinine and PROP intensity. Hence, for the majority of bitter compounds (and likely for other tastes as well) spatial summation over distinctly varying FP densities is likely to account for variation in intensity. This finding parallels the earlier work of Fischer and Griffin (1963) who reported the existence of unimodally-ditributed variations in taste sensitivity (primarily thresholds for the bitterness of quinine) that were independent of PROP sensitivity. They found that PROP thresholds of both tasters and non-tasters increased with increasing quinine thresholds suggesting that PROP sensitivity overlays a more general variation in taste sensitivity.

14.5 PROP taster differences and flavour perceptions

As noted earlier, individual differences in the perception of both tastes and somatosensory qualities related to PROP/FP density are also evident in the few studies that have used real foods or beverages. In addition, however, it might be expected that such variations have an overall impact on flavour, which reflects the integration of sensory information from taste, somatosensory and olfactory afferents in producing an overall sensory quality (Prescott 2004). Within such synthetic percepts, overall intensity and quality of flavour quality is determined by the interactions that occur between these different sensory inputs. For example, the effect of 'sweet-smelling' odours in enhancing the overall sweetness of a flavour is well known (Frank and Byram 1988). It is thus highly likely that the perception of the characteristic olfactory component of a food or beverage flavour in the mouth is strongly influenced by concurrent taste information. Responses to tastes will, in turn, be subject to somatosensory influences. In addition to the role of somatosensory afferents in partly determining taste quality (see above), oral irritants such as capsaicin have been shown to suppress sweet tastes (Prescott, Allen and Stephens 1993) as well as flavours that have a prominent sweet component (Prescott 1999). Because of the substantial variations in responses to irritation among PROP taster groups, these interactions and the consequent profile of the flavour might also be influenced by taster status. Such hypotheses have received recent support through evidence that variations in oral anatomy may have an impact on overall flavour. Duffy, Chapo *et al.* (2003) showed, first, that taste intensities, FP density and, to a lesser extent, PROP bitterness predicted retronasal olfactory intensity of a simple sweet food (jelly beans) and, second, that the overall flavour was best predicted from the intensity of the sweet taste and number of FP. It thus appears that individual variations in FP density–taste intensity will be directly reflected more generally in overall flavour perception.

An even broader source of individual differences in taste and flavour perception has recently been described. Green and George (2004) showed systematic individual differences in generalised responsiveness to a range of chemosensory qualities in groups defined on the basis of whether or not they experienced tastes in response to thermal stimulation of papillae (Cruz and Green 2000). These groups showed systematic differences in ratings of the intensity of a range of tastes, as well as retronasal odours. Interestingly, these differences appeared to be independent of FP density (*chorda tympani n.*), since the differences were present when taste intensities were measured at the back of the tongue (*glossopharyngeal n.*) and on the soft palate (*superficial petrosal n.*). These data raise the possibility that there exists a central process that regulates the overall intensity of stimulation for tastes and other chemosensory qualities, and that individuals differ in the extent to which such regulation occurs. This notion bears considerable similarity to earlier concepts of central perceptual regulation in which individuals maintain an optimum level of stimulation through enduring 'perceptual styles' that either maximise or limit sensory intensity (Buchsbaum 1976; Petrie 1967). As yet, there are no indications whether or not these individual differences reflect genetic diversity.

14.6 Conclusion

Unsurprisingly, perhaps, recent research into taste genetics, together with psychophysical studies of variations in sensitivity to PROP and other compounds, have combined to produce an increasingly complex picture of taste perception. Odours have traditionally received emphasis as the defining and distinguishing characteristic of flavours. Nevertheless, we are beginning to see that genetic variations in the form of both multiple receptors and receptor polymorphisms, as well as individual differences in taste anatomy, are important influences on the perception of flavours and perhaps, as a consequence, a strong determinant of food and flavour preferences.

14.7 References

ANLIKER, J.A., BARTOSHUK, L.M., FERRIS, A.M. and HOOKS, L.D. (1991) Children's food preferences and genetic sensitivity to the bitter tastes of 6-n-propylthiouracil (PROP). *Am. J. Clin. Nutr.* 54, 316–320.

BARTOSHUK, L.M. (1979) Bitter taste of saccharin related to the genetic ability to taste the bitter substance of 6-n-Propylthiouracil. *Science*, 205, 934–935.

BARTOSHUK, L.M. (1993) The biological basis of food perception and acceptance. *Food Qual. Pref.*, 4, 21–32.

BARTOSHUK, L.M. (2000). Comparing sensory experiences across individuals: Recent psychophysical advances illuminate genetic variation in taste perception. *Chem. Senses*, 25, 447–460.

BARTOSHUK, L.M., FAST, K., KARRER, T.A., MARINO, S., PRICE, R.A. and REED, D.A. (1992) PROP supertasters and the perception of sweetness and bitterness. *Chem. Senses*, 17, 594.

BARTOSHUK, L.M., RIFKIN, B., MARKS, L.E. and HOOPER, J.E. (1988) Bitterness of KCl and benzoate, related to genetic status for sensitivity to PTC/PROP. *Chem. Senses*, 13, 517–528.

BARTOSHUK, L.M., CONNER, E., GRUBIN, D., KARRER, T., KOCHENBACH, K., PALSCO, M., SNOW, D., PELCHAT, M. and DANOWSKI, S. (1993) PROP supertasters and the perception of ethyl alcohol. *Chem. Senses*, 18, 526–527.

BARTOSHUK, L.M., DUFFY, V.B., LUCCHINA, L.A., PRUTKIN, J. and FAST, K. (1998) PROP (6-n-propylthiouracil) supertasters and the saltiness of NaCl. *Ann. N.Y. Acad. Sci.*, 855, 793–796.

BARTOSHUK, L.M., DUFFY, V.B. and MILLER, I.J. (1994) PTC/PROP tasting: Anatomy, psychophysics, and sex effects. *Physiol. Behav.*, 56(6), 1165–1171.

BARTOSHUK, L.M., DUFFY, V.B., REED, D. and WILLIAMS, A. (1996) Supertasting, Earaches and Head Injury: Genetics and Pathology Alter our Taste Worlds. *Neurosci. Biobehav. Rev.*, 20(1), 79–87.

BARTOSHUK, L.M., FAST, K., SNYDER, D. and DUFFY, V.B. (2004a) Genetic differences in human oral perception. Advanced methods reveal basic problems in intensity scaling. In Prescott, J. and Tepper, B. (eds) *Genetic Variation in Taste Sensitivity*. New York: Marcel Dekker, 1–41.

BARTOSHUK, L.M., DAVISSON, A., KIDD, J., KIDD, K.K., SPEED, W., PAKSTIS, A., REED, D., SNYDER, D. and DUFFY, V.B. (2004b) Supertasting is not explained by the PTC/PROP gene. *Paper presented at the 26th Association for Chemoreception Sciences conference*. April 21–25, 2004, Sarasota, USA.

BELL, G. and SONG, H-J. (2004) Genetic basis for 6-n-propylthiouracil taster and supertaster status determined across cultures. In Prescott, J. and Tepper, B. (eds) *Genetic Variation in Taste Sensitivity*. New York: Marcel Dekker, 137–154.

BLAKESLEE, A.F. and SALMON, T.N. (1935) Genetics of sensory thresholds: Individual taste reactions for different substances. *Proc. Nat. Acad. Sci.*, 21, 84–90.

BUCHSBAUM, M. (1976) Self-regulation of stimulus intensity: Augmenting/reducing and the average evoked response. In G.E. Schwartz and D. Shapiro (eds) *Consciousness and Self Regulation* (Vol. 1, 101–135). New York: Plenum Press.

BUFE, B., BRESLIN, P.A.S., KUHN, C., REED, D.R., THARP, C.D., SLACK, J.P., KIM, U-K., DRAYNA, D. and MEYERHOF, W. (2005) The molecular basis of individual differences in phenylthiocarbamide and propylthiouracil bitterness perception. *Current Biology*, 15, 322–327.

CARDELLO, A.V. (1981) Comparison of taste qualities elicited by tactile, electrical and chemical stimulation of single human taste papillae. *Percept. Psychophys*, 29, 163–169.

CHAUDHARI, N., LANDIN, A.M. and ROPER, S.D. (2000) A metabotropic glutamate receptor variant functions as a taste receptor. *Nature Neurosci.*, 3(2), 113–119.

CRUZ, A. and GREEN, B.G. (2000) Thermal stimulation of taste. *Nature*, 403, 889–892.

DELWICHE, J.F., BULETIC, Z. and BRESLIN, P.A.S. (2001a) Covariation in individuals' sensitivities to bitter compounds: Evidence supporting multiple receptor/transduction mechanisms. *Percept. Psychophys.*, 63(5), 761–776.

DELWICHE, J.F., BULETIC, Z. and BRESLIN, P.A.S. (2001b) Relationship of papillae number to bitter intensity of quinine and PROP within and between individuals. *Physiol. Behav.*, 74, 329–337.

DREWNOWSKI, A. (2004) 6-n-propylthiouracil sensitivity, food choices, and food consumption. In Prescott, J. and Tepper, B. (eds) *Genetic Variation in Taste*

Sensitivity. New York: Marcel Dekker, pp. 179–194.

DREWNOWSKI, A., HENDERSON, S.A. and SHORE, A.B. (1997) Taste responses to naringin, a flavonoid, and the acceptance of grapefruit juice are related to genetic sensitivity to 6-n-propylthiouracil. *Am. J. Clin. Nutr.*, 66, 391–397.

DREWNOWSKI, A., HENDERSON, S. A. and BARRATT-FORNELL, A. (1998) Genetic sensitivity to 6-n-Propylthiouracil and sensory responses to sugar and fat mixtures. *Physiol. Behav.*, 63(5), 771–777.

DREWNOWSKI, A., HENDERSON, S.A., LEVINE, A. and HANN, C. (1999) Taste and food preferences as predictors of dietary practices in young women. *Pub. Health Nutr.*, 2, 513–519.

DREWNOWSKI, A. and ROCK, C.L. (1995) The influence of genetic taste markers on food acceptance. *Am. J. Clin. Nutr.*, 62, 506–511.

DUFFY, V.B. and PETERSON, J.M. (2000) Genetic variation in taste: Associations with alcohol sensation and intake. *Chem. Senses*, 25, 638.

DUFFY, V.B., BARTOSHUK, L.M., LUCCHINA, L.A., SNYDER, D.J. and TYM, A. (1996) Supertasters of PROP (6-n-propylthiouracil) rate the highest creaminess to high-fat milk products. *Chem. Senses*, 21(5), 598.

DUFFY, V., CHAPO, A.K., HUTCHINS, H.L. and BARTOSHUK, L.M. (2003) Retronasal olfactory intensity: Associations with taste. *Paper presented at the 25th Association for Chemoreception Sciences conference*, Sarasota, FL, USA.

DUFFY, V.B., LUCCHINA, L.A. and BARTOSHUK, L.M. (2004) Genetic variation in taste: Potential biomarker for cardiovascular disease risk? In Prescott, J. and Tepper, B. (eds) *Genetic Variation in Taste Sensitivity*. New York: Marcel Dekker, 195–228.

ESSICK, G., CHOPRA, A., GUEST, S. and McGLONE, F.P. (2003) Lingual tactile acuity, taste perception, and the density and diameter of fungiform papillae in female subjects. *Physiol. Behav.*, 80(2–3), 289–302.

FARBMAN, A.I. and HELLEKANT, G. (1978) Quantitative analyses of the fiber population in rat chorda tympani nerves and fungiform papillae. *Am. J. Anat.*, 153, 509–522.

FAST, K., DUFFY, V.B. and BARTOSHUK, L.M. (2002) New psychological insights in evaluating genetic variation in taste. In C. Rouby, B. Schaal, D. Dubois, R. Gervais and A. Holley (eds) *Olfaction, Taste and Cognition*. Cambridge: Cambridge University Press.

FINGER, T.E. and SIMON, S.A. (2000) Cell biology of the taste epithelium. In T.E. Finger, W.L. Silver and D. Restrepo (eds) *The Neurobiology of Taste and Smell*. 2nd edn, New York, Wiley-Liss, 287–314.

FISCHER, R. and GRIFFIN, F. (1963) Quinine dimorphism: A cardinal determinant of taste sensitivity. *Nature*, 200, 343–347.

FISCHER, R., GRIFFIN, F., ENGLAND, S. and GARN, S.M. (1961) Taste thresholds and food dislikes. *Nature*, 191, 1328.

FOX, A.L. (1931) Six in ten 'tasteblind' to bitter chemical. *Sci. News Lett.*, 9, 249.

FRANK, R.A. and BYRAM, J. (1988) Taste-smell interactions are tastant and odorant dependent. *Chem. Senses*, 13(3), 445–455.

GENT, J.F. and BARTOSHUK, L. (1983) Sweetness of sucrose, neohesperidin dihydrochalone, and saccharin is related to genetic ability to taste the bitter substance 6-n-propylthiouracil. *Chem. Senses*, 7(3/4), 265–272.

GILBERTSON, T.A. (1998) Gustatory mechanisms for the detection of fat. *Curr. Opin. Neurobiol.*, 8, 447–452.

GILMORE, M.M. and GREEN, B.G. (1993) Sensory irritation and taste produced by NaCl and citric acid: effects of capsaicin desensitization. *Chem. Senses*, 18(3), 257–272.

GLANVILLE, E.V. and KAPLAN, A.R. (1965) Food preference and sensitivity of taste for bitter compounds. *Nature*, 205, 851–853.

GLENDINNING, J.I., CHAUDHARI, N. and KINNAMON, S.C. (2000) Taste transduction and molecular biology. In T.E. Finger, W.L. Silver and D. Restrepo (eds) *The Neurobiology of Taste and Smell*. 2nd edn, New York: Wiley-Liss, 315–351.

GREEN, B.G. (2003) Studying taste as a cutaneous sense. *Food Qual. Pref.*, 14, 99–109.

GREEN, B.G. and GEORGE, P. (2004) 'Thermal taste' predicts higher responsiveness to chemical taste and flavor. *Chem. Senses*, 29, 617–628.

GREEN, B.G. and HAYES, J.E. (2003) Capsaicin as a probe of the relationship between bitter taste and chemesthesis. *Physiol. Behav.*, 79, 811–821.

GUO, S.-W. and REED, D.R. (2000) The genetics of phenylthiocarbamide perception. *Ann. Hum. Biol.*, 28(2), 111–142.

HALL, M.J., BARTOSHUK, L.M., CAIN, W.S. and STEVENS, J.C. (1975) PTC taste blindness and the taste of caffeine. *Nature*, 253, 442–443.

HARRIS, H. and KALMUS, H. (1949) The measurement of taste sensitivity to phenylthiourea (P.T.C.). *Ann. Eugen.*, 15, 24–31.

HLADIK, C.M., PASQUET, P. and SIMMEN, B. (2002) New perspectives on taste and primate evolution: The dichotomy in gustatory coding for perception of beneficent versus noxious substances as supported by correlations among human thresholds. *Am. J. Phys. Anthrop.*, 117, 342–348.

HUTCHINSON, I., SHAHBAKE, M., LAING, D.G. and JINKS, A. (2002) Taste pore density on the tongue and PROP sensitivity in children. *Paper presented at the 24th Association for Chemoreception Sciences conference*, Sarasota, FL, USA.

KAMINSKI, L.C., HENDERSON, S.C. and DREWNOWSKI, A. (2000) Young women's food preferences and taste responsiveness to 6-n-propylthiouracil (PROP). *Physiol. Behav.*, 68, 691–698.

KARRER, T., BARTOSHUK, L.M., CONNER, E., FEHRENBAKER, S., GRUBIN, D. and SNOW, D. (1992) PROP status and its relationship to the perceived burn intensity of capsaicin at different tongue loci. *Chem. Senses*, 17, 649.

KIM, U.-K., BRESLIN, P.A.S., REED, D.R. and DRAYNA, D. (2004) Genetics of human taste perception. *J. Dent. Res.*, 83(6), 448–453.

KIM, U.-K., JORGENSON, E., COON, H., LEPPERT, M., RISCH, N. and DRAYNA, D. (2003) Positional cloning of the human quantitative trait locus underlying taste sensitivity to phenylthiocarbamide. *Science*, 299, 1221–1225.

KIRKMEYER, S.V. and TEPPER, B.J. (2004) A current perspective on creaminess perception and 6-n-propylthiouracil taster status. In Prescott, J. and Tepper, B. (eds) *Genetic Variation in Taste Sensitivity*. New York: Marcel Dekker, 117–135.

LAUGERETTE, F., PASSILLY-DEGRACE, P., PATRIS, B., NIOT, I., FEBBRAIO, M., MONTMAYEUR, J-P. and BESNARD, P. (2005) CD36 involvement in orosensory detection of dietary lipids, spontaneous fat preference and digestive secretions. *The Journal of Clinical Investigation*, 115(1), 3177–3184.

LEACH, E.J. and NOBLE, A.C. (1986) Comparison of bitterness of caffeine and quinine by a time-intensity procedure. *Chem. Senses*, 11, 339–345.

LI, X., STASZEWSKI, L., XU, H., DURICK, K., ZOLLER, M. and ADLER, E. (2002) Human receptors for sweet and umami taste. *Proc. Nat. Acad. Sci. USA*, 99, 4692–4696.

LIAO, J. and SCHULTS, P.G. (2003) Three sweet receptor genes are clustered in human chromosome 1. *Mamm. Genome*, 14, 291–301.

LOOY, H. and WEINGARTEN, H.P. (1992) Facial expressions and genetic sensitivity to 6-n-Propylthiouracil predict hedonic response to sweet. *Physiol. Behav.*, 52, 75–82.

LUGAZ, O., PILLIAS, A.-M. and FAURION, A. (2002) A new specific ageusia: Some humans cannot taste l-glutamate. *Chem. Senses*, 27, 105–115.

MATTES, R.D. (2004) 6-n-propylthiouracil taster status: Dietary modifier, marker, or misleader? In Prescott, J. and Tepper, B. (eds) *Genetic Variation in Taste Sensitivity*. New York: Marcel Dekker, 229–250.

MELA, D.J. (1989) Bitter taste intensity: The effect of tastant and thiourea taster status. *Chem. Senses*, 14, 131–135.

MILLER, I.J. and BARTOSHUK, L.M. (1991) Taste perception, taste bud distribution, and spatial relationships. In T.V. Getchell, L.M. Bartoshuk, R.L. Doty and J.B. Snow (eds), *Smell and Taste in Health and Disease* (205–233). New York: Raven Press.

MILLER, I.J. and REEDY, F.E. (1990) Variations in human taste bud density and taste intensity perception. *Physiol. Behav.*, 47, 1213–1219.

MONTMAYEUR, J.-P. and MATSUNAMI, H. (2002) Receptors for bitter and sweet taste. *Curr. Opin. Neurobiol.*, 12(4), 366–371.

NELSON, G., CHANDRASHEKAR, J., HOON, M.A., FENG, L., ZHAO, G., RYBA, N.J.P., *et al.* (2002) An amino-acid taste receptor. *Nature*, 416, 199–202.

PANGBORN, R.M. (1970) Individual variation in affective responses to taste stimuli. *Psychon. Sci.*, 21(2), 125–126.

PETRIE, A. (1967) *Individuality in Pain and Suffering*. Chicago: University of Chicago Press.

PFAFFMAN, C. and BARTOSHUK, L.M. (1989) Psychophyscial mapping of a human case of left unilateral ageusia. *Chem. Senses*, 14, 738.

PICKERING, G.J., SIMUNKOVA, K. and DiBATTISTA, D. (2004) Intensity of taste and astringency sensations elictied by red wines is associated with sensitivity to PROP (6-n-propylthiouracil). *Food Qual. Pref.*, 15(2), 147–154.

PRESCOTT, J. (1998) Comparisons of taste perceptions and preferences of Japanese and Australian consumers: overview and implications for cross-cultural sensory research. *Food Qual. Pref.*, 9(6), 393–402.

PRESCOTT, J. (1999) Introduction to the trigeminal sense: The role of pungency in food flavours. In G.A. Bell and A.J. Watson (eds), *Tastes and Aromas. The Chemical Senses in Science and Industry* (58–49). Sydney: University of New South Wales Press.

PRESCOTT, J. (2004) Psychological processes in flavour perception. In A.J. Taylor and D. Roberts (eds), *Flavour Perception* (256–277). London: Blackwell Publishing.

PRESCOTT, J. and SWAIN-CAMPBELL, N. (2000) Responses to repeated oral irritation by capsaicin, cinnemaldehyde and ethanol in PROP tasters and non-tasters. *Chem. Senses*, 25, 239–246.

PRESCOTT, J., RIPANDELLI, N. and WAKELING, I. (2001) Intensity of tastes in binary mixtures in PROP non-tasters, medium-tasters and super-tasters. *Chem. Senses*, 26, 993–1003.

PRESCOTT, J., BARTOSHUK, L.M. and PRUTKIN, J. (2004a) PROP tasting and the perception of non-taste oral sensations. In Prescott, J. and Tepper, B. (eds) *Genetic Variation in Taste Sensitivity*. New York: Marcel Dekker, 89–104.

PRESCOTT, J., SOO, J, CAMPBELL, H. and ROBERTS, C. (2004b) Responses of PROP taster groups to variations in sensory properties in foods and beverages. *Physiol. Behav.*, 82 (2–3), 459–469.

PRESCOTT, J., ALLEN, S. and STEPHENS, L. (1993) Interactions between oral chemical irritation, taste and temperature. *Chem. Senses*, 18(4), 389–404.

PRUTKIN, J. (2002) *Mechanoreception and Nociception on the Anterior Tongue*, MSc thesis, Yale University School of Medicine.

PRUTKIN, J.M., FAST, K., LUCCHINA, L.A. and BARTOSHUK, L.M. (1999) Prop (6-*n*-propyl-thiouracil) genetics and trigeminal innervation of fungiform papillae. *Chem. Senses*, 24, 243.

REED, D.R. (2004) Progress in human bitter genetics. In Prescott, J. and Tepper, B. (eds) *Genetic Variation in Taste Sensitivity*. New York: Marcel Dekker, 43–61.

REED, D.R., BARTOSHUK, L.M., DUFFY, V., MARINO, S. and PRICE, R.A. (1995) Propylthiouracil tasting: Determination of underlying threhold distributions using maximum likelihood. *Chem. Senses*, 20(5), 529–534.

RICHARDSON, N.J. and BOOTH, D.A. (1993) Multiple physical patterns in judgements of the creamy texture of milks and creams. *Acta Psychol.*, 84, 93–101.

ROZIN, P. (1982) Human food selection: The interaction of biology, culture and individual experience. In L.M. Barker (ed.) *The Psychobiology of Human Food Selection* (225–254). Westport, CT: AVI Publishing Co.

SCHIFFERSTEIN, H.N.J. and FRIJTERS, J.E.R. (1991) The perception of the taste of KCl, NaCl, and quinine HCl is not related to PROP sensitivity. *Chem. Senses*, 16(4), 303–317.

SCOTT, T.R. (1992) Taste, feeding, and pleasure. In Epstein, A.N. and Morrison, A.R. (eds) *Progress in Psychobiology and Physiological Psychology* (Vol. 15, 231–291): Academic Press.

SCOTT, T.R. and MARK, G.P. (1987) The taste system encodes stimulus toxicity. *Brain Res.*, 414, 197–203.

SHI, P., ZHANG, J., YANG, H. and ZHANG, Y.P. (2003) Adaptive diversification of bitter taste receptor genes in mammalian evolution. *Mol. Biol. Evol.*, 20, 805–814.

SMAGGHE, K. and LOUIS-SYLVESTRE, J. (1998) Influence of PROP-sensitivity on taste perceptions and hedonics in French women. A study performed without retronasal olfaction. *Appetite*, 30, 325–339.

SMITH, D.V. (1971) Taste intensity as a function of area and concentration: Differentiation between compounds. *J. Exp. Psychol.*, 87(2), 163–171.

SNYDER, L.H. (1931) Inherited taste deficiency. *Science*, 74, 151–152.

STEINER, J.E., GLASER, D., HAWILO, M.E. and BERRIDGE, K.C. (2001) Comparative expression of hedonic impact: affective reactions to taste by human infants and other primates. *Neurosci. Biobehav. Rev.*, 25, 53–74.

TEPPER, B.J. and NURSE, R.J. (1997) Fat perception is related to PROP taster status. *Physiol. Behav.*, 61(6), 949–954.

TEPPER, B.J. (2004) 6-n-propylthiouracil as a genetic taste marker for fat intake, obesity and chronic disease risk. Current evidence and future promise. In Prescott, J. and Tepper, B. (eds) *Genetic Variation in Taste Sensitivity*. New York: Marcel Dekker, 155–177.

TODRANK, J. and BARTOSHUK, L.M. (1991) A taste illusion: Taste sensation localised by touch. *Physiol. Behav.*, 50, 1027–1031.

TURNBULL, B. and MATISOO-SMITH, E. (2002) Taste sensitivity to 6-n-propylthiouracil (PROP) predicts acceptance of bitter-tasting spinach in 3–6 year old children. *Am. J. Clin. Nutr.*, 76, 1101–1105.

WHITEHEAD, M.C., BEEMAN, C.S. and KINSELLA, B.A. (1985) Distribution of taste and general sensory nerve endings in fungiform papillae of the hamster. *Am. J. Anat.*, 173, 185–201.

YACKINOUS, C. and GUINARD, J.-X. (2001) Relation between PROP taster status and fat perception, touch, and olfaction. *Physiol. Behav.*, 72, 427–437.

YIEE, J.H., DUFFY, V.B. and BARTOSHUK, L.M. (2003) Analytic nature of taste mixture interactions: Contribution of PROP status. *Paper presented at the 25th Association for Chemoreception Sciences conference*, Sarasota, USA.

YOKOMUKAI, Y., COWART, B.J. and BEAUCHAMP, G.K. (1993) Individual differences in sensitivity to bitter-tasting substances. *Chem. Senses*, 18(6), 669–681.

15

Texture–aroma interactions

S. Lubbers, INRA, France

15.1 Introduction

The comprehension of the phenomena which modify the availability of aroma compounds within the food matrices is a very important field in the food science world. Many teams of researchers have sought to identify the mechanisms which explain at the physicochemical level how the compounds responsible for olfactive stimuli are available for perception in the foodstuffs. The evolution of the research in this field tends to an increased desire for knowledge on the phenomena of interactions between the two major components of the acceptability of food for consumers: the aromatic component and the textural component. From the physicochemical point of view, this research was carried out, in the setting into prospect for data resulting from the flavour release analysis in the headspace of foodstuffs, the structure characterisation of matrices and their rheological behaviours. The instrumental analyses of flavour should reflect the flavour component, perceptible by the individual. The rheological analysis should allow the characterisation of the perceived texture.

During the last few decades the studies on the interactions between non-volatile components and flavour compounds were carried out in systems presenting an increasing complexity of organisation. Thus the behaviour of aroma compounds in aqueous solutions, protein and polysaccharide colloidal suspensions, and in viscous and/or gelled systems was studied. It appears that the organisation of molecules into the matrix, in which these studies were undertaken, is particularly significant. The assumption of the influence of the structure and resulting texture of the foodstuff on its flavour was made.

To illustrate this assumption, the precursory work of Pangborn and Szczesniak (1974) is frequently cited in literature, because it poses the bases

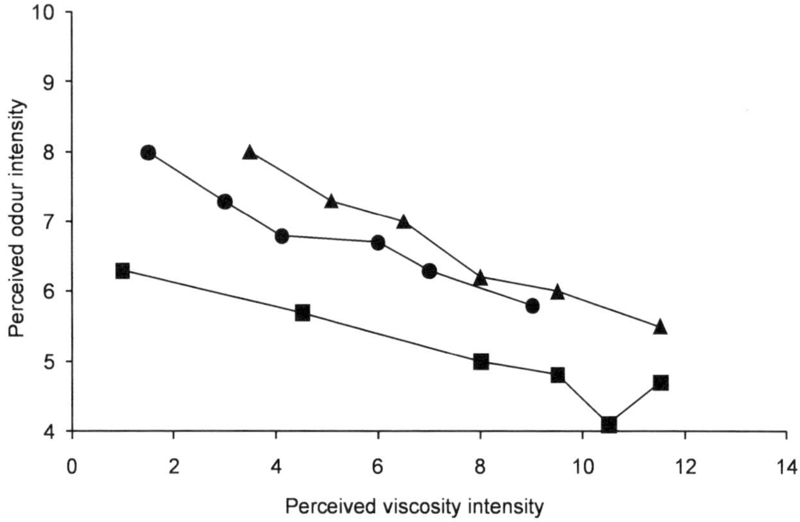

Fig. 15.1 Intensity of perceived odour as a function of perceived viscosity in mouth. ■ coffee odour, ● orange odour, ▲ tomato odour. Adapted from Pangborn *et al.* (1978).

of the influence of the viscosity of the matrix on the perception of the odour and the flavour. The perceived intensity of the odour of coffee aroma, orange aroma and tomato aroma, as a function of the viscosity of the matrices, was evaluated by panellists (Fig. 15.1, adapted from Pangborn *et al.* 1978). Thus, the perception of the odour decreased with an increase in the perception of the viscous character by the panellist. The intensity of the perceived flavour from the solution (in the mouth) is also modified according to the viscosity of the medium. This reduction depended on the nature of the thickener used, the nature of the flavour and the sensory method tested (nose or mouth).

Thus, the major question is by which mechanisms the increase in viscosity or texture of foodstuff induced a reduction in the perception of flavour or odour. Taking into account the influence of the chemical nature of the macromolecules, which conditions the structure and so the perceived texture of the foodstuff, and the nature of aroma compounds, it is necessary to consider various levels of investigation to highlight the appropriate mechanisms. First, from the physico-chemical point of view, a certain number of studies made it possible to understand, at a molecular and macroscopic level, how the structure and the rheological behaviour of the matrix affected the release of aroma compounds. Second, the influence of the perceived texture on the flavour perception was determined in numerous studies but the explanations of the phenomena involved were more complex. The temporal aspect and kinetic dimension appeared as major parameters and the modifications, which the matrices underwent, when taken into the mouth had to be considered. Furthermore, interactions at the psychophysical level had also to be considered because the perception of texture and flavour were made simultaneously.

15.2 Influence of rheological behaviour on flavour release

15.2.1 Effect of structure and composition of matrix on flavour release

The impact of the texture of the product on olfactive perception is generally indirectly examined by studying the effect of factors modifying the rheological properties of the matrix on the relative volatility of aroma compounds from this product. The studies were carried out with model systems in more the share of cases. For works carried out with complex foodstuffs (several thickening agents and flavour compounds), the majority of authors agreed on an impact of the addition of thickeners on the aroma release. The effect observed, varied with the couple thickeners/flavour compounds (Table 15.1).

The increase in the apparent viscosity of a solution of macromolecules, (polysaccharides or proteins), with the increase of their concentration, involved a reduction in the quantity of flavour compounds released in the headspace of the solution in comparison with water.

Two phenomena have been proposed to explain the behaviour of flavour compounds in these systems. It can demonstrate that the nature of the macromolecule is involved, first, either directly by engaging interactions with the aroma compound, and second, either by modifying the conditions of mass transfers within the system. The study of the system, under thermodynamic equilibrium conditions and kinetic conditions, permitted us, partly, to distinguish between the phenomena.

Indeed, the increase in macromolecule concentration conditions the structure of the product, turning it from a solution to a solid system. The molecular organisation can be approached by rheological analyses, allowing, in particular, to describe changes in the state of the system; the transition from a solution towards a gel is a significant mechanism in the modifications of the transfers of aroma compounds in the system.

However, direct interaction, at a molecular level, was not systematically demonstrated. Indeed, taking into account the diversity of chemical classes and physicochemical properties of aroma compounds, it was difficult to generalise. For a great number of compounds, a combination between the effects of molecular interactions and, the rheology of the matrix, remained the most probable assumption.

At thermodynamic equilibrium, the partition coefficients between the two phases can be estimated and thus, interactions between the polymer and the volatile compound can be supposed. Molecular interactions between macromolecules, in particular, polysaccharides, proteins and aroma compounds were presented in Chapters 9 and 10.

These interactions can explain the differences in flavour release, between solutions with the same viscosity, but composed of different polysaccharides (Roberts *et al.* 1996). The effects of isoviscous solutions (sucrose, guar and CMC) on dynamic flavour release was tested by Roberts *et al.* (1996). Highly volatile aroma compounds such as ethyl 2-ethylbutyrate showed a large decrease in mass spectrometer signal (TIC) as viscosity increased (Fig. 15.2, adapted

Table 15.1 Impact of the texture on flavour perception into complex foodstuffs

Authors	Products	Thickeners	Flavours	Measurements	Flavour perception
Pangborn *et al.* (1978)	tomato juice orange juice coffee	HPC, Xanthan, CMC-L, alginate	tomato orange coffee	η_{app} viscosity in mouth flavour intensity taste	effect of the thickener nature
Guichard *et al.* (1991)	strawberry jam	HM pectin LM pectin	strawberry	HS thickness in mouth taste flavour intensity	effect of the thickener nature
Wendin *et al.* (1997)	acidified milk	gelatine, xanthane, pectin	maltol methyl butyrate	η_{app} thick, smooth flavour intensity: fruit, maltol sugar, acidity	$[T] \rightarrow I_{maltol} \searrow$ no effect on fruit attribute
Cayot *et al.* (1998)	custard dessert	potato starch (1) corn starch (2) waxy starch (3) modified starch (4)	isoamyl acetate	η_{app}, strength gel HS flavour intensity	effect of the thickener nature $I_{(1)} > I_{(2)} > I_{(3)} > I_{(4)}$
Kälviainen *et al.* (2000)	gelled candies	pectin (1) gelatine (2) starch (3) [2+3]	strawberry	texture attributes acidity flavour intensity	effect of the thickener nature $I_{(2)} < I_{(3)} < I_{(2+3)} < I_{(1)}$

Walker and Prescott (2000)	apple juice	CMC, xanthane, pectin	apple	HS flavour intensity sugar, acidity thickness astringency	[pectine] $\rightarrow$ $I_{cooked\ apple}$, I_{honey} $\searrow$
Brennan *et al.* (2002)	stirred yoghurt	modified starch	strawberry green lemon	G', G'' thickness flavour intensity colour	no significant effect of perceived texture on flavour perception
Paci Kora *et al.* (2004)	stirred yoghurt	modified waxy starch+HM pectin	amyle acetate ethyl pentanoate hexanal octalactone diacetyl trans-2-hexenal	η_{app} HS flavour intensity thickness sugar, acidity	[T] low effect on [HS] [T] $\rightarrow$ $I_{green\ apple}$ $\searrow$
Decourcelle *et al.* (2004)	stirred yoghurt	corn starch (1) pectin (2) locus bean gum (3)	strawberry	η_{app} HS flavour intensity thickness	[1], [2] $\rightarrow$ [HS] $\searrow$ [3] $\rightarrow$ [HS] $\nearrow$ [T] $\rightarrow$ $I_{strawberry}$ $\searrow$

HS: headspace
[ ]: concentration in
T: thickeners
η_{app}: apparent viscosity
I: sensory intensity
$\nearrow$: increase
$\searrow$: decrease
$\rightarrow$: to cause

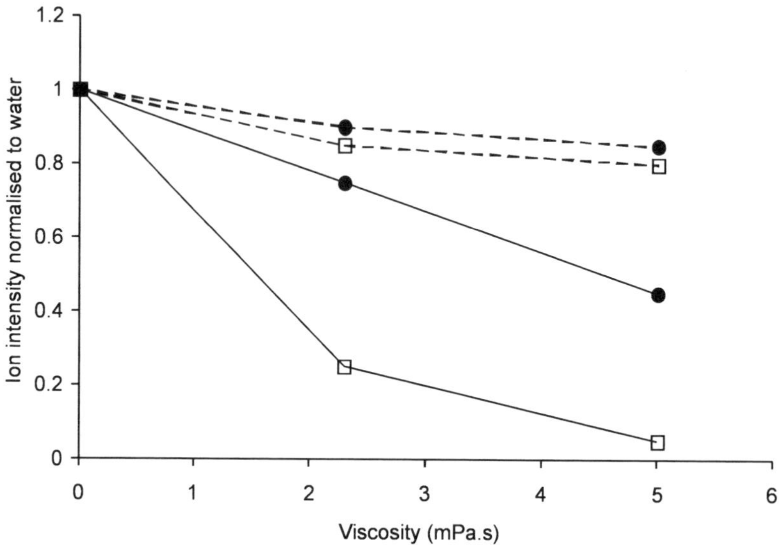

Fig. 15.2 Effect of viscosity on the release of flavour compounds from model solution of sucrose and CMC. –●– maltol/CMC, –□– maltol/sucrose, –●– ethyl 2-methylbutyrate/CMC, –□– ethyl 2-methylbutyrate/sucrose, adapted from Roberts *et al.* (1996).

from Roberts *et al.* 1996). Less volatile compounds such as maltol showed less of an effect. Using isoviscous solutions of CMC and guar, Roberts *et al.* showed that the flavour release was significantly higher with CMC solution than guar mixture.

In the polysaccharide mixture, it appeared that the modifications of the ratio between macromolecules used significantly modified the behaviour of flavour (Secouard *et al.* 2003, Lubbers and Guichard 2003). As shown in Fig. 15.3 (adapted from Secouard *et al.* 2003), the lower the inverse of viscosity of the xanthan/guar mixture, the lower was the limonene release. The linonene retention in the mixture of polysaccharides is mainly influenced by the viscosity of the matrix. A comparison of the limonene release from different matrices with the same viscosity showed that the retention was not only due to the viscosity of the media but may also have resulted from specific limonene–matrix interaction.

In the case of polysaccharides such as carboxy-methyl cellulose (CMC), pectin, guar, locus bean gum, the critical concentration C*, corresponds to the beginning of the entanglement of macromolecules as defined by Baines and Morris (1987). Below the C* value, the presence of macromolecules did not affect the release of flavour in the headspace (Fig. 15.4). Beyond the value of C*, the density of macromolecules seemed to be sufficient to modify the behaviour of the aroma compounds.

The increase in the polymer concentration beyond the critical concentration C*, generated the establishment of a three-dimensional network, as a function of the nature of the polymer. When this type of organisation is set up within the

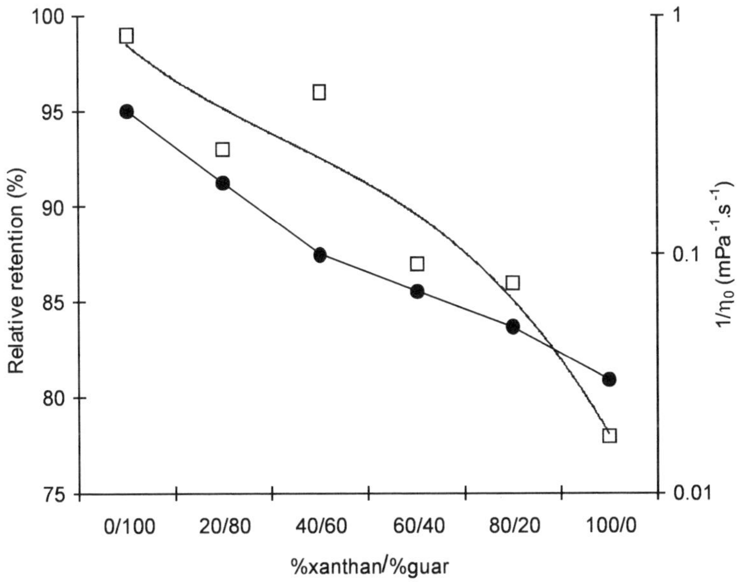

Fig. 15.3 Influence of the xanthan/guar ratio on limonene release (□) and the inverse viscosity of the solutions (●). Adapted from Secouard *et al.* (2003).

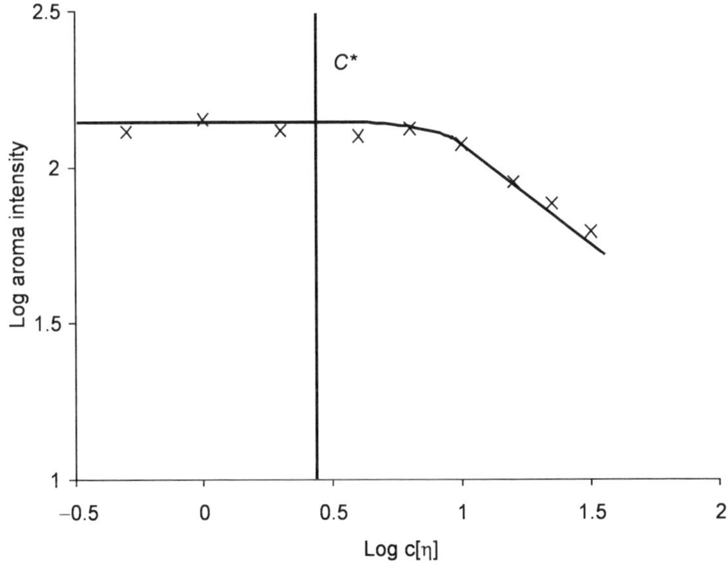

Fig. 15.4 Variation of perceived flavour intensity with degree of coil overlap for gum samples. Adapted from Baines and Morris (1987).

matrix, the system presents intermediate rheological characteristics between a liquid and a solid, defined as viscoelastic. The macroscopic state of the system is a gel.

These various levels of organisation can be obtained with pectin, one of the polysaccharides most used as a thickener in the foodstuffs. Rega *et al.* (2002) showed that at the C* in pectin concentration, the decrease in the headspace concentration of flavour was significant compared with water (medium of reference). All the authors agreed on the effect of the nature of the macromolecule on the phenomenon observed.

The state of dispersion of the pectin molecule and the nature of sugars, which are associated within the system, were significant factors in the mechanisms of flavour release. In matrices with a low water content (great amount of sugar), it appeared that the properties of the gel obtained affected significantly the release of aroma compounds from these matrices (Table 15.2, adapted from Lubbers and Guichard 2003). In syrups, it is well known that water activity more than viscosity of the matrix, induced changes in the diffusion of aroma compounds which explained the variation in the partition coefficients. The pectin addition, which gelled the syrups, again caused a decrease in the flavour release. A higher decrease was observed with the gel, which presented the higher gel hardness. The compressions until fracture, adhesion strengths of gel, and the sensory evaluations of texture (hardness, sticky in mouth) were successfully correlated with the release of flavour in these kinds of product (Lubbers and Guichard 2003, Lubbers *et al.* 2004, Boland *et al.* 2005).

The perception of the odour and the flavour are a dynamic phenomenon, so dynamic methods of analysis of the headspace were used by various authors (Cayot *et al.* 1998, Juteau *et al.* 2004, Roberts *et al.* 1996, Seuvre *et al.* 2004), in order to provide physicochemical data, which might connect significantly to the sensory data. Nevertheless detailed attention should be given to the time scale between the various phenomena. The studies in model systems with instrumental analysis are generally carried out over a long period of observation: i.e. several minutes to hours. On the other hand, time involved in sensory analysis is that of

Table 15.2 Vapour–matrix partition coefficients (K × 1000) for aroma compounds in sugar solutions and pectin gels, adapted from Lubbers and Guichard (2003)

Product	Rheological properties	Methyl 3-butanol	Phenyl 2-ethanol	Ethyl phenyl glycidate
Syrups	**Viscosity** (Pa.s)			
Glucose	0.196	220 a	6.4 a	33 a
Corn syrup DE40	0.507	180 b	4.4 b	25 b
Pectin gels	**Gel hardness** ($N.m^{-2}$)			
Glucose	17 500	200 a	3.7 c	31 a
Corn syrup DE40	20 000	133 c	3.4 d	16 c

Note: Products with the same letter are not significantly different.

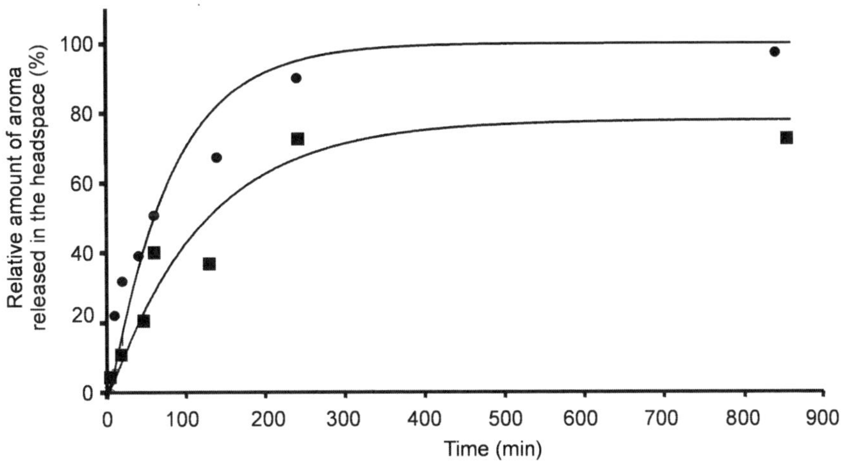

Fig. 15.5 Isoamyl acetate exhaustion from starch gel (■) and water (●) followed by dynamic cumulative headspace analysis. Adapted from Juteau *et al.* (2004).

the perception of the stimulus or of the consumption of food; i.e. of around a millisecond to a minute.

In the following, the time of the observation is more than 850 seconds. Figure 15.5 represents the experimental values obtained in dynamic mode for isoamyl acetate, in a gelled matrix of corn starch, and in water (Juteau *et al.* 2004). The curves represent the release of flavour from the matrices, as a function of time. In water, the total amount of flavour added was released in 850 minutes. In the gel, 27% of the initial quantity of aroma was not released in 850 minutes; corresponding to the quantity of the compound engaged in an irreversible binding with the molecule of amylose (Arvisenet *et al.* 2002). In this work, the interaction between the macromolecule and the aroma compound and the effect of the organisation of this macromolecule on the flavour release were highlighted.

15.2.2 Involved mechanisms

The state of organisation of the matrix, caused by macromolecules, results in a more or less dense entanglement, or a three-dimensional network, established between a more or less significant number, of macromolecule chains. These two organisations, although very different from the rheological point of view, can trap either, a continuous phase, which contains the compounds of flavour. The presence of the macromolecules limits the exchange between the various phases: liquid or solid, and vapour.

From a dynamic point of view, the transfer of the compounds is conditioned by the renewal in volatile compounds at the food–air interface, itself related to the migration of volatile compounds within the product. In a viscous system, friction strength can be estimated between the macromolecules and the aroma compounds. In a gelled system, this limitation of the exchanges was characterised by a restriction on the diffusion of aroma compounds within the food matrix. The presence of a three-dimensional network can generate the reduction in the diffusivity of volatile compounds, slowing down their migration to the matrix–air interface (Voilley *et al.* 1998, Rega *et al.* 2002, Lubbers and Guichard 2003). For these physical phenomena, it is necessary to add the possibility of molecular interactions between the aroma compound and the macromolecule, which also generates a limitation in the exchanges between phases, until the trapping of part of the volatile compound in the product.

Several mathematical models were worked out to account for the release of flavour compounds from viscous, or gelled matrices. The model of Harrison and Hills (1997) integrates the effect of interaction between components, viscosity of the system and the mass transfer coefficient of aroma compound. Thus, the mass transfer coefficient is inversely proportional to the square root of the viscosity of the solution. This model applied by Bakker *et al.* (1998) significantly fitted the release of the diacetyl from solutions of gelatine in stirred condition. However, the models adapted from the Harrison's model, supposed a homogeneity of the phases and, thus a system under agitation. This condition is not obtained in the gelled system. In this case the diffusivity of aroma compounds within the matrix have often been discussed. However, difficulties were noted in obtaining diffusivity coefficients of small volatile solutes as aroma. The importance of the parameter 'diffusivity of the volatile compounds' in the establishment of a mathematical model for flavour release from gelled systems was partly called into question by Juteau *et al.* (2004). The reduction in the time of residence of volatile at the air–gel interface would allow a better explanation on the decrease in the mass transfer coefficients in the studied gel. However, the determination of the apparent kinetic orders of the release in different samples tended also to confirm the presence of a gradient of concentration between the surface layer and the bulk of the matrices.

To summarise, Table 15.3 presents a digest of knowledge on the effect of the proteins and polysaccharides, in increased concentrations, on flavour release.

15.3 Texture measurements and perceived intensity of aroma

15.3.1 Perception of texture and flavour

When a food is taken into the mouth, a large number of stimuli are generated, the majority of which interact. For a long time, the concept of perceptive interactions have been ignored by the physicochemical analysts due to the complexity of the studied systems. Moreover, model systems were used frequently, and thus, the conclusions on the relationships between a molecule or a property of the food and a sensory stimulus, were simplified artificially. The dimension of the texture in mouth is terribly complex, and concepts in the physics of materials or fluid engineering are necessary to understand what occurs during consumption of a spoon of yoghurt, for example.

Table 15.3 State of the art on the effects of proteins and polysaccharides on flavour release and flavour perception in model systems

	Conc	Structure	Rheology	Texture	Flavour release	Flavour perception	Molecular interaction	Transfer
Protein	Low[ ] $\nearrow$	'solution'	$\eta_{app} \nearrow_L$	viscous	cst $\rightarrow \searrow$	cst $\rightarrow \searrow$	+/−, +	+/−
Polyoside	Low[ ] $\nearrow$	'solution'	$\eta_{app} \nearrow_L$	viscous	cst $\rightarrow \searrow_L$	cst $\rightarrow \searrow_L$	no, +/−	+/−
	Low[ ] > c*	soft network	$G' \nearrow$	soft gel	$\searrow$	$\searrow$	no, +/−	+/−
Protein	[ ] $\nearrow$	hard network	$G' \nearrow$	hard gel	$\searrow$	$\searrow$	+	+
Polyoside	[ ] $\nearrow \gg$ c*	hard network	$G' \nearrow$	hard gel	$\searrow$	$\searrow$	no, +	+
Protein	[ ] $\nearrow$	entangled	$\eta_{app} \nearrow_H$	thickened	$\searrow$	$\searrow$	+	+
Polyoside	[ ] $\nearrow$	entangled	$\eta_{app} \nearrow_H$	thickened	$\searrow$	$\searrow$	no	+

[ ] $\nearrow$: increased concentration
c*: critical concentration from Baines and Morris (1987)
η_{app}: apparent viscosity
G': storage modulus
$\nearrow_L$: low increase; $\nearrow_H$: high increase
cst $\rightarrow \searrow$: constant until a decrease
$\searrow$: decrease
+/− possible
+ measurable

From liquids to gels, there is a wide range of rheological properties generating complex feelings in the mouth, such as stickiness, thickness, smoothness, slipperiness. Nevertheless, there is still much work to be done to connect these sensory attributes of texture with some rheological parameters measurable on the product. Moreover, destructuration phenomena take place on a sample of food during its consumption. The texture of the product can change from solid to liquid during swallowing. Many authors showed that the texture perceived by the panellist could influence the sensory response to an attribute of flavour (Pangborn and Szczesniak 1974, Baines and Morris 1987, Guinard and Marty 1995, Lundgren *et al.* 1986, Chai *et al.* 1991, Jaime *et al.* 1993, Yven *et al.* 1998, Hollowood *et al.* 2002, Weel *et al.* 2002, Cook *et al.* 2003, Paci Kora *et al.* 2004). Amongst other things, Mälkki *et al.* (1993) showed that the perception of aroma compounds in oat gum, guar gum and CMC gel of the same viscosity, was different as a function of polysaccharide used. Paci Kora *et al.* (2004) obtained a significant decreasing effect of thickening agents on apple sensory scores associated with hexanal in stirred yoghurts. However, these authors concluded that the sensory effect of thickening agents may be due to sensory interactions between perceptions rather than physicochemical interactions.

On the other hand, depending on the structure of the foodstuff, the time for food remaining in the mouth can vary significantly. For example, orange slices are chewed for a distinct period of time while orange juice stays in contact with the oral cavity just for a few seconds. Complete equilibration of the flavours between the food material and the air of the mouth during chewing can not be achieved immediately after the food is introduced into the mouth (Buettner and Schieberle 2000). Consequently, the composition of the flavours released from the foodstuff changes significantly during mastication not only resulting in higher amounts of the flavours with longer chewing but also in a different flavour profile. Mastication of model solutions containing hexanal, octanal and decanal, respectively, revealed not only lower amounts of the three aldehydes being retained in the mouth after mastication of short duration. The flavour profile changed significantly with increasing duration of mastication resulting in the relative number of the longer-chain aldehydes increasing with longer mastication times (Fig. 15.6, adapted from Buettner and Schieberle 2000).

During mastication the texture of the food material changes more and more to liquid due to salivation and mastication, resulting in a reduced possibility of odorant transfer to the nasal cavity (Buettner *et al.* 2001). When swallowing liquid material, retronasal aroma perception will more or less be reduced to one main aroma flash associated with the swallowing itself (Buettner *et al.* 2001, 2002, Hansson *et al.* 2003). When masticating solid foods, there is a series of retronasal aroma perceptions, mainly related to the swallowing of small portions of the foodstuff.

In summary, it was shown that, once food has been introduced into the oral cavity, no continuous aroma release into the nasal cavity is possible and distinct physiological actions such as swallowing are necessary to allow or increase

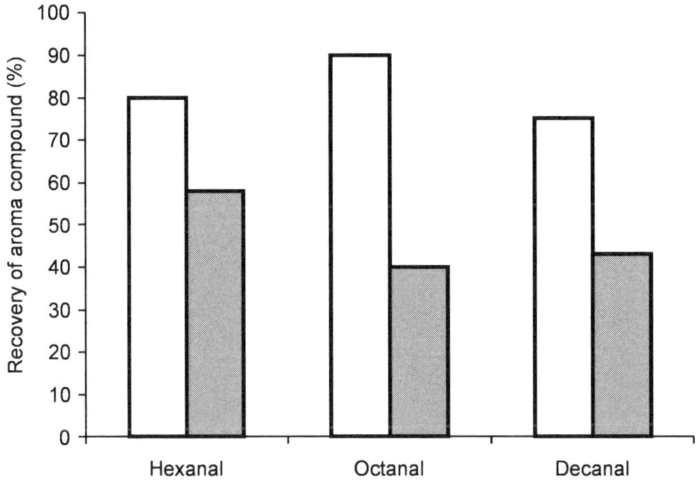

Fig. 15.6 Influence of the duration of mastication. 5 s (white bar) and 1 min (grey bar) on the remaining quantities of aldehydes in spat-off aqueous solutions (initial concentration 100 μg/l water). Adapted from Buettner and Schieberle (2000).

aroma perception (Buettner *et al.* 2002, Hansson *et al.* 2003, Buettner and Montserrat 2004). Thus, the temporal dimension of aroma perception should be carefully taken into account.

15.3.2 Psychophysical interactions versus texture impact on flavour perception

Taking into account the importance of the skew from the possible perceptive interactions, the experimenters developed techniques allowing them to analyse directly in the mouth or nose of a subject, during eating, the quantity of aroma compounds released from food (Overbosch 1987, Soeting and Heridema 1988, Taylor 1996, Taylor *et al.* 2000). This quantity of aroma compounds is supposed to be proportional to the intensity of the aromatic note perceived by the subject. Thus, the techniques of 'nosespace' or 'mouthspace' were used. Sensory measurements in time–intensity carried out in parallel, made it possible to compare the flavour perception during time and the quantity of aroma compounds analysed in the nasal cavity.

The phenomena connecting the differences in the perceived texture and the intensity of the aromatic note seem obvious, when very different products are compared. Thus the nosespace concentration measured in the nasal cavity of a subject, eating a viscous pectin solution, is higher than in the case of the gel, obtained with the same polysaccharide (Fig. 15.7, adapted from Hansson *et al.* (2003)). The sensory perception of the flavour in the two products is in agreement with the nosespace result. The subjects perceived a less significant intensity of flavour in the gel than in the viscous solution. The mechanisms exposed above were consistent with explanations of the present phenomena.

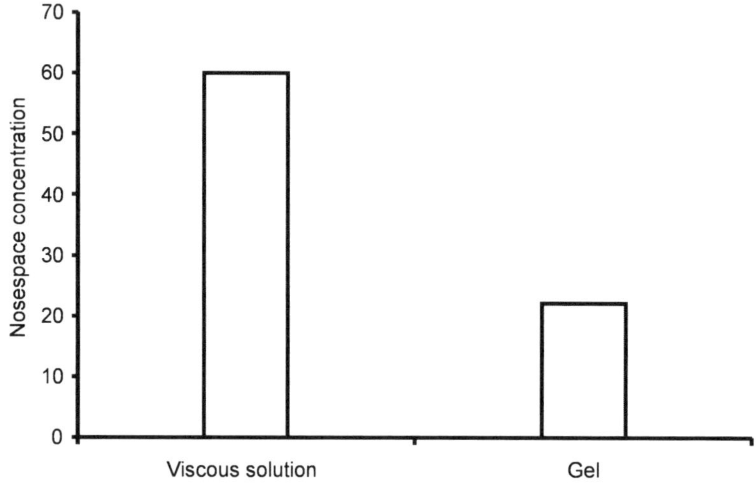

Fig. 15.7 Maximal nosespace concentration for 2-butanoate after swallowing when the assessor ate a viscous solution and a gel. Adapted from Hansson *et al.* (2003).

The recent data in literature on this phenomenon agree that the difference in the flavour perception in matrices with variable textures was not always explained by a difference in the amounts of aroma compounds released. Several authors showed that the relationship between the intensity of flavour perception and the nosespace concentration in aroma compounds was not correlated (Baek *et al.* 1999, Linforth *et al.* 1999, Weel *et al.* 2002, Hollowood *et al.* 2002, Cook *et al.* 2003, Paci Kora *et al.* 2004).

Thus, Baek *et al.* (1999) did not highlight a relation between the maximum intensity of flavour perception and the maximum concentration in furfuryl acetate from nosespace measurement, but they showed that the differences in sensory perception can be explained by different flavour release velocities, as a function of matrices.

Results from time intensity measurements and the nosespace analysis on protein gels were compared (Weel *et al.* 2002). The panellists perceived a reduction in the intensity of the flavour with an increase in the firmness of the gel (Fig. 15.8). However, measurements in concentrations of the aroma compounds in the nasal cavity showed that there was no difference between the samples. Cook *et al.* (2003) confirmed the reducing effect of the thickener concentration beyond C* on the perception of the banana note. However, the concentration of isoamyl acetate, the aroma compound corresponding to the banana odour, in the nasal cavity during the consumption of the solutions, appeared relatively independent of the nature and the concentration of thickeners. On the other hand, these authors observed a significant correlation between olfactive perception and the oral shear stress.

Weel *et al.* (2002), Cook *et al.* (2003) and Paci Kora *et al.* (2004) supposed that psychophysical interactions (texture in mouth and flavour) formed the basis

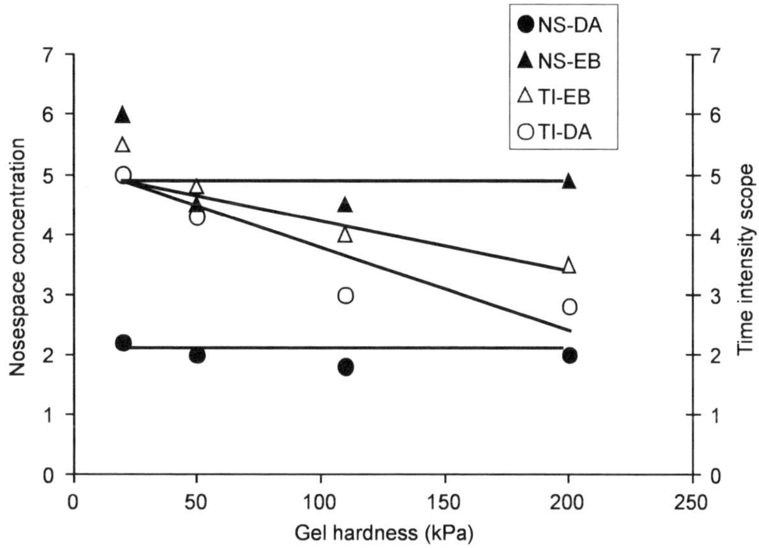

Fig. 15.8 Relation between gel hardness and overall average values of Cmax of nosespace concentration (●,▲) and time intensity (△,○) for diacetyl (DA) and ethyl butyrate (EB). Adapted from Weel *et al.* (2002).

of their results. The influences of texture on flavour perception could be explained by integration of the signals from various senses that reach the brain simultaneously during the eating of a gel.

It should nevertheless be highlighted that in all the publications, the results were treated in terms of average curves of release and perception, from all the individuals. Thereby, it can suppose that all absence of effect could be concluded, whereas an individual analysis of data from each panellist could be profitable. Indeed, all the authors have agreed on the significant variation between panellists, probably due to anatomical differences. It was shown that specific chewing patterns play a major role on flavour release patterns (Lethuaut *et al.* 2004, Pionnier *et al.* 2004, Buettner and Montserrat 2004). Based on recent physiological observations obtained during food consumption (Buettner and Montserrat 2004), new approaches to interpreting time–intensity and nosespace data will be proposed by the experts of this field. Thus we can hope clearer conclusions on the respective impact of physicochemical (texture) and the psychophysical (cognitive interaction) factors on the release of flavour from foodstuffs will be forthcoming.

15.4 References

ARVISENET, G., LE BAIL, P., VOILLEY, A. and CAYOT, N. (2002) Influence of physicochemical interactions between amylose and aroma compounds on the retention of aroma in food-like matrices, *J Agric Food Chem*, 50, 7088–7093.

BAEK, I., LINFORTH, R. S. T., BLAKE, A. and TAYLOR, A. J. (1999) Sensory perception is related to the rate of change of volatile concentration in-nose during eating of model gels, *Chem Senses*, 24, 155–160.

BAINES, Z. V. and MORRIS, E. R. (1987) Flavour/taste perception in thickened systems: the effect of guar gum above and below c*, *Food Hydrocolloid*, 1, 197–205.

BAKKER, J., BOUDAUD, N. and HARRISON, M. (1998) Dynamic release of diacetyl from liquid gelatin in the headspace, *J Agric Food Chem*, 46, 2714–2720.

BOLAND, A. B., DELAHUNTY, C. M. and VAN RUTH, S. M. (2005) Influence of the texture of gelatin gels and pectin gels on strawberry flavour release and perception, *Food Chem*. In press.

BRENNAN, E., SESTER, C. and SCHMIDT, K. (2002) Yogurt thickness: effects on flavour perception and liking. *J Food Sci*, 67, 2785–2789.

BUETTNER, A., BEER, A., HANNIG, C. and SETTLES, M. (2001) Observation of the Swallowing Process by Application of Videofluoroscopy and Real-time Magnetic Resonance Imaging – Consequences for Retronasal Aroma Stimulation, *Chem. Senses*, 26, 1211–1219.

BUETTNER, A., BEER, A., HANNIG, C., SETTLES, M. and SCHIEBERLE, P. (2002) Physiological and analytical studies on flavor perception dynamics as induced by the eating and swallowing process, *Food Qual Prefer*, 13, 497–504.

BUETTNER, A. and MONTSERRAT, M. (2004) In *State of the art in flavour chemistry and biology* (Hoffman, T., Rothe, M. and Schieberle, P., eds) Deutsche Forschungsanstalt für Lebensmittelchemie, Garching, 3–11.

BUETTNER, A. and SCHIEBERLE, P. (2000) In *Flavor Release* (Roberts, D. D. and Taylor, A. J., eds) American Chemical Society, Washington, 87–98.

BRENNAN, E., SESTER, C. and SCHMIDT, K. (2002) Yogurt thickness: effects on flavor perception and liking, *J Food Sci*, 67, 2785–2789.

CAYOT, N., TAISANT, C. and VOILLEY, A. (1998) Release and perception of isoamyl acetate from a starch-based food matrix, *J Agric Food Chem*, 46, 3201–3206.

CHAI, E., OAKENFULL, D. G., MCBRIDE, R. L. and LANE, A. G. (1991) Sensory perception and rheology of flavoured gels, *Food Aust*, 43, 256–261.

COOK, D. J., LINFORTH, R. S. T. and TAYLOR, A. J. (2003) Effects of hydrocolloid thickeners on the perception of savory flavors, *J Agric Food Chem*, 51, 3067–3072.

DECOURCELLE, N., LUBBERS, S., VALLET, N., RONDEAU, P. and GUICHARD, E. (2004) Effect of thickeners and sweeteners on the release of blended aroma compounds in fat-free stirred yoghurt during shear conditions, *Int Dairy J*, 14, 783–789.

GUICHARD, E., ISSANCHOU, S., DESCOURVIERES, A. and ETIÉVANT, P. (1991) Pectin concentration, molecular weight and degree of esterification: influence on volatile composition and sensory characteristics of strawberry jam, *J Food Sci*, 56, 1621–1627.

GUINARD, J. X. and MARTY, C. (1995) Time–intensity measurement of flavor release from a model gel system: effect of gelling agent type and concentration, *J Food Sci*, 60, 727–730.

HANSSON, A., GIANNOULI, P. and VAN RUTH, S. (2003) The influence of gel strength on aroma release from pectin gels in a model mouth and in vivo, monitored with proton-transfer-reaction mass spectrometry, *J Agric Food Chem*, 51, 4732–4740.

HARRISON, M. and HILLS, B. P. (1997) Mathematical model of flavor release from liquids containing aroma-binding macromolecules, *J Agric Food Chem*, 45, 1883–1890.

HOLLOWOOD, T. A., LINFORTH, R. S. T. and TAYLOR, A. J. (2002) The effect of viscosity on the perception of flavour, *Chem Senses*, 27, 583–591.

JAIME, I., MELA, D. J. and BRATCHELL, N. (1993) A study of texture flavor interactions using free choice profiling, *J Sens Stud*, 8, 177–188.

JUTEAU, A., CAYOT, N., CHABANET, C., DOUBLIER, J. L. and GUICHARD, E. (2004) Flavour release from polysaccharide gels: different approaches for the determination of kinetic parameters, *Trends Food Sci Technol*, 15, 394–402.

KÄLVIAINEN, N., ROININEN K. and TUORILA, H. (2000) Sensory characterization of texture and flavor of high viscosity gels made with different thickeners, *J Sens Stud*, 31, 407–419.

LETHUAUT, L., WEEL, K. G. C., BOELRIJK, A. E. M. and BROSSARD, C. D. (2004) Flavor perception and aroma release from model dairy desserts, *J Agric Food Chem*, 52, 3478–3485.

LINFORTH, R. S. T., BAEK, I. and TAYLOR, A. J. (1999) Simultaneous instrumental and sensory analysis of volatile release from gelatine and pectine/gelatine gels, *Food Chem*, 65, 77–83.

LUBBERS, S., DECOURCELLE, N., VALLET, N. and GUICHARD, E. (2004) Flavor release and rheology behavior of strawberry fat free stirred yogurt during storage, *J Agric Food Chem*, 52, 3077–3082.

LUBBERS, S. and GUICHARD, E. (2003) The effects of sugars and pectin on flavour release from a fruit pastille model system, *Food Chem*, 80.

LUNDGREN, B., PANGBORN, R. M., DAGET, N., YOSIDA, M., LAING, D. G., McBRIDE, R. L., GRIFFITHS, N., HYVOENEN, L., SAUVAGEOT, F., PAULUS, K. and BARYLKO PIKIELNA, N. (1986) An interlaboratory study of firmness, aroma, and taste of pectin gels, *Lebensm-Wiss Technol*, 19, 66–76.

MÄLKKI, Y., HEINÏO, R. L. and AUTIO, K. (1993) Influence of oat gum, guar gum and carboxymethyl cellulose on the perception of sweetness and flavour, *Food Hydrocolloid*, 6, 525–532.

OVERBOSCH, P. (1987) Flavour release and perception, in *Flavour Science and Technology* (Martens, M., Dalen, G. A. and Russwurm, H., eds) John Wiley & Sons, Chichester, 291–300.

PACI KORA, E., SOUCHON, I., LATRILLE, E., MARTIN, N. and MARIN, M. (2004) Composition rather than viscosity modifies the aroma compound retention of flavored stirred yoghurt, *J Agric Food Chem*, 52, 3048–3056.

PANGBORN, R. M., MISAGHI GIBBS, Z. and TASSAN, C. (1978) Effect of hydrocolloids on apparent viscosity and sensory properties of selected beverages, *J Texture Stu*, 9, 416–436.

PANGBORN, R. M. and SZCZESNIAK, A. S. (1974) Effect of hydrocolloids and viscosity on flavor and odor intensities of aromatic flavor compounds, *J Texture Stu*, 4, 467–482.

PIONNIER, E., CHABANET, C., MIOCHE, L., LE QUÉRÉ, J. L. and SALLES, C. (2004) *In Vivo* Aroma Release during Eating of a Model Cheese: Relationships with Oral Parameters, *J Agric Food Chem*, 52, 557–564.

REGA, B., GUICHARD, E. and VOILLEY, A. (2002) Flavour release from pectin gels: effects of texture, molecular interactions and aroma compounds diffusion, *Sci Aliment*, 22, 235–248.

ROBERTS, D. D., ELMORE, J. S., LANGLEY, K. R. and BAKKER, J. (1996) Effects of sucrose, guar gum, and carboxymethylcellulose on the release of volatile flavor compounds under dynamic conditions, *J Agric Food Chem*, 44, 1321–1326.

SECOUARD, S., MALHIAC, C., GRISEL, M. and DECROIX, B. (2003) Release of limonene from polysaccharide matrices: viscosity and synergy effect, *Food Chem*, 82, 227–234.

SEUVRE, A. M., DIAZ, M., CAYOT, P. and VOILLEY, A. (2004) Influence of the composition and the structure of different media on the release of aroma compounds, *Lait*, 84, 305–316.

SOETING, W. and HERIDEMA, J. (1988) A mass spectrometric method for measuring flavour concentration/time profiles in human breath, *Chem Senses*, 13, 607–617.

TAYLOR, A. J. (1996) Volatile flavor release from foods during eating, *Crit Rev Food Sci Nut*, 36, 765–784.

TAYLOR, A. J., LINFORTH, R. S. T., HARVEY, B. A. and BLAKE, A. (2000) Atmospheric pressure chemical ionisation mass spectrometry for *in vivo* analysis of volatile flavour release, *Food Chem*, 71, 327–338.

VOILLEY, A., FOUSSE, C. and MARTIN POLO, M. (1998) In *Interactions of food matrix with small ligands influencing flavour and texture*, Vol. 5 (Escher, F. and Nüssli, J., eds) European Communities, Luxembourg, Zurich, Switzerland, 69–71.

WALKER, S. and PRESCOTT, J. (2000) The influence of solution viscosity and different viscofying agents on apple juice flavor, *J Sens Stud*, 15, 285–307.

WEEL, K. G. C., BOELRIJK, A. E. M., ALTING, A. C., VANMIL, P. J. J. M., BURGER, J. J., GRUPPEN, H., VORAGEN, A. G. J. and SMIT, G. (2002) Flavor release and perception of flavored whey protein gels: Perception is determined by texture rather than by release, *J Agric Food Chem*, 50, 5149–5155.

WENDIN, K., SOLHEIM, R., ALIMERE, T. and JOHANSSON, L. (1997) Flavour and texture in sourmilk affected by thickeners and fat content, *Food Qual Prefer*, 8, 281–291.

YVEN, C., GUICHARD, E., GIBOREAU, A. and ROBERTS, D. D. (1998) Assessment of interactions between hydrocolloids and flavor compounds by sensory, headspace, and binding methodologies, *J Agric Food Chem*, 46, 1510–1514.

16

Odour–taste interactions in flavour perception

C. Salles, INRA, France

16.1 Introduction

Flavour perception is essentially a multi-modal phenomenon within which aroma and taste are considered to play a prominent role. The perception of flavour thus involves integration of the separate sensory modalities and is influenced by their interactions. These interactions are multi-factorial in nature. Physicochemical interactions, including interactions with the food matrix, physiological (peripheral) interactions, and cognitive and psychological aspects thus contribute to the complexity of the flavour perception. Within each modality, many contributing factors and their intramodal interactions, such as enhancing or masking effects, have, moreover, been described.

The perception of flavour is a dynamic phenomenon because the stimuli changes continuously as food is eaten. Measurements of the changes in stimuli with time *in vivo* are essential to the understanding of the relationships between stimuli and perception, but these add some complexity. In the perception of odours, there is a highly integrative interaction between olfaction and other sensory modalities, such as taste or tactile sensations. Emotional and affective aspects of sensory modalities, and even mood, have also to be considered. For instance, affective aspects of tactile stimuli influence the other sensory modalities such as taste and smell. However, if the overall perception of flavour is considered as an integration of simultaneous perceptions of sensory modalities that include not only taste and smell but also chemesthesis and 'mouthfeel' sensations, the word 'flavour' cannot be understood without considering first taste and smell, and their interactions. Thus it has been recognised that some tastes can increase the perceived intensity of aromas and, conversely, that the perceived intensity of tastes may be increased by odours, especially when they are associated within congruent mixtures (such as classical sweetness and fruitiness). For some years, sensory scientists have studied these interactions between odour and taste to learn more on their mechanisms themselves and on the locations where these interactions take place and are processed. These studies will help in understanding not only taste and smell perceptions, but also more generally the processing of information that occurs in the brain when an overall flavour is perceived.

Odour and taste interactions, like all the other sensory interactions, result from physicochemical, physiological and psychological effects, where cognitive aspects are certainly of prime importance. After reviewing the state of the art in the studies on odour–taste interactions, this chapter will focus on the possible origins of these interactions at these different levels, to end with perspectives of applications of the current research.

16.2 Odour–taste interactions

Several studies reported relationships between odour and taste perception. Most of them were conducted using model and simple solutions. We attempted to report the main previous works showing odour–taste interaction in Table 16.1. It is generally difficult to make comparison between different approaches or to draw general and clear conclusions because different sensory methods and different data treatments were used. According to Table 16.1, we can observe that although most studies were carried out on the influence of taste on aroma, some studies also report on the influence of aroma on taste. The most studied interaction is sweetness–aroma. Murphy and Cain (1980) reported an increase of sweetness of a sucrose solution when a citral solution was smelt simultaneously. Cliff and Noble (1990) noted the same effect with a glucose solution and a peach aroma, but including a study of the evolution of the perception with time during the solution intake, showed the influence of the concentration of implied taste and aroma components on the importance of the observed interaction. The increase of sweetness was also observed for a sucrose solution (Frank *et al.* 1989) and for yoghurt containing soybean proteins (Drake *et al.* 2001) flavoured with strawberry aroma. The condition to observe consistently and clearly such interaction is that aroma and taste association are congruent. For example, Frank and Byram (1988) observed an increase of sweetness by adding strawberry aroma to beaten cream. However, these authors suggested that the effect may vary with the taste–aroma association as they did not observe this phenomenon with peanut butter aroma, since strawberry aroma was not able to enhance saltiness (Fig. 16.1).

Stevenson *et al.* (1999) investigated the effects of twenty aromatic notes on sweetness and sourness of sucrose and citric acid solutions. Among aromatic notes, they made a distinction between food-like and not food-like odours. For food-like odours, they found that caramel, lychee, strawberry and maracuja

Table 16.1 Examples of taste–aroma interactions

Product	Taste	Aroma	Effect	Reference
MS	sweet	citrus	AT+	Murphy and Cain (1980)
MS	sweet	peach	AT+	Cliff and Noble (1990)
MS	sweet	strawberry	AT+	Franck *et al.* (1989)
Favoured cream	sweet	strawberry	AT+	Franck and Byram (1988)
Flavoured yaourt (+soybean proteins)	sweet	strawberry	AT+	Drake *et al.* (2001)
MS	sweet	caramel	AT+	Stevenson *et al.* (1999)
MS	sweet	strawberry	AT+	Stevenson *et al.* (1999)
MS	sweet	maracuja	AT+	Stevenson *et al.* (1999)
MS	sweet	lychee	AT+	Stevenson *et al.* (1999)
MS	sweet	maltol	AT–	Stevenson *et al.* (1999)
MS	sweet	damascone	AT–	Stevenson *et al.* (1999)
MS	sweet	angelique	AT–	Stevenson *et al.* (1999)
MS	sour	caramel	AT–	Stevenson *et al.* (1999)
MS	sweet	orange	TA+	Bonnans and Noble (1993)
MS	sour	orange	TA+	Bonnans and Noble (1993)
MS	sweet	citrus	TA+	Kuo *et al.* (1993)
MS	sour	citrus	TA+	Kuo *et al.* (1993)
MS	sweet	vanille	TA+	Kuo *et al.* (1993)
MS	sour	vanille	TA–	Kuo *et al.* (1993)
MS	sweet	fruity	TA+	Stampanoni (1993)
MS	sour	fruity	TA+	Stampanoni (1993)
Chewing-gum	sweet	mint	TA+	Davidson *et al.* (1999)
Hydrocolloid sweeteners	salty	garlic	AT+	Cook *et al.* (2003)
Model cheese	salty	cheese	AT+	Pionnier *et al.* (2004c)
Olive oil	bitter	cut grass	AT+	Caporale *et al.* (2004)
MS	salty	Soy sauce	AT+	Djordjevic *et al.* (2004)

MS: model solution; AT: effect of aroma on taste; TA: effect of taste on aroma; + enhancing effect; – masking effect

significantly enhanced the sweetness of sucrose when compared with pure sucrose whilst maltol suppressed it and mango had no significant effect. For food-unlike odours, they found that damascone, angelica oil, and cedryl acetate suppressed sweetness ratings when compared with pure sucrose whilst acetyl methyl carbamol and eucalyptol showed no significant effect on sweetness ratings. So, in some cases, these interactions are dependant on the concentration of stimuli but the mechanism seems complex. Additional studies should be conducted to clarify these effects.

Additionally, Stevenson *et al.* (1999) showed that the sourness of a citric acid solution could be lowered by caramel aroma. However, the authors also observed an increase of sweetness which could be at the origin of the observed decrease in sourness.

In connection with lychee, enhancement was found significant at the lowest concentration tested but not at the highest. For water chestnut, no significant

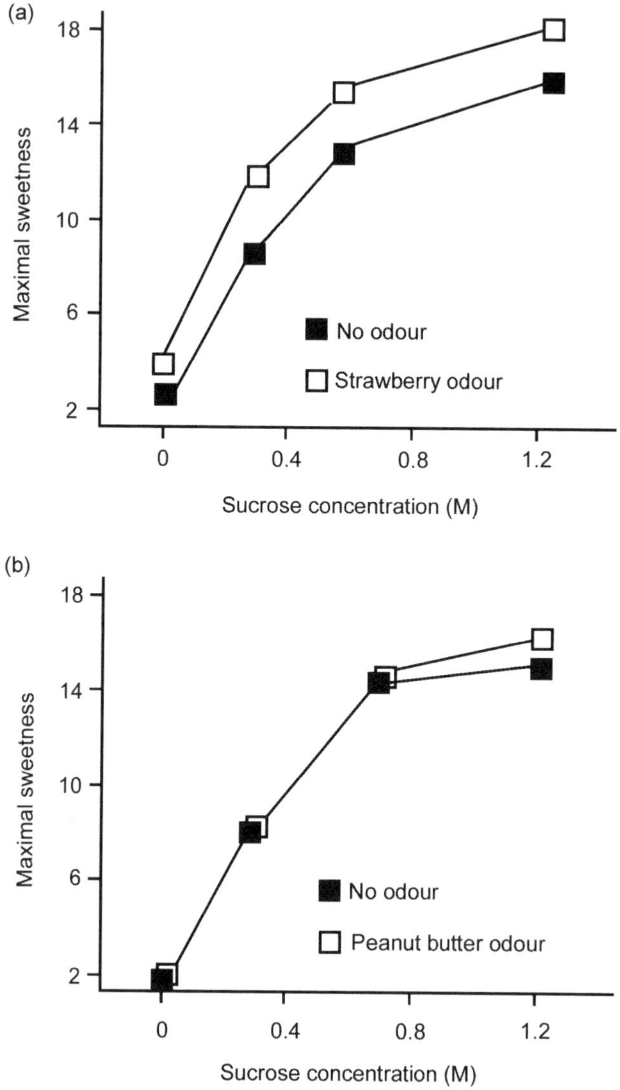

Fig. 16.1 Mean maximum sweetness ratings for sucrose and strawberry odour (a), sucrose and peanut butter odour (b) (Franck and Byram 1988, with the permission of Oxford University Press).

difference was observed between the different aroma concentrations for any effect on sweetness. Effect of concentration was also studied for peach (Cliff and Noble 1990) and strawberry (Schifferstein and Verlegh 1996) with an increase of sweetness enhancement when aroma concentration increased. Few other studies reported observation of possible interaction between all other basic tastes and aroma perception. In tomato, positive correlation between sweetness and cis-3-hexenal, trans-2-hexenal, hexanal, cis-3-hexenol, geranylacetone, 2+3-methylbutanol, trans-2-heptenal, 6-methyl-5-hepten-2-one and 1-nitro-2-phenylethane, and between sourness and acetaldehyde, acetone, 2-isobutyl-thiazole, geranylacetone, beta-ionone, ethanol, hexanal and cis-3-hexenal were reported by Baldwin *et al.* (1998). They hypothesised probable taste–aroma interaction and particularly the associations of fruity notes with sweetness and green note with sourness.

Observations were made also for the influence of taste on aroma perception (Table 16.1). Bonnans and Noble (1993) reported an increase of the fruitiness of a water solution flavoured with orange when sourness and sweetness were perceived. On the contrary, Cliff and Noble (1990) did not report such an observation of the effect of sweetness of a glucose solution on the fruitiness of a water solution flavoured with peach. Retronasally, citric acid and sucrose were able to contribute to aroma intensity of a citral solution (Kuo *et al.* 1993) for a part of the panel, the other part being not able to detect any effect. Additionally, these authors showed an exhausting effect of sucrose on the intensity and persistence of a vanilla note and a suppressive effect of citric acid and sodium chloride on a vanilla note. Stampanoni (1993) studied the influence of the concentration of sucrose and citric acid on the flavour of drinks and sherbets containing orange and lemon aromas. She reported a positive effect of citric acid and sucrose on all the rated descriptors: fruitiness, freshness, juiciness and global impact but not on a peely note. In the case of orange sherbets, sourness is correlated with freshness and peeliness while in the case of lemon sherbets, sourness is correlated with freshness and juiciness. So, the effects of sugar and acid are flavour-dependant for peeliness, juiciness and freshness. These results could also suggest taste-odour interactions due to physico-chemical as well as psychological effects. Davidson *et al.* (1999) reported temporal analyses of flavour release in the retronasal cavity, taste compound present in saliva and aroma perception during chewing gum. They showed clearly that the intensity of mint aroma perception is, surprisingly, related to the release kinetic of saccharose and not to the kinetic release of menthone. This experiment showed the existence of interaction between sweetness and mint aroma due to cognitive effects, as physicochemical effects were not significant in this system.

To point out such cross-modal interactions from a model or natural mixture, the most frequent, simplest and more common technique used is to stop the retronasal airway flow with nose-pinches or to generate an air counter-current in the nostril, flowing back aroma compounds in the mouth. The comparison of results between free and blocked retronasal airflow may supply evidence for the interaction of aroma perception with taste perception. In the same way, to show the interaction of taste perception with aroma perception, it should be necessary to mask taste receptors located on the tongue which seems rather difficult and to avoid contact between saliva and the taste stimulus. In an experiment aiming to show the interaction of sweetness on strawberry aroma, Delarue (2002) used a taste inhibitor, commonly called 'lactisole' (2-(4-methoxyphenoxy)propanoic acid), which reversibly inhibits sweet perception (Johnson *et al.* 1994). In some case, the delivery of taste and aroma compounds changes with time. The temporal delivery of flavour may therefore be important and investigation was made to find a technique to deliver the components to panellists in controlled conditions. Cook *et al.* (2003) and Hort and Hollowood (2004) developed a technique called 'dynataste' or multichannel flavour delivery system, enabling the simultaneous monitoring of the temporal delivery of proximal taste and the aroma stimuli (Taylor and Hort 2004).

16.3 Origin of odour–taste interactions

Odour and taste interactions may originate from physicochemical, physiological or psychological effects (Fig. 16.2).

16.3.1 Physicochemical interactions

'Salting out' effects

Non-volatile compounds, which can be responsible for taste, can play directly or indirectly a role in the volatility of aroma compounds. The effect of these non-volatile compounds on the aroma compounds in the vapour phase has been extensively studied (Nelson and Hoff 1968, Voilley *et al.* 1977, Dubois *et al.* 1995). Salt and acid were found to increase the headspace concentration of polar volatile compounds (Nawar 1971). Jennings (1965) showed that radioactivity due to the quantity of C^{14}-labelled ethyl acetate increased in headspace in accordance with sodium chloride concentration. Schinneler *et al.* (1972) observed an increase of headspace concentration of octanol in presence of monophosphate inositol in the liquid phase. Voilley *et al.* (1977) reported three types of behaviour of model volatiles with sucrose, calcium chloride and citric acid, three non-volatile solutes, in aqueous solutions. The activity coefficients increased, decreased or remained the same. In a general manner, the volatility of aroma compounds increases with the presence of salts in the media. This 'salting-out' phenomenon is due to the mobilisation of solvation water molecules by salt molecules, leading to the repulsion of volatile molecules. Therefore, their volatility increase compared to a media without salt. The importance of volatile compound release depends on its nature and on the nature and salt concentration in the media (Dubois *et al.* 1995).

Apart from salts, several organic small molecules are found to be able to modify volatility of volatile compounds. Intense sweeteners such as aspartame and neohesperidine dihydrochalcone interact with volatile compounds and modify the intensity of flavour attributes (Lindley *et al.* 1993, Nahon *et al.* 1998). This phenomenon can lead to a selective release of these aroma compounds during consumption and it changes the flavour quality of a drink for example.

As already described for salts, mono- and disaccharides also affect the volatility by altering the activity coefficients of volatile compounds (Land 1978).

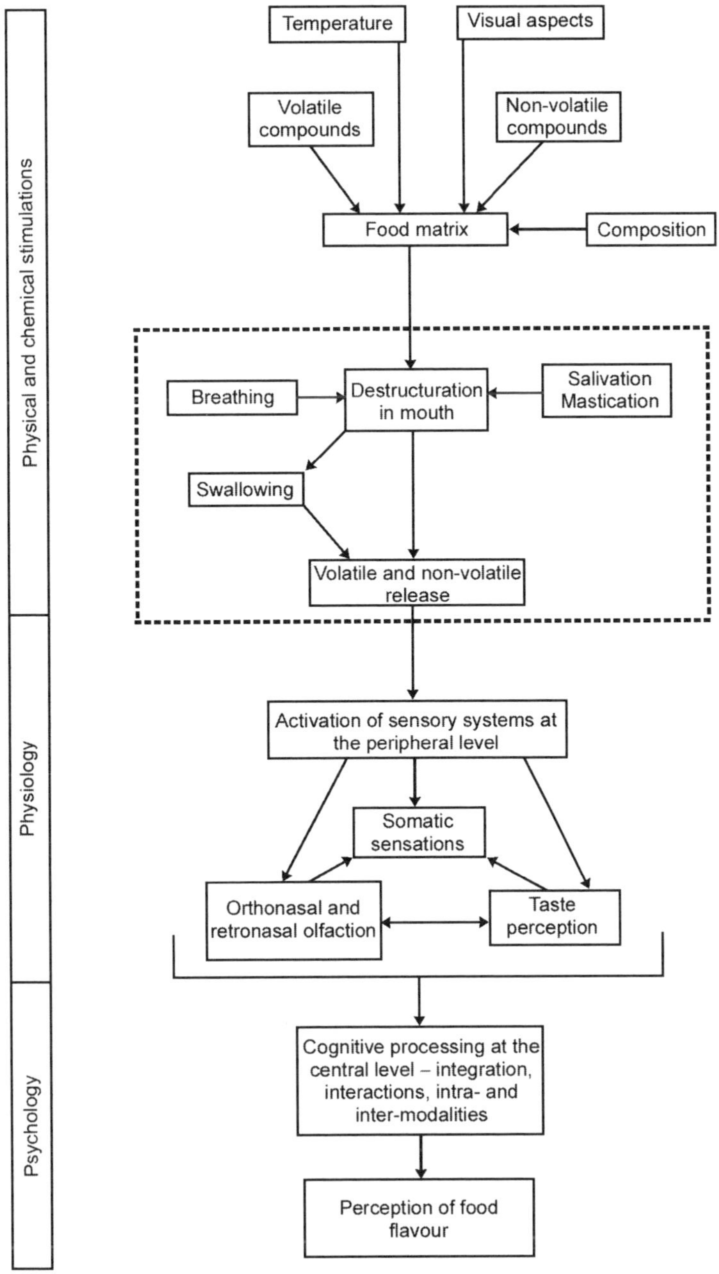

Fig. 16.2 Main factors influencing flavour perception.

At a sufficient concentration, they lower the amount of bulk water by structuring water, increasing their effective concentration and consequently enhancing their volatility (Nawar 1971). It was also shown that chemical reactions may occur between sweeteners and some volatile compounds (Hussein *et al.* 1984). For example, Le Quéré *et al.* (1994) found a decrease in some aldehydes concentration in diet orange soft drinks containing aspartame and a formation of new volatile compounds in diet orange drinks containing cyclamate. Larger carbohydrates such as polysaccharides often contribute to the increase in viscosity of a beverage, which influences the diffusion of small molecules.

Dufour and Bayonove (1999) reported interaction between wine polyphenols and aroma compounds. In a hydroalcoholic solution, isoamyl acetate, ethyl hexanoate, and benzaldehyde appeared to be more retained than limonene at low catechin concentrations (0–5 g/l). The tannin fraction induced a slight decrease of benzaldehyde volatility, a salting-out effect on limonene and had no effect on the two esters. On the basis of investigations at the molecular level, they hypothesised a hydrophobic driving force between catechins and benzaldahyde. Concerning ethanol, Conner *et al.* (1998) reported a decrease in activity coefficients of ethyl esters above 17% ethanol, corresponding to a progressive aggregation of alcohol molecules reducing the hydrophobic hydration of the alkyl chains.

A more systematic study on these effects was reported on a complex model food product, considering all the small molecules identified in the product. Delahunty and Piggott (1995) and Solms (1986) suggested that interaction between small water-soluble molecules and aroma compounds may occur in cheese. Studying a model solution representing a water extract of Camembert cheese, Pionnier *et al.* (2002) reported, beside classic salting-out effects mainly due to the presence of sodium chloride, the effect of some organic compounds. These components such as peptides had an effect on the retention or release of volatile compounds which can either decrease or increase headspace concentration of aroma. In particular, the presence of water-soluble extract compounds increase the headspace concentration of 2-heptanone, 1-octen-3-ol and 3-methylbutanol. For other volatiles, no effect was globally observed but omission tests made on the model water solution containing soluble fraction components showed that mineral salts lead to a release of 2-nonanol while peptides lead to its retention in the solution.

Aroma–matrices interactions (proteins, lipids, polysaccharides)

The chemical composition of the matrix, and consequently its structure, influences release and perception of flavour. The main components of food matrix are primarily lipids, proteins, carbohydrate and water. In food, lipids are mainly triglycerides that increase solubility of aroma substances (Allaneau 1979). The volatility of most compounds is lower in lipids than in water (Le Thanh *et al.* 1992) and, consequently, flavour threshold concentrations determined in oil are generally higher than in water (Jo and Ahn 1999). A decrease of the concentration of volatile compounds in the vapour phase was observed when the quantity of lipids increased in the food matrix. For example, Lubbers *et al.* (1994) showed that the retention of 28% of beta-ionone on yeast walls, in a model wine, was due to endogenous lipids. The quantity of sorbed compounds depends on the length of triglyceride fatty acid chains and retention is more important for a higher unsaturation degree of lipids (Maier 1975). The addition of a small amount of fat in an aqueous medium results in a partition equilibrium between the two in favour of the lipid phase due to its hydrophobicity (Bakker 1995). Changes in fat affect significantly the flavour release, according to volatile compound lipophilicity and lipid type (Roberts *et al.* 2003). Brauss *et al.* (1999) showed that low-fat yoghurts released volatiles more quickly and at a higher intensity but with less persistence than yoghurts containing higher fat proportion. More information about lipid–flavour and emulsion–flavour interactions can be found in Chapters 7 and 8 written by Ollivon and Dumont, respectively.

As macromolecules, proteins and carbohydrates are well known to interact with aroma compounds. These aspects will not be developed in this chapter as Chapters 9 and 10 written by Tromelin *et al.* and Delarue and Giampaoli, respectively, are mainly dedicated to protein–flavour and carbohydrate–flavour interactions. The existence of interactions between aroma compounds and proteins or carbohydrates were mainly studied in rather simple model systems. However, when fat is added in the system, the presence of proteins or saccharides no longer influences flavour release (Roberts and Pollien 2000).

Flavour release in mouth conditions

Flavour release in the mouth is affected by several parameters such as the structure of the food matrix, which is indirectly responsible for texture, and oral parameters: efficiency, duration and strength of chewing, mastication rate, salivary flow and composition, swallowing rate, temperature of the food when placed in the oral cavity. These factors are determinant for the food concentration of taste and aroma compounds released in the mouth during the chewing process. These compounds then reach odour receptors at the same time to produce flavour perception, including perceptual interaction effects.

Saliva is a complex aqueous medium containing several inorganic salts and organic compounds such as organic acids, sugars, glycoproteins (mucin), alpha-amylase, several other enzymes, antibodies .... The functions of saliva are multiple: dilution, hydration of food, lubrication of the oral mucus. Salivary components can influence volatile partitioning from solutions. Saliva components may interact with flavour components of food. Friel and Taylor (2001) showed that volatile component partition between aqueous and gaseous phases is unequally affected by artificial saliva according to their physical and chemical properties. They noted three types of behaviour; compounds not affected by the presence of mucin, compounds showing a decrease of partition in presence of mucin and compounds showing also a decrease with mucin that was modulated by the presence of salivary salts and sugars. In this last case, some competitions of binding were observed, as the final headspace concentration was dependent on the order of incorporation of mixture ingredients. In the same way, van Ruth *et al.* (2001) showed different types of interactions of salt and protein of artificial saliva with some aroma compounds, depending on saliva composition and saliva/water and oil/saliva ratio of the medium. In particular, saliva proteins lowered retention of highly volatile compounds such as dimethyl sulfide, 1-propanol, diacetyl, 2-butanone and ethyl acetate and increased retention of less volatile hydrophobic compounds. With oil/saliva systems, the volatile compounds were also found affected by the saliva composition. In particular, proteins of saliva were shown to increase salting-out effect for the more polar compounds which were distributed in a large proportion in the water phase. Concerning the ratio level, for the water/saliva system, the effect on the volatile compounds was limited to an increase of the salting-out effect observed with the higher saliva ratios for only eight compounds. For the oil/saliva system, all compounds but 1-butanol were significantly influenced by the ratio, due to the difference of solubility and affinity of volatile compounds for oil and aqueous phases. Most of the tested compounds were more retained in the system with high oil ratio, except for 1-propanol and diacetyl, which were less retained.

Other factors relative to saliva, during the time the food is in the mouth, may affect the concentration of aroma compounds. During ingestion in the mouth, aroma compounds can be adsorbed on the oral mucous membrane according to their chemical properties and released slowly in saliva for a longer time than the time residence of food in the oral cavity (Buettner and Welle 2004, Buettner 2004). This phenomenon is probably at the origin of persistence perception. Some compounds such as esters or sulphur compounds can also be degraded by enzymatic activities in saliva, thereby reducing persistence or generating other potent odorants (Buettner 2002a,b).

Beside salivary composition and volume, mastication is another important factor in flavour release in mouth (van Ruth and Roozen 2000), increasing the release of aroma compounds with progressive destructuration of the food matrix which is the first transformation process of food during feeding. Mastication involves the breakdown of a solid food into smaller particles, the incorporation of saliva, the agglomeration and shaping of the resulting mixture into a cohesive bolus, and finally the transport of the bolus to the pharynx. These functions correspond to a highly complex sensory-motor activity integrating various components of the masticatory system, such as teeth, jaw muscles, lips, cheeks, tongue and the production of salivary secretion. The resulting pattern of chewing is adjusted to the food bolus properties at any time during the chewing sequence according to a precise peripheral feedback. The chewing behaviour adapts itself to the initial structure of food and the evolution of texture during the in-mouth chewing process (Lucas *et al.* 2002, Bourne 2004, Mioche 2004). During this process, texture and released flavour are perceived, and these perception determinants have a large impact on food acceptability and choice. The process occurring from aroma compounds release to their passage through the retronasal airway where receptors are located are very complex. During liquid ingestion, it seems that aroma compounds are able to pass through the retronasal airway only during the swallowing phases (Linforth *et al.* 2002) because the mouth works as a closed system. It seems that the barrier between the mouth and the pharynx is opened intermittently only during these phases, leading aroma pulses to be carried by retronasal airflow in the throat to reach aroma receptors (Buettner *et al.* 2002). During mastication, the air is pumped out of the mouth into the throat with each chew and aroma release can be associated with the resulting pulses of air pumped from the mouth (Hodgson *et al.* 2003). So, in the case of consumption of solid foods, both mastication and swallowing contribute to aroma delivery to receptors. In the case of taste compounds, the process seems simpler. As they are generally water soluble, these compounds are extracted by saliva which allows them to be readily in contact with taste receptors located on the tongue.

Interindividual differences are very important in flavour release and perception patterns (Brown *et al.* 1996), that is, mainly related to different chewing, salivary and airflow parameters which vary between individuals (Mioche *et al.* 1999, Mathoniere *et al.* 2000, Pionnier *et al.* 2004a). Moreover, jaw muscle activity is known to change with age, leading to changes in several parameters such as bite force, number of chews and chewing duration (Mioche *et al.* 2004). Pionnier *et al.* (2004a,b,c) used these interindividual differences to show relationships between aroma and taste compound release, aroma and taste perception, and oral parameters. These authors, working on a flavoured model cheese, showed in particular that for aroma compounds the maximum concentration reached and the total amount released during chewing are related to respiratory and masticatory parameters. For the non-volatile taste compounds, a maximum concentration detected late and a low total taste compound release are related to long chewing time and low saliva flow rate, low chewing rates, low masticatory performances and low swallowing rates. Concerning the relationships with perception, they found that the time necessary to reach the maximal intensity of salty, sour and mouldy attributes of the model cheese is positively related to the time to reach the maximal concentration of sodium, citric acid and heptan-2-one and to physiological parameters. The role of saliva flow in the release and perception of salt was also studied by Neyraud *et al.* (2003). In particular, these authors showed the impact of salt released from a matrix and of the chewing action on saliva flow rate, pH and sodium chloride concentration. They also described the cyclic swallowing of saliva which is replaced progressively by newly secreted saliva of low salt content, leading to further extraction of salt from the gum matrix. For more information, see Chapters 12 and 13 written by Boelrijk, Smit and Weel, and Linforth and Taylor, respectively, which are dedicated to these aspects.

16.3.2 Physiological interactions

Aroma and taste receptors are respectively located on the nasal mucous and on the tongue, mainly. Despite this separation at the peripheral level, many people frequently confuse taste and aroma perception. In some case, particular flavour molecules may have a multimodal action. In the case of cheeses, ammonia or volatile acids may be present at sufficient concentration to influence the pH and, consequently, sourness (McSweeney and Fox 1997) while they also influence the global aroma. Menthol is also an interesting molecule with a well known peppermint aroma but it also causes a cold sensation (chemesthetic) and bitterness. In these cases, the same molecule can access and stimulate both aroma and taste receptors. The number of such examples are rather limited and most reported taste–aroma interactions have to be related to other phenomenon. The hypothesis of competition between aroma and taste molecules for taste receptors located on the tongue were considered as improbable (Noble 1996). Rolls (1995), using primates, showed that amongst 112 single neurons responding to taste, olfactory and visual modalities, many of them were unimodal. Some single neurons showed convergence, responding, for example, to taste and visual inputs (13%), and olfactory and visual inputs (5%). Some of them had corresponding sensitivities in the two modalities. For example, neurons responded best to sweet taste or in an olfactory discrimination task to fruit odour, or neurons responded best among the tastants to sodium chloride and best among the odours to onion odour and well also to salmon. The olfactory input to these neurons was measured while the monkey performed olfactory discrimination tasks. The authors hypothesised that the multimodal representations were being formed from unimodal inputs to the orbitofrontal region. So, the area where odour and taste projections converge could be the integration and interaction site of odour and taste modalities. The neurophysiological origin of odour–taste interactions is also supported by the results reported by Portman (1999) who summarises anatomic and physiological concepts of perception.

Measuring thresholds for benzaldehyde, saccharin and monosodium glutamate in single and summation conditions, Dalton *et al.* (2000) showed that subliminal concentration of saccharin was able to modify the threshold level for olfactive detection of benzaldehyde (Fig. 16.3). The detection threshold of benzaldehyde was, in particular, significantly increased (28%) by the presence of a sub-threshold concentration of saccharin in mouth. The detection threshold of the combination benzaldehyde + saccharin (0.049 mM) ranked between that for benzaldehyde alone (0.071 mM) and that for saccharin alone (0.026 mM) showing summation but not complete additivity of taste and smell perceptions. This phenomenon was not observed when pairing benzaldehyde with deionised water and monosodium glutamate, providing support for the congruency between benzaldehyde note and sweetness. The authors suggested that cross-modal response summation requires the existence of a central point of intermodal convergence containing neurons responsive to the combined inputs: insular cortex, orbitofrontal cortex and the amygdala. Moreover, the specificity of saccharin–benzaldehyde integration may be specific to previously encountered combinations. The authors suggested that the amygdala could lower the detection threshold of one element of a congruent stimulus pair if this pair had been previously encountered. However, these psychophysical data and physiological hypotheses requested electrophysiological and neuroimaging studies to be tested.

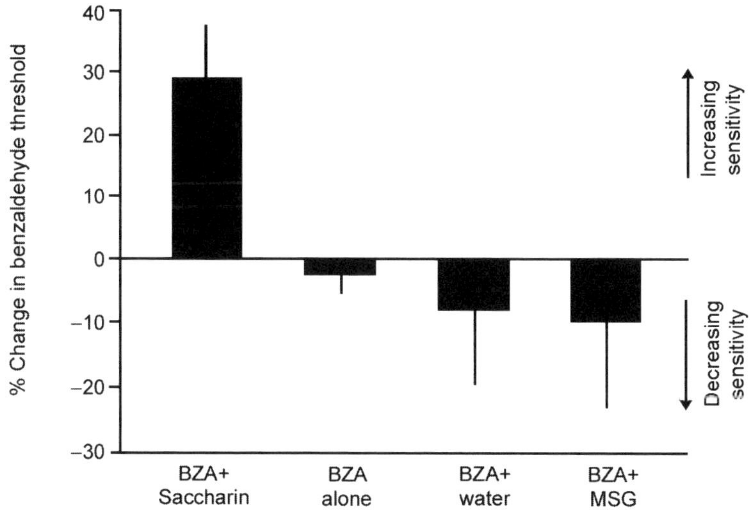

Fig. 16.3 Changes in threshold level for benzaldehyde (BZA) odour when different tastants such as saccharine or monosodium glutamate (MSG) are in the mouth (Dalton *et al.* 2000 with the permission of Nature Publishing Group).

To state precisely how taste-odour interactions and thus flavour perception are implemented, de Araujo *et al.* (2003) used functional magnetic resonance imaging (fMRI) to investigate where these interactions may occur in the human brain. They showed, using sucrose as a taste stimulus and strawberry aroma as an olfactive stimulus, that common brain areas activated by both taste and smell are parts of the caudal orbitofrontal cortex, amygdala, insular cortex and adjoining areas, and adjoining insular cortex. A small part of the anterior insula responded to unimodal taste and to unimodal olfactory stimuli, and a part of the anterior frontal operculum was a unimodal taste area not activated by olfactory stimuli. To reveal brain areas corresponding to positive interactions between taste and olfactory stimuli, de Araujo *et al.* (2003) tested whether the activation produced by the mixture of sucrose and strawberry is significantly greater than the sum of any activations produced by the two unimodal components presented alone. Significant activations showed an area of the left anterior orbitofrontal cortex that was activated more by the combined odour and taste stimuli than by the sum of any activations produced by each one separately where there was little or no activation, providing evidence for interactions between taste and olfactory inputs. Multisensory integration seems not to be restricted to olfactory–taste stimuli. Moreover, the results obtained by de Araujo *et al.* (2003) suggested, compared with similar works, in which volatile compounds were orthonasally delivered (Small *et al.* 1997), that taste–smell integration is dependent upon the mode of olfaction delivery.

Other studies showed that facilitation of odour detection by visual cues depended on object congruency and is associated with enhanced activity in the superior temporal sulcus and a region of the orbitofrontal cortex (Small 2004). However, it seems that whatever the implied senses, spatial concordance is a key factor for integration.

Djordjevic *et al.* (2004) confirmed that taste–smell interactions occurred at the central level by delivering odorants nasally and tastants orally, minimising the possibility of interactions taking place in the mouth by eventual peripheral mechanisms. This procedure avoided physical contact between odorants and tastants, and between odorants and taste receptors. They examined the influence of strawberry and soy bean odours on perceived sweetness and saltiness, using three levels of tastant concentration: zero, weak and strong (Fig. 16.4). They found the sweetness enhancement effect induced by strawberry odour and saltiness enhancement induced by soy sauce odour, although these effects vary with the taste stimulus concentration. Integration of perceptions occurring at the central level, the authors tried to explore whether the effect of imagined odours

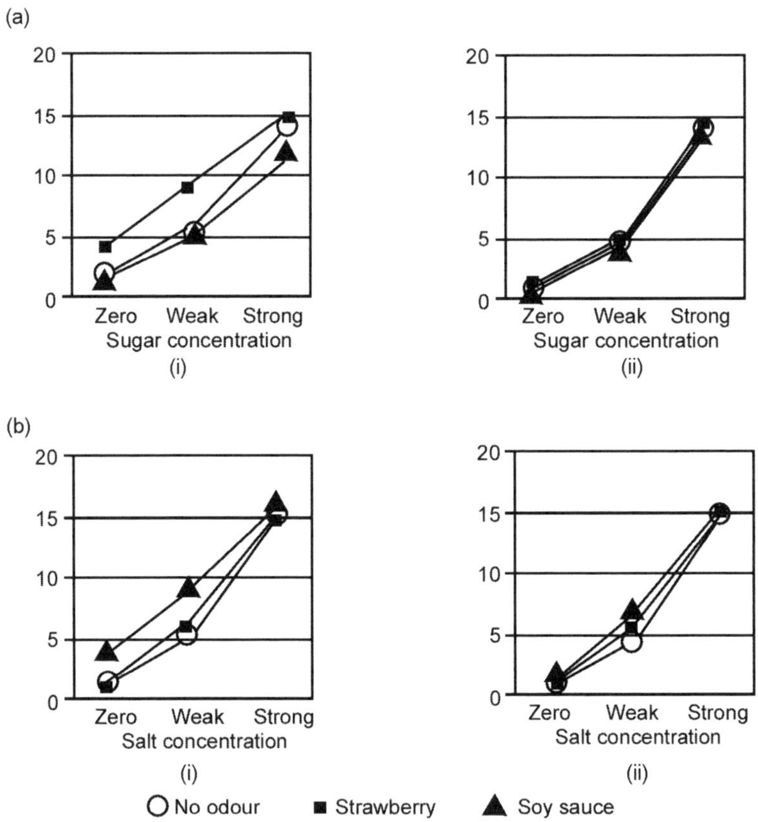

Fig. 16.4 Perceived sweetness (a) and saltiness (b) in function of presented (i) or imagined (ii) odours and taste compound concentrations (Djordjevic *et al.* 2004 with the permission of Springer-Verlag GmbH).

were comparable to those of physically presented odours. For perceived sweetness induced by imaging strawberry odour, no significant effect was observed but the authors noted that 20% of the panel showed significant sweetness enhancement. Concerning imagined odour of soy sauce and saltiness, a significant effect was observed for the weak salt concentration. These reported results show very modest global effects of the imagined odours. The important individual difference existing in odour imagery ability may explain these limited results. Functional magnetic resonance imaging experiments should allow us to visualise and to confirm the existence and the level of such effects.

The previous experience of taste–odour pairing seems of high importance on the integration process of taste–odour. Small *et al.* (2004) used fMRI to evaluate brain response to a retronasally presented odour (vanilla) in combination with a congruent taste (sweet) and with an uncongruent taste (salty). They tried, in particular, to determine whether super-additive responses, occurring when the neural activity evoked in response to bimodal stimulation is greater than the sum of activities evoked by independent and unimodal presentation of stimulus, were specific to the congruence of taste–smell stimuli. They observed super-additive responses during the perception of congruent taste and aroma (sweet/vanilla) in the anterior cingulate cortex, dorsal insula, anterior ventral insula extending into the caudal orbitofrontal cortex, frontal operculum, ventral lateral prefrontal cortex and posterior parietal cortex whereas these regions were not activated in the case of incongruent mixture. They found also that the anterior cingulate, posterior parietal cortex, frontal operculum and ventral insula/caudal orbitofrontal cortex were more active in comparison with the sum of the activities obtained using vanilla and salt independently or the incongruent mixture flavour. This suggested that the delivery of an uncongruent or unfamiliar taste–odour combination may lead to neural response suppressions and that insula, orbitofrontal and anterior cingulate cortex are important regions of the network underlying flavour perception.

16.3.3 Psychological interactions

Odour–taste integration

Psychological processes involved in flavour perception were recently reported in detail by Prescott (2004). Various sensory information perceived during the ingestion of food or drink through the mouth or only by smelling give us important information about identification of the product. These are olfactory, gustatory, visual and tactile sensations. The combination of these or some of these sensations, when we ingest a food or a drink are generally integrated and it is difficult to separate (or discriminate) spontaneously the global perception (Rozin 1982). When we perform sensory analysis, to be able to rate different descriptors from one food sample, concentration is needed. Training sessions are therefore frequently requested before a good performance is obtained to recognise and to rate, for example, several tastes and aromas in a model mixture or in a true food. In the recognition task, the first tendency is to give the global perception to a known and memorised object. This integrated global perception of physiologically distinct sensory modalities is what we learnt to recognise and what we are used to perceive when we eat a particular food. Like and dislike or acceptance and rejection are associated with this integrated perception (Prescott 2004).

Odour–taste integration may be illustrated by some odours defined by taste descriptors, in particular sweetness and sourness, and by odours enhancing tastes. Frequently, particular odours are described by basic taste qualities which are usually congruent (Burdach *et al.* 1984). Co-occurrence seems related to combinations we are used to find in the current food and beverages we are used to consume, for example, strawberry or vanilla odour and sweetness, cheese or meat aroma with saltiness, but the couple strawberry/saltiness is much less familiar (Frank and Byram 1988). In fact, a stimulus in one perception modality elicits a consistent corresponding stimulus in another modality (Martino and Marks 2001). Ratings of taste descriptors for odours were found to be as reliable as ratings for odour quality over several tests (Stevenson and Boakes 2003).

These odour–taste combinations should therefore be specific of cultures. For example, French and Vietnamese vary in their judgement on the congruence of odour–taste couples. Nguyen *et al.* (2002) observed a stronger strawberry–sweetness enhancement for American subjects than for French subjects and a stronger effect of vanilla on sourness for Vietnamese subjects than for French subjects.

Stevenson *et al.* (1995, 1998) tried to elucidate the origin of these taste–odour associations. They repeatedly paired novel odours such as lychee and water chestnut with either sucrose or citric acid in solution. These two odours were rated as low in 'sweetness or sourness odour' when sniffed before the pairing experiment with the sweet or sour tastes. After an exposure phase, these novel odours were rated significantly higher in 'sweetness or sourness odour' according to the taste with which they were paired. They concluded that these associative processes are some examples of learned synesthaesia (Stevenson *et al.* 1998). On a hedonic point of view, no significant change in rating or liking was noted when increases in the perceived sweetness or sourness of the novel odours were observed. Stevenson (2001a,b) showed that qualities of odours were influenced by pairing with other odours. For example, when mushroom odour is repeatedly paired with cherry odour, an enhancement of cherry odour is observed. These observations showed a certain 'malleability' of odour quality through association with another flavour quality.

As already seen, odours described with odour attributes when smelled are also able to influence taste intensity evaluation. For example, strawberry or vanillin odours enhance sweetness when added to sucrose solutions. Closing of the retronasal airway stops the effect, demonstrating that chemical interactions between aroma and taste compounds are not responsible for it. It has been stated already that these effects are rather specific to the taste and odour qualities: sweetness but not peanut butter enhances strawberry odour; saltiness is not enhanced by strawberry odour (Frank and Byram 1988). Conversely, odours added to taste compound solutions could suppress taste intensity. Sweet odours at low intensity such as peanut butter odour suppressed sweetness when added to a sucrose solution while raspberry odour enhanced it. Stevenson *et al.* (1999) showed that the sweet characteristic of a complex caramel odour similarly enhanced the sweetness of sucrose and suppressed the sourness of citric acid solution. These observations fit with real perceptual properties arising from odour–taste interactions.

Another important factor to study is the level concentration of flavour compounds necessary to observe such effects. Dalton *et al.* (2000) showed that the detection threshold for benzaldehyde odour which has a cherry–almond quality, increased when subjects held a saccharin solution with a sweet taste in their mouth compared with benzaldehyde alone or in combination with a non-sweet solution. This observation suggests that the sweetness of the benzaldehyde odour and the sweetness of saccharin are probably integrated at the sub-threshold levels. Such effects on sensitivity were obtained with the pineapple–aspartam sweet odour–taste pair, but also with monosodium glutamate as a taste compound, which is not sweet at all (Delwiche and Heffelfinger 2003). Conversely, an increase of taste threshold by odours was also reported. The detection of a sucrose solution at threshold level was increased when a sweet odour such as strawberry was compared with a non-sweet odour such as ham (Djordjevic *et al.* 2004).

White and Prescott (2001) reported experiments taking into account reaction times to identify a sweet or a sour taste at supra-threshold levels. This lapse of time was shorter for congruent pairs (cherry odour–sucrose, for example) than for uncongruent pairs (cherry odour–citric acid). To summarise, the main observations for a congruent aroma–taste pair are an increase of perception intensity, a decrease of identification time and a decrease of detection threshold value.

Cognitive aspects in flavour perception

'Sweet-smelling' odours are able to enhance the sweetness of sucrose in solution as reported previously. However, in some cases, these odours failed to do so. In fact, it seems that enhancement of taste by an odour and consequently the degree of integration of the two modalities is dependent on the task asked of the judges. For example, when judges are asked to rate only sweetness of a mixture sucrose–strawberry aroma, strawberry odour is found to enhance the sweetness of sucrose. This enhancement is no longer observed when judges were asked to rate the sourness and fruitiness of mixtures (Frank *et al.* 1993). These authors also reported that these effects occur for some taste mixtures containing compounds judged as similar (sour–bitter mixtures) but not for other mixtures containing dissimilar components (sweet–bitter mixtures). It seems that enhancement effects depend on the characteristics which have to be rated in the sensory procedure.

The halo-dumping effect (Clark and Lawless 1994) can be an explanation for such a phenomenon. This occurs when a possibility of an alternative response is not provided for a particular perception, for example, when fruitiness is not provided as a descriptor of a sweet–fruity mixture. So judges dump the values for the fruitiness attribute onto other provided rating scales (sweetness rating), thereby producing enhancement.

Such an enhancing effect can also be a function of the odour quality itself. This is illustrated by the work reported by Stevenson *et al.* (1999) (Table 16.1), showing an enhancement or suppression of sweetness of a sucrose solution according to the sweet-smelling of the odours tested. Nguyen *et al.* (2002) showed that the dumping effect due to rating scale number can occur but it is not the only mechanism responsible for the enhancement phenomenon of a taste by an odour.

Integration of perceptually similar dimensions is determined by the attention requested in a task. The perceptual process of taste enhancement effects suggests that the apparent influence of a number of rating scales on odour–taste interactions results from the impact of these scales on how the odour and taste are perceived.

Multiple ratings of appropriate attributes force an analytical approach whereas a single rating of a sensory quality encourages the synthesis of sensory modalities (Prescott 1999, Prescott *et al.* 2004). When instruction requires separation of the components, the components of a flavour are evaluated individually, and taste enhancement is eliminated. So, rating requirements lead to different perceptual approaches: an analytical approach when appropriate response alternatives are provided, a synthetic approach when response alternatives are limited, which influence the degree of perceptual integration occurring.

Odour–taste interactions can therefore be influenced by the extent to which an analytical or synthetic perceptual approach is taken during rating. As an illustration, van der Klaauw and Frank (1996) were able to eliminate taste enhancement by focusing the attention of judges to the appropriate attributes of a taste–odour mixture, even when they were only required to rate sweetness.

These results show the importance of strategy of encouraging either synthesis or analysis of the taste–aroma couple in taste–aroma interaction effect study.

16.4 Future trends

Knowledge about the nature of aroma- and taste-active compounds, their concentration in food and their complex release process in the mouth from a food matrix to reach receptors at the peripheral level are insufficient to explain the global perception of flavour. Recent studies show that perceptual information is integrated at the central level to give a global representation of the food flavour that could be modulated by psychological or cognitive processes. In particular, whether the odour and taste are perceived as a single entity or as multiple elements is an important determinant of the perceptual effects. New techniques for imaging such as functional MRI allowed an important advance in the understanding and knowledge of the physiological aspects of flavour perception, in particular where and how the sensory information is treated. The combination of multimodal psychophysics experiments with spectral or imaging techniques in future should give important additional information about the complex flavour perception processes. To reach this goal, now, fRMI, a non-invasive technique of choice for studying the functional areas involved in taste and odour perception is a performance tool that could be applied to the study of other perceptual interactions. In particular, studies by fRMI of interactions between taste and somesthetic sensations have already been reported by Cerf-Ducastel *et al.* (2001).

In a more practical point of view, this kind of interaction could be beneficial for the food industry, in particular, in flavour formulation. The possibility of using aroma to enhance taste perception will be of great interest in the reduction of taste compound concentration in food, i.e. sodium chloride and sucrose which, when consumed to excess, may lead to serious diseases. In the case of sodium chloride, an excessive intake of sodium is a cause of illness. So, health authorities recommend progressively decreasing salt content in food products. The consequence could be an increase in blandness that could result in a loss of food acceptability with a negative economical impact. The main challenge is to reduce sodium content without decreasing acceptability. Among the possible solutions, an enhancement of the perception of salt by perceptual interaction with aroma seems a very promising way. An aroma note such as 'smoked' has already shown a significant impact on saltiness (Renou, personal communication). Concerning sweetness, in the same way, sugar which is responsible for diabetes or, indirectly, for obesity can be reduced in foods and sweetness could be reinforced by several congruent aroma notes.

16.5 Sources of further information

Flavour Perception, edited by A.J. Taylor and D.D. Roberts (2004), Blackwell Publishing Ltd.

16.6 References

ALLANEAU B (1979) 'Influence des lipides sur la rétention des constituants volatils – Aspects thermodynamiques', PhD thesis, E.N.S.I.A.A.

BAKKER J (1995) 'Flavor interactions with the food matrix and their effects on perception', in Gaonkar A G, *Ingredient Interactions. Effect of Food Quality*, New York, Marcel Dekker, 411–438.

BALDWIN E A, SCOTT J W, EINSTEIN M A, MALUNDO T M M, CARR B T, SHEWFELT R L and TANDON K S (1998) 'Relationship between sensory and instrumental analysis for tomato flavor', *Journal of American Society for Horticultural Science*, 123, 906–915.

BONNANS S and NOBLE A C (1993) 'Effect of sweetener type and of sweetener and acid levels on temporal perception of sweetness, sourness and fruitiness', *Chemical Senses*, 18, 278–283.

BOURNE M (2004) 'Relation between texture and mastication', *Journal of Texture Studies*, 35, 125–143.

BRAUSS M S, LINFORTH R S T, CAYEUX I, HARVEY B and TAYLOR A J (1999) 'Altering the fat content affects flavor release in a model yogurt system', *Journal of Agricultural and Food Chemistry*, 47, 2055–2059.

BROWN W E, LANGLEY K R, MIOCHE L, MARIE S, GÉRAULT S and BRAXTON D (1996) 'Individuality of understanding and assessment of sensory attributes of foods, in particular, tenderness of meat', *Food Quality and Preference*, 7, 205–216.

BUETTNER A (2002a) 'Influence of human saliva on odorant concentrations. 2. aldehydes, alcohols, 3-alkyl-2-methoxypyrazines, methoxyphenols, and 3-hydroxy-4,5-dimethyl-2(5H)-furanone', *Journal of Agricultural and Food Chemistry*, 50, 7105–7110.

BUETTNER A (2002b) 'Influence of human salivary enzymes on odorant concentration changes occurring *in vivo*. 1. Esters and thiols', *Journal of Agricultural and Food Chemistry*, 50, 3283–3289.

BUETTNER A (2004) 'Investigation of potent odorants and afterodor development in two Chardonnay wines using the buccal odor screening system (BOSS)', *Journal of Agricultural and Food Chemistry*, 52, 2339–2346.

BUETTNER A and WELLE F (2004) 'Intra-oral detection of potent odorants using a modified stir-bar sorptive extraction system in combination with HRGC-O known as the buccal odor screening system (BOSS)', *Flavour and Fragrance Journal*, 19, 505–514.

BUETTNER A, BEER A, HANNIG C, SETTLES M and SCHIEBERLE P (2002) 'Physiological and analytical studies on flavor perception dynamics as induced by the eating an swallowing process', *Food Quality and Preference*, 13, 497–504.

BURDACH K J, KROEZE J H A and KOSTER E P (1984) 'Nasal, retronasal and gustatory perception: an experimental comparison', *Perception and Psychophysics*, 36, 205–208.

CAPORALE G, POLICASTRO S and MONTELEONE E (2004) 'Bitterness enhancement induced by cut grass odorant (cis-3-hexen-1-ol) in a model olive oil', *Food Quality and Preference*, 15, 219–227.

CERF-DUCASTEL B, VAN DE MOORTELE P-F, MACLEOD P, LE BIHAN D and FAURION A (2001) 'Interaction of gustatory and lingual somatosensory perceptions at the cortical level in the human: a functional magnetic resonance imaging study', *Chemical Senses*, 26, 371–383.

CLARK C C and LAWLESS H T (1994) 'Limiting response alternatives in time-intensity scaling: an examination of the halo-dumping effect', *Chemical Senses*, 19, 583–594.

CLIFF M and NOBLE A C (1990) 'Time–intensity evaluation of sweetness and fruitness and their interaction in a model solution', *Journal of Food Science*, 55, 450–454.

CONNER J M, BIRKMYRE L, PATTERSON A and PIGGOTT J (1998) 'Headspace concentrations of ethyl esters at different alcoholic strengths', *Journal of the Science of Food and Agriculture*, 77, 121–126.

COOK D J, DAVIDSON J M, LINFORTH R S T and TAYLOR A J (2003) 'Measuring the sensory impact of flavour mixtures using controlled delivery', in Deibler K D and Delwiche J, *Handbook of Flavor Characterization: Sensory Analysis, Chemistry and Physiology*, New-York, Marcel Dekker, 135–150.

DALTON P, DOOLITTLE N, NAGATA H and BRESLIN P A S (2000) 'The merging of the senses: integration of subthreshold taste and smell', *Nature Neuroscience*, 3, 431–432.

DAVIDSON J M, LINFORTH R S T, HOLLOWOOD T A and TAYLOR A J (1999) 'Effect of sucrose on the perceived flavor intensity of chewing gum', *Journal of Agricultural and Food Chemistry*, 47, 4336–4340.

DE ARAUJO I E T, ROLLS E T, KRINGELBACH M L, McGLONE F and PHILLIPS N (2003) 'Taste–olfactory convergence and the representation of the pleasantness of flavour in the human brain', *European Journal of Neuroscience*, 18, 2059–2068.

DELAHUNTY C M and PIGGOTT J R (1995) 'Current methods to evaluate contribution and interactions of components to flavour of solid foods using hard cheese as an example – Review', *International Journal of Food Science and Technology*, 30, 555–570.

DELARUE J (2002) 'Influence de l'acide et du sucré sur la perception de la flaveur. Application au cas du yaourt à la fraise', PhD Thesis, E.N.S.I.A.

DELWICHE J F and HEFFELFINGER A L (2003) 'Crossmodal activity of taste and smell', presented at Association for Chemoreception Sciences, Sarasota, USA.

DJORDJEVIC J, ZATORRE R J and JONES-GOTMAN M (2004) 'Odor-induced changes in taste perception', *Experimental Brain Research*, 159, 405–408.

DRAKE M A, GERARD P D and CHEN X Q (2001) 'Effects of sweetener, sweetener concentration and fruit flavor on sensory properties of soy fortified yogurt', *Journal of Sensory Study*, 16, 393–405.

DUBOIS C, LUBBERS S and VOILLEY A (1995) 'Revue bibliographique: interactions arômes autres constituants', *Cahier scientifique*, 112, 186–193.

DUFOUR C and BAYONOVE C L (1999) 'Interactions between wine polyphenols and aroma substances. An insight at the molecular level', *Journal of Agricultural and Food Chemistry*, 47, 678–684.

FRANK R A and BYRAM J (1988) 'Taste–smell interactions are tastant and odorant dependent', *Chemical Senses*, 13, 445–455.

FRANK R A, DUCHENY K and MIZE S J S (1989) 'Strawberry odor, but not red color, enhances the sweetness of sucrose solutions', *Chemical Senses*, 14, 371–377.

FRANK R A, VAN DER KLAAUW N J and SCHIFFERSTEIN H N J (1993) 'Both perceptual and conceptual factors influence taste–odor and taste–taste interactions', *Perception & Psychophysics*, 54, 343–354.

FRIEL E N and TAYLOR A J (2001) 'Effect of salivary components on volatile partitioning from solutions', *Journal of Agricultural and Food Chemistry*, 49, 3898–3905.

HODGSON M, LINFORTH R S T and TAYLOR A J (2003) 'Simultaneous real-time measurements of mastication, swallowing, nasal airflow, and aroma release', *Journal of Agricultural and Food Chemistry*, 51, 5052–5057.

HORT J and HOLLOWOOD T A (2004) 'Controlled continuous flow delivery system for investigating taste–aroma interactions', *Journal of Agricultural and Food Chemistry*, 52, 4834–4843.

HUSSEIN M M, D'AMELIA R P, MANZ A L, JACIN H and CHEN W-T C (1984) 'Determination of reactivity of aspartame with flavor aldehydes by gas chromatography, HPLC and GPC', *Journal of Food Science*, 49, 520–524.

JENNINGS W G (1965) 'Influence of temperature and salt addends on vapor equilibration of headspace', *Food Science*, 30, 445–449.

JO C and AHN D U (1999) 'Fat reduces volatiles production in oil emulsion system analysed by purge and trap dynamic headspace/gas chromatography', *Journal of Food Science*, 64, 641–643.

JOHNSON C, BIRCH G G and MacDOUGALL D B (1994) 'The effect of the sweetness inhibitor 2(-4-methyloxyphenoxy)propanoic acid (sodium salt) (Na-PMP) on the taste of bitter–sweet stimuli', *Chemical Senses*, 19, 349–358.

KUO Y L, PANGBORN R M and NOBLE A C (1993) 'Temporal patterns of nasal, oral, and retronasal perception of citral and vanillin and interaction of these odourants with selected tastants', *International Journal of Food Science and Technology*, 28, 127–137.

LAND D (1978) 'Some factors influencing the perception of flavour-contributing substances in food', in Land D G and Nursten H E, *Progress in Flavour Research*, London, Applied Science Publishers Ltd, 53–66.

LE QUÉRÉ J L, LESSCHAEVE I, DEMAIZIÈRES D, ISSANCHOU S and DELACHE R (1994) 'Chemical and sensory effects of intense sweeteners on the flavour of diet orange soft drinks', in Maarse H and van der Heij D G, *Trends in Flavour Research*, Amsterdam, Elsevier, 387–393.

LE THANH M, PHAM THI S T and VOILLEY A (1992) 'Influence of the presence of vegetable oil in emulsion on the volatility of aroma substances during the course of the extraction', *Sciences des Aliments*, 12, 587–592.

LINDLEY M G, BEYTS P K, CANALES I and BORREGO F (1993) 'Flavor modifying characteristics of the intense sweetener neohesperidin dihydrochalcone', *Journal of Food Science*, 58, 592.

LINFORTH R, MARTIN F, CAREY M, DAVIDSON J and TAYLOR A J (2002) 'Retronasal transport of aroma compounds', *Journal of Agricultural and Food Chemistry*, 50, 1111–1117.

LUBBERS S, CHARPENTIER C, FEUILLAT M and VOILLEY A (1994) 'Influence of yeast walls on the behavior of aroma compounds in a model wine', *American Journal of Enology and Viticulture*, 45, 29–33.

LUCAS P W, PRINZ J F, AGRAWAL K R and BRUCE I C (2002) 'Food physics and oral physiology', *Food Quality and Preference*, 13, 203–213.

MAIER H G (1975) 'Binding of volatile aroma substances to nutrients and foodstuffs', in Maarse H and Groenen P J, *International Symposium on Aroma Research*, Zeist, The Netherlands: Podoc, Wageningen, 143–157.

MARTINO G and MARKS L E (2001) 'Synesthesia: strong and weak.', *Current Directions in Psychological Science*, 10, 61–65.

MATHONIERE C, MIOCHE L, DRANSFIELD E and CULIOLI J (2000) 'Meat texture characterisation: comparison of chewing patterns, sensory and mechanical measures', *Journal of Texture Studies*, 31, 183–203.

McSWEENEY P L H and FOX P F (1997) 'Chemical methods for the characterization of proteolysis in cheese during ripening', *Le Lait*, 77, 41–76.

MIOCHE L (2004) 'Mastication and food texture perception: variation with age', *Journal of Texture Studies*, 35, 145–158.

MIOCHE L, BOURDIOL P, MARTIN J F and NOËL Y (1999) 'Variations in human masseter and temporalis muscle activity related to food texture during free and side-imposed mastication', *Archives of Oral Biology*, 44, 1005–1012.

MIOCHE L, BOURDIOL P, MONIER S, MARTIN J F and CORMIER D (2004) 'Changes in jaw muscles activity with age: effects on food bolus properties', *Physiology and Behavior*, 82, 621–627.

MURPHY C and CAIN W S (1980) 'Taste and olfaction: independence vs interaction', *Physiology & Behavior*, 24, 601–605.

NAHON D F, ROOZEN J P and DE GRAAF C (1998) 'Sensory evaluation of mixture of sodium cyclamate, sucrose and orange aroma', *Journal of Agricultural and Food Chemistry*, 46, 3426–3430.

NAWAR W W (1971) 'Some variables affecting composition of headspace aroma', *Journal of Agricultural and Food Chemistry*, 19, 1057–1059.

NELSON P E and HOFF J E (1968) 'Food volatiles: gas chromatographic determination of partition coefficients in water–lipid systems', *Journal of Food Science*, 33, 479–482.

NEYRAUD E, PRINZ J and DRANSFIELD E (2003) 'NaCl and sugar release, salivation and taste during mastication of salted chewing gum', *Physiology & Behavior*, 79, 731–737.

NGUYEN D H, VALENTIN D, LY M, CHREA C and SAUVAGEOT F (2002) 'When does smell enhance taste? Effect of culture and odorant/tastant relationship'. Paper presented at European Chemoreception Research Organisation conference, Erlangen, Germany.

NOBLE A C (1996) 'Taste–aroma interaction', *Food Science and Technology*, 7, 439–444.

PIONNIER E, CHABANET C, MIOCHE L, LE QUÉRÉ J L and SALLES C (2004a) '*In vivo* aroma release during eating of a model cheese: relationships with oral parameters', *Journal of Agricultural and Food Chemistry*, 52, 557–564.

PIONNIER E, CHABANET C, MIOCHE L, TAYLOR A J, LE QUÉRÉ J L and SALLES C (2004b) '*In vivo* nonvolatile release during eating of a model cheese: relationships with oral parameters', *Journal of Agricultural and Food Chemistry*, 52, 565–571.

PIONNIER E, ENGEL E, SALLES C and LE QUÉRÉ J L (2002) 'Interactions between non-volatile water-soluble molecules and aroma compounds in Camembert cheese', *Food Chemistry*, 76, 13–20.

PIONNIER E, NICKLAUS S, CHABANET C, MIOCHE L, TAYLOR A J, LE QUÉRÉ J L and SALLES C (2004c) 'Flavor perception of a model cheese: relationships with oral and physico-chemical parameters', *Food Quality and Preference*, 15, 843–852.

PORTMAN M (1999) 'Anatomie des organes des sens', *Journal International des Sciences de la Vigne et du Vin*, Hors Série, 7–13.

PRESCOTT J (1999) 'Flavour as a psychological construct: implications for perceiving and measuring the sensory qualities of foods', *Food Quality and Preference*, 10, 349–356.

PRESCOTT J (2004) 'Psychological processes in flavour perception', in Taylor A J and Roberts D D, *Flavour Perception*, Oxford, Blackwell Publishing, 256–277.

PRESCOTT J, JOHNSTONE V and FRANCIS J (2004) 'Odor–taste interactions: Effects of attentional strategies during exposure', *Chemical Senses*, 29, 331–340.

ROBERTS D D and POLLIEN P (2000) 'Relative influence of milk component on flavor compound volatility', in Roberts D D and Taylor A J, *Flavor Release: Linking Experiments, Theory and Reality*, Washington DC, American Chemical Society, 312–332.

ROBERTS D D, POLLIEN P, ANTILLE N, LINDINGER C and YERETZIAN C (2003) 'Comparison of nosespace, headspace, and sensory intensity ratings for the evaluation of flavor absorption by fat', *Journal of Agricultural and Food Chemistry*, 51, 3636–3642.

ROLLS E T (1995) 'Olfactory and taste processing in the brain', in Salvadori G, *Olfaction and Taste*, Geneva 8, Switzerland, Allured Publishing Corporation, 149–189.

ROZIN E (1982) 'The structure of cuisine', in Barker L M, *The Psychobiology of Human Food Selection*, Chichester, England, Ellis Horwood, 189–203.

SCHIFFERSTEIN H N J and VERLEGH P W J (1996) 'The role of congruency and pleasantness in odor-induced taste enhancement', *Acta Psychologica*, 94, 87–105.

SCHINNELER D J, DOUGHERTY R H and BIGG R H (1972) 'Influence of selected 5′-nucleotides on flavor threshold of octanal', *Journal of Food Science*, 37, 935–937.

SMALL D M (2004) 'Crossmodal integration – insights from the chemical senses', *Trends in Neurosciences*, 27, 120–123.

SMALL D M, JONES-GOTMAN M, ZATORRE R J, PETRIDES M and EVANS A C (1997) 'Flavor processing: more than the sum of its parts', *NeuroReport*, 8, 3913–3917.

SMALL D M, VOSS J, MAK Y E, SIMMONS K B, PARRISH T R and GITELMAN D R (2004) 'Experience-dependent neural integration of taste and smell in the human brain', *Journal of Neurophysiology*, 92, 1892–1903.

SOLMS J (1986) 'Interactions of non-volatile and volatile substances in foods', in Birch G G and Lindley M G, *Interactions of Food Components*, London, Elsevier Applied Science Publishers, 189–210.

STAMPANONI C R (1993) 'The "quantitative flavor profiling" technique', *Perfumer & Flavorist*, 18, 19–24.

STEVENSON R J (2001a) 'The acquisition of odour qualities', *The Quarterly Journal of Experimental Psychology. A, Human Experimental Psychology*, 54A, 561–578.

STEVENSON R J (2001b) 'Is sweetness taste enhancement cognitively impenetrable? Effects of exposure, training and knowledge', *Appetite*, 36, 241–242.

STEVENSON R J and BOAKES R A (2003) 'Sweet and sour smells: the acquisition of taste-like qualities by odors', in Spence C, Calvert G and Stein B, *Handbook of Multisensory Processes*, Boston, MIT Press.

STEVENSON R J, BOAKES R A and PRESCOTT J (1998) 'Changes in odor sweetness resulting from implicit learning of a simultaneous odor–sweetness association: an example of learned synesthesia', *Learning and Motivation*, 29, 113–132.

STEVENSON R J, PRESCOTT J and BOAKES R A (1995) 'The acquisition of taste properties by odors', *Learning and Motivation*, 26, 433–455.

STEVENSON R J, PRESCOTT J and BOAKES R A (1999) 'Confusing tastes and smell: how odours can influence the perception of sweet and sour tastes', *Chemical Senses*, 24, 627–635.

TAYLOR A J and HORT J (2004) 'Measuring proximal stimuli involved in flavour perception', in Taylor A J and Roberts D D, *Flavor Perception*, Ames, Blackwell Publishing, 1–38.

VAN DER KLAAUW N J and FRANK R A (1996) 'Scaling component intensities of complex stimuli: the influence of response alternatives', *Environmental International*, 22, 21–31.

VAN RUTH S M, GROSSMANN I, GEARY M and DELAHUNTY C M (2001) 'Interactions between artificial saliva and 20 aroma compounds in water and oil model systems', *Journal of Agricultural and Food Chemistry*, 49, 2409-2413.

VAN RUTH S M and ROOZEN J P (2000) 'Influence of mastication and saliva on aroma release in a model mouth system', *Food Chemistry*, 71, 339–345.

VOILLEY A, SIMATOS D and LONCIN M (1977) 'Gas phase concentration of volatiles in equilibrium with a liquid aqueous phase', *Lebensmittel – Wissenschaft & Technologie*, 10, 45–49.

WHITE T and PRESCOTT J (2001) 'Odors influence speed of taste naming', *Chemical Senses*, 6, 1119.

17

The learning of human flavour preferences

A. Blake, Firmenich SA, Switzerland

17.1 Introduction

Flavours are the most recognised quality characteristic of our foods and each of us has very personal opinions about the ones we choose and prefer. We can enjoy some foods to the extent that we find them exquisitely delicious and comforting but there are those we find so unpleasant and conceptually unacceptable that we are disgusted or feel nauseous when we see other people eating them. It may be the appearance, the texture or the source of the food that creates such strong feelings but often it is simply the flavour. Why and how we come to have such personal and yet individually different likes and dislikes for food flavour is the subject of this chapter which will review some of the recent work showing what exactly flavour is and why it is so important to us. Before we start doing this, however, it is important to make some definitions about the words we use when we refer to the flavour of food.

17.1.1 The linguistic confusion between flavour and taste

One source for great confusion in the English language is our understanding of the word taste, which is used casually and interchangeably to describe several quite different things which relate to food. Leaving aside the non-food uses of the word to describe general attributes of quality (a person who has good taste in clothes) or experience (they developed a taste for the outdoor life) we also use taste to describe quite different aspects and activities related specifically to food and drinks. We can use it as a transitive verb to assess the flavour quality of something (I tasted the soup and found it quite delicious), an intransitive verb to describe a quality of something (the butter tasted rancid), a noun to describe total flavour quality (the wine had a taste that reminded me of apricots) or a noun which describes a single sensory characteristic (lemons have a sour taste). This chapter will argue that flavour is a multisensory experience created by our brain from all the sensory inputs it receives during eating and drinking; moreover because our mouth is the focus of eating and drinking it is here that we experience the overall and composite sensation we call flavour. This is essentially why we confuse flavour with our sense of taste, the latter contributing only one aspect of the totality. There are many publications which continue this confusion, e.g. Korsmeyer (1999) and Roudot (2004), but throughout this chapter we will use 'taste' specifically to mean the single sense detected in the mouth and limited to the qualities of sweet, salt, sour, bitter and umami (Ikeda 1909), the last being the taste of mono-sodium glutamate. There is still debate about whether we have yet more taste receptors for qualities such as fatty or astringent. The word 'flavour' will be used here exclusively to describe the totality of the perceived food quality that will be shown to rely predominantly on our sense of smell and yet also on all our other senses including taste.

17.2 The relationship between cooking and flavour

Human beings have the ability to cook and process their food whereas all other living creatures must eat the things they can forage for or kill in the state that they are found. Granted that some animals will hide or bury food and return to eat it later by which time it will often have changed, becoming more or less edible in the process; insects such as honeybees convert nectar and store it as honey but such examples are a long way from the deliberate and systematic blending, processing and cooking of raw materials that human societies have practised since they first invented tools and discovered how to control and use fire. The development of cookery was an event of profound importance to human beings since it greatly expanded the range of things that could be eaten. The key objective of all living things is to obtain enough nutrients and energy to survive and to avoid being eaten at least until reproduction has taken place. Prior to the emergence of homo sapiens animals had found two different solutions to this challenge, either by evolving a physiology which would allow them to live on various plant materials or to develop hunting skills which enabled them to catch and consume other animals. For their part the food sources, whether plant or animal, often evolved protective defences that made them less attractive as a source of food or even in some cases completely inedible. There are many examples of both plants and animals which are poisonous. Many plants evolved a dual strategy whereby some parts of them became attractive as foods but the rest of the plant remained inedible, for example, ripened fruits have odours and tastes to encourage their consumption and help distribute their seeds, but the leaves and roots either taste unpleasant or are difficult to digest. It is also a fact that many plant seeds are not easily digested, birds have evolved the appropriate enzymes which allow them to eat these as food but they need to be cooked before they can be safely eaten in quantity by mammals. The ability of human societies to do this allowed them to change from existing exclusively as hunter-gatherers; by becoming farmers they were able to secure a more reliable and plentiful food supply based on cereals and grains. The ability to cook not only expanded the range of basic foodstuffs available to mankind but it also brought the ability to preserve and keep foods which would otherwise have perished. Knowing how to process and store raw materials in ways that would keep or even improve their nutritional value were skills that led to many of the fermentation-based foods we still eat today – cheese, pickles, wine, beer, tempeh, miso. Knowing how to treat perishable meats and fish such that they could be kept for long periods gave us the prototypes, amongst others, for kippers, cured ham and beef jerky. These and similar innovations not only greatly extended the range of foods that humans could eat but also created an interesting and inevitable challenge – how to learn, remember, recognise and communicate that these new foods, which had never before existed in nature, could be safely eaten. Until this point in time all the foods eaten by all species had evolved slowly; in parallel with these gradual changes the consumers had evolved the ability to detect and recognise the chemicals which identified them as being fit to eat, in other words the molecules which gave them taste and odour.

17.2.1 Cooking creates new flavours

When humans developed cooking they introduced molecules that had never existed in the food chain before. Fortunately nature gave them a sensory system that was supremely adaptable and that allowed them to recognise and learn the odours of the new molecules and through them to recognise what could or could not be eaten, then finally to make judgements about the flavour of the food and its quality. From this they developed concepts of deliciousness that were created during and because of the cooking process. Although there is an ongoing discussion in today's consumer society about foods and whether their flavours are more or less natural this usually overlooks the fact that most of these are produced during cooking and are in this sense manmade. It is obviously not natural for a cow to spontaneously jump onto a bonfire in order to give us the pleasure of a roast beef sandwich.

The nature of individual human societies and the fact that they had migrated over most of the earth's surface with different sources of food and different cooking practices inevitably meant that the range of flavours enjoyed by human beings was very varied. Different patterns of cuisine had emerged and each individual experienced the food of the society into which it was born and presumably it learned to enjoy and accept the flavours of its tribe's foods in preference to those of other groups. This raises questions that are as relevant in today's world as they were then:

- When and how do human infants learn which flavours to accept and reject?
- Are some food flavours inherently more acceptable than others?

- To what extent are the flavours learned in infancy still liked in maturity and how do these preferences change during life?
- When preferences change what are the factors that drive and influence this?

These are some of the questions that we will explore in this chapter but first we need to consider what exactly this sensation is that we call flavour.

17.3 Flavours and how we experience them

Until recently the scientists who are interested in the subject of 'food flavour' conducted their research within very specialised and quite different disciplines. Flavour chemists are generally interested in the structure and quantity of the molecules that are present in foods and drinks and that give them taste and odour. Biochemists seek to understand how the cells in the sensory systems detect these molecules and convert this information to signals that can be transmitted to the brain. Neurologists and experimental psychologists examine how and where the nerve impulses from the sensory systems are processed, converge in the brain and give us conscious sensations. Behavioural psychologists study our liking for flavour sensations and how we develop preferences for different foods. However, in recent years it has been realised that to fully understand what flavour is we need to integrate many aspects of all this work and look at the problem from a multidisciplinary perspective (Blake 2004).

17.3.1 The human brain and its relationship with flavour

Perhaps the single key realisation that changed the understanding of flavour was when scientists investigating its perception became aware of the then comparatively new field of experimental psychology dealing with the multisensory processing of signals to the brain. Knowledge of the brain and how it processes information has grown dramatically with the development of non-invasive imaging techniques and in particular magneto-encephalography (MEG), positron emission tomography (PET) and functional magnetic resonance imaging (fMRI). These techniques allow images of a functioning brain to be created and the last two, in particular, highlight all regions of the brain where neural activity is taking place. The modern concept of the human brain is now one of a hugely interconnected and inherently holistic information processing system and although we have specific areas of the brain that are primarily concerned with a particular task, for example, the processing of vision or hearing, these regions do not work in isolation but are interconnected via neurons that transfer information between them.

17.3.2 Multisensory perception

It is a fact that an input into one sensory channel can affect the perception of events in another. Perhaps the most dramatic and demonstrable example of this is the McGurk effect (McGurk and MacDonald 1976) where the sight of a face mouthing a sound can change the actual perception of the sound. Some seven years ago workers at the University of Oxford showed from fMRI scans of people engaged in silent lip-reading that neuronal activity occurred not only in the visual but also in the auditory processing regions of their brains, thus demonstrating the existence of the interconnections between these two sensory channels (Calvert, Bullmore *et al.* 1997). The convergence between the senses of taste and odour have also been demonstrated and the orbitofrontal cortex appears to be involved in this process (Rolls and Baylis 1994). There is a growing body of evidence that much of our perception of the external world is multi-sensory and that the interactive nature of these processes develops according to previous experience and practice; it appears that neurons which are stimulated by the same event but through different sensory channels develop lasting interconnections; for recent reviews see Duran and Costell (1999) and Spence and Driver (2004).

In 2000 a multisensory link between taste and smell (Dalton, Doolittle *et al.* 2000) was reported by scientists working at the Monell Chemical Senses Center in Philadelphia. This pivotal paper showed that the presence of a sub-threshold level of sweetness in the mouth changed the perception of an odorant (benzaldehyde) that was itself presented at a sub-threshold level. This was the first time that a neurological link between a taste stimulus and an olfactory perception was directly demonstrated. Further work at the University of Nottingham has since confirmed these findings and has shown that our perception and recognition of flavours although largely based on olfaction is strongly influenced and reinforced by our sense of taste. It has been demonstrated that the perception and strength of peppermint flavour in a chewing gum is strongly correlated with the residual sweetness of the gum and not with the presence of peppermint volatiles released to the nose (Davidson, Hollowood *et al.* 1999). This was an important finding not only because it explained that flavour fade in chewing gum has more to do with residual sweetness than with reduction in flavour volatiles but also raised the question of why we associate mint aroma with sweetness, since mint leaves themselves are not actually sweet. As an extension to this work the same group (Hort and Hollowood 2004) has shown that when a solution containing an olfactory stimulus, e.g. banana aroma and a taste cocktail of sweetness and acidity is steadily fed by tube into the mouths of panellists they have an overall perception of banana flavour. If after several seconds the level of tastant is reduced and replaced by water (with no change in flow rates) then the perception of banana flavour is also reduced even though the same olfactory stimulus is present; bringing the taste back to the level it was at the beginning also restores the overall strength of banana flavour; the identity of the flavour depends on olfaction but the strength is modulated by taste. What is even more interesting is the experiment that starts with the same conditions, flows of taste and olfactory stimulus, but after several seconds the latter is stopped while the taste stimulus continues. With this protocol many people fail to notice that the banana aroma has disappeared and they continue to perceive banana flavour even though they only have a sweet and acidic solution in their mouth. Very recent work suggests that people may differ in their ability to dissociate taste and olfactive signals and it could be that the ability to do this can be trained, or to put it another way we may be able to separate learned associations with sufficient practice. The fact that taste signals can change olfactive perception also gives rise to the fact that single olfactive stimulants may have different flavours depending on what taste they are presented with; thus phenyl propionic acid has a flavour of honey when presented in a sweet solution but has a flavour which more resembles cheese when the solution is salty.

A complete understanding of what we know as 'flavour' thus requires a consideration of all our senses and how these contribute to what is probably the most multisensory experience we have on a regular basis every few hours. It is equally important to keep in mind the fact that eating is an intrinsically hazardous activity. Each time we consume food or drink we take into our bodies part of the external world in order to provide the energy and raw materials that are needed to keep our system running and in good repair, but on each occasion that we do this we risk ingesting things that are harmful or dangerous. It should be no surprise that along with other animals we have evolved an effective monitoring and control system that minimises the risks inherent in this necessary activity; our assessment of the flavour of the things which go into our mouth is key to this monitoring system and is what decides whether we carry on with the process of eating and drinking them or whether we spit them out. It is this important role of the mouth that has decided why we experience flavour there, even though its overall quality depends on sensory pathways that have nothing at all to do with the mouth. Professor John Prescott from the University of Otago in New Zealand says that flavour is a psychological construct of the brain (Prescott 1999). He points out that:

> since the mouth acts as a gateway to the gut, our chemical senses can be seen as part of a defence system to protect our internal environment.

17.3.3 The role of all the senses in flavour perception

Vision

For us humans sight is the most far reaching sense that we have and we usually first become aware of potential food sources because of it. It may seem unusual that we discuss vision in the context of flavour but there is now convincing evidence that the appearance of an object can influence our perception, recognition and liking of its flavour (Zellner and Kautz 1990). There is also the much publicised study that showed that experienced wine tasters were quite unable to correctly describe and identify white wines which had been coloured red with flavourless food colouring (Morot, Brochet *et al.* 2001). When a food is eaten to satiety the appearance of the food stops being attractive and becomes repellent (Rolls 1996). It is reported that diners who were happily eating a steak meal under monochromatic light were repulsed when the lighting was brought back to white and they realised that the meat was in fact green (Wheatley 1973). It seems that even before we take food into our mouths the visual appearance of it is already influencing the way that we will perceive its flavour (Gottfried and Dolan 2003) and we also appear to learn associations between colours and specific flavours (Calvert and Osterbauer 2002).

Hearing
If the linkage between flavour and vision might have seemed tenuous the relationship with hearing might appear to be even more obscure. Nevertheless we do learn to expect that when we eat celery it will crunch and the appropriate sound of breakfast cereals even features as part of the selling strategy for them. Very recently it has been demonstrated that the perception of the texture and mouthfeel of a snack food can be changed by relaying the sound generated during chewing to the ears by way of an amplifier and headphones. Changing the quality of the sound by means of a frequency filter alters the judgement of crispness of the product (Zampini and Spence 2004).

Touch
Information about the temperature and texture of what we have in our mouth passes to the brain via the trigeminal nerve. In recent years the mechanisms by which we know whether food is hot in the sense of spicy or piquant or cooling as with menthol have also been elucidated. The receptors that detect the burn of chilli or pepper are now known to be exactly the same as those which sense the temperature of the food (Caterina, Schumacher *et al.* 1997) so they tell the brain that a very spicy curry is burning the mouth even though no physical damage is being done. The subtleties of how the mouth assesses the texture of food are still not fully understood (Dobraszczyk and Vincent 1999), certainly it is very sophisticated in the way it detects characteristics such as firmness, crispness, sliminess, smoothness, greasiness, graininess, thickness and creaminess – characteristics which we use to describe food but that cannot all be adequately measured in the laboratory. The attribute of crispness has been studied in detail and some progress made to relate this sensation to the sudden drop in load on the jaw muscles when food breaks during biting (Norton, Mitchell *et al.* 1998). Likewise the origins of creaminess have been investigated and discussed at length, for a review see Kilcast and Clegg (2002), but usually several different parameters need to be measured and integrated to adequately explain a single perceived quality attribute (Bourne 2002). Other textural aspects are even less well understood, for example, there is an ongoing discussion about what exactly causes the sensation we know as astringency, whether we have specific receptors for the molecules that cause it or whether it is a physical effect. Recent work suggests it is caused by changes in the epithelial lining of the mouth (Prinz and deWijk 2004). It is also still unclear whether we have specific receptors for the presence of fat (or fat-related molecules) or whether we sense fat because of the effects it has on food texture. Much of what is perceived in the mouth as aspects of taste and texture themselves appear to be multisensory events that are integrated within the brain to form a unitary perceived quality of the foodstuff (Green 2001) and fMRI has confirmed that taste and touch stimuli activate

common areas of the brain (Cerf-Ducastel and Murphy 2001). One issue that has been studied in some depth in recent years is the multisensory link between the texture of food in our mouths and our perception of flavour strength. It has been known for some time that as the viscosity of liquid foods and drinks increases the apparent strength of the flavour diminishes (Baines and Morris 1989). This was first explained by the proposal that a more viscous product would release fewer volatile molecules for detection by the nose than would be the case for a less viscous system. This has now been shown to be untrue and the development of the real-time dynamic monitoring of flavour release from foods (Taylor, Linforth *et al.* 2000) has clearly demonstrated that viscosity has little or no effect on the transfer of volatiles from food to the breath and that other mechanisms are responsible for this phenomenon (Hollowood, Linforth *et al.* 2000). The group of Professor Andrew Taylor at the University of Nottingham has studied many aspects of this and it has been demonstrated that increased viscosity reduces both flavour and taste perception for both sweet and salty foods but apparently not for bitter and sour foods (Cook, Hollowood *et al.* 2003). They have also demonstrated that there is a good correlation between reduced flavour perception and the pressure that the tongue puts on the soft palate of the roof of the mouth during the chewing of food (see also Elejalde and Kokini 1992). There is no direct evidence but it is tempting to speculate that our brains learn to associate specific textural characteristics of food with flavour and taste sensations and that we may make these connections very early in life, for example, during suckling after birth. Certainly most baby foods are salty or sweet but are seldom sour or bitter.

Taste

There has never been any doubt that our sense of taste is extremely important to our perception of flavour. The common misconception seems to be that this sense is of paramount importance when, in fact, it provides only a limited part of the total information. It is ironic that while much of the elementary teaching about our sense of taste is understated and often incorrect many people nevertheless make the mistake of ascribing it with discriminating abilities it does not have because, in fact, our mouths contain receptors for only a limited number of water soluble molecules. It is now accepted that we have five families of receptors that respectively detect molecules that are sweet, salty, sour, bitter or have the savoury characteristic of mono-sodium glutamate which is described by the Japanese word 'umami' (Ikeda 1909). Receptors of all five types are distributed over the entire surface of the tongue and although the distribution is uneven they are not localised and separated into discrete regions as is often wrongly shown in elementary textbooks on sensory science. The total number of receptors differs from individual to individual and many people can have receptor deficiences without even being aware of this. I once sat next to an individual during a demonstration of how salt could reduce bitterness and thus enhance the apparent sweetness of tonic water. This person could only perceive the drink as being saltier and after some discussion and further tasting it was realised that he had no perception of bitterness whatsoever; for the first time in fifty years he started to understand why he found that a Campari cocktail tasted like just another sweet, red fruit juice. Most of us, however, appear to have several receptors for the detection of bitterness that perhaps reflects the evolutionary importance of being able to detect the presence of many bitter and potentially toxic ingredients in foods. Human babies are born with an inherent liking for sweetness and some time after birth also develop preferences for saltiness and umami character; by contrast bitterness and sourness (acidity) are disliked initially by babies but after some years they often learn to like these tastes too. Indeed many children around the age of eight to ten have a delight in eating sour foods that are disliked by adults (Liem and Menella 2003) while adolescents gradually develop a liking for the bitter things such as coffee and beer enjoyed by their parents.

Olfaction

We now arrive at our fifth sense and the one that provides most of the information to our brain about the flavour of the food in our mouth. It has been estimated (Murphy, Cain *et al.* 1977) that over 80% of the sensory signals that allow the brain to form the conscious impression of flavour arrive via the olfactory epithelium which is high in the nose and close to the brain; it can even be considered an extension of the brain since the neurons within this organ project directly into the olfactory bulb.

The detailed anatomy of the olfactory system has been studied in great detail in recent years and for a recent review see (Pernollet and Briand 2004). In brief the organ of smell is capable of recognising many thousands of odours and has enormous adaptability in doing this. The olfactory epithelium in humans contains several million neurons but these are thought to fall into only some hundreds of different types each of which expresses a different olfactory receptor protein. The nerve impulses from these neurons pass into the brain via the cribriform plate, a bony structure between the base of the brain and the nasal cavity with tiny holes to let the nerve fibres pass into the latter; remarkably each type of neuron in the olfactory epithelium connects to the same glomerular cell within the olfactory bulb. The different olfactory receptors can interact with more than one odorant and any odorant can bind to several receptors, although with different levels of affinity (Malnic, Hirono *et al.* 1999, Touhara 2002). When an odour is bound to several receptors it creates a pattern of neuronal activity in the olfactory bulb and subsequently within the olfactory cortex and it is this pattern that is the basis for the olfactory memory (Fig. 17.1).

There are several consequences to this system, most importantly it is highly adaptable (Araneda, Kini *et al.* 2000) since any odorant, even one that has never been experienced before, will set up a specific pattern of olfactory reception and there is no requirement for a specific receptor for that stimulus, a fact that would have been extremely important for an animal that was busily engaged in cooking food and creating smells that never existed in nature. It also means that different concentrations of the same odorant may create different patterns of receptor

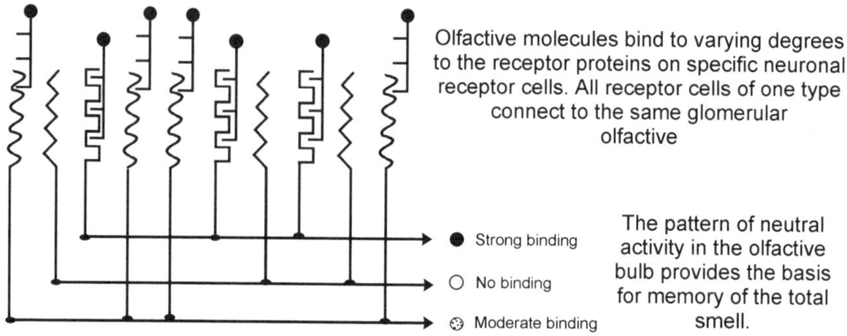

Fig. 17.1 Schematic diagram showing how a single odour molecule can bind to specific receptor proteins to different degrees, which then sets up a remembered pattern of neural activity in the olfactive bulb.

activation with the consequence that the character of an odorant can change with concentration (Kajiya, Inaki *et al.* 2001); another feature is that whether the stimulus is a single odorant or a blend of odorants it is remembered as a single pattern of neuronal activity and it is not possible to re-interpret this pattern in terms of detailed information about the composition of the stimulus. For example, it has been shown that even highly trained perfumers and flavourists are unable to identify the components of a mixture containing more than a few individual molecular species (Livermore and Laing 1996). It has been reported that single odorants presented in mixtures with other odorants start to acquire the perceptual qualities of the odour with which they were mixed (Stevenson 2001a,b), which again becomes explicable in terms of neuronal patterns being the basis of olfactive memory rather than the detection of a specific stimulus. It is claimed that perceptual learning and the association of other events with an odour plays a fundamental role in recalling that odour (Wilson and Stevenson 2003). Other work points out that people are best able to remember specific odours, e.g. of wine, if they relate them to verbal descriptors (Hughson and Boakes 2001). For further discussion about odour perception and its recall see also Stevenson and Boakes (2003). It should be noted that just as tastes can change olfactive perception the reverse is also true. A number of workers have investigated and reported the way that smells can change the perception of taste, especially sweetness (Harper, Land *et al.* 1968, Dravnieks 1985, Frank and Byram 1988). Other authors have proposed that perceptions of sweetness are not inherent to the olfactant but are caused by associative learning of the taste and olfactive stimuli (Stevenson, Boakes *et al.* 1998) – in other words they develop a multisensory relationship

Although the olfactive signal is supremely important to the overall flavour quality of food we are curiously unaware of this. When we sniff and smell an object we breathe air in and the molecules it carries enter the nose orthonasally, the brain senses this and gives us a conscious impression of the smell having originated from a source external to the body, i.e. we associate the smell with the air breathed in or the object in front of the nose. However, when we are eating or drinking, volatile molecules from the food pass over the olfactory epithelium only when we breathe out, in other words retronasally, we no longer recognise the stimulus as a smell but it is now integrated into the total perception of flavour in the mouth. It might seem a simple process for the brain to receive olfactory information about the food in our mouth from the air which passes from the throat to the nasal cavity but this event appears to be far more complex than had previously been realised and it is intimately linked to the mechanical activities associated with chewing and swallowing. Because our mouths have to both break down food by chewing and also deliver it to the stomach via the oesophagus without it getting into our lungs our olfactory experiences during eating and drinking are discontinuous. They are sensed as intermittent bursts of information whereas, by contrast, taste and touch give steady streams of information. Some authors go further and propose that the dynamic nature of olfaction may also play a direct part in the memory and identification of a specific odour (Laurent, Stopfer *et al.* 2001, Friedrich and Laurent 2001). An extreme difference between the phasing and continuity of olfactive and taste stimuli occurs when we drink continuously, as, for example, when 'downing' a pint of beer. The first sip delivers volatiles to the throat that allow the consumer to recognise that he or she is drinking beer but during the emptying of the glass no breathing takes place until after the final swallow which reconfirms that the drink is still beer. During the actual drinking process the only information about what is in the mouth comes from the senses of taste and touch, yet the drinker experiences beer throughout. Viewed in this light we can perhaps start to understand why the brain ignores the discontinuity when an olfactive signal is removed during constant drinking as reported in the previous section. At the time of writing, however, there has been little work done to consider the dynamics and weighting factors involved in the integration of the individual sensory inputs to give the final conscious sensation of flavour that for the reasons already discussed, is perceived in our mouths.

17.3.4 The relevance of propriosense to flavour learning

Before we leave the discussion about our five senses we should mention the other very important flow of information to our brain that probably plays the single most important part in whether we remember the flavour of the food or drink just consumed as good or bad. I refer to what is sometimes described as propriosense that incorporates all the information received by the brain about the state of the body. Many sensory channels are involved in this, both nervous, for example, the information from our stomach and gut carried to the brain by the vagus nerve, or chemical, such as the release of insulin to adjust circulating glucose levels in the blood by controlling the movement of glucose into cellular tissues. A detailed discussion of the mechanisms by which the brain regulates food intake in general and more specifically the satiety which occurs during the eating of a meal lies outside the scope of this chapter (for further reading see Panksepp (1998), Rolls (1999), and Seeley and Woods (2003)); however, this is obviously important to the appreciation and enjoyment of the flavour of food and fMRI has been used to investigate the changes in brain activity as a single food (chocolate) is consumed beyond satiety, when the experience changes from being pleasant to being aversive (Small, Zatorre *et al.* 2001). One very well known phenomenon that dramatically changes the liking of a flavour and one that has been extensively researched in animals is post-ingestive nausea. This has been most studied in rats which develop an extreme aversion to the flavours of foods that have been eaten prior to gastric distress, even if this has been induced other than by the food itself, e.g. by direct intubation of gastric irritants into the stomach (Sclafani 2001). Humans also show such behaviour and learn to dislike a flavour particularly if it is associated with nausea or vomiting and in extreme cases they may then reject foods with that flavour throughout life (Kalat 1985). Less easy to understand and predict are the psychological links between foods, their flavours and the experiences of eating via associative learning. This will be discussed in greater depth in the section on how we learn to like flavours but it is becoming increasingly clear that the joint brain/body responses to foods by humans and their future liking or rejection of these is extremely subtle and will lie outside the scope of animal experiments, especially given the introspection and imagination that humans have. Professor Jaak Panksepp stresses that while a detailed study of mammalian brains such as that of the rat has helped in the understanding of the functioning of the human brain at the level of basic instinctive behaviour this becomes impossible when considering activities such as introspection, imagination and creativity (Panksepp 1998). In the chapter in his book which deals with Neurostatics, the anatomy of the brain/mind he says:

> attempts to span cognitive issues, by trying to relate the higher psychic functions of humans to animal brain circuits, will be vastly more difficult and perhaps impossible when it comes to our highest cortical abilities, the four R's – reading writing arithmetic and rational thought.

To these we can add two more R's – restaurants and their ratings. It is this argument which leads to the conclusion that while studies of animal feeding behaviour may be useful for understanding instinctive behaviours such as the development of aversive responses to those foods which cause nausea, they will be unlikely to help in explaining the peculiarities humans show in their broad spectrum of food and flavour preferences. As support for this view it is worth mentioning that while animals can show instinctive reactions of fear to certain predator smells, humans do not apparently do this. Thus, the odour of cats causes rats to crouch and freeze (Morrow, Redmond *et al.* 2000, Blanchard, Yang *et al.* 2001) and several species (deer, rabbits, hare and zebra) are terrified by the smell of lion dung (Boag and Mlotkiewicz 1994) but humans do not react in this way.

Panksepp makes the observation:

> Similarities in cortical interconnectivities (from species to species) diminish markedly as one begins to compare the more complex

> secondary and tertiary association cortices where perceptions, as well as most cognitive and rational processes, are generated. In short, multimodal association areas of the cortex, where information from different senses is combined to yield concepts and ideas, are structurally similar in microscopic detail, but because of the types of exchange of information among an increasing number of areas, similarities between humans and other animals begin to diminish.

The multimodal aspects of flavour perception, particularly as they relate to the quality and the emotional context of a meal clearly fall into this category. Figure 17.2 illustrates in extremely simplified form how the conscious impression of flavour is a consequence of all the sensory inputs to the brain.

Because the brain is central to the construction of flavour memory and the judgement of whether food or drink can be safely consumed we need to introduce some of the most recent thinking about how this organ develops, especially how the human brain acquires its unique capabilities. Obviously a subject as vast as this lies largely outside the scope of this chapter but some basic knowledge of brain growth and function is essential to a proper understanding of flavour and what it is. For a more complete discussion of brain anatomy and development together with some of the latest thinking on this subject the reader is referred to specific books (Panksepp 1998, Scientific American 1999, Brown, Keynes *et al.* 2001) and the review by Clifford (1999).

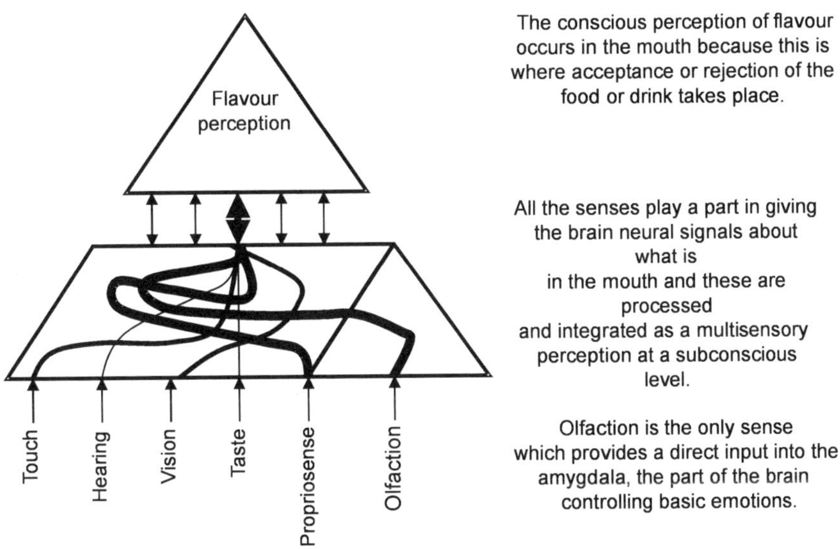

Fig. 17.2 Highly simplified representation of the processing by the brain of the sensory signals from which flavour becomes a conscious and perceived quality of what is being eaten or drunk.

17.3.5 The development of the human brain in infancy

In comparison with all other mammals the human infant is essentially born too soon in the sense that it is quite helpless for a long time and relies for survival totally on adults for several years. It has been argued that this essentially premature delivery is a consequence of the unusually large size of human brains and heads, if fully developed before birth they would require the mother to have a cervix impossibly wide to support her weight when standing on two legs. The growth and development of the human brain is certainly different from all other animals and this is particularly true for the formation of its higher cognitive abilities such as those which lead to language and conscious thought. An alternative explanation for the birth of a human when its brain is still at an early stage of development, however, might be that experiences and sensations from the external world are essential for it to develop the potential which it has. The human brain may simply be incapable of developing its higher cognitive functions without the appropriate sensory stimuli which come after it has been born. It is estimated that at birth a baby has some 10^{11} neurons and although during the development of the brain to maturity the total number of neurons does not change greatly their population does, some of them die while new ones are created. More importantly, the connections between the surviving neurons increase dramatically through the growth of dendritic links (King 2003). It is estimated that by the time a brain reaches maturity any one neuron may be connected to several thousand other neurons resulting in a vast interconnectivity ($> 10^{14}$ dendritic links) and that the extent and detailed structure of this are largely determined by the frequency, consistency and coherency of neuron activation. Much of this is associated with spontaneous brain activity, termed experience-independent plasticity, but sensory input from the external world also plays a critical role. External influences that mould the brain fall into two categories – experience-expectant and experience-dependent plasticity. The former relates to external stimuli that direct normal development of brain function, a good example is the requirement of vision for a proper development of the visual cortex. On the other hand experience-dependent plasticity occurs in those operations of the brain which are changed and modified by external sensory stimuli that are essentially unpredictable; in other words it is a learning process dependent on the experiences of life (Black and Greenough 1998). It is well known that the brains of young children can show remarkable degrees of plasticity (Cleaver and Derbyshire 2002) and even in adults the brain can recover from minor damage or injury, for example, after a stroke (Levin and Grafman 2000). There is an ongoing discussion about the extent to which stem cells can migrate and create neurons and glial cells in the brain after it has matured and while some research says that this cannot happen in the human brain (Rakic 2004, Sanai, Tramontin *et al.* 2004) other workers report evidence for the development of new brain activity as a consequence of the learning of new skills (Draganski, Gaser *et al.* 2004). To summarise the key points of modern neuroscience which impinge on a proper understanding of what flavour is and how we develop our preferences for it:

- The brain of a human being has a general structure which is determined by the genetic make-up passed on from its parents but which evolves according to experience.
- When a human child is born its brain lacks much of the detailed inter-connectivity between the neurons, this is subsequently laid down in infancy and is moulded by sensory inputs from the outside world, from within its own body and from emotional responses generated within the brain itself.
- Much of the experience of the world which influences the behaviour of the brain is perceived simultaneously through more than one sensory channel and the convergence of this forms links between neurons that provide a multi-sensory processing of future inputs.
- The brains of animals lack the higher functions possessed by humans and the extrapolation of animal behaviours, that relate to food preferences, to human behaviour should be made with caution for those behaviours that are not purely instinctive.

17.4 How we learn flavour

With these facts in mind let us consider when and how an individual human gains experience of foods and drinks from which it will learn to form concepts such as delicious or disgusting.

17.4.1 The learning of flavour by the neonate

About eleven weeks after a baby is conceived it has developed an olfactory epithelium (Doty 1992). At this stage the only source of any olfactive stimulus is from the amniotic fluid surrounding the baby, however, this is in contact with the mother's blood and there is considerable evidence that volatile flavour molecules from the mother's diet can reach the baby *in utero* (Schaal, Marlier *et al.* 1998, 2000). As the baby grows and its brain develops it experiences more and more olfactive stimuli because it regularly takes amniotic fluid into its lungs which passes over the olfactive epithelium. Especially in the last few months before birth the food eaten by the mother has a strong influence on those flavours which will be preferred by the baby after birth (Menella, Jagnow *et al.* 2001, Schaal, Soussignan and Marlier 2002) though hardly any courses in pre-natal instruction appear to mention this fact to mothers-to-be. After the baby has been born many new sensations become part of its life. It already knows and can recognise its mother's voice (De Casper and Fifer 1980, De Casper and Spence 1986) but now it can also see and has direct contact with others, especially its mother. Of particular relevance to the present discussion is the fact that it starts to feed; it experiences not only the flavours that are in the milk but also, and as importantly, it learns to associate these with the other physical and emotional aspects of the feeding occasion: nourishment, security, contact, warmth and attention. There is an odour characteristic of human milk which makes it particularly appealing to babies (Schaal, Soussignan and Marlier 2002) but it has also been shown that odours combined with thirty seconds of massage of one-day old babies condition them positively for that particular odour (Sullivan, Taborsky-Barba *et al.* 1991). In addition to the intrinsic appeal of human milk a baby can recognise the smell of its own mother's milk and is aware of changes in its flavour; when nursing mothers eat either garlic (Menella and Beauchamp 1991), or vanilla (Menella and Beauchamp 1996) then in both cases there is a change in the sucking rate of the baby. Breast feeding in particular gives the benefit that the baby's diet is not monotonous since the foods eaten by the mother will affect the flavour of the milk and provide variety, it has been shown that mothers who consume specific flavours during breast feeding will enhance acceptance of those flavours in their child's diet at the time of weaning (Menella, Jagnow *et al.* 2001). Breast fed infants are also reported to be more willing to accept a novel vegetable on first presentation than are formula fed infants (Sullivan 1994).

Other work at Monell has shown that when babies are fed on formula feed from a bottle then the flavour of this can have a marked effect on food choice several years later (Menella and Beauchamp 2002). Children who had been bottle fed with three different types of compound feed during their first year were subsequently compared for eating behaviour some three to four years later. One of the feeds was derived from milk, one was based on soya and another formulated from hydrolysed proteins; the last was described by the researchers as bitter and sour and as being unpalatable to themselves. The children under study who were now about four years older than when they were first fed the formula feed showed clear differences in their preferences for apple juices, these were either presented unchanged, were acidified with lemon juice or were modified by the addition of naringin, the molecule that gives the bitter taste in grapefruit. Milk fed children preferred the untreated apple juice but soy fed children preferred the bitter apple juice and the hydrolysate fed children preferred the acidified juice. These children were also offered the original three formula feeds and it was found that those which had been fed hydrolysate or soy in infancy were most likely to judge the flavour of hydrolysate formula as pleasant when compared with those originally fed milk. Those fed hydrolysate also judged the odour of hydrolysate as more pleasant than those fed milk. A further interesting observation was that the mothers which had fed hydrolysate or soya feeds to their babies were more likely to rank broccoli as one of their children's favourite foods in comparison with those who had relied on milk-based feeds. Introducing children to a range of flavours at the time of weaning has also been shown to improve their subsequent acceptance of novel foods (Gerrish and Menella 2001). It is clear from these studies that exposure of very young children to differently flavoured foods whether at the breast or the bottle is already forming patterns of acceptability that will last into childhood and probably into adult life. This is certainly a fact that should be known by all mothers but one that does not appear in books on child rearing.

17.4.2 The role of carers in flavour learning by the child

The Department of Human Development and Family Studies at The Pennsylvania State University have shown another interesting consequence of bottle feeding on later food consumption (Fisher, Birch *et al.* 2000). In this case comparisons were made between toddlers which had either been breast fed or bottle fed earlier in life. One of the main conclusions of this work was that breast feeding in the first year showed benefits on food intake and feeding style which persisted into the toddler phase. However, a noteworthy aspect of this work was the conclusion that these benefits were indirectly derived from the mother–child relationship that developed during breast feeding and that this had more to do with the behaviour of the mother than the child; those who breast fed their babies could not see how much milk they provided and they developed a relaxed attitude to the fact that their infant was getting enough to eat, the mothers who bottle fed their children were able to see how much the babies drank and became concerned about how well they were feeding, this anxiety persisted beyond the bottle feeding stage with consequences during weaning. The bottle fed children were put under pressure to eat more but perversely ate less than the breast fed children who had more relaxed mothers. It is frequently found that children coerced to eat one particular food rather than another will develop eating preferences which are quite the opposite to those wanted by the parent. Much work has been done on what is needed to get a small child to try and then come to like a new food. Familiarity and repeated exposure appear to be the key requirements to this (Sullivan and Birch 1994, Gunder *et al.* 1998) and peer pressure appears to play a role even in pre-school life (Birch 1980). However, it is not sufficient for a child to see a food being consumed by others, it has actually to be tried by the child on several occasions (Birch 1987, 1990), and will then gradually be accepted. Some work has related the nutritional value of the food to the flavour and there is evidence that flavour acceptance is positively associated with energy-rich foods (Birch 1990), especially those containing fat (Johnson 1991). However, it appears that many parents fail to get their children to like new foods because their strategies often create a stressful meal occasion and the parent's mood simply becomes associated with the food and more importantly its flavour. The effects of presenting foods as rewards has also been studied in 3–5-year-old children (Birch, Zimmermann *et al.* 1980). The food was either presented as a reward for specific behaviour, paired with extra attention from an adult or simply presented in a non-social context, such as being offered at lunch time. The foods given as a reward or paired with attention got enhanced preferences. However, the idea of enhancing preference for a food by offering it as a reward need not always work. The same group (Birch, Marlin and Romer 1984) showed that when children were allowed to play but only if they consumed a flavoured drink then this led to a decreased acceptance of the flavour, i.e. play was the reward but consuming the drink was the condition. Indeed the offering of one food as a reward for eating another less popular choice simply gives positive reinforcement for the reward food and negative feelings about the target food (Newman and Taylor 1992), a better parental strategy for encouraging the consumption of vegetables might be 'If you don't finish all your ice cream you won't be able to have some of this lovely cabbage'.

17.4.3 A unique study of spontaneous food selection

There is evidence that children left to their own choice will select foods which their bodies need. In a classic study which was reported 65 years ago and which would be virtually impossible to repeat today, 15 infants from 6 to 11 months of age were allowed to self-select their diets over a period of six years (Davis 1939). Each day there were three or four meal occasions when the children were offered the range of 34 different foods shown in Table 17.1.

The supervisory staff who were present at the meal had strict instructions not to influence in any way the infants' choice of food but only to give practical help with feeding and only when required. What is interesting is that the list includes a wide selection of fruit and vegetables and several types of offal that most children and a high proportion of young adults would probably refuse to eat today. However, all the children thrived on their self-chosen diets and instinctively met energy and nutritional needs without guidance. They were regularly checked for health and were judged above average in this respect. Not surprisingly some of the food combinations selected were unusual, one example given is a breakfast consisting of a pint of orange juice and a bowl of minced liver followed by a supper of eggs (not included in the table but referred to in the text), bananas and milk. In the early stages of the experiment all the children tried virtually all of the foods presented, often several times, but after a few weeks they settled on the few which they preferred although these differed from child to child. Sometimes patterns of choice altered and particularly after illness when it was noted that children

Table 17.1 Foods available to children during the self-selection trials of Dr Clara Davis

1. Water	18. Potatoes
2. Sweet milk	19. Lettuce
3. Sour (lactic) milk	20. Oatmeal
4. Sea salt	21. Wheat
5. Apples	22. Corn meal
6. Bananas	23. Barley
7. Orange juice	24. Ry-krisp
8. Fresh pineapple	25. Beef
9. Peaches	26. Lamb
10. Tomatoes	27. Bone marrow
11. Beets	28. Bone jelly
12. Carrots	29. Chicken
13. Peas	30. Sweetbreads
14. Turnips	31. Brains
15. Cauliflower	32. Liver
16. Cabbage	33. Kidneys
17. Spinach	34. Fish (haddock)

consumed more raw beef, carrots and beets. In this study the children were given no guidance on what they should eat but mothers today would probably be delighted if their children only ate the type of foods which were available for the duration of this study.

To the best of my knowledge no other study of this nature has ever been made and for most children the people with which they have contact will play a very important part in influencing their food choices. Particularly in early infancy the bonding between carer and child must play a uniquely important role that will influence the future food preferences of the child far more than has been generally assumed. Some years ago Professor Leann Birch (1998) wrote:

> While children do not need to learn to like sweet or salty foods, and will readily accept even novel sweets, there is substantial evidence that children's preferences for the majority of other foods are strongly influenced by learning and experience, especially via the impact of associative learning.

Learning how to connect incidents which are associated in time and context is something that is central to the learning of evaluative responses (Martin and Levey 1978). This behaviour is not exclusive to humans; it is well known that monkeys show fear of snake-shaped objects and it was once believed that this was an instinctive behaviour but it is now thought that this fear response is learned by observation of the mother's reaction to the sight of a snake, even if this occurs only once (Mineka, Davidson *et al.* 1984). The role of child–carer relationships and observational learning by the child from siblings and other adults (for reviews see Dunn and Munn (1985) and Lazarus (1991)) appear to have been only rarely investigated in the case of eating habits but this may prove to be one of the most important influences on food acceptance and hence flavour learning in the early years of life. A mother's expression of disgust at something a young child is attempting to eat may well provide a powerful and negative association for the future acceptability of that item as a food. Conversely it has been reported that the sight of a mother eating a particular food can positively influence a child to try that food (Harper and Sanders 1975).

17.4.4 Do feeding experiences in early life modify brain structure?

The relationship between the emotional content of an eating situation and the flavour of the food has never, to the knowledge of the author, been considered in terms of the direct part it may play in the development of the baby's brain. There are, however, several things which lead to the conclusion that this could be an important and neglected influence on neonatal development. First, we have the well known fact that among all our senses it is only smell which directly influences the amygdala of the brain, the part that is the source of emotions. Second, the human brain is particularly plastic after birth and its growth and neural interconnectivity, especially in relation to the higher cognitive functions, is strongly influenced by sensory stimuli. Third, and most obviously, feeding takes up a significant part of a newborn baby's life when it is not sleeping. It is hard to see why these three associated facts could not have a lasting influence on the mental processes and emotions of that infant thereafter and particularly as they relate to the eating and enjoyment of food.

17.4.5 Neophobia and disgust in humans

As children reach the age of about two they start to develop their sense of independence and their wish to control their own environment; the children express themselves both vocally and physically and now show strong preferences or dislikes for foods. They increasingly show distrust for any new food or flavour, a phenomenon described as food neophobia. A conclusion from the previous two paragraphs is that many food prejudices are actually formed during eating experiences in the first two years of life and these may prove very difficult to change. It is known that individuals differ greatly in their degrees of neophobia and while some people show a real pleasure in trying new foods others are much more conservative in this respect (Pliner and Hobden 1992, Pliner 1994). Several authors (Rozin and Schulkin 1990, Birch and Marlin 1982) have written extensively on the subject of neophobia but population groups as a whole show dramatic and substantial shifts in food preferences that undermine arguments that this is a natural behaviour of human beings. For some animals neophobia is clearly a key factor in food selection after infancy but given appropriate motivation human societies appear to be much more flexible in changing food preferences than is generally assumed. There are nevertheless reasons why we should be more concerned about this issue because if at an early age we fail to develop likings for foods that are nutritionally important then we may create problems for future health. This is a topic to which we will return later but first let us look at the general issue of disgust and how it relates to food.

If one considers the wide range of foods eaten by all human groups on earth one must ask the question whether any edible material that provides nourishment without any ill effects can be intrinsically disgusting. If presented at a sufficiently early age with positive reinforcement from the child carer it would become an accepted part of the diet; it is possible that all notions of disgust are, in fact, learned. There are enormous differences between population groups and the flavours and foods that are accepted and liked within them; sometimes these are relatively trivial but they can also be so deeply felt that one group of people finds the food choice of another quite offensive. To illustrate how widespread these prejudices are we can take some examples. At the level of the trivial we can mention the difference in the liking of wintergreen flavour (methyl salicylate) from one side of the Atlantic to the other. Although both Americans and Europeans are largely derived from the same genetic pool the former enjoy this flavour in many confectionery products and drinks, especially in root beer, by contrast Europeans associate the flavour with ointments and embrocation but certainly not food and drinks. At a narrower geographic divide we see big differences in the liking of the flavour of autolysed yeast which is sold as a nutritional spread, rich in B vitamins. The British (Marmite®), Australians (Vegemite®) and Swiss (Cenovis®) all like these products and they are fed to children at an early age but most other nationalities find the flavour of these products quite disgusting. Nowhere are these differences seen more strikingly than across the Swiss–French border where children born within a few kilometres of each other fall into either the like or the hate groups. So deeply ingrained are these early flavour experiences that they usually remain throughout life. Moving from the trivial to the more seriously prejudiced is the eating of horse meat. Whereas in many parts of Europe a horse steak is well appreciated, most Americans find the concept quite disgusting. I once sat at a table where an American visitor unknowingly started to eat his steak with obvious enjoyment until he was told what it was; he left the table deeply distressed and nauseous. Both Americans and Europeans, however, are equally revolted by the idea of eating dog in a Korean restaurant. Similarly there are many different taboos between religious practices; many foods enjoyed and appreciated by one group are avoided by another with no sound nutritional or health logic but with deeply held convictions.

It has been suggested that disgust for foods is related mainly to animal products (Rozin and Fallon 1987) but there are also examples of extreme prejudice against plant products. The Durian fruit is loved by many people in South East Asia but most Europeans and Americans dislike its smell of fried onions with faecal overtones. From personal experience it is, however, easy to turn repulsion into enjoyment and the key to this, as for young children, is repeated exposure and encouragement to try it; in my case six to ten eating experiences were sufficient to start enjoying this fruit. I have found that I can now eat and appreciate the flavours of many foods that I initially found unpalatable (raw fish, soft shell crabs, pickled jelly fish, kimchi, and caterpillar tortillas being a few examples, but I have never had the courage to try Balut – the Philippino delicacy of a boiled embryonic duckling still in its shell. Nor, I suspect, would I be willing to try Okah, a food flavouring used in the Sudan and with a strong resemblance to soya sauce but made from cows' urine (Dirar 1993)). However, the reasons why I have or had these prejudices were probably formed in my early childhood when I learned from my mother's expressions that eggs with partly developed chicks in them or anything to do with the lavatory should be considered disgusting as food. Had I been born in the Philippines I would no doubt enjoy Balut as much as it is prized by the locals.

It may seem that such examples merely reflect curious differences between people and their choice of foods but there are some issues that relate to the ways in which children form concepts of the acceptable and digusting that may need to be considered more seriously. For most of human history food sources were dependent on supply and season, often there was simply not enough food to eat and what was available had to be consumed; snails may now be regarded as something of a delicacy on restaurant menus but we should not ignore the fact that they were eaten in years gone by simply because there was a shortage of proteins in the diets of the poor. Only 50 years ago, even in the most wealthy countries, when an animal was slaughtered most of it went into foods which were eaten and enjoyed by people. Offals were not set aside for use in pet foods, children were introduced to products made from them at an early age and they developed a liking for their flavours.

17.4.6 Should we be concerned about how children learn to like foods?

In the last 50 years there has been a dramatic change not only in the selection and choice of foods available within the developed world but also in the structure of the family unit. First, there is a substantial surplus of food which means that people can be very selective about what they eat – they can pick and choose foods that may be delicious to eat but not necessarily good nutritionally, especially if they are eaten to the exclusion of everything else. The second point to make is that in many countries the family meal has essentially disappeared and children no longer eat with their parents. A generation or two ago most children were raised in larger groups where grandparents, aunts and uncles played an important part in family life and meal-times were social occasions too. One of the most obvious demographic trends in modern industrialised societies is the decrease in the size of the family unit. The influence this change has had on the eating habits of young children does not appear to have been studied in depth but it is very likely that reducing the family meal to a trivial occasion where the child does not interact with adults will inevitably have important consequences on the flavour preferences it will develop as it matures. Many children now largely self-select their diets but from a much wider range of foods and certainly not from a range as nutritionally acceptable as that in the Davis study reported earlier. Given the general concern that exists at the present time in many countries about obesity and health problems associated with the poor diets of young people, especially children, we may need to rethink our attitudes to the importance of teaching them to eat healthy foods and to learn to like the flavours of these at an early stage in life. At least one author has expressed concern about a serious deficiency in key components of the modern diet that has essentially been created by the luxury of being able to survive on a very narrow choice of over-refined foods (Clayton 2002). If the flavours of these foods become the preferred flavours at an early age then other foods which are nutritionally important may be judged to have unacceptable or even disgusting flavours.

17.4.7 Changes in food preferences after infancy

Up until now we have discussed how and when young children form their likes and dislikes for foods. We now need to consider what happens after childhood and how adults change food preferences during life. It is not difficult to understand why neophobia in humans should exist, what is much more difficult to understand is the fact that human beings both individually and collectively change their food preferences throughout their lives. Even for young children we see behaviour which is hard to understand, for example, they eat and apparently enjoy confectionery products with levels of sourness that are totally unacceptable to their parents. Darwin in 1877 commented on his own children's preferences for rhubarb, unripe gooseberries and cooking apples which he found disgusting. In a recent study carried out at Monell it is reported that more than 30% of five- to nine-year-old children show preferences for highly acidic confections and that this correlates with lower levels of neophobia (Liem and Menella 2003). What is puzzling is why children should show this behaviour at all since it is hard to explain it in terms of any real benefits other than the gratification of being more daring than their peers, shocking their elders or of seeking novelty.

For adults it is equally hard to see why a society with an adequate and nutritious diet would spontaneously change its food and flavour preferences from generation to generation. Population groups have their own cuisine styles which have been described as culturally transmitted food practices (Rozin 1973). It has been argued that flavours (herbs, spices, condiments) are added to basic foodstuffs to satisfy the human desire for variety within what would otherwise be a monotonous diet that is dependent on a limited range of staple ingredients. It is claimed that the important function of flavourings is to label foods as familiar and thus safe (Rozin 1977) and neophobia encourages the conservatism to stay within this range of proven foodstuffs. In spite of this the history of food acceptance patterns in many cultures does not fit with this logic. When I was a child the most popular food in the United Kingdom was 'Fish and Chips', today it is 'Chicken Tikka Masala'. Who could have predicted 20 years ago that Americans would now be eating raw fish and that sushi bars would be found there in most small townships? Shifts in food preferences are not new although in the last century they have happened faster because of our global awareness; the flavours of tomatoes, chillis, peanuts, vanilla, potatoes and maize were unknown outside the Americas until the fifteenth century but became totally assimilated into other countries, sometimes to the extent that the locals often believe that these foods are indigenous to their country.

Shifts in eating habits are often driven more by health concerns, fashion and status than real nutritional considerations or availability of the foodstuff. It has been said that our food preferences are defined by socio-cultural rules rather than being influenced by physiological need (Rozin 1982) and the changes in these rules that are peculiarly human, make us behave very differently from a truly neophobic animal. There are several reasons why human beings change their diets during their lives. Nowadays the driving force for change is often health-related and a food or a drink is perceived as being more or less beneficial to general well-being or having a specific health benefit. The growing liking for cranberry juice in the Western world has little to do with the organoleptic qualities of the juice but is much more related to its claimed health benefits. The dramatic shift from high to low fat dairy products was driven by a general and popular concern about blood cholesterol levels. The recent craze for the Atkins diet with its emphasis on foods that are high in proteins and fats but low in carbohydrates is a good example, where motivation to lose weight and follow the eating habits of a few celebrities caused a dramatic shift in food consumption patterns for basic foodstuffs.

By becoming unfashionable and no longer considered appropriate for human consumption a perfectly good food can disappear. This was the fate of the parsnip in France which was once widely eaten but is now largely unknown; nevertheless it remains very popular in other European countries. The generally accepted explanation is that although well known in the fourteenth century (Pichon 1847) parsnips, known as escheroys or panais, were used less and less while carrots grew in popularity, finally the parsnip was cultivated only as a fodder crop for animals and gradually it was perceived as being suitable only for this. As other crops replaced it in this respect, especially the higher yielding varieties of beet, the parsnip finally disappeared. The perception of a food as being mainly something eaten by the poor can also lead to its demise; in 1920 there were more than 200 shops preparing and selling tripe in Manchester and now there are essentially none (Mason 2002). In contrast to these examples some foods can suddenly become very fashionable, the enormous popularity of Italian Balsamic vinegar is an example of one that was driven mainly by TV chefs who promoted its special flavour qualities to great effect. All the evidence points to the fact that human beings, given motivation and justification, like to try new foods and flavours just as much as they like to change their styles in dress sense, music and art. It seems that we simply get bored with food that is repetitive. There has been a limited amount of work on the reactions of people fed a monotonous but nutritionally adequate diet; twenty-year-old volunteers on such a regime reported a dramatic increase in their food cravings that diminished once they returned to a varied diet, whereas older people in their seventies did not show this (Pelchat 1997, 2000). Other studies also report an age-related decline in food cravings (Hill, Weaver and Blundell 1991, Basdevant 1993). One is left with the conclusion that in spite of some aspects of neophobia, we show a unique human behaviour in that we like to try new foods and flavours but that perhaps our inclination to do this declines as we get older. Ironically olfactive discrimination also decreases with age (Schiffman and Pasternak 1979) so that elderly people in spite of a reduction in food cravings are nevertheless more willing to try novel foods (Pelchat and Schaefer 2000) because they are less influenced by the odour of them. Some very recent work has used fMRI to study the brains of people who experience food craving and there are some very preliminary suggestions that the brain regions which are active during this are also those involved with drug addiction (Pelchat 2002, Pelchat, Johnson *et al.* 2004).

17.4.8 Preferences for certain popular flavours can be difficult to explain

Many of the flavours we find most attractive are first experienced and become accepted in childhood. Vanilla is the most popular flavour tonality worldwide and this is thought to be a result of its widespread use in baby-food formulations and that children's early exposure to flavours is the key to their acceptance. As adults, however, we often show preferences for flavours that are much harder to understand, we like flavours that are bitter and it is an intriguing and puzzling aspect of human behaviour that the most popular beverages in the world are tea, coffee and beer. We also like spicy foods and there has been much written on the curiously popular appeal of chilli (Rozin and Schulkin 1990). It is generally accepted that a liking for sweet things is the only innate flavour preference that we have at birth, so what are the reasons for people eventually learning to like flavours that, according to all the accepted rules, should be rejected? The drinking of coffee is an interesting subject of study, not only because of the popularity of this drink even though it has intrinsically unpleasant flavour characteristics, but also because it is served in so many different versions, an Italian ristretto in Rome is a very different drink from a café latté served in San Francisco. Several groups have investigated whether the stimulant action of caffeine could explain our liking for coffee (Yeomans, Jackson *et al.* 2000, Rogers, Martin *et al.* 2003). These workers were able to show that when flavours were paired with drinks which contained caffeine then some people showed enhanced preferences for these flavours in comparison with drinks that did not contain the stimulant. However, this only occurred for those people that were already consumers of coffee but who had been deprived of caffeine prior to the trials; people that were not dependent on caffeine showed no enhanced liking for the flavours paired with it, so a post-ingestive boost from caffeine could not explain why people learned to like the flavour of coffee before they became caffeine dependent. This therefore leaves the question of why we persist in drinking coffee completely open. One possible explanation that does not appear to have been investigated is that as children we learn to associate coffee, and in particular its very recognisable smell, with pleasant memories associated with adult company. Thus the smell of freshly brewed coffee at the start of a new day when breakfast is being prepared, being with adults enjoying coffee in a happy social situation, such as at the end of a meal, might form a learned association between this easily identifiable smell and feelings of contentment and maturity (see also Rozin (1982)). Has the association of coffee odour with things that are intrinsically adult made the eventual liking of its flavour a rite of passage from childhood to maturity?

As adults we may experience several life changes that can directly influence our food choices and, to a greater or lesser degree, modify our flavour preferences. Usually when people change foods or drinks they gradually come to prefer their new choice over the original. This is what happens when people who normally sweeten their coffee decide to avoid sugar; initially the unsweetened coffee is not really liked but eventually they will prefer it and often will then dislike coffee that is sweetened. Relocation to a new country is a situation where social pressures usually force a change in food choice and there is also a reported shift towards a liking for odours found in the host country (Hudson and Distel 2002). One situation which has been studied in some detail is the way that marriage requires a shift in food acceptability for both partners towards a jointly accepted diet (Bove, Sobal *et al.* 2003). Seen in the context of flavour learning and how with repeated exposure to new foods the brain gradually integrates sensory signals to create a new pattern of neuronal activity, one that is associated with satisfying food, can explain many of our observed behaviours. It seems inevitable that a deeper understanding of flavours and the human preferences for them will rely increasingly on neuroscience and experimental psychology as key scientific disciplines.

Let us briefly review the main points made in the preceeding paragraphs before we move to the final section:

- Flavour is a multi-sensory experience which is constructed by the brain from the totality of sensory experiences during eating and drinking and one that is then perceived in the mouth.
- The flavour memories of a human baby start to be laid down before birth from the foods eaten by the mother, after birth they are greatly influenced by feeding occasions when the emotional content of the experience also influences what becomes liked.
- Associative learning by the young infant probably plays a very important role in developing attitudes towards flavours that are liked or disliked.
- There is probably no food which is inherently disgusting and prejudice against any specific food is a learned behaviour.
- Any attempt by the carer to influence preferences for food or flavour by the developing child will fail if these are associated with stressful situations although they may be unrecognised as such by the carer.
- Repeated exposure to a novel food or flavour is important if it is to be eventually accepted and liked.
- Although young children may show patterns of neophobia, human beings in general are much more curious to try new foods and flavours than one might expect.
- The gradual acceptance of new foods and drinks by the adolescent is probably driven by the wish to adopt the habits of and to participate in adult society.
- Many human beings generally become bored with eating the same foods and seek variety just as they do in other pleasure seeking activities such as fashion, art or music. This appears to be true more for younger than older people.
- An issue for concern is that in affluent societies with a surplus of food we may inadvertently be allowing young children to self-select foods and to develop patterns of flavour preference which could discourage future liking for nutritionally essential components of the diet.

17.5 Future trends

There is no doubt that advances in brain scanning techniques have greatly improved our knowledge of how the brain develops and functions. The present generation of body scanners have, of course, been designed with medical diagnostic applications as the first priority but they are not wholly suitable for studies on eating and drinking. Most fMRI scanners are configured such that the subject lies prone and is introduced into the magnet coil horizontally. Consuming foods and beverages when lying flat on one's back is not very easy and the brain under study may well be using a lot of effort to prevent the subject from inadvertently choking. The likely future development of head scanning fMRI which will allow the subject to sit upright will significantly improve this situation. Many groups are seeking to increase the speed and precision of scanning techniques and there is a general desire to observe brain activity in real-time and thus understand the temporal flow of its neuronal activity. There is evidence that oscillatory patterns set up between neurons may be an important aspect of brain function (Freeman 1991); this author has also examined the olfactory bulbs of thirsty rabbits and has reported that when odours are associated with significant events, such as the availability of water, then oscillatory firing of neurons occurs which is related to the animals' understanding that water may be forthcoming (Freeman 1995).

Scientists are also trying to understand brain processes from a mathematical perspective and theories of chaos and attractor states (Gleick 1987, Kauffmann 1995) seem highly relevant to a better understanding of the dynamic changes in brain states which can lead to transient but stable patterns of neuronal activity. Chaos theory may also give clues about how small changes in the environment can have dramatic effects on the higher brain states which affect mood, an important aspect of observational learning (Elbert, Ray *et al.* 1994).

Many research groups are studying the unsolved questions related to conscience, feelings and emotions; we all know what these are but from where they come and how a vastly interconnected set of neurons within our heads can create them still remains largely a mystery. In this chapter we have digressed a long way from a preoccupation with flavour molecules and their chemistry. Nevertheless food and the pleasure of its flavours play an important part in our lives and are intimately linked to many of our deeply satisfying experiences. Marcel Proust famously commented on this in his 'Remembrance of things past' when he described the memory and emotions associated with eating a madeleine cake dipped in lime-blossom tea (Proust 1913). The pleasures of eating start when we are born and if we are lucky only stop when we die; a fact that cannot be claimed even for sex, although the latter seems to pre-occupy people far more than gastronomy does. The fact that we have meals several times each day means that most people take eating for granted and this may obscure an important role it might play in the development of the human brain and its complex emotions. Many schools of philosophy have linked the foods we eat to the vitality of the body, Hippocratic doctrine laid down ideas of health and diet which have reappeared with various modifications from the time of the Golden Age of Greece. The Chinese concepts of yin and yang and bodily harmony reflect these principles as well. New theories are emerging that link feelings and emotions to the state of the body in general (Ledoux 1996). The neurologist Professor Antonio Damasio (2003) develops his theory of feelings, emotions and their origins in his book *Looking for*

Spinoza. In summary, he believes that there is a constant dialogue between the brain and the body with the objective of maintaining homeostasis and if possible to optimise the quality of the body state. This is achieved by the brain constructing neuronal maps of all aspects of the body's physiology and any departure of the body-in-mind from the optimum leads to feelings and actions which seek to restore the system. This might involve externally directed activities such as eating and drinking, changes in behaviour such as sleeping, or brain directed changes of the body itself, for example, the control of blood glucose levels. Feelings are the perceptions derived from the representation of the body within the brain; for example, a non-optimum state of our internal body chemistry could appear in the map of body-in-mind as a composite feeling of hunger. In the case of humans the situation is more complicated because our brains are capable of recreating situations from memory or indeed of creating mental situations which are entirely fictional. Human beings not only have maps of the body state within their heads but also have imagination which creates perceptions that need not relate to actual events. The feelings that we have are not only derived from actual body states but also all the associated mind states. A feeling for a food such as disgust or, conversely, one of delight, as in the case of Proust, not only comes from the food itself through our senses (which are all mapped into the brain as changes in body states) but are also derived from mind states which can relate to real or imagined situations. The survival of an organism depends on a rapid development of brain maps which will be guided by the experience of feeding events. Monitoring of a food before and during eating depends on all the senses and from these a composite sensation of flavour is derived. The post-ingestional consequences are stored for future reference as maps of the body-state and hence as feelings – satiation, comfort, happiness, pain, nausea, guilt, sadness etc. As humans we are unique in that our mental state can be modified because of internally generated beliefs and convictions. These can be social in origin (religion, fashion, situation) or self-generated (self-esteem, joy, anxiety, sadness) but all are important drivers of eating habits. In creating these feelings humans can also project into the future the consequences of an activity such as eating. Thus a person seeking to lose weight may well want to accept the offer of another helping of dessert but because he or she can relate this to their longer-term health concerns they may well decline; conversely if he or she accepts they will experience not only the immediate pleasure of eating the dessert but also feelings of guilt. Work has been published that aims to understand how music affects our brain activity and changes our emotional state (Blood and Zatorre 2001). In the next decade we will no doubt learn much more about what is going on inside our heads when we sit down to a favourite meal and experience the anticipation and pleasure of the occasion.

17.6 References

ARANEDA, R. C., A. D. KINI *et al.* (2000) 'The molecular receptive range of an odorant receptor.' *Nature Neuroscience* 3: 1248–1255.

BAINES, Z. V. and E. R. MORRIS (1989) Suppression of perceived flavour and taste by food hydrocolloids. *Food Colloids*. R. D. Bee, P. Richmond and J. Mingens. Cambridge, Royal Society of Chemistry 75: 184–192.

BASDEVANT, A. *et al.* (1993) 'Snacking patterns in obese French women.' *Appetite* 21: 17–23.

BIRCH, L. L. (1980). 'Effects of peer model's food choices and eating behaviors on preschoolers' food preferences.' *Child Development* 51: 489–496.

BIRCH, L. L. (ed.) (1987) *Children's Food Preferences: developmental patterns and environmental influences*. Annals of Development. Greenwich, JAI Press.

BIRCH, L. L. (1990) The control of food intake by young children: the role of learning. *Taste, Experience and Feeding*. D. Capaldi and T. L. Powley. Washington D.C., American Psychological Association: 116–135.

BIRCH, L. L. (1998) 'Development of Food Acceptance Patterns in the First Years of Life.' *Proceedings of the Nutrition Society* 57: 617–624.

BIRCH, L. L., L. GUNDER *et al.* (1998) 'Infants' consumption of a new food enhances acceptance of similar foods.' *Appetite* 30(3): 283–295.

BIRCH, L. L. and D. W. MARLIN (1982) 'I don't like it; I never tried it: effects of exposure on two-year-old children's food preferences.' *Appetite* 3: 353–360.

BIRCH, L. L., S. I. ZIMMERMANN *et al.* (1980) 'The influence of social-affective context on the formation of children's food preferences.' *Child Development* 51: 856–861.

BIRCH, L. L., MARLIN D.W. and ROMER, J. (1984). 'Eating as the "means" activity in a contingency: effects on young children's food preference.' *Child Development* 55: 432–439.

BLACK, J. E. and W. T. GREENOUGH (1998) Developmental approaches to the memory process. *Neurobiology of Learning and Memory*. J. L. J. Martinez and R. P. Kesner. San Diego, Academic Press: 55–88.

BLAKE, A. (2004) Flavour perception and the learning of food preferences. *Flavour Perception*. A. J. Taylor and D. D. Roberts. Oxford, Blackwell Publishing.

BLANCHARD, R. J., M. YANG *et al.* (2001). 'Cue and context conditioning of defensive behaviors to cat odor stimuli.' *Neuroscience and Behavioural Reviews* 25: 587–595.

BLOOD, A. J. and R. J. ZATORRE (2001) Intensely pleasurable responses to music correlates with activity in brain regions implicated in reward and emotion. *Proc. Natl. Acad. Sciences* 98: 11818–11823.

BOAG, B. and J. MLOTKIEWICZ (1994) Effect of odour derived from lion faeces on behaviour of wild rabbits. *J. Chem. Ecol.* 20: 631–637.

BOURNE, M. C. (2002) *Food Texture and Viscosity*. San Diego, Elsevier.

BOVE, C. F., J. SOBAL *et al.* (2003) 'Food choices among newly married couples: convergence, conflict, individualism, and projects.' *Appetite* 40: 25–41.

BROWN, M., R. KEYNES *et al.* (2001) *The Developing Brain*, Oxford, Oxford University Press.

CALVERT, G. A., E. T. BULLMORE *et al.* (1997) 'Activation of auditory cortex during silent lipreading.' *Science* 276: 593–596.

CALVERT, G. A. and R. OSTERBAUER (2002) 'The Scent of Color.' *The Aroma-Chology Review* 11(1): 1–2, 6.

CATERINA, M. J., M. A. SCHUMACHER *et al.* (1997) 'The capsaicin receptor: a heat-activated ion channel in the pain pathway.' *Nature* 389: 816–824.

CERF-DUCASTEL, B. and C. MURPHY (2001). 'fMRI activation in response to odorants orally delivered in aqueous solutions.' *Chemical Senses* 26(6): 625–637.

CLAYTON, P. (2002) Health Defense in *Textbook of Men's Health*, Lunenfeld, B. and

Gooren, L. New York, Parthenon Press.

CLEAVER, H. and D. DERBYSHIRE (2002) Girl with half a brain becomes fluent in two languages. *Daily Telegraph*. London.

CLIFFORD, E. (1999) Neural plasticity. *Harvard Brain*. Merzenich, Taub and W. T. Greenough. 16: 16–20.

COOK, D. J., T. A. HOLLOWOOD *et al.* (2003) 'Oral shear stress predicts flavour perception in viscous solutions.' *Chemical Senses* 28: 11–23.

DALTON, P., N. DOOLITTLE *et al.* (2000) 'The merging of the senses: integration of subthreshold taste and smell.' *Nature Neuroscience* 3(5): 431–432.

DAMASIO, A. (2003) *Looking for Spinoza – Joy, Sorrow and the Feeling Brain*. Orlando, Harcourt Inc.

DAVIDSON, J. M., T. A. HOLLOWOOD *et al.* (1999) 'The effect of sucrose on the perceived flavour intensity of chewing gum.' *Journal of Agricultural and Food Chemistry* 47: 4336–4340.

DAVIS, C. M. (1939) 'Results of the self-selection of diets by young children.' *J. Canadian Medical Assoc.* 41: 257–261.

DE CASPER, A.-J. and W. P. FIFER (1980) 'Of human bonding. Newborns prefer their mother's voices.' *Science* 208: 1174–1176.

DE CASPER, A.-J. and M. J. SPENCE (1986) 'Prenatal material speech influences newborn's perception of speech sounds.' *Infant Behaviour & Development* 9: 133–150.

DIRAR, H. A. (1993) Flavours and Substitutes (chapter 9). *The Indigenous Fermented Foods of the Sudan: A Study in African Food & Nutrition*. Wallingford, UK, CAB International: 469–472.

DOBRASZCZYK, B. J. and J. F. V. VINCENT (1999) Measurement of mechanical properties of food materials in relation to texture. *Food texture: Measurement and perception*. A. J. Rosenthal. Gaithersburg, Aspen Publishers Inc.: 99–151.

DOTY, R. L. (1992) Olfactory function in neonates, in *The Human Sense of Smell*. D. G. Laing, R. L. Doty and W. Breipohl, Springer-Verlag, Berlin/Heidelberg, 155–165.

DRAGANSKI, B., C. GASER *et al.* (2004) 'Changes in grey matter induced by training.' *Nature* 427: 311–312.

DRAVNIEKS, A. (1985) *Atlas of Odour Character Profiles*. Philadelphia, American Society for Testing of Material.

DUNN, J. and P. MUNN (1985) 'Becoming a family member: Family conflict and the development of social understanding in the second year.' *Child Development* 56: 480–492.

DURAN, L. and E. COSTELL (1999) 'Perception of taste physiochemical and psychophysical aspects.' *Food Science and Technology International* 5(4): 299–309.

ELBERT, T., W. J. RAY *et al.* (1994) 'Chaos and Physiology: Determining chaos in excitable cell assemblies.' *Physiology Review* 74: 1–47.

ELEJALDE, C. C. and J. L. KOKINI (1992) 'The psychophysics of pouring, spreading and in-mouth viscosity.' *Journal of Texture Studies* 23: 315–336.

FISHER, J. O., L. L. BIRCH *et al.* (2000) 'Breast-feeding through the first year predicts maternal control in feeding and subsequent toddler energy intakes.' *Journal of the American Dietetic Association*. 100: 641–646.

FRANK, R. A. and J. BYRAM (1988) 'Taste–smell interactions are tastant and odorant dependent.' *Chemical Senses* 13(3): 445–455.

FREEMAN, W. J. (1991) 'The physiology of perception.' *Scientific American*. 264 (Feb): 78–85.

FREEMAN, W. J. (1995) *Societies of Brain: A study in the Neuroscience of Love and Hate*.

Hillsdale, New Jersey, Laurence Erlbaum.

FRIEDRICH, R. W. and G. LAURENT (2001) 'Dynamic organisation of odour representations in the olfactory bulb by slow temporal patterning of mitral cell activity.' *Science* 291: 889–894.

GERRISH, C. J. and J. A. MENELLA (2001) 'Flavor variety enhances food acceptance in formula-fed infants.' *American Journal of Clinical Nutrition* 73: 1080–1085.

GLEICK, J. (1987) *Chaos: making a new science*. New York, Penguin Books.

GOTTFRIED, J.A. and R. J. DOLAN (2003) 'The nose smells what the eye sees: crossmodal visual facilitation of human olfactory perception.' *Neuron* 39: 375–386

GREEN, B. G. (2001) Psychophysical measurement of oral chemesthesis. *Methods in chemosensory research*. S. A. Simon and M. A. L. Nicolesis. Boca Raton, CRC Press: 3–19.

HARPER, L. and K. M. SANDERS (1975) 'The effect of adult's eating on young children's acceptance of unfamiliar foods.' *Journal of Experimental Child Psychology* 20: 206–214.

HARPER, R., D. G. LAND *et al.* (1968). 'Odour qualities: a glossary of usage.' *British Journal of Psychology* 59: 231–252.

HILL, A. J., C. F. L. WEAVER and J. E. BLUNDELL (1991) 'Food craving, dietary restraint and mood.' *Appetite* 17: 187–197.

HOLLOWOOD, T. A., R. S. T. LINFORTH *et al.* (2000). The relationship between volatile release and perception. *Flavor release: Linking experiments, theory and reality*. D. D. Roberts and A. J. Taylor. Washington DC, American Chemical Society. 763: 370–380.

HORT, J. and T. A. HOLLOWOOD (2004) 'Controlled continuous flow delivery system for investigating taste-aroma interactions' *Journal of Agricultural and Food Chemistry*. 52: 4834–4843.

HUDSON, R. and H. DISTEL (2002) The Individuality of Odor Perception. *Olfaction, Taste and Cognition*. Rouby, C., Schaal, B., Dubois, D., Gervais, R. and Holley, A. Cambridge University Press, Cambridge, UK.: 408–420

HUGHSON, A. L. and R. A. BOAKES (2001) 'Perceptual and Cognitive Aspects of Wine Expertise.' *Australian Journal of Psychology* 53(2): 103–108.

IKEDA, K. (1909). 'About new seasoning.' *Journal of the Tokyo Chemical Society* 30: 820–836.

JOHNSON, S. L. A. A. (1991) 'Conditional Preferences: Young Children Prefer Flavors Associated with High Dietary Fat.' *Physiology & Behavior* 50: 1245–1251.

KAJIYA, K., K. INAKI *et al.* (2001) 'Molecular bases of odour discrimination: reconstruction of olfactory receptors that recognise overlapping sets of odorants.' *Journal of Neuroscience* 21: 6018–6025.

KALAT, J. W. (1985) Taste-aversion learning in ecological perspective. *Issues in the Ecological Study of Learning*. T. Johnston and A. Pietrewicz. Hillsdale, NJ, Erlbaum: 119–141.

KAUFFMANN, S. (1995) *At Home in the Universe: the search for the laws of self-organisation and complexity*, Oxford, Oxford University Press.

KILCAST, D. and S. CLEGG (2002) 'Sensory perception of creaminess and its relationship with food structure.' *Food Quality and Preference* 13(7–8): 609–623.

KING, A. (2003). *Development of Multi-sensory Spatial Integration*, Oxford University Press, Oxford.

KORSMEYER, C. (1999) *Making Sense of Taste: food and philosophy*. New York, Cornell University Press.

LAURENT, G., M. STOPFER *et al.* (2001) 'Odor encoding as an active dynamical process.' *Ann. Rev. Neuroscience* 24: 263–297.

LAZARUS, R. S. (1991) Individual development. *Emotion and Adaptation*. New York, Oxford University Press: Chapter 8.

LEDOUX, J. E. (1996) *The Emotional Brain*. New York, Simon & Schuster Inc.

LEVIN, H. S. and J. GRAFMAN (2000) *Reorganisation of Function after Brain Damage*. Oxford, Oxford University Press.

LIEM, D. G. M. and J.A. MENELLA (2003) 'Heightened sour preferences during childhood.' *Chem. Senses* 28: 173–180.

LIVERMORE, A. and D. G. LAING (1996) 'Influence of training and experience on the perception of multicomponent odor mixtures.' *Journal of Experimental Psychology – Human Perception and Performance* 22(2): 267–277.

MALNIC, B., J. HIRONO *et al.* (1999) 'Combinatorial receptor codes for odors.' *Cell* 96: 713–723.

MARTIN, I. and A. B. LEVEY (1978) 'Evaluative conditioning.' *Advances in Behaviour Research and Therapy* 1: 57–102.

MASON, L. (2002) 'Mancunian Foods.' *The International Herald of Taste* (24): 36.

MCGURK, H. and J. MACDONALD (1976) 'Hearing lips and seeing voices.' *Nature* 264: 746–748.

MENELLA, J. A. and G. K. BEAUCHAMP (1991) 'Maternal diet alters the sensory qualities of human milk and the nursling's behavior.' *Pediatrics* 88: 737–744.

MENELLA, J. A. and G. K. BEAUCHAMP (1996) 'The human infants' responses to vanilla flavors in human milk and formula.' *Infant Behaviour & Development* 19: 13–19.

MENELLA, J. A. and G.K. BEAUCHAMP (2002) 'Flavor experiences during formula feeding are related to preferences during childhood.' *Early Human Development* 68: 71–82.

MENELLA, J. A., C. P. JAGNOW *et al.* (2001) 'Prenatal and postnatal flavor learning by human infants.' *Pediatrics* 107(E88).

MINEKA, S., M. DAVIDSON *et al.* (1984) 'Observational conditioning of snake fear in rhesus monkeys.' *Journal of Abnormal Psychology* 93: 355–372.

MOROT, G., F. BROCHET *et al.* (2001) The color of odors. *Brain and Language*, New York, Academic Press.

MORROW, B. A., A. J. REDMOND *et al.* (2000) 'The Predator Odour, TMT, displays a unique, stress-like pattern of dopaminergic and endocrinological activation in the rat.' *Brain Research* 864: 146–151.

MURPHY, C., W. S. CAIN *et al.* (1977) 'Mutual Action of Taste and Olfaction.' *Sensory Processes* 1: 204–211.

NEWMAN, J. and A. TAYLOR (1992) 'Effect of a means-end contingency on young children's food preferences.' *Journal of Experimental Child Psychology* 64: 200–216.

NORTON, C. R. T., J. R. MITCHELL *et al.* (1998) 'Fractal determination of crisp or crackly textures.' *Journal of Texture Studies* 29(3): 239–253.

PANKSEPP, J. (1998) *Affective Neuroscience – The foundations of human and animal emotions*. New York, Oxford, Oxford University Press.

PELCHAT, M. L. (1997) 'Food cravings in young and elderly adults.' *Appetite* 28: 103–113.

PELCHAT, M. L. (2000) 'You can teach an old dog new tricks: olfaction and responses to novel foods by the elderly.' *Appetite* 35(2): 153–160.

PELCHAT, M. L. (2002) 'Of human bondage: Food craving, obsession, compulsion, and addiction.' *Physiology and Behaviour* 76: 347–352.

PELCHAT, M. L., A. JOHNSTON *et al.* (2004) 'Images of Desire: Food Craving Activation During fMRI.' *NeuroImage* 23, 1486–1493.

PELCHAT, M. L. and S. SCHAEFER (2000) 'Dietary monotony and food cravings in young and elderly adults.' *Physiology and Behaviour* 68: 353–359.

PERNOLLET, J. C. and L. BRIAND (2004) Structural recognition between odorants, olfactory-binding proteins and olfactory receptors, first events in odour coding. *Flavour Perception*. A. J. Taylor and D. D. Roberts. Oxford, Blackwell Publishing.

PICHON, J. (1847) Menagier de Paris. L'imprimerie de Crapelet, rue de Vaugirard 9, Paris.

PLINER, P. (1994) 'Development of Measures of Food Neophobia in Children.' *Appetite* 23(2): 147–163.

PLINER, P. and K. HOBDEN (1992) 'Development of a scale to measure the trait of food neophobia in humans.' *Appetite* 19: 105–120.

PRESCOTT, J. (1999) 'Flavour as a psychological construct: implications for perceiving and measuring the sensory qualities of food.' *Food Quality and Preference* 10: 349–356.

PRINZ, J. F. and R. A. DEWIJK (2004) 'The role of oral processing in flavour perception,' Chapter 2 in *Flavour Perception*, Taylor, A.J. and D.D. Roberts, Blackwell Publishing, Oxford, 45–46

PROUST, M. (1913) *Du côté de chez Swann (Swann's Way)*. New translation into English (2003) by Lydia Davis, New York, Viking Press.

RAKIC, P. (2004) 'Immigration denied.' *Nature* 427: 685–686.

ROGERS, P. J., J. MARTIN *et al.* (2003) 'Absence of reinforcing, mood and psychomotor performance effects of caffeine in habitual non-consumers of caffeine.' *Psycho Pharmacology* 167: 54–62.

ROLLS, E. T. (1996) 'A theory of hippocampal function in memory.' *Hippocampus* 6: 601–620.

ROLLS, E. T. (1999) *The Brain and Emotion*. Oxford, Oxford University Press.

ROLLS, E. T. and L. L. BAYLIS (1994) 'Gustatory, olfactory, and visual convergence within the primate orbitofrontal cortex.' *Journal of Neuroscience* 14(9): 5437–5452.

ROUDOT, A. C. (2004) 'Food science and consumer taste.' *Gastronomica – the Journal of Food and Culture* 4(1): 41–46, University of California

ROZIN, E. (1973) *The Flavor Principle Cookbook*. New York, Hawthorn.

ROZIN, E. (1982) The structure of cuisine. *The Psychobiology of Human Food Selection*. L. M. Barker. Westport, AVI Publishing Co. Inc.

ROZIN, P. (1977) The use of characteristic flavorings in human culinary practice. *Flavor: Its Chemical, Behavioral and Commercial Aspects*. C. M. Apt. Boulder, CO, Westview Press.

ROZIN, P. (1982) Human food selection: the interaction of biology, culture and individual experience. *The Psychology of Human Food Selection*. L. M. Barker. Chichester, Wiley: 225–254.

ROZIN, P. and A. E. FALLON (1987) 'A perspective on disgust.' *Psychological Review* 94(1): 23–41.

ROZIN, P. and J. SCHULKIN (1990) Food selection. *Handbook of Behavioral Neurobiology*, volume 10: *Food and Water Intake* E. M. Stricker (ed.). Plenum, New York, 297–328.

SANAI, N., A. D. TRAMONTIN *et al.* (2004) 'Unique astrocyte ribbon in adult human brain contains neural stem cells but lacks chain migration.' *Nature* 427: 740–744.

SCHAAL, B., L. MARLIER *et al.* (1998) 'Olfactory function in the human fetus: evidence from selective neonatal responsiveness to the odor of amniotic fluid.' *Behavioural*

Neuroscience 112: 1438–1439.

SCHAAL, B., L. MARLIER *et al.* (2000) 'Human fetuses learn odors from their pregnant mother's diet.' *Chemical Senses* 25: 729–737.

SCHAAL, B., R. SOUSSIGNAN and L. MARLIER (2002) Olfactory cognition at the start of life: the perinatal shaping of selective odour responsivess. *Olfaction, Taste and Cognition*, Rouby, C., Schaal, B., Dubois, D., Gervais, R. and Holley, A. Cambridge University Press, Cambridge, UK: 421–440.

SCHIFFMAN, S. S. and M. PASTERNAK (1979) 'Decreased discrimination of food odours in the elderly.' *Journal of Gerontology* 34(1): 73–79.

SCIENTIFIC AMERICAN (1999) *The Scientific American Book of the Brain*. New York, Scientific American.

SCLAFANI, A. (2001) 'Post-ingestive positive controls of ingestive behaviour.' *Appetite* 36: 79–83.

SEELEY, R. J. and C. WOODS (2003) 'Monitoring of Stored and Available Fuel by the CNS: Implications for Obesity.' *Nature Neuroscience Reviews* 4: 901–909.

SMALL, D. M., R. J. ZATORRE *et al.* (2001) 'Changes in brain activity related to eating chocolate: from pleasure to aversion.' *Brain* 124 (Pt 9): 1720–1733.

SPENCE, C. and J. DRIVER (2004) *Crossmodal Attention*. Oxford, Oxford University Press.

STEVENSON, R. J. (2001a) 'Associative learning and odour quality perception: How sniffing an odour mixture can alter the smell of its parts.' *Learning & Motivation* 32: 154–177.

STEVENSON, R. J. (2001b) 'The acquisition of odour qualities.' *The Quarterly Journal of Experimental Psychology* 54: 561–577.

STEVENSON, R. J. and R. A. BOAKES (2003) 'A mnemonic theory of odour perception.' *Psychological Review* 110(2): 340–364.

STEVENSON, R. J., R. A. BOAKES *et al.* (1998) 'Changes in odor sweeteners resulting from implicit learning of a simultaneous odor-sweetness association: an example of learned synethesia.' *Learning & Motivation* 29: 113–132.

SULLIVAN, R. M., TABORSKY-BARBA, S. *et al.* (1991) 'Olfactory classical conditioning in neonates.' *Pediatrics* 87: 511–517.

SULLIVAN, S. A. and L. L. BIRCH (1994) 'Infant dietary experience and acceptance of solid foods.' *Pediatrics* 93: 271–277.

TAYLOR, A. J., R. S. T. LINFORTH *et al.* (2000) 'Atmospheric pressure chemical ionisation mass spectrometry for in vivo analysis of volatile flavour release.' *Food Chemistry* 71: 327–338.

TOUHARA, K. (2002) 'Odor discrimination by G protein-coupled olfactory receptors.' *Microscopy Research and Technique* 58: 135–141.

WHEATLEY, J. (1973) 'Putting Colour into Marketing.' *Marketing* 67: 24–29.

WILSON, D. A. and R. J. STEVENSON (2003) 'The fundamental role of memory in olfactory perception.' *Trends in Neuroscience* 26(5): 243–247.

YEOMANS, M. R., A. JACKSON *et al.* (2000) 'Expression of flavour preferences conditioned by caffeine is dependent on caffeine deprivation state.' *Psychopharmacology* 150: 208–215.

ZAMPINI, M. and C. SPENCE (2004) 'The role of auditory cues in modulating the crispness and staleness in crisps.' *Journal of Sensory Studies* 19: 347–363.

ZELLNER, D. and M. KAUTZ (1990) 'Color affects perceived odor intensity.' *Journal of Experimental Psychology: Human Perception and Performance* 16(2): 391–397.

18

The development of flavour perception from infancy to adulthood

B. Schaal, Centre Européen des Sciences du Goût, France

18.1 Introduction

Our current knowledge of human olfaction derives mainly from research carried out with samples of adult participants. What is more, this knowledge often taps on biased sampling of more easily accessible age groups in the adult population, i.e. young adults in their twenties. Recently, due to growing preoccupations regarding the well-being of elderly people in Western countries, the public health authorities have sponsored research into the ageing of chemical senses. This has stimulated the design of standardised chemo-sensometric instruments and to the establishment of corresponding sensitivity norms, and has resulted in improved understanding of some processes affecting olfactory perception and cognition throughout the *adult* lifespan. However, at the other end of the life cycle, during the first 15 years, our knowledge of smell still remains fragmentary. In fact, the amount of developmental data is dependent on the accessibility of paediatric populations. It is fairly in-depth for easily reachable newborns, but then extremely scarce between 2 weeks and 3 years of age, after which it again becomes more abundant between the age of 4 and puberty. Thus, our overall appreciation of the development of odour sensitivity remains poor during the first years of life (cf. Beauchamp *et al.* 1991, Doty 1992, Schaal 1988, 1999, for reviews), and the data on when and how odour and flavour preferences are generated is both limited and still contradictory.

Increased research effort in the domain of olfactory development during the newborn, infant and juvenile periods would be important for several reasons. First, from a basic science standpoint, such research would illuminate more general issues on the biological and psychological mechanisms of human perception: how does perception change according to age, brain maturation, experience, cognitive growth and language acquisition? Which processes modulate early fluctuations in preferences and cognition? Do certain early sensory skills result from specialisations evolved to optimise responses at the times of the challenging transitions of childhood (birth, weaning, puberty)? Second, from more applied perspectives, some rare findings suggest that infantile experience with specific odours or flavours can canalise lifelong perceptual abilities and subsequent attitudes. These long-term effects of early experience relate to notions such as sensitive period, cerebral plasticity and memory, and also, more importantly, to the early programming of food liking, addictive habits, and social choices.

The aim of this chapter is to point out the importance of a developmental approach in understanding the functions and functioning of olfaction in our own species. It is mainly devoted to the earliest odour and flavour sensations in the newborn and young infant, and how these may mediate subsequent preferences. The first part will briefly summarise a number of observations and experiments often inspired from animal research and which situate the roles of smell in functional contexts met in the subjects' everyday or habitual life. The second part provides a (non-exhaustive) review of results obtained using experimental techniques focusing on meticulous stimulus monitoring and response recording at multiple levels of integration. Although both of these approaches were designed after theoretical options – ethology and experimental psychology – that were divergent at times, we will see that they complement each other. A final section will illustrate the environmental susceptibility of olfactory functioning in very early life and its long-term, sometimes lifelong, consequences.

18.2 Functional value of olfaction in early development

The application of various methods designed in animal ethology (Tinbergen 1951) has greatly fuelled the research on the development of smell in our own species. One experimental method most employed in ethology has been to analyse the spontaneous responses of an organism when exposed to a given conspecific, or to a more or less complex set of representative stimulations from it but in its physical absence. This method relying on models or dummies has been successfully used to understand which key stimulations control and regulate inter-individual exchanges. In 1975, A. Macfarlane came up with the idea of applying this principle to studying olfaction in human infants. His method is based on two simple principles, particularly innovative in a scientific context dominated at the time by 'hard-line' experimental psychology: present an odour which had acquired psychological salience for the infant, viz. an ecological stimulus, and then measure responses which are normally expressed in the interactions with the mother or, more generally, with the wider social environment or physical environment. Macfarlane's now classical contribution initiated a series of research projects on infant reactivity to ecological odours, and on the implications of such odours in the advent of adaptive development. Using this ethological approach, the involvement of olfaction was explored in more detail in two areas, early social interactions and the acquisition of food preferences/aversions.

In both of these functional domains, the responses elicited by odorants reflect the generally dichotomous organisation of behaviour in terms of approach/withdrawal (Schneirla 1965). These dichotomous responses can be operationally defined by opposite changes in either vigilance state (high vs low arousal states) and related attentional capacities (interest vs disinterest), exploratory and directional movements (moving towards vs away), optimisation of information-intake (intensified vs inhibited inhalation), lingual movements (increased contact vs refusal), and lastly ingestion decisions (intake vs rejection).

18.2.1 Behavioural activation

Odours can markedly affect the activational state of newborn infants. For example, infants between 2 and 10 days old, when fussing before nursing, reduce the amplitude of head and arm movements when exposed to the odour of a gauze pad previously worn on the breast or neck of a nursing mother (Schaal *et al.* 1980). Their motor activity diminishes more when these odours come from their own mother rather than from another nursing mother or from an odourless gauze pad (Schaal 1986). Such a soothing effect was also reported for the mother's overall odour as opposed to the homologous odour from another woman (Sullivan and Toubas 1998). When such stimulations are administered to calm or crying infants aged under 24 hours, those crying are soothed when presented with a maternal odour (irrespective of its source), while those who are calm display increased mouthing-licking-sucking responses (more to their mother's odour than to the odour of an unfamiliar woman).

The odour of human milk appears to be especially effective in modulating the activational state of infants. As further detailed below, milk odour elicits oral and general movements in calm infants. In certain conditions, it can also bring them to a calm state when they are aroused. For example, when three different groups of 3-day-olds were exposed to the odour of their familiar milk during a painful heel prick (breast-fed infants to their mother's milk, bottle-fed infants to their formula milk, and a group composed equally of breast- and bottle-feeders to a blank) those exposed to the odour of maternal milk reduced their crying more quickly than those in the other two groups, which did not differ from each other (Mellier *et al.* 1997). Before concluding from the prior study that human milk has any particular calming effect, however, the differential effects caused by conspecific milk, by cow-based artificial milk and by a control odour have to be compared in the same infants. Recently, experiments with preterm and full-term infants indicated that an odour (viz. vanilla) that had become familiar through mere diffusion in the incubator or in the crib was sufficient to reduce expressed pain reactions during blood sampling (Goubet *et al.* 2003, Rattaz *et al.* 2005). But when, in this context, the effect of the familiar vanilla odour was compared with the effect of the familiar mother's milk odour, the latter appeared more efficient in reducing motor agitation during the painful heel stick (Rattaz *et al.* 2005), suggesting that human milk may carry compounds which are especially reactogenic to infants. Finally, the physiological response (cortisol level in saliva) to the stress provoked by a painful procedure can be attenuated by presenting non-biological odours (such as lavender, or γ-dodecalactone thought to mimic milk odour) (Kawakami *et al.* 1997).

All in all, these results suggest that different maternal body odours, or odorants that have gained positive meaning through familiarisation, can bring about adjustments in the activation state of a very young infant and induce a behavioural state compatible with the expression of coordinated motor activities.

18.2.2 'Exploratory' and directional behaviours

Olfactory attenuation of crying and distress in infants is often followed by low amplitude movements, such as slow cephalic orientation, eye opening and 'gazing', activation of oral and lingual movements. These are interpreted as indicative of information seeking, attention and attraction. According to Macfarlane, mentioned in the Introduction, cephalic orientation movements are also held to indicate differential processing of simultaneously-presented olfactory stimulations. In the choice test developed by him, newborns lay on their backs in their cradles, two odorised gauze pads hanging one on each side of their faces, in direct contact with their cheeks (Fig. 18.1a). The stimulation side was controlled by inverting the stimuli between the two one-minute trials composing a test. These tests were video-recorded and then analysed by an observer who was unaware of which stimulation was presented and from which side. Using this procedure, Macfarlane (1975) conducted two experiments to appraise the absolute and relative orientation responses of infants confronted

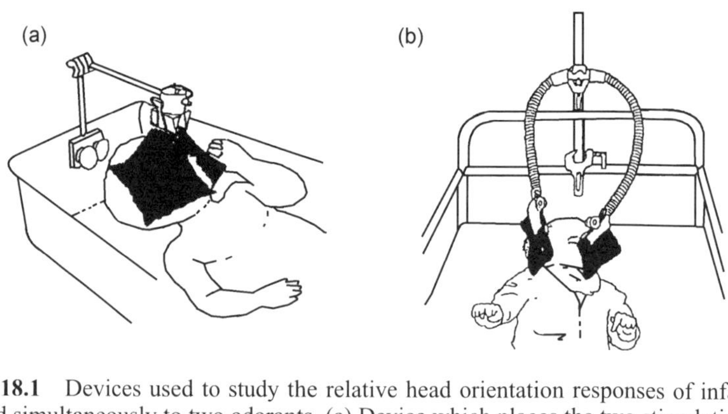

Fig. 18.1 Devices used to study the relative head orientation responses of infants exposed simultaneously to two odorants. (a) Device which places the two stimulations, in direct contact with the cheeks, like two sides of a roof above the face of the supinely lying infant (Macfarlane 1975); (b) Device in which the stimulations are initially distant from the infant's face, avoiding tactile elicitation of orientation (Schaal *et al.* 1980).

with an odour acquired in the context of nursing. In the absolute response test, infants aged 2–7 days faced the choice between the odour of their mother's breast and an odourless stimulus. The majority of newborns turned longer towards their mother's odour, thus indicating that they were capable of detecting it, and that it was attractive. In the relative choice experiment, infants aged 2, 6 or 8–10 days were confronted simultaneously with two odour sources, namely the odour of their mother's breast and that of another nursing mother's breast. Although 2-day-olds did not display any discrimination between the two stimulations, 6-day-olds showed a response in favour of the mother's breast odour, and this pattern was even clearer in infants aged 8–10 days. Thus, as early as the end of the first week of life, infants are capable of distinguishing odours of low intensity.

Subsequent experiments have confirmed and refined this result, but in using odour-presentation devices which minimise the impact of tactile stimulation on the initial orientation of head movements (Fig. 18.1b). Such experiments have further investigated the development of infants' responses to the olfactory surroundings of the mother's breast. Thus, 15-day old infants (breast- and bottle-fed) display greater attraction towards the breast odour of an unfamiliar nursing mother than towards the breast odour of a non-nursing mother or the axillary odour of a nursing mother (Cernoch and Porter 1985). Although this result is not surprising in infants regularly exposed to their mother's breast, it is particularly interesting in bottle-fed infants, who in principle have never been exposed to it. These infants are, in fact, more attracted by the unfamiliar breast odour of a nursing mother than by the familiar odour of their milk formula. This result suggests two hypotheses. First, a milk-producing breast may emit a particularly attractive odour. Second, this olfactory attraction is independent of postnatal familiarisation (further detailed below).

The attractive odorant factor of the lactating breast appears also to be present in the milk itself which seems to convey at least two kinds of olfactory information that newborn infants seem able to pick up. First, breast-fed infants are capable of discriminating between the odour of their mother's milk and that of another mother (Marlier and Schaal 1997). They can thus detect a first element of olfactory individuality in milk. Second, they turn more insistently towards the odour of human milk taken from an unfamiliar woman rather than towards another iso-intensitive, unfamiliar odour (Marlier and Schaal 1999, 2005). As this attraction towards human milk has also been revealed in infants who had never been exposed to it (bottle-fed since birth), the hypothesis can be made that human milk in general conveys an attractive factor, which potency does not depend on postnatal reinforcement (see below).

Odours can activate the cephalic orientation of infants when they are released 1–2 cm from their nostrils, but in addition, they can apparently motivate the infants to move closer by themselves to the odour source when it is placed further away (10–15 cm). Recent observations indicate that newborn infants, laid on their stomachs on the mother's chests, are capable of reaching and orally grasping a breast after coordinated crawling (Righard and Alade 1990). This highly ecological performance has provided the opportunity to examine the power of attraction of breast odour. When placed on the body of their reclining mother, and given the choice between a washed breast and an unwashed breast, newborns (aged 5 to 13 minutes) move more often towards the former than the latter (Varendi *et al.* 1994). However, newborns attracted to the unwashed breast manage to grasp it orally after 22–100 minutes, whilst those who are attracted to the washed breast grasp it after 22–73 minutes, which suggests that when a particular direction has been taken, their movements are not much affected by the odour. Also, in these experimental conditions, the infants do not seem to make a real choice between the differentially treated breasts since they do not display prior bilateral head movements, or one sole cephalic orientation movement in the direction of this breast. The authors of this experiment have nevertheless interpreted this result in terms of newborn aptitude in detecting olfactory differences at a distance of 10–15 cm.

18.2.3 Odour sampling responses

In olfaction, the sampling of information is expressed as sniffing the source of an odorant, i.e. eliciting a pattern of respiration which improves the airflow over the nasal mucosa (Laing 1983). Sniffing varies with the nature and intensity of the odorant (Bensafi *et al.* 2005). The motor act of breathing is part of the olfactory percept, as the inhalation of odourless air triggers brain regions involved in the processing of odours, and conversely the mental evocation of smelly objects in the absence of any odour elicits respiratory change (Bensafi *et al.* 2003).

Sniffing is strongly influenced by the hedonics of odour stimuli in adults (e.g., Schwartz *et al.* 1986), and though volitional sniffing remains poor in infants, their respiratory patterns are nevertheless also affected by hedonics. For example, breast-fed infants increase their respiratory rates to the odour of breast milk, whereas they reduce it when exposed to the odour of formula milk. Bottle-fed infants display the reverse pattern to either stimuli (Soussignan *et al.* 1997). Such hedonically induced breathing variations are also seen in response to odours that are hedonically contrasted to adults, such as vanilla and butyric acid. Term and preterm infants show accelerated and decreased respiratory rates to vanilla and butyric acid, respectively, even when these stimulations are highly diluted to match the odour of human milk or amniotic fluid (Soussignan *et al.* 1997, Marlier *et al.* 2001). Thus, infants tend to adjust their nasal airflow depending on whether it carries a pleasant/familiar or unpleasant/unfamiliar odorant.

18.2.4 Appetitive movements of face, mouth and tongue

Maternal odour, whilst polarising newborn cephalic and general body activity, also stimulates oral movements (mouth opening, lip protrusion, chewing), and lingual movements (licking, sucking). Sleeping newborns display these responses when exposed to the odour of mother's milk, but not that of cow's milk, nor water (Russell 1976). Similarly, wakeful infants (4 days old) more frequently display anticipatory oral responses when they are turned towards the odour of their mother's milk rather than when they are turned towards the odour of another mother's milk in a two-choice test (Marlier and Schaal 1997, 2005); or when they are turned towards the odour of their amniotic fluid compared with water (Schaal *et al.* 1995a).

Lastly, from the very first day after birth, the mother's general body odour (gathered on clothing) stimulates oro-lingual responses in newborns more efficiently than another mother's odour (Sullivan and Toubas 1998). Oral movements can thus be highly selective, and can express neonatal motivation to explore and orally grasp the offered stimulus. The hypothesis that these oral movements anticipate feeding is suggested by the comparison of responses to different odours in the tested newborns, before and after nursing (see below).

In addition to their ability to analyse an odorant administered nasally, infants have shown themselves capable of detecting odorants that reach them by the retronasal pathway during feeds. The modification of the infant's ingestive response has been studied during nursing, using experimental manipulation of the dominant aroma carried in the mother's milk. Ingestion by the lactating mother of garlic, alcohol, or vanilla, considerably affect the olfactory properties of mother's milk for an evaluation panel of adults (Mennella and Beauchamp 1991a, 1996). It also does so in young infants. Nurslings aged 3–4 months respond to the introduction of the garlic aroma in the milk by increasing their sucking rhythm (compared with a test where they ingest milk lacking this aroma), but without a noticeable effect on the quantity of milk ingested. Although they are capable of distinguishing the 'garlicky' milk from the normal milk, this change does not seem to affect its global palatability (Mennella and Beauchamp 1991a). On the other hand, when mothers ingest vanilla, the nurslings stay longer on the breast and the quantity of milk ingested increases. The same effect has been observed in bottle-fed infants who suck the 'vanillised' formula milk more avidly than the normal formula milk. However, this increase in ingestion of recently vanillised milk does not last long and subsides after several occurrences (Mennella and Beauchamp 1996). These results indicate that infants detect the olfactory variations in their feed and that these variations govern in part their ingestive decisions. On this subject, Hall (1975) has suggested that infant perception of spontaneous changes in milk flavour during nursing could be one of the mechanisms governing appetite adjustment.

18.2.5 Physiological effects of odours

Little information is available about the physiological responses of infants to odours emitted by conspecifics or by foods. In the domain of digestive processes, the hypothesis can be made that visceral reflexes probably function very early in life. In adults, a cascade of nervous, endocrine and metabolic mechanisms is elicited by the sole perception of food odours. For example, exocrine digestive secretions (salivary, gastric, hepatic) and endocrine digestive secretions (gastro-intestinal and pancreatic hormones) are activated by food odours independent of any corresponding visual or tactile perception (Brand *et al.* 1982, Feldman and Richardson 1986). This sort of olfactory mechanism which anticipates the digestive process might be of high adaptive value from the neonatal period onwards.

Conversely, the metabolic stage of infants, roughly operationalised as the prandial state, has a marked influence on their responses to odours. This has been shown in an experiment which consisted in exposing bottle-fed infants (3 days old) to five odorants one hour before and one hour after a feed (Soussignan *et al.* 1999). The odour substrata were (1) usual milk formula, (2) unfamiliar milk formula, but qualitatively similar to the previous one, (3) milk formula very different to the first, (4) unfamiliar odour (vanillin), and (5) control stimulus. Detailed analysis of facial responses (using the Facial Action Coding System developed by Ekman and Friesen) revealed that muscular actions of aversion and disgust are more frequently provoked during the post-prandial stage by the odour of familiar milk, and to a lesser extent, by the similar odour of unfamiliar milk. The other stimulations, which clearly differed qualitatively from the usual food, did not affect the occurrence of negative facial responses. This fluctuation in hedonic responsiveness, connected with prandial state, is reminiscent of the phenomenon of negative alliesthesia[1] described in adults. In newborns, as in adults, this response is specific to the dominant sensory quality associated with satiation. It is therefore probable that the post-ingestive or post-absorptive mechanisms underlying this motivational change function from the first days following birth.

Considered collectively, the findings drawn from ethological and psychobiological approaches reveal particularly subtle sensory abilities in very early life, and differentiated capacities in the affective and cognitive processing of olfactory information emitted by (or associated with) the social and food environments (which, in fact, are inseparable in early life). These approaches are precious sources of observations and hypotheses, awaiting further investigation with regard to the perceptual mechanisms involved and to their susceptibility to change during development. A survey of such data is provided in the following sections.

18.3 Early functioning of olfaction

18.3.1 Sensitivity of early olfactory detection

Which nasal chemoreceptive systems are involved?

In humans, at least four chemoreceptive systems are located in the nasal cavities: the main olfactory, the trigeminal, the accessory olfactory or vomeronasal, and lastly the terminal systems (cf. Doty 2003). The functional status of vomeronasal

1. Alliesthesia defines the change in the level of pleasure associated with sensory stimulation, depending on the physiological and metabolic state of the organism (Cabanac 1971). Negative alliesthesia denotes the decrease in pleasure of a stimulus after a meal, when this stimulus was dominant in the satiating food.

and terminal structures remains debated in our species, and so it will not be discussed further here. As for the two other subsystems, it is usually acknowledged that the neuroreceptors of the main olfactory system are the best sensors of very low intensity volatile stimulations, and that the trigeminal fibres react to more intense stimulation, including tactile or irritant compounds. This dual nature of nasal chemosensory detection was, until the 1960s, behind the opinion that infant olfaction was a result of trigeminal activation rather than olfaction *per se* (cf. Schaal 1988, for review). But more recent experiments have demonstrated that infants are capable of detecting odours of weak-to-very-weak intensity, which in certain cases are even imperceptible for adults, and thus suggest that this performance is accomplished by the olfactory system rather than the trigeminal system. However, this conventional distribution of the sensory duties in the nasal sensorium will move further as our knowledge increases. One can already mention a study indicating that anosmic adults are perfectly capable of subtle discrimination regarding low intensity odorants, probably thanks to trigeminal input (Laska *et al.* 1997). Therefore, 'pure' olfactory stimulations do not exist, and any olfactory image derives from the simultaneous activation of the two categories of nasal chemoreceptors. For this reason, in the rest of this chapter, olfaction will be used to denote nasal chemoreception in the wider sense, without further specifying the receptive subsystem involved.

Stimuli can reach the nasal chemoreceptive subsystems by two different routes, the orthonasal and the retronasal pathways. The former is used when we direct a sniff to an object and inhale it through our external nares, and the second pathway is taken by the volatile molecules released in the mouth from chewed food and aspired up into the nasal cavities through the choanas. Conventionally the orthonasal access of odorants gives rise to the odour sensation, while the retronasal access is part of flavour experience, the mixed sensation including cues from olfaction, taste, somesthesis and irritation. Although olfaction has mostly been the focus of developmental research summarised below, some data also pertain to flavour sensation.

Inception of nasal chemoreception

Nasal chemosensory reactivity in the human fetus *in utero* has not, to our knowledge, been the subject of any direct experimental studies. However, two indirect experimental approaches, the only two compatible with ethical requirements, have been followed. The first consisted in examining the responses of *ex utero* fetuses to odours when gestation was interrupted before term (cf. Schaal *et al.* 2004, for a review). The second analysed the responses of full-term newborns when they were confronted with odours or odorant mixtures that could only be encountered before birth.

Investigation on premature newborns has enabled us to demonstrate that, as early as 31 to 38 weeks of gestation, the nasal chemoreceptor systems are capable of detecting odours that contrast both in quality and in trigeminal properties (i.e. nonanoic acid and cineol; Pihet *et al.* 1997). Prior work had suggested that these subjects are able to display behavioural reactivity to intense olfacto-trigeminal stimulations (menthol) as early as 28 weeks of gestation (Sarnat 1978). Even when odorants of much weaker intensity are used, after having matched them in subjective intensity, premature newborns remain capable of differentiating among them, and between them and a control scentless stimulus (Pihet *et al.* 1996, Goubet *et al.* 2002). Finally, stimulations of very weak intensity (matched in subjective intensity with the odour of amniotic fluid) were revealed to elicit reliable responses in 28–33-week old premature infants (Marlier *et al.* 2001). These functional data indicate the nasal system's capacity to respond to chemosensory stimulations well before gestational term, as soon as the start of the third trimester of gestation. Even though commencing aerial life prematurely can accelerate the maturation of certain mechanisms governing odour reception, it cannot be excluded that a fetus of an equivalent gestational age is capable of a similar sensory performance to that of a premature newborn. This hypothesis is backed up by recent neuroanatomical and neurochemical findings (described in Schaal *et al.* 1995c, 2005).

The course of olfactory sensitivity in early development

Research referred to in the first section indicates that olfactory discrimination capacity improves with postnatal age. Self *et al.* (1972) analysed the evolution of olfactory reactivity over the three first days after birth in terms of variations in respiratory rhythm and various behavioural indices. To that aim, she exposed newborns to constant, supra-liminal artificial odorants selected on the basis of their presumed lack of trigeminal elements (extracts of aniseed, *Asa foetida*, lavender and valerian) and to control stimulations (dry and humid blanks). As early as the first day after birth, over two-thirds of infants were capable of detecting most of these odorants. The percentage of subjects reactive to *Asa foetida*, lavender, or valerian, tends to increase between days 1 and 3. Is this increased reactivity connected with improved sensitivity of the olfactory system, or with functional changes in the effectors on which the detection indices are based?

Examining in more detail the postnatal variations in olfactory acuity, Lipsitt *et al.* (1963), exposed newborns aged 1–4 days to solutions of *Asa foetida* in increasingly stronger concentrations, and at the same time, recorded general motor modifications and variations in respiratory rhythm (which they aggregated in a compound response index). They noted a marked decrease in the average concentration of *Asa foetida* necessary to provoke a response, which according to the authors reveals a drop in the threshold of olfactory detection over the first four days. However, the use in this study of a lengthy procedure involving frequent stimulus repetition does not allow the various possible causes of functional improvement to be differentiated. This improvement can be interpreted either in terms of overall gain in olfactory sensitivity or in terms of induction by repeated exposure of sensitivity to the experimental odorant. A more recent study (Shimada *et al.* 1987), using a series of five odorants (phenylethylic alcohol, cyclotene, isovaleric acid, undecalactone and skatole) reports no correlation between postnatal age and the thresholds of olfactory detection in newborns. The drop in liminal value observed in the study carried out by Lipsitt *et al.* could thus result from an effect of repeated olfactory exposure. Such cases of facilitating or inducing detection of an odour, initially neutral or unperceived, using repeated exposure have been examined several times in adults (e.g., Rabin and Cain 1986, Wysocki *et al.* 1989).

Specificity of early olfactory detection?

The results presented in section 18.2 show that neonatal aptitude in detecting odours is undeniably comprehensive. Infants respond to diverse odorant mixtures including those emitted by various maternal exocrine secretions (areolar, axillary, lacteal), by food aromas or by artificial odorants associated with nursing. Since the first attempts to examine newborn olfaction in 1850 and up until now, up to 80 artificial odorant stimuli, variable in quality, intensity, complexity, or trigeminal properties have been shown to provoke diverse behavioural reactions in infants.

However, parallel to these attempts to comprehensively assay olfactory detection in early life, the existence of perceptual specialisations can be questioned for certain odorants or groups of odorants. Various mechanisms may explain such olfactory specialisations. First, given the role of genetic and developmental determinants, molecular reception mechanisms for certain odorants can be missing at certain ages (e.g., age-dependent specific anosmia). Second, the maturative chronology of various categories of olfactory neuroreceptors could render individuals more receptive to certain stimulations than to others at given points in development (compare with heterochronic development of sweet and salty tastes). Lastly, olfactory experience common to all individuals of the species might determine long-term perceptual salience, and hence differentiate the sensory or affective value of given odorants against the olfactory background. For example, electric cerebral responses to odours indicate that certain odours, such as musk, are inefficient in eliciting a cerebral response in newborns (Yasumatsu *et al.* 1992), suggesting that selectivity phenomena could limit the spectrum of odours detected in very early life.

The hypothesis regarding early olfactory specialisations has been examined in more detail using a series of very weak-intensity odorants, comprising homospecific psychobiological substrata (human amniotic fluid and milk), heterospecific biological substrata (two different milks of cow and plant origin emitting very different odours), pure odorants without any particular biological significance (vanillin, butyric acid), and a blank stimulation (distilled water; Soussignan *et al.* 1997). In this study, care was taken to match the pure odorants with each of the biological substrata in intensity and trigeminal potency, thus constituting a series of 12 iso-intense stimulations (comprising four triplets of odorants: each biological odorant and the two pure odorants in corresponding intensities). The subjects were exposed for 10 seconds to each of the 12 odorants and to the blank, and at the same time their respiratory and facial responses were recorded. Analysis of these two response criteria allows the following conclusions to be drawn. First, 3-day old infants are capable of detecting all the odours (compared with the blank stimulus), those produced by homospecific or heterospecific mixtures, as well as pure odorants. Second, this comprehensive reactivity is qualified by exposure conditions, especially for lacteal odours: whilst breast-fed infants manifest greater respiratory activation when exposed to the odour of human milk compared with that of milk formula, on the contrary bottle-fed infants respond more readily to the odour of milk formula. Therefore, the respiratory indicator of the olfactory process appears to be significantly influenced by prior exposure of the subject to the odour. Third, given the conditions of this experiment, the analysis of facial indicators does not enable any conclusions to be drawn regarding the capacities of infants to differentiate between biological stimulations and pure iso-intensive odours. However, using a more 'active' behavioural indicator (head orientation in a two-choice test), newborns several days old are shown to respond in a different way to the odour of human milk as opposed to the odour of artificial milk. The importance of this observation will be discussed below.

18.3.2 The discriminative power of early olfaction

Discrimination level of early olfaction

Several experimental strategies have appraised the capacities of newborns to segregate olfactory qualities. The most global, used by early experimenters, consisted in merely spotting different motor responses following presentation of distinct odorants. They noticed that the frequency, amplitude, latency and duration of the odour-induced movements vary for contrasting olfactory qualities (Stirnimann 1936, Engen *et al.* 1963, Self *et al.* 1972). More sophisticated approaches have produced more dependable data, but also contradictory at times, concerning early discriminative abilities of olfaction.

The first study is based on electroencephalographic responses in sleeping infants. A former study indicated that the odours of coffee, citral, vanillin and pyridine, do not cause any significant alteration in the EEG of infants aged less than 6 days; repeatable variations were reported only on infants between 6 and 90 days old (Fusari and Pardelli 1962). This negative result for the newborn period is surprising given highly discriminative behavioural findings presented above. A comparable discordance between behavioural and electrophysiological indices has been noted in human adults (Perbellini and Scolari, 1966) and in young animals. So, for example, even though no spontaneous electric activity has been detected in the olfactory bulb of newborn rats before 4 days (Salas *et al.* 1969), they rely exclusively on olfactory indices in heading for and localizing the nipple from the first postnatal hours. This incoherence between different levels of functional analysis should probably be put down to methodological problems rather than to a lack of an electric cerebral response to odorants. Furthermore, more recent research, involving imaging techniques for blood flow variations using infrared spectroscopy, reveals cortical activation in newborns as early as the first hours after birth, caused by odours of human origin (human milk, axillary odour), as well as by artificial odorants (vanilla) (Bartocci *et al.* 2000). EEG results in infants of 3–4 months also reveal inconsistent results. While Kendal-Reed and Van Toller (1992) report undifferentiated reactogenic potency for various complex food odours in wakeful infants, other studies reveal differentiated cerebral reactions when contrasting olfactory qualities are presented to sleeping infants. Thus, at 3–4 months, maternal milk odour causes a bilateral reduction in the amplitude of δ and θ band waves recorded from frontal and central electrodes; on the other hand, the odour of orange caused an increase in δ waves in the central and right parietal positions (Yasumatsu *et al.* 1994). This difference in cerebral response can be attributed to qualitative and/or intensitive differences between the two stimulations. Recording conditions and choice of stimulations appear thus critical in revealing olfactory brain response.

The second strategy utilises the habituation paradigm. This consists of repeatedly presenting odour A until response is extinguished or reaches a minimum criterion (habituation), then inducing the subject to inhale odour B, whilst the differences with odour A are monitored. If the level of response initially noted for odour A is restored (dishabituation) following this second stimulation, the two stimulations are considered to be processed as different. This procedure has been successfully applied to demonstrate the ability of infants between three days and one month to discriminate pure odorants (Engen *et al.* 1963, Guillory *et al.* 1980), or mixtures and their component odorants (Engen and Lipsitt 1965; cf following subsection). Recently, this paradigm was used in preterm infants (Goubet *et al.* 2002).

A third strategy consisted in associating odour A with a rewarding event and then studying the time necessary for establishing classical conditioning, using odour A, then odour B, which was never rewarded, as the conditional stimuli. This type of experiment was carried out by exposing newborns, for at least one month, to the odour of mint present on their bottle teats (Irzhanskaia and Felberbaum 1954). The palpebral reflex (eyelid closing in response to tactile stimulation of the conjunctiva) was used in order to determine the time lapse necessary for establishing classical conditioning. After repeated association of mint with the tactile stimulus (an air-puff to the eyes), the odour of mint alone switched on the palpebral reflex. The authors compared the time necessary for acquiring this association regarding the familiar odour of mint, with an unfamiliar odour of anise. The conditioning reflex was created for both odours, but appeared quicker for mint than for anise. Although the methodology is flawed, this experiment suggests that an odorant is more easily conditionable when it is familiar to the subject.

Other studies, described in more detail in the third section, used more ecological conditioning procedures. For example, Schleidt and Genzel (1990) and Sullivan (1990) associated a novel odorant with either nursing or bottle-feeding. The application of such associative procedures led to very early (over the first two days), quick (30 minutes) and more or less stable acquisition of olfactory discrimination.

Dimensions of olfactory stimulation discriminable to newborns

The perceived olfactory environment is multidimensional, and the scope of this 'multi' remains poorly delimited. Psychophysicists classically retain three main characteristics on which individuals rely to discriminate olfactory stimulations: quality (jasmine vs vanilla odour, for example); intensity (strong vs weak odour), and the subject's prior experience with the odour. The hedonic value of the corresponding odour perception derive from the individuals' history with odorants (pleasant vs unpleasant odour), or its judgement of familiarity or other kinds of categories (e.g., edibility). These dimensions are interdependent, however. It is well known that modifying the intensity of one stimulus can, for example, influence its quality and, hence, its hedonic evaluation. Although they are clearly non-exclusive, one may ask whether these sensory 'dimensions' are relevant for the early functional capacities of the olfactory system?

Quality

The experiments summarised above suggest that, given equal subjective intensities, contrasting olfactory qualities can be discriminated by infants. A judicious demonstration of the early discrimination potency of olfaction was carried out by Engen and Lipsitt (1965) using the habituation paradigm. If a mixture of anise and Asa foetida odours (dominated nonetheless by the anise note) is used as the habituation stimulus of the respiratory response, the infants recover their initial reactivity response to Asa foetida alone, but not to anise alone. Similarity assessments made by a group of adults revealed indeed that the odour of the anise component is more difficult to differentiate in the odour mixture than that of the Asa foetida component. If the same habituation paradigm is used with a mixture of odours of equal subjective intensity, and which contribute equally to the mixture (case of an acetate of amyl-heptane mixture), the two components are discriminated in the dishabituation test. Also, the effectiveness of dishabituation is proportional to the dissimilarity of the olfactory qualities of the mixture compared to those of the elementary compounds (as evaluated by adults). This experimental approach has led to the conclusion that 3-day-olds not only discriminate, but also appear to evaluate olfactory similarities almost like adults. Since recovery of the habituated response depends on the qualitative distance between the habituation stimulus and the dishabituation stimulus, it can be inferred that the 3-day-old brain is capable of evaluating qualitative resemblance between odours in a similar way to the adult brain.

Intensity

Early capacity to process intensity information from olfactory stimuli is contradictory. Rovee (1969) thought that this was already possible in 3-day-olds because they manifested motor responses of increasing amplitude with increasing concentration of different aliphatic alcohols. It is nonetheless probable that the increasing recruitment of trigeminal chemoreceptors is responsible for this effect, as alcohols are known to carry notable trigeminal potency which also increases with concentration. On the other hand, the presentation of increasing concentrations of vanilla vapour or butyric acid does not greatly affect latency, amplitude or duration regarding variations of respiratory rhythm or facial response (Soussignan *et al.* 1997).

Familiarity

Several other factors play a decisive role in discrimination when simple sensory skills combine with more complicated psychological processes. In particular, the subject's prior experience with odours affects familiarity, i.e. the aptitude in detecting recurrence of a given stimulus. In adults, familiarity with an odour noticeably increases its power of discrimination (Rabin 1988). Familiarity effects have been noted from the foetal-neonatal period onwards. As mentioned in section 18.2, infants display more insistent orientation or mouthing towards stimuli which have become familiar during interactions with the mother (e.g., Macfarlane 1975, Schaal *et al.* 1980, Delaunay-El Allam *et al.* 2006a). Odorant stimuli can also be recorded as familiar when merely diffused in the air surrounding infants while sleeping. Given the choice between an odorant with which they have been familiarized in their crib over the previous 24 hours and a new odorant, 2-day-old infants (especially girls) manifest a preference for the familiar odour (Balogh and Porter 1986). In the same way, preterm infants retain an odour introduced in their incubator's atmosphere (Goubet *et al.* 2002).

Hedonic value

Numerous experiments have sought to determine the influence on early reactivity of the affective impact inherent to odours. But such studies are easily flawed by confounded variables (either of sensory nature, such as disparity in intensity levels between the stimulations under test; or of psychological nature, such as different degrees of familiarity). The criteria selected for examining odour differentiation from the hedonic point of view are generally behavioural or psychophysiological, covering either the degree of interest shown by the subject for a certain stimulation, the facial actions conveying affective connotations, or the level of motor mobilisation invested to get nearer to one or another stimulation offered in a choice test.

Some of these studies suggest that humans of all ages and origins use similar hedonic criteria in dividing up their olfactory universe. Thus, from the first moments following birth, orange and geranium odours, considered as pleasant by experimenters, were reported to evoke positive oral activation responses, whilst unpleasant-to-adult odours (*Asa foetida*) cause expressions of disgust and head turning in infants aged between 15 minutes and 6 hours (Peterson and Rayney 1910–1911). Also, Stirnimann (1936) established a listing of expressions emitted by newborns in response to odours considered pleasant or unpleasant by adults. This pioneering work lacks, however, the methodological refinements that are considered today to be basic. Resumed by Steiner (1977, 1979) in a more systematic way, this research confirms that newborns under 12 hours old, before any postnatal ingestion, differ in their responses when tested with various olfactory stimulations. The range of stimulations used is justified by the hedonic evaluations made by adults and represents pleasant and unpleasant extremes (respectively, banana, vanilla, milk, vs shrimp and rotten egg). The method consisted in presenting these odours under the noses of newborns as their faces are photographed. These images were then shown to an evaluation panel unaware of the nature of the tested stimulus and their task was to assess the hedonic value of the odour-induced facial expressions. Generally speaking, odours considered pleasant by adults induced facial responses thought to express contentment and acceptance (relaxed facial muscles, raising of mouth corners, licking and sucking), whilst odours considered to be unpleasant by adults provoked expressions interpreted to be rejection and disgust (lowering of mouth corners, lip and tongue protrusion, gaping). Agreement among members of the evaluation panel was greater for infant responses to unpleasant odours (particularly for the odour of rotten egg), than responses to pleasant or test odours, suggesting that either newborns exhibit less ambiguous facial expressions in response to unpleasant odours, or adult humans are more sensitive to negative facial emotions in general.

Considering that morphological analogy of facial expressions between newborns and adults may be taken to imply an analogy of underlying emotional states, Steiner postulated that certain olfactory stimulations are more acceptable than others for the newborn organism. According to him, the perceptual attribute used by the infant to differentiate artificial odours is its hedonic connotation, and he posits the existence of a hard-wired brain mechanism controlling the hedonic facial reactivity of newborns to odours by reflex action (Steiner 1979; Steiner *et al.* 2001). He supports this hypothesis by the fact that newborns suffering from a severe cortical deficit (anencephaly, hydro-anencephaly) respond to odours through the same sort of facial expressions as normal infants.

However, one can argue that the experimental conditions used by Steiner did not allow full examination of the hypothesis regarding automatic and pre-programmed olfactory reactivity in infants. Several weaknesses in the method used can be pointed out:

1. The odorants used have no obvious ecological validity for newborns and their parameters were imprecise in terms of intensity and duration.
2. Actual detection of stimulation was not verified, so that the absence of facial reaction could have been interpreted as hedonic indifference.
3. The stage of infant wakefulness was not defined.
4. Newborn responses were recorded using one sole image (photograph) taken at a poorly defined moment during stimulation (at least in Steiner's initial work).
5. The coding technique of facial expressions was indirect and imprecise.
6. The hedonic significance of newborn expressions was extrapolated from similar expressions in adults.

With a view to re-examining olfacto-facial response in newborns, Soussignan *et al.* (1997) took these various methodological biases into consideration. The procedure they used was explained above as regards the nature and intensity of stimulations: we recall here that 12 stimulations are involved (comprising four unfamiliar biological odours (amniotic fluid,[2] human milk and two artificial milks emitting very different odours) and corresponding intensities of vanillin and butyric acid), as well as an odourless control. Recorded responses were respiratory rhythm, to attest odour detection, and oro-facial expression, to examine hedonic polarity of responses. Facial responses were videotaped and then decoded using the Facial Action Coding System developed by Ekman and Friesen (1982, infant version; Oster and Rosenstein, in press) and interpreted using facial data previously obtained from newborns and older children.

Systematic analysis of facial actions emitted in response to the different odorants indicates that these are not automatic as predicted by Steiner's reflex model. However, it was possible to differentiate the spatial and temporal configurations of facial responses according to the hedonic value of the odour inhaled by the newborn. An odour deemed unpleasant by adults (butyric acid) provoked significantly more negatively-valenced responses (screwing up of nose and lowering of mouth corners). On the other hand, an odour deemed pleasant by adults (vanilla) did not generate any more positive facial response than butyric acid. Relatively reliable hedonic responses therefore seem more easily activated on the face by negative stimulations than by positive stimulations (in adults). As the two odours were equalized in intensity and irritating power, negative facial response in newborns is based on their aptitude to differentiate odours in qualitative and/or hedonic terms. Thus, these results confirm, in part, those obtained by Steiner.

In the present conditions, the hedonic polarity of facial responses set on by biological substrata remains equivocal. Among these substrata, only amniotic fluid (unfamiliar – see footnote below) evokes more frequent negative facial responses than the control stimulus. Odours of unfamiliar milk formula, although perceived by adults as highly unpleasant, do not seem to be associated with negative responses in infants. Therefore, the hedonic facets of infant and adult odour universes cannot be totally superimposed.

The keenness of hedonic discrimination in newborn olfaction is also corroborated by experiments involving an active choice between two odours presented simultaneously (cf section 18.2). Thus, when infants are confronted with two attractive odours, they are capable of showing which one they prefer by the relative duration of head orientation or their differential mouthing responses.

In conclusion, the research presented in this second section clearly demonstrates the capacities of newborns to detect, discriminate, and to a certain extent, categorise olfactory stimulations. On top of this, their aptitude for subdividing the olfactory universe is organised selectively: infants can show by their behaviour that they prefer one odour to another. The dimensions at the root

2. The samples of amniotic fluid and breast milk were 'unfamiliar' to the infants in that they were not collected from their own amnion or mother.

of these very early olfactory preferences remain poorly explored, but any moderately intense olfactory quality, to which the infant has previously been familiarised, is almost certain to have a higher reinforcing value compared with any new odour. This hypothesis attributes a major role to direct experience with odours in the acquisition of olfactory preferences. In the next section, we shall see that the mechanisms behind chemosensory acquisition are numerous and varied, and that radical empiricist positions, according to which odour preferences are absent at birth and acquired with age (e.g., Engen 1982), cannot be defended any more.

18.4 Memory and plasticity of olfactory function in early life

18.4.1 Early postnatal shaping of odour preferences

The fact that human infants can be familiarised with a novel odorant by mere exposure, i.e. without any apparent reinforcement, has already been mentioned above (see page 416, Dimensions of olfactory stimulation discriminable to newborns). This form of passive acquisition seems to prepare more complex and more specific olfactory acquisitions. Thus, the repeated pairing of an initially neutral odorant with maternal contact leads to the progressive development of a preference. Delaunay-El Allam *et al.* (2006a) showed that the circumstantial odorisation of the maternal breast with a chamomile-scented salve induces the rapid formation of a significant preference for this odorant. This result corroborates a previous study by Schleidt and Genzel (1990) who also asked breastfeeding mothers to odorise their breasts (with rose essence) for the first two weeks following the birth of their infant. During preference tests carried out after 1 week and 2 weeks, these newborns displayed more insistent orientation towards the odour associated with nursing compared with a novel odour. In this study, the remanence of the odour acquisition appears to be limited, however. When the mothers stopped odorising their breasts after 2 weeks, the infants no longer displayed any preferential orientation 2 weeks later. Thus, infants learn about odour qualities in connection with nursing and related reinforcing agents (comfort contact; mother's voice; milk, gastric filling, and post-ingestive and post-absorptive processes), but seem also capable of unlearning (or deactivating) this association when it is no longer reactivated. However, recent data indicate that an odour linked with breastfeeding can be retained for much longer periods after exposure interruption (Delaunay-El Allam *et al.* 2004, 2006b).

These associative capacities are apparent even earlier (Sullivan 1990) and in the absence of any reinforcement linked with food. Between 2 and 10 days, maternal neck odour induces a reduction of the infant's motor activity (Schaal 1986); thus contingency between a breast-unrelated odour and simple, non-feeding contact with the mother was efficient in assigning a specific meaning to the odour. Also, a brief session pairing massage and odour (for 10 periods of 30 seconds) is sufficient for newborns aged 4 to 16 hours to manifest a preference for this odour 24 hours later. On the other hand, infants of the same age exposed at different times either to massage or to the odour alone show no changes in behaviour (Sullivan 1990).

Such mechanisms also occur in the acquisition of natural odours emitted by the mother. According to whether they are breast- or bottle-fed, newborns are introduced to contrasting olfactory learning situations. It has been shown that, when given the choice between their mother's axillary odour or that of an unfamiliar mother, breast-fed infants turn towards their mother's odour for a longer time, whilst bottle-fed infants show no selective reactivity (Cernoch and Porter 1985). This difference is interpreted in terms of amount of direct exposure to maternal skin odour between these two groups of infants.

Also, monitoring, between postnatal days 1 and 5, of the development of relative attraction towards two odour substrata, to which the infants were exposed just before or just after birth (amniotic fluid and milk), confirms the importance of repeated exposure in establishing infant discrimination and hedonic aptitudes. During days 1 to 3, the infants did not respond to the two stimulations in differential ways, indicating equal sensory and/or motivational treatment of them (Marlier *et al.* 1997). However, with increasing age and nursing experience, they displayed an enhanced attraction for the milk odour compared to amniotic fluid odour, and this differentiation was significantly apparent after day 3. Progressively divergent responses towards the two odours can be explained by two mechanisms. The first phase (days 1–3 with 0–12 feeds) seems to reflect newborn preference for an odour acquired in the womb, as well as equivalent sensory and/or hedonic responses to milk and amniotic odours. The second phase (days 4–5 and more than 12 feeds) seems to reflect perception, at the time of milk inflow, of the qualitative modification of milk odour. This sequence illustrates the development of a preference *in vivo* and suggests that, in the earliest phases of life, postnatal olfactory development depends on current and prior influences.

The ontogeny of olfactory control of behaviour reveals high plasticity since a variety of odorants can acquire affective significance, not only those emitted naturally by preferred individuals – breast, neck, axillary odours – but also artificial odours chosen arbitrarily. At the same time, the processes involved in the acquisition of the hedonic properties of odours are versatile and extremely sensitive in certain contexts. Mere exposure, as well as association with highly reinforcing events – nursing, massage or comfort contact with the mother – can determine the perceptual salience of an odour and confer strong affective meaning on it. However, the plasticity of early postnatal learning may be constrained at a certain level by prior influences from previous chemosensory experience: the postnatal repetition of odour exposure is not always sufficient to favour the establishment of an associative acquisition. If, between birth and 4 days, breast-fed infants rapidly develop a preference for the odour of mother's milk (see above), bottle-fed infants (as deduced from an equivalent choice test between the odour of their milk formula and their amniotic fluid) do not display this trend over the same period. The odour of cow-based formula milk, to which they have been exposed 5–6 times a day since birth, remains less attractive than the amniotic odour (Marlier *et al.* 1998b). In this case, the odour acquired *in utero* appears to be more reinforcing than the odour acquired *ex utero*. Consequently, odours may not be all equally 'learnable', and one may hypothesise that the postnatal learning of an odour is facilitated when it has already been encountered *in utero*.

Finally, it must be borne in mind that developmental time is not linear. Some periods are more favourable than others for the more or less permanent establishment of sensory or cognitive processes. Such periods, denoted as 'critical' or 'sensitive', have been well documented in animal models regarding the development of taste. For example, the reduction of environmental sodium in the pregnant rat before and on embryonic day 8 is followed by a permanent, non-recoverable reduction in peripheral taste responses in the young rat (Hill and Przekop 1988, Hill and Vogt 1989). Another such prenatal effect has been reported after experimental extra-cellular dehydration in pregnant rats, resulting in enhanced salt preference in the offspring (Nicolaïdis *et al.* 1990). Comparable endocrine-mediated mechanisms exist in women in whom dehydration can be caused by severe gestational vomiting. Children born to mothers who suffered severe episodes of vomiting reported higher levels of salt use and greater preference for salty foods than peers born to women with mild, or without, vomiting (Crystal and Bernstein 1995). Comparable sensitive periods for the acquisition of odour or flavour preferences are abundantly attested in the animal literature, but they have been less investigated in humans. One case is suggested in the acceptability changes of particular milk formulae made of casein or whey proteins which have been hydrolysed to improve digestibility. The side-effect of this milk treatment is a strongly offensive odour and a taste with distinctive bitter and sour notes for adult judges. Before the age of 2 months, infants readily accept to ingest such formulae to satiety; but after 7–8 months, they generally reject it (Mennella and Beauchamp 1996). This phenomenon possibly relates to age or maturation-dependent changes in odour/taste sensitivity, reactiveness, and/or rejection pattern of novelty. Being exposed before 4–5 months to protein hydrolysate (PH) formulae, and associated sourness-bitterness, determines their later acceptability over long periods, and a stable transfer of acceptability to real foods containing sour/bitter components. For example, 4–5-year-old children fed PH formulae early in life were more acceptant of acidified apple juice and of the PH formula they had consumed as compared to matched children fed different kinds of formulae. They were also reported to like foods with distinctive bitter compounds, such as broccoli (Mennella and Beauchamp 2002). The analysis of early feeding with differently flavoured formulae helped to unveil that, roughly during the first semester of life, human infants are open recipients of any flavour quality afforded by the mother. Thus, the flavour experience mothers build up for their 'captive' infants during this period of neurocognitive openness, strongly contributes to shaping chemosensory phenotypes that will like and want similar foods or beverages.

18.4.2 Olfactory preferences established independently of postnatal experience

Inborn unconditional odorants?

The demonstration of the remarkable malleability of postnatal olfactory acquisitions in no way attenuates the importance of the results presented above and which are recalled here. First, certain unfamiliar smells appear to be immediately more pleasant than others for newborn babies (Steiner 1979, Soussignan *et al.* 1997). Second, the 'naïve newborn' model, which consists in analysing the responses of subjects confronted for the first time with a given odorant, reveals that infants deprived of any direct exposure to the breast (bottle-fed since birth) are attracted by its odour (Makin and Porter 1989); and that bottle-fed newborns prefer the odours of a nursing mother's breast or of human milk compared with the odour of their familiar artificial milk (Marlier and Schaal 2005). Mothers' breasts and milk seem therefore to emit one or several odorant factors to which newborns react with no prior experience. This response to mother's milk is evident even in premature newborns (29 to 36 weeks of gestation; Bingham *et al.* (2003)). Lastly, responses showing such unconditional attraction have been noted in the newborns of several other mammalian species, especially towards volatile compounds of milk (reviewed in Schaal 2005). The most remarkable case concerns the reaction of newborn rabbits to the odour of homospecific milk even though they had never been exposed to it (Schaal *et al.* 2003). This specific reactivity in newborn rabbits to rabbit milk is carried in the fraction of odorant volatiles which develop over standing milk within 30 minutes of milking. From this headspace a single compound was identified (2-methyl-but-2-enal) which has the same behavioural activity as whole milk. This compound having extremely repeatable behavioural activity on rabbit pups and being emitted in milk in the distal part of the mammary tract (Moncomble *et al.* 2005), it was called 'mammary pheromone'. Remarkably, this pheromonal compound was not present in the prenatal environment, and it is thus reasonable to assume that pup responsiveness to it cannot be induced by foetal experience. This mammary pheromone represents a case of unconditional odorant which elicits activation, approach and oral grasping responses in the naïve newborn and this effect is lasting for the first postnatal weeks until weaning is engaged. But in addition to its behaviour-releasing effect, this pheromone promotes the learning of any associated odorant (Coureaud *et al.* 2005), and so plays a direct role in the rapid expansion of the stimuli that are meaningful in the newborn organism's environment. The generality of such milk-borne pheromones remains to be determined in other mammals, including in our own species.

As a whole, studies using odour cues carried in biological substrata suggest that animal and human newborns manifest preferential responses towards certain odours even in the absence of direct exposure to them in the postnatal environment. The apparently unconditional nature of these positive responses suggests either perceptual capacities which derive from a genetic coupling of given odour stimuli and behaviour, or the possibility of acquisitions that are predetermined by foetal experience.

Foetal learning and memory

Foetal learning is supported by several arguments, primarily those provided by animal research (reviewed in Schaal *et al.* 1999). First, the chemoreceptive mechanisms in rat or sheep fetuses function during the last days of gestation (Smotherman and Robinson 1995, Schaal *et al.* 1991). Aversive conditioning carried out on embryonic day 17 in rats, associating mint with a motor inhibition, is retained until day 19. Thus, the chain for encoding, retaining and retrieving chemosensory information appears to function perfectly in the fetuses of placental mammals. Second, the manipulation of the olfactory quality of amniotic fluid can modify newborn reactivity to odours; the effect of this prenatal practice is apparent either because it directly induces olfactory attractions or aversions active from birth and which can persist into adulthood (Smotherman 1982, Hepper 1988, Schaal *et al.* 1995b), or because it facilitates postnatal learning of odours encountered *in utero* (Pedersen and Blass 1982). Third, transnatal retention of olfactory acquisition is actually expressed in the natural environment; this has been proven in the absence of any experimental manipulation, since newborn rats and lambs show a preference for the odour of the amniotic fluid in which they developed, compared with the odour of an unfamiliar amniotic fluid (Hepper 1987, Schaal *et al.* 1995b).

Other points specific to our own species support the hypothesis that nasal chemosensory detection does function in the foetus. Maturation of the chemoreceptor system is sufficiently advanced at the end of gestation to make the processing of chemical information possible (Schaal *et al.* 1995c, 2005). Chemical analysis has revealed the presence of numerous potentially odorous components in amniotic fluid (Schaal 2005). Lastly, fetal brain aptitude in detecting intra-uterine sensory information and in proving capable of retaining it after birth has been demonstrated in audition (DeCasper and Spence 1986, Granier-Deferre *et al.* 2005).

For obvious ethical reasons, the human foetus is not accessible for experimentation. Accordingly, *a posteriori* approaches have to be devised to examine the aptitude of the foetal brain to memorise odours. These approaches examined responses of newborns to prenatal odours. It was first proven that newborns remain attracted to the odour of amniotic fluid during the first days after birth, as opposed to water (Schaal *et al.* 1995a). Then, facing a choice between an olfactorily neutral maternal breast and the other breast which was smeared with amniotic fluid, newborns more often turn towards the latter (Varendi *et al.* 1996). Thus, amniotic odour is detected by newborns, it conveys attractive properties, but in addition, this response is selective: 3-day-old infants exposed to the odours of both their own amniotic fluid and an unfamiliar amniotic fluid turned their heads for a longer time towards the familiar fluid (Schaal *et al.* 1998). Since this preference was displayed by bottle-fed infants, who had not been in contact with the components common to both amniotic fluid and mother's milk, it is highly probable that this is the result of prenatal experience alone (Schaal *et al.* 1998).

Psychobiological processes involved in foetal chemosensory acquisitions
Two experiments have sought to specify whether this prenatal preference for amniotic odour is the result of complex olfactory recognition (human amniotic fluid contains more than 390 potentially odorous components) or of the perceptual extraction of a dominant aroma. The first experiment suggests that infants born of mothers who had eaten garlic during pregnancy show less aversion to the odour of allyl sulphide on the first postnatal day (Hepper 1995). The second study compared different behavioural indicators of newborns regarding appetence (oral movements, head turning) and aversion (facial expressions of disgust), their mothers having or not having consumed foods rich in anise aroma (Schaal *et al.* 2000). Blind analysis of the responses of these infants showed that those exposed to the anise aroma at the end of gestation display greater appetitive response and less aversive response to anethole than those who were not exposed to it. The nasal chemoreceptive system of the human foetus is thus capable of extracting, and its brain capable of memorising, a dominant odour in its prenatal environment. This capacity to retain a prenatal odour remains active for at least four days after birth, the first days during which many novel situations (e.g., first searching and suckling of a nipple, first ingestion of colostrum) have to be worked out to facilitate neonatal adaptive beginnings in life.

The stimulations afforded by the prenatal chemosensory environment can have neurocognitive effects at different levels in the chain of odour information processing. Several hypotheses can be put forward that are not mutually exclusive. At the most peripheral level, this plasticity can entail:

1. selective production of olfactory binding proteins, which convey odours through the mucus bathing the olfactory cilia
2. differential expression of neurons carrying a certain type of olfactory receptor proteins on their cilia
3. selective apoptosis phenomena that tend to bias the olfactory neuroreceptor populations (Najbauer and Leon 1995).

Such peripheral plasticity regulated by the chemical environment has been revealed at the level of the olfactory neuro-epithelium (e.g., Wang *et al.* 1993). The chemosensory profile of the amniotic milieu can also impact on higher-level cognitive processes. Associative or non-associative olfactory experience can affect synaptic organisation within and between the different nervous relays, as shown by Leon, Wilson and Sullivan and their associates (Wilson and Sullivan 1994). This odour environment-induced moulding of the corresponding neuronal network organisation is thought to be the cause of one or a category of olfactory images that newborns will specifically search for (notion of 'search image'). Also, since the intensity level of intra-amniotic chemical stimulations is probably weak, it is possible that this induces after birth a non-specific preference for low-intensity odours. These two mechanisms relative to qualitative and intensitive selectivity are valid in rat foetuses (Smotherman and Robinson 1987).

The most direct explanation of prenatal acquisition *in vivo* is that of simple familiarisation with the chemosensory properties of the amniotic environment. This process is thought to trigger a specific development in processing certain olfactory indices thus facilitating postnatal conditioning of them. The presence of these same olfactory indices in the postnatal environment is thought to confirm their salience in view of reinforcing events associated with parental care. This 'foetal preparation' process is confirmed in infant rats: when sensitised to citral *in utero*, they only develop a preference for citral if they are re-exposed to it *ex utero* in particular stimulation conditions (Pedersen and Blass 1982). Under normal conditions, amniotic fluid plays the role of an olfactory link between amniotic and post-amniotic milieus (Blass and Teicher 1980, Schaal 2005). But apart from this sequence linking prenatal sensitisation to postnatal reinforcement of a specific odour, associative learning processes occur from foetal life onwards. Thus, each maternal meal provides the conditions for establishing an association between aroma flow and metabolite flow which stimulate in a short time window chemoreception, high foetus activation level, and the stimulation of opioidergic pathways.

18.4.3 Early odour acquisitions can persist in later childhood and adulthood

Several studies carried out on animals indicate the possibility of long-term memorisation of odours acquired during infancy. Infantile olfactory memory has been examined particularly in relation to the orientation of social and sexual preferences. In domestic dogs, maternal odour is retained by puppies, and even after two years of total separation, they remain capable of recognising their mother's odour (Hepper 1994). This 'olfactory imprinting' process has been analysed more thoroughly in rodents. Young female mice raised in the absence of adult males do not display, as adults, a normal attraction for male mice (Mainardi 1963). This effect is due to an olfactory mechanism: female mice whose mothers were odorised with Parma violet odour throughout the nursing period, when sexually mature, manifest a preference for males with this odour compared with test males. On the other hand, females raised in a normal environment prefer non-odorised males (Mainardi *et al.* 1965). This early effect of maternal odorising is not however to be found in young male mice: in the experiments conducted by Mainardi *et al.*, young male mice do not seem to be differentially affected by the experimental odour. However, in rats, an olfactory experience associated with suckling modifies male reactivity (in terms of reducing ejaculation latency) towards a receptive female carrying the same odour (Fillion and Blass 1986). This effect appears to be specific to the odour-suckling association because an odour connected indirectly with suckling (applied to the back of the nursing female or in the nest) entails no modification of adult sexual behaviour. Control tests further reveal that acquisition of the odour-suckling association has no influence on sexual performance of males exposed to non-receptive females or females without the experimental odour; in addition, males not exposed to the odour-suckling association show no changes in sexual behaviour in the presence of an artificially odorised female. In these species, mouse and rat, the olfactory canal plays a predominant role in early behaviour and the olfactory imprint can have long-term effects. The validity of these mechanisms for humans has not been examined in depth, and the differences between rodents themselves (see the case of male mice and rats described above) calls for prudence regarding inter-specific extrapolation.

However, some findings that should be treated with caution suggest that infant olfactory learning could last until adulthood. For example, among 92 young adults (20–35 years old) surveyed using a questionnaire on the positive and negative odours that have affected them most, 38% declared the most pleasant olfactory memories go back to childhood (Lenti-Boero 1994). The olfactory ambience associated with the mother is mentioned by 37%, whilst only 27% refer to present-day close relationships, and 6.5% to sexual relationships. These long-lasting childhood olfactory memories are most often reported in relation to food perception (Lenti-Boero 1994). Another research work traces the aversions acquired during childhood (between 0 and 5 years) and which persist 50 years later (Garb and Stunkard 1974); among subjects aged 13–20 when responding to the questionnaire, 12% trace their aversion to the 0–5 age period, 58% to the 6–12 age period, and 30% to the 13–20 age period.

The early milk-based feeding context, either at breast or on bottle, can clearly be causal in the developmental trajectory of odour or flavour preference phenotype. Infants who have sucked a breast scented with chamomile during the first postnatal months are more attracted by a toy odorised with the same note, and suck more actively from a chamomile-scented bottle, when they are re-tested 21 months later (Delaunay-El Allam *et al.* 2006b). Another element in favour of persistent effects of early experience in milk, but this time in formula milk, has already been discussed above (Mennella and Beauchamp 2002). It is further consolidated by the fact that children with phenylketonuria, who received from the first months phenylalanine-free protein hydrolysate formulas with distinctive flavours, made up of unpleasant smell and sweet–sour taste, are not reluctant to consume such formulae as adolescents, after years of exposure discontinuation. Lastly, a quasi-experimental trial took advantage of the flavouring of infant milk formula. Before 1992, most milk formulas were flavoured with vanilla in Germany. Thus, subjects having been either exclusively breast-fed or bottle-fed before that date could be considered free of exposure to the vanilla flavour during nursing or to have massively experienced it in formula, respectively. Subjects (aged between 12 and 59 years, average age 28.8) were then asked to evaluate standard ketchup and the same ketchup slightly flavoured with vanilla (Haller *et al.* 1999). The first group preferred normal ketchup to vanilla-flavoured ketchup (70.9% compared to 29.1%), while those exposed to vanilla through bottle-feeding responded the other way round (40% compared to 60%). Although this approach did not account for exposure to vanilla since weaning, it does underline the fact that infantile memory for odours and flavours may remain accessible and then possibly influence the attitudes and decisions of older children, juveniles and adults.

Another research considered the problem from the opposite perspective: guessing the infancy environment of adults from 19 countries by direct examination of their pattern of flavour preferences and aversions. Teerling *et al.* (1994) endeavoured to determine the food context to which these persons had been exposed during the first 5 years of life. To accomplish this, a series of eight contrasting flavours with strong cultural connotations was defined.[3] Based on the positive, negative and neutral responses to this series of aromatic indicators in national or ethnic flavour principles, each person was assigned to an infantile culinary context. Among the 74 tested subjects, 35% were correctly assigned from a cultural point of view, which is a significantly different proportion as compared to random distribution.

18.5 Conclusions and future trends

In humans, as in other mammals, nasal chemoreception is already functional before birth and it can then contribute to the behavioural and physiological adjustments necessary to adaptive transitions in the postnatal period (cf. Schaal 2005). Chemosensory systems are comprehensive in nature, i.e. they can detect a large spectrum of odorants and flavorants, and are open to early influences, as attested by the variety of stimuli influencing behaviour and the variety of rewarding means and contexts which can boost up learning processes. However, some findings suggest that olfactory learning in early life may be paralleled with cognitive mechanisms that are more or less pre-functional. Certain odorants, often carried in homospecific milk, have particular reactogenic properties which are revealed in the absence of any prolonged exposure after birth. Such inborn predispositions may result either from sensory images shaped by pre-birth interaction between olfactory neurons and the environment, or from hard-wired processes unaffected or minimally affected by the environment. These two parallel pathways to environmentally-attuned chemosensation and related behaviours have been well investigated in young animals, and should be better explored in human infants.

In the above sections, we have mainly addressed quality-specific perception and learning of odours and flavours, so that being exposed to stimulus Z facilitates later acceptance of stimulus Z or any substrate containing stimulus Z in a distinctive manner (e.g., foetal or neonatal exposure to carrot flavour *in utero* or *in lacto* leads to preference for carrot-flavoured cereals in 5-month-olds; Mennella *et al.* 2001). But our senses also afford amodal information about stimuli. For instance, tactile, visual, and auditory stimuli may share common

3. These culturally tainted odorants were cinnamon, ginger, liquorice, wintergreen, blue cheese, spicy or hot-spicy, brown-sugar biscuit (speculaas), sweet/acid or sweet/sour. It was assumed, for example, that wintergreen is highly appreciated in North America but disliked everywhere else; or the odour of blue cheese which causes aversion in Japan, in Indonesia and in South Africa but is appreciated elsewhere.

traits of intensity, noisiness/clarity, complexity, rhythm/tempo and variety. In the chemical senses, stimuli can have intensity, variety or complexity properties in common. Such amodal qualities may have strong influences on subsequent responsiveness. For example, early exposure to low intensity odour stimulations, which is typically the case in the amnion or milk, may be followed by a general trend of preferential responses to low intensity odorants. Otherwise, early experience of chemosensory variety has been shown to condition later acceptance of chemosensory novelty in animal infants (Kuo 1967). In rats, an effect of flavour variety experience on novelty acceptance could be ascertained in immature animals, but not in adult animals (Capretta *et al.* 1975), suggesting the existence of a sensitive period for the impact of chemosensory variety. Human cultures have fashioned distinct circumstances of early exposure to chemosensory variety: breastfeeding in which infants are exposed to chemosensory variability within and between feeds, and bottle-feeding in which infants are exposed to monotonous food within and between feeds. When this is put in line with the evidence presented above, that the chemosphere of early feeding can promote lasting odour/flavour impressions, phenotypic differences at the sensory, neurocognitive and behavioural levels may be reasonably expected between both subgroups. For example, due to chemosensory similarity between amniotic fluid and colostrum, both groups of infants are exposed to contrasted conditions of perinatal continuity in the flavour of ingested fluids; although the consequence of this disparity is unknown, it seems to affect postnatal responses (Schaal 2005). Breast- and bottle-fed infants, in line with their respective experiences of flavour variety in their lacteal diets, also differ in their responses to flavour novelty in foods at the time of weaning (e.g., Sullivan and Birch 1994, Maier *et al.* 2005). Although both subgroups of infants are certainly not homogenous (in terms of duration and exclusive character of early feeding practices), and may be exposed to chemosensory variety by other means than through food (e.g., cosmetics), this source of individual differences deserves more intensive research.

Mother-to-infant transfer of food flavours *in utero* and *in lacto*, infant learning of such information and use of it in the first selective responses to food, and stable influences into later childhood and adulthood of such early flavour experience should logically concur in the intergenerational reproduction of locally preferred foods. Indeed, mother food use during pregnancy and lactation is prolonged in the foods given to infants during the food diversification period (e.g., Bril *et al.* 2001, Mennella *et al.* 2005). However, analysing the similarity of favoured food patterns between parents and children raises a paradox – the so-called 'family paradox' (Rozin 1991) – according to which offspring preferred foods do not strongly correlate with parents' preferred foods (at least in late childhood and early adolescence). Moreover, mother–child preferences surprisingly do not correlate much higher than father–child preferences (Burt and Hetzler 1980, Birch 1980, Pliner and Pelchat 1986, Rozin 1991). This reveals that flavour preferences and related food liking are not solely and linearly programmed by early experience. Instead they are enriched and further re-organised by multiple and idiosyncratic influences outside the restricted mother–infant context, in the nuclear family, in peer groups evolving with age, and under the pressure of incentives afforded by the surrounding culture at large. The rate at which early flavour acquisitions enacted in the maternal niche are progressively interwoven with active, individual flavour experiences, and then self-organise into stable flavour preference phenotypes remains a vast, fallow area for future *developmental* biologists, psychologists, and clinicians.

18.6 Acknowledgements

This chapter has benefited greatly from numerous discussions with E. Hertling, R. Soussignan and T. Jiang, and from the linguistic skills of P. Brossier.

18.7 References

BALOGH, R.D. and PORTER, R.H. (1986) Olfactory preferences resulting from mere exposure in human neonates. *Infant Behavior and Development*, 9, 395–401.

BARTOCCI, M., WINBERG, J., RUGGIERO, C. *et al.* (2000) Activation of olfactory cortex in newborn infants after odor stimulation: a functional near-infrared spectroscopy study. *Pediatric Research*, 48, 18–23.

BEAUCHAMP, G.K., COWART, B.J. and SCHMIDT, H.J. (1991) Development of chemosensory sensitivity and preference. In T.V. Getchell, R.L. Doty, L.M. Bartoshuk and J.B. Snow (eds), *Smell and Taste in Health and Disease*, 405–416. Raven Press, New York.

BENSAFI, M., PORTER, J., POULIOT, S., MAINLAND, J., JOHNSON, B., ZELANO, B., YOUNG, N., BREMNER, E., AFRAMIAN, D., KAHN R. and SOBEL, N. (2003) Olfactomotor activity during imagery mimics that during perception. *Nature Neuroscience*, 6, 1142–1144.

BENSAFI, M., ZELANO, C., JOHNSON, B., MAINLAND, J. and SOBEL, N. (2005) Olfaction: Where there is a sniff there is a smell. In M.S. Gazzaniga (ed.), *The New Cognitive Neurosciences*, MIT Press, New York.

BINGHAM, P.M., ABASSI, S. and SILVERI, E. (2003) A pilot study of milk odor effects on non-nutritive suckling by premature infants. *Arch. Pediatr. Adolesc. Med.*, 157, 72–75.

BIRCH, L.L. (1980) The relationship between children's food preferences and those of their parents. *Journal of Nutritional Education*, 12, 14–18.

BLASS, E.M. and TEICHER, M.H. (1980) Suckling. *Science*, 210, 15–22.

BRAND, J.G., CAGAN, R.H. and NAIM, M. (1982) Chemical senses in the release of gastric and pancreatic secretions. *Annual Review of Nutrition*, 2, 249–276.

BRIL, B., HOMBESSA-NKOUNKOU, E., BOUVILLE, J.F. and OCAMPO, C. (2001) From milk to adult diet: A comparative study on the socialization of food. *Food and Foodways*, 9, 155–186.

BURT, J.V. and HETZLER, A.A. (1980) Parental influence on the child's food preference. *Journal of Nutritional Education*, 10, 127–128.

CABANAC, M. (1971) Physiological role of pleasure. *Science*, 173, 1103–1107.

CAPRETTA, P.J., PETERSIK, J.T. and STEWART, D.J. (1975) Acceptance of novel flavours is increased after early experience of diverse tastes. *Nature*, 254, 689–691.

CERNOCH, J.M. and PORTER, R.H. (1985) Recognition of maternal axillary odours by infants. *Child Development*, 56, 1593–1598.

COUREAUD, G., MONCOMBLE, A.S., MONTIGNY, D., PERRIER, G. and SCHAAL, B. (2005) Instant learning promotion: a new feat for mammalian pheromones. Submitted.

CRYSTAL, S.R. and BERNSTEIN, I.L. (1995) Morning sickness: Impact on offspring salt preference. *Appetite*, 25, 231–240.

DECASPER, A.J. and SPENCE, M.J. (1986) Prenatal maternal speech influences newborn's perception of speech sounds. *Infant Behavior and Development*, 9, 133–150.

DELAUNAY-EL ALLAM, M., MARLIER, L. and SCHAAL, B. (2004) Human infants' memory for an odour acquired during breast-feeding: Positive retention tests at 7 and 21 months. 16th Biennial Meeting of the European Chemoreception Research Organisation (Dijon, France).

DELAUNAY-EL ALLAM, M., MARLIER, L. and SCHAAL, B. (2006a) Learning at the breast: Preference formation for an artificial scent and its attraction against to odor of maternal milk. *Infant Behavior and Development*, in press.

DELAUNAY-EL ALLAM, M., SOUSSIGNAN, R., MARLIER, L., PATRIS, B. and SCHAAL, B. (2006b) A longitudinal analysis of memory for an odour associated with breast-feeding. In preparation.

DOTY, R.L. (1992) Olfactory function in neonates. In D.G. Laing, R.L. Doty and W. Breipohl (eds), *The Human Sense of Smell*, 155–165. Springer, Berlin.

DOTY, R.L. (ed.) (2003) *Handbook of Olfaction and Gustation*. Marcel Dekker, New York.

EKMAN, P. and FRIESEN, W. (1982) *Facial Action Coding System: A Technique for Measurement of Facial Movement*. Consulting Psychologist Press, Palo Alto, CA.

ENGEN, T. (1982) *The Perception of Odours*. Academic Press, New York.

ENGEN, T. and LIPSITT, L.P. (1965) Decrement and recovery of responses to olfactory stimuli in the human neonate. *Journal of Comparative and Physiological Psychology*, 59, 312–316.

ENGEN, T., LIPSITT, L.P. and KAIE, H. (1963) Olfactory response and adaption in the human neonate. *Journal of Comparative and Physiological Psychology*, 50, 73–77.

FELDMAN, M. and RICHARDSON, C.T. (1986) Role of thought, sight, smell and taste of food in the cephalic phase of gastric acid secretion in humans. *Gastroenterology*, 90, 428–433.

FILLION, T.J. and BLASS, E.M. (1986) Infantile experience with suckling odours determines adult sexual behavior in male rats. *Science*, 231, 729–731.

FUSARI, C. and PARDELLI, C. (1962) L'olfattometria elettroencefalografica nel lattante, *Bolletino delle Mallatie del'Orechio Gola Naso*, 80, 719–734.

GARB, J.L. and STUNKARD, A.J. (1974) Taste aversion in man. *American Journal of Psychiatry*, 131, 1204–1207.

GOUBET, N., RATTAZ, C., PIERRAT, V., ALLÉMAN, V., BULLINGER, A. and LEQUIEN, A. (2002) Olfactory familiarisation and discrimination in preterm and full-term newborns. *Infancy*, 3, 53–75.

GRANIER-DEFERRE, C., SCHAAL, B. and DE CASPER, A. (2004) Les prémisses ftales de la cognition. In R. Lécuyer (ed.), *Traité de Psychologie du Développement du Nourrisson*, 57–94. Dunod, Paris.

GUILLORY, A.W., SELF, P.A., FRANCIS, P.L. and PADEN, L.Y. (1980a) Habituation in studies of neonatal olfaction. *Biennial Meeting of the South-western Society for Research in Human Development*, Lawrence, KS.

HALL, B. (1975) Changing composition of human milk and early development of an appetite control. *Lancet*, 1, 779–781.

HALLER, R., RUMMEL, C., HENNEBERG, S., POLLMER, U. and KÖSTER, E. P. (1999) The influence of early experience with vanillin on food preference later in life. *Chemical Senses*, 24, 465-467.

HEPPER, P.G. (1987) The amniotic fluid: an important priming role in kin recognition. *Animal Behavior*, 35, 1343–1346.

HEPPER, P.G. (1988) Adaptive fetal learning: prenatal exposure to garlic affects postnatal preferences. *Animal Behavior*, 36, 935–936.

HEPPER, P.G. (1994) Long-term retention of kinship recognition established during infancy in the domestic dog. *Behavioural Processes*, 33, 3–14.

HEPPER, P.G. (1995) Human fetal 'olfactory' learning. *International. Journal of Prenatal and Perinatal Psychology and Medicine*, 7, 147–151.

HILL, D.L. and PRZEKOP, P.R. (1988) Influences of dietary sodium on functional taste receptor development: A sensitive period. *Science*, 241, 1826–1828.

HILL, D.L. and VOGT, M.B. (1989) Behavioral taste responses in rats 'recovered' from sodium deprivation instituted in early development. *Chemical Senses*, 14, 710.

IRZHANSKAIA, K.N. and FELBERBAUM, R.A. (1954) 'Conditioned reflex activity in premature children' (in Russian). *Fisiologikeskii. Zhurnal SSSR*, 40, 668–672.

KAWAKAMI, K., TAKAI-KAWAKAMI, K., OKAZAKI, Y., KURIHARA, H., SHIMIZU, Y. and YANAIHARA, T. (1997) The effect of odor on human newborn infant under stress. *Infant Behavior and Development*, 20, 531–535.

KENDAL-REED, M. and VAN TOLLER, S. (1992) Brain electrical activity mapping: an exploratory study of infant response to odours. *Chemical Senses*, 17, 765–777.

KUO, Z.Y. (1967) *The Dynamics of Behavior Development: An epigenetic perspective*. New York, Random House.

LAING, D.G. (1983) Natural sniffing gives optimum odour perception for humans. *Perception*, 12, 99–117.

LASKA, M., DISTEL., H. and HUDSON, R. (1997) Trigeminal perception of odorant quality in congenitally anosmic subjects. *Chemical Senses*, 22, 447–456.

LENTI-BOERO, D. (1994) Sedici odori per molti ricordi: una ricerca preliminare sull'olfatto e la vita quotidiana. *Studi Urbinati (Urbino)*, 5, 167–181.

LIPSITT, L.P., ENGEN, T. and KAIE, H. (1963) Developmental changes in the olfactory threshold of the neonate. *Child Development*, 34, 371–376.

MACFARLANE, A.J. (1975) Olfaction in the development of social preferences in the human neonate. *Ciba Foundation Symposia*, 33, 103–117.

MAIER, A.S., LEATHWOOD, P., CHABANET, C., ISSANCHOU, S. and SCHAAL, B. (2005) Exposure to different regimen of food variety influences the acceptance of new flavors by weanling infants. *6th Pangborn Conference*, Harrogate, UK.

MAINARDI, D. (1963) Eliminazione della barriera etologica all'isolamento riproduttivo tra Mus musculus domesticus e M.m. bactrianus mediante azione sull'apprendimento infantile. *Rendiconti dell'Instituto Lombardo di Scienze e Lettere, B: Scienze Biologische e Medische*, 97, 291–299.

MAINARDI, D., MARSAN, M. and PASQUALI, A. (1965) Causation of sexual preferences of house mouse. The behaviour of mice reared by parents whose odour was artificially altered. *Atti della Societa Italiana de Scienze Naturali*, 104, 325–338.

MAKIN, J.W. and PORTER, R.H. (1989) Attractiveness of lactating females' breast odors to neonates. *Child Development*, 60, 803–810.

MARLIER, L. and SCHAAL, B. (1997) Familiarité et discrimination olfactive chez le nouveau-né: influence différentielle du mode d'alimentation? In B. Schaal (ed.), *L'Odorat chez l'Enfant, Perspectives croisées*, Paris, Presses Universitaires de France, Enfance, 1, 47–61.

MARLIER, L. and SCHAAL, B. (1999) Feeding-experience dependent and feeding-experience independent odour preference responses in human infants: in search of an inborn attractive factor in human milk. *1st European Symposium on Olfaction and Cognition*, Lyon, France.

MARLIER, L. and SCHAAL, B. (2005) Human newborns prefer human milk: Conspecific milk odor is attractive without postnatal exposure. *Child Development*, 76, 155–168.

MARLIER, L., SCHAAL, B. and SOUSSIGNAN, R. (1997) Orientation responses to biological odours in the human newborn. Initial pattern and postnatal plasticity. Comptes Rendus de l'Académie des Sciences (Paris) *Life Sciences*, 320, 999–1005.

MARLIER, L., SCHAAL, B. and SOUSSIGNAN, R. (1998a) Neonatal responsiveness to the odor of amniotic and lacteal fluids: A test of perinatal chemosensory continuity. *Child Development*, 69, 611–623.

MARLIER, L., SCHAAL, B. and SOUSSIGNAN, R. (1998b) Bottle-fed neonates prefer an odor experienced *in utero* to an odor experienced postnatally in the feeding context. *Developmental Psychobiology*, 33, 133–145.

MARLIER, L., SCHAAL, B., GAUGLER, C. and MESSER, J. (2001) Olfaction in premature newborns: Detection and discrimination abilities two months before gestational term. In A. Marchlewska-Koj, J.J. Lepri and D. Müller-Schwarze (eds), *Chemical Signals in Vertebrates*, Volume 9, 204–208. Plenum/Kluwer Publishers, New York.

MELLIER, D., BÉZARD, S. and CASTON, J. (1997) Etudes exploratoires des relations intersensorielles olfaction-douleur. In B. Schaal (ed.), *L'Odorat chez l'Enfant, Perspectives croisées*, Paris, Presses Universitaires de France, Enfance, 1, 98–111.

MENNELLA, J.A. and BEAUCHAMP, G.K. (1991a) Maternal diet alters the sensory qualities of human milk and the nursling's behavior. *Pediatrics*, 88, 737–744.

MENNELLA, J.A. and BEAUCHAMP, G.K. (1991b) The transfer of alcohol to human milk, effects on flavor and infant's behavior. *New Engl. J. Med.*, 325, 981–985.

MENNELLA, J.A. and BEAUCHAMP, G.K. (1996) Developmental changes in the acceptance of protein hydrolysate formula. *Developmental and Behavioral Pediatrics*, 17, 386–391.

MENNELLA, J.A. and BEAUCHAMP, G.K. (2002) Flavor experiences during formula feeding are related to preferences during childhood. *Early Human Development*, 68, 71–82.

MENNELLA, J.A. and BEAUCHAMP, G.K. (1996) The human infants' response to vanilla flavors in mother's milk and formula. *Infant Behavior and Development*, 19, 13–19.

MENNELLA, J.A., JAGNOW, C.P. and BEAUCHAMP, G.K. (2001) Prenatal and postnatal flavor learning by human infants. *Pediatrics*, 107, 1–6.

MENNELLA, J.A., TURNBULL, B., ZIEGLER, P.J. and MARTINEZ, H. (2005) Infant feeding practices and early flavor experiences in Mexican infants: an intra-cultural study. *Journal of the American Dietetic Association*, 105, 908–915.

MONCOMBLE, A.S., COUREAUD, G., QUENNEDEY, B., LANGLOIS, D., PERRIER, G. and SCHAAL, B. (2005) The Mammary pheromone of the rabbit: From where does it come? *Animal Behaviour*, 69, 29–38.

NAJBAUER, J. and LEON, M. (1995) Olfactory experience modulates apoptosis in the developing olfactory bulb. *Brain Research*, 674, 245–251.

NICOLAÏDIS, S., GALAVERNA, O. and METZLER, C.H. (1990) Extracellular dehydration during pregnancy increases salt appetite of offspring. *American Journal of Physiology*, 258, R281–R283.

OSTER, H. and ROSENSTEIN, D. (in press) *Baby-FACS: Analysing facial movements in infants*. Palo Alto, CA, Consulting Psychology Press.

PEDERSEN P.A. and BLASS, E.M. (1992) Prenatal and postnatal determinants of the 1st suckling episode in albino rats. *Developmental Psychobiology*, 15, 349–355.

PERBELLINI, D. and SCOLARI, R. (1966) L'ellettroencefalo-olfattometria. *Ann Laryngol Otol Rin Far*, 65, 421–429.

PETERSON, F. and RAYNEY, L.H. (1910–11) The beginnings of mind in the newborn. *Bulletin of the Lying-in Hospital of NY City*, 7, 99–122.

PIHET, S., MELLIER, D., BULLINGER, A. and SCHAAL, B. (1997) Réponses comportementales aux odeurs chez le nouveau-né prématuré: étude préliminaire. *Enfance*, 1, 33–46.

PIHET, S., SCHAAL, B., BULLINGER, A. and MELLIER, D. (1996) An investigation of olfactory responsiveness in premature newborns. *International Conference on Infant Studies*, Providence, RI.

PLINER, P. and PELCHAT, M.L. (1986) Similarities in food preferences between children and their siblings and parents. *Appetite*, 7, 333–342.

RABIN, M.D. (1988) Experience facilitates olfactory quality discrimination. *Perception and Psychophysics*, 44, 532–540.

RABIN, M.D. and CAIN, W.S. (1986) Determinants of measured olfactory sensitivity. *Perception and Psychophysics*, 39, 281–286.

RATTAZ, C., GOUBET, N. and BULLINGER, A. (2005) The calming effect of a familiar odor on full-term newborns. *Journal of Developmental and Behavioral Pediatrics*, 26, 86–92.

RIGHARD, L. and ALADE, M.O. (1990) Effect of delivery routines on success of first breast-feed. *Lancet*, 336, 1105–1107.

ROVEE, C.K. (1969) Psychophysical scaling of olfactory response to the aliphatic alcohols in human neonates. *Journal of Experimental Child Psychology*, 7, 245–254.

ROZIN, P. (1991) Family resemblance in food and other domains: the family paradox and the role of parental congruence. *Appetite*, 16, 93–102.

RUSSELL, M.J. (1976) Human olfactory communication. *Nature*, 260, 520–522.

SALAS, M., GUZMAN-FLORES, C. and SHAPIRO, S. (1969) An ontogenetic study of olfactory bulb activity in the rat. *Physiology and Behavior*, 4, 699–703.

SARNAT, H.B. (1978) Olfactory reflexes in the newborn infant, *Journal of Pediatrics*, 92, 624–626.

SCHAAL, B. (1984) La compétence olfactive du nouveau-né humain: aspects fonctionnels. *Psychologie Médicale*, 16, 305–314.

SCHAAL, B. (1986) Presumed olfactory exchanges between mother and neonate in humans. In J. Le Camus and J. Cosnier (eds), *Ethology and Psychology*, Toulouse, Privat (101–110).

SCHAAL, B. (1988) Olfaction in infants and children: developmental and functional perspectives. *Chemical Senses*, 13, 145–190.

SCHAAL, B. (1999) Le développement de la sensibilité olfactive: de la période foetale à la puberté. In Y. Christen, L. Collet and M.T. Droy-Lefaix (eds), *Recontres IPSEN en Oto-Rhino-Laryngologie*, Vol. 3, Irvinn Editions, Neuilly sur Seine.

SCHAAL, B. (2005) From amnion to colostrum to milk: Odour bridging in early developmental transitions. In B. Hopkins and S. Johnson (eds), *Prenatal Development of Postnatal Functions. Advances in Infancy Research*, Volume 2, 52–102. Praeger, Westport, CT.

SCHAAL, B., MONTAGNER, A., HERTLING, E., BOLZONI, D., MOYSE, A. and QUICHON, R. (1980) Les stimulations olfactives dans les relations entre l'enfant et la mere. *Reproduction, Nutrition, Développement*, 20, 843–858.

SCHAAL, B., ORGEUR, P., LECANUET, J.P., LOCATELLI, A., GRANIER-DEFERRE, C. and POINDRON, P. (1991) Nasal chemoreception *in utero*: first data in the fetal sheep. Comptes Rendus de l'Académie des Sciences, *Life Sciences*, 113, 319–325.

SCHAAL, B., MARLIER, L. and SOUSSIGNAN, R. (1995a) Responsiveness to the odour of amniotic fluid in the human neonate. *Biology of the Neonate*, 67, 397–406.

SCHAAL, B., ORGEUR, P. and ARNOULD, C. (1995b) Olfactory preferences in newborn lambs: possible influence of prenatal experience. *Behaviour*, 132, 351–365.

SCHAAL, B., ORGEUR, P. and ROGNON, C. (1995c) Odor sensing in the human fetus: anatomical, functional and chemo-ecological bases. In Lecanuet, J.P., Krasnegor, N.A., Fifer, W.A. and Smotherman, W. (eds), *Prenatal Development, A Psychobiological Perspective*, 205–237. Lawrence Erlbaum: Hillsdale, NJ.

SCHAAL, B., MARLIER, L. and SOUSSIGNAN, R. (1998) Olfactory function in the human fetus: evidence from selective neonatal responsiveness to the odor of amniotic fluid. *Behavioral Neuroscience*, 112, 1438–1449.

SCHAAL, B., LECANUET, J.P. and GRANIER-DEFERRE, C. (1999) Sensory and integrative development in the human fetus and perinate: the usefulness of animal models. In M. Haug and R.E. Whalen (eds), *Animal Models of Human Cognition and Emotion*, 119–142. American Psychological Association, Washington, DC.

SCHAAL, B., MARLIER, L. and SOUSSIGNAN, R. (2000) Human foetuses learn odours from their pregnant mother's diet. *Chemical Senses*, 25, 729–737.

SCHAAL, B., COUREAUD, G., LANGLOIS, D., GINIES, C., SÉMON, E. and PERRIER, G. (2003) Chemical and behavioural characterization of the mammary pheromone of the rabbit. *Nature*, 424, 68–72.

SCHAAL, B., HUMMEL, T. and SOUSSIGNAN, R. (2004) Olfaction in the fetal and premature infant: functional status and clinical implications. *Clinics in Perinatology*, 31, 261–285.

SCHLEIDT, M. and GENZEL, C. (1990) The significance of mother's perfume for infants in the first weeks of their life. *Ethology and Sociobiology*, 11, 145–154.

SCHNEIRLA, T.C. (1965) Aspects of stimulation and organization in approach/withdrawal processes underlying vertebrate behavioral development. In D.S. Lehrman, R.A. Hinde and E. Shaw (eds), *Advances in the Study of Behavior*, Vol. 1, 1–74. New York, Academic Press.

SCHWARTZ, G.E., WHITEHORN, D., HERNON, J.C. and JONES, M. (1986) Subjective and respiratory differentiation of fragrances: Interactions with hedonics. *Psychophysiology*, 23, 460.

SELF, P.A., HOROWITZ, F.D. and PADEN, L.Y. (1972) Olfaction in newborn infants. *Developmental Psychology*, 7, 349–363.

SHIMADA, M., TAKAHASHI, S., IMURA, S. and BABA, K. (1987) Olfaction in neonates. *Chemical Senses*, 12, 518.

SMOTHERMAN, W.P. (1982) Odor aversion learning by the rat fetus. *Physiology and Behavior*, 29, 769–771.

SMOTHERMAN, W.P. and ROBINSON, S.R. (1987) Psychobiology of fetal experience in the rat. In Krasnegor, N.E., Blass, E.M., Hofer, M.A. and Smotherman, W.P. (eds), *Perinatal Development. A Psychobiological Perspective*, 39–60. Academic Press, Orlando.

SMOTHERMAN, W.P. and ROBINSON, S.R. (1995) Tracing development trajectories into the prenatal period. In Lecanuet, J-P., Fifer, W.P., Krasnegor, N.A. and Smotherman, W.P. (eds), *Fetal Development: A Psychobiological Perspective*, 15–32. Hillsdale, NJ, Lawrence Erlbaum.

SOUSSIGNAN, R., SCHAAL, B., MARLIER, L. and JIANG, T. (1997) Facial and autonomic responses to biological and artificial olfactory stimuli in human neonates: Re-examining early hedonic discrimination of odours. *Physiology and Behavior*, 62, 745–758.

SOUSSIGNAN, R., SCHAAL, B. and MARLIER, L. (1999) Olfactory alliesthesia in human neonates: Prandial state modulates facial and autonomic responses to milk odors. *Developmental Psychobiology*, 35, 3–14.

STEINER, J.E. (1977) Facial expressions of the neonate infant indicating the hedonics of food-related stimuli. In Weiffenbach, J.M. (ed.), *Taste and Development, The Genesis of Sweet Preference*, NIH-DHEW, Bethesda, MD.

STEINER, J.E. (1979) Human facial expressions in response to taste and smell stimulations. In Lipsitt, L.P. and Reese, H.W. (eds), *Advances in Child Development*, Vol. 13, 257–295. Academic Press, New York.

STEINER, J.E., GLASER, D., HAWILO, M.E. and BERRIDGE, K. (2001) Comparative evidence of hedonic impact: affective reactions to taste by human infants and other primates. *Neuroscience and Biobehavioral Reviews*, 25, 53–74.

STIRNIMANN, F. (1936) Versuche über Geschmack und Geruch am ersten Lebenstag. *Jahrbuch für Kinderheilkunde*, 146, 211–227.

SULLIVAN, R.M. (1990) Newborn infants exhibit CR's to an odor previously paired with either breast or bottle feeding. *Annual Meeting of International Society for Developmental Psychobiology*, Cambridge, UK.

SULLIVAN, R.M. and TOUBAS, P. (1998) Clinical usefulness of maternal odor in newborns: soothing and feeding preparatory responses. *Biology of the Neonate*, 74, 402–408.

SULLIVAN, R.M., TABORSKY-BARBA, S., MENDOZA, R., ITANO, A., LEON, M., COTMAN, C.W., PAYNE, T. and LOTT, I. (1991) Olfactory classical conditioning in neonates. *Pediatrics*, 87, 511–517.

SULLIVAN, S.A. and BIRCH, L.L. (1994) Infant dietary experience and acceptance of solid foods. *Pediatrics*, 93, 271–277.

TEERLING, A., KÖSTER, E.P. and VAN NISPEN, V. (1994) Early childhood experiences and future preferences. *11th Congress of the European Chemoreception Research Organization*, Blois.

TINBERGEN, N. (1951) *The Study of Instinct*. Oxford University Press, Oxford.

VARENDI, H., PORTER, R.H. and WINBERG, J. (1994) Does the newborn baby find the nipple by smell? *Lancet*, 344, 989–990.

VARENDI, H., PORTER, R.H. and WINBERG, J. (1996) Attractiveness of amniotic fluid odor: evidence of perinatal learning? *Acta Paediatrica*, 85, 1223–1227.

WANG, H., WYSOCKI, C.J. and GOLD, G. (1993) Induction of olfactory receptor sensitivity in mice. *Science*, 260, 998–1000.

WILSON, D.A. and SULLIVAN, R.M. (1994) Neurobiology of associative learning in the neonate: early olfactory learning. *Behavioral and Neural Biology*, 61, 1–18.

WYSOCKI, C.J., DORRIES, K.M. and BEAUCHAMP, G.K. (1989) Ability to perceive androstenone can be acquired by ostensibly anosmic people. *Proceeding of the National Academy of Sciences of the USA*, 81, 7976–7978.

YASUMATSU, K., UCHIDA, S., SUGANO, H. and SUZUKI, T. (1992) The effect of fragrance on spectral power of EEG in human infants. *Kyushu N- Psychology*, 38, 287–294.

YASUMATSU, K., UCHIDA, S., SUGANO, H. and SUZUKI, T. (1994) The effect of the odour of mother's milk and orange on the spectral power of EEG in infants. *Journal of UOEH*, 16, 71–83.

Index

absorption, in the lungs and nose 295–6
acetone 212, 263–4
acid–base equilibria 233
acid sensing ion channels (ASIC) 5–6
across-fibre pattern model 15–18, 39, 45
activational state of neonates 405–6
activity coefficients 118–19
adaptation 38
affective component of taste 47
affinity chromatography 183
agarose 289–90
age 392
aggregates/aggregation
  β-lactoglobulin 176
  lipid 134–6, 139–40
    formation 140–3
    micelles and related aggregates 143–5
aggregation number 145
air–product partition coefficient 229, 230–4, 240
air–water partition coefficients 294–5
alcohols 246–7
  β-lactoglobulin stability and denaturation 177
aldehydes 192, 338, 339
  aldehyde–protein interactions 275
alkenals 192
α-gustducin 12–15
α-lactalbumin 177–8
amiloride-insensitive salt taste 7
amino acids 185
amniotic fluid 383, 419, 421, 424, 425, 426
amphiphilic molecules 134, 141–2, 148
amygdala 41, 43, 52, 356–7
amylaceous matrices 218–22
amylopectin 214
amylose 214, 215
amylose complexes 220–2
analysis of variance (ANOVA) 71, 100
analytical intensity rating tests 64–8
  descriptive analysis 65–7
  time-intensity 67–8
  types of scales 64
anatomical barrier 262, 263, 293, 296–7
animals 370, 389–90
anise 415, 416
anosmia 88
anterior cingulate cortex 43, 52, 359
apparent concentration 158–60
apparent diffusivity 125
apparent mass transfer coefficient 240
appetitive movements 408–9
aroma 163–4, 373–4
  modelling aroma interactions *see* modelling aroma interactions
  modelling persistence of aroma after swallowing 249–50
  taste–aroma interactions *see* taste–aroma interactions
  texture–aroma interactions *see* texture–aroma interactions

aroma analysis 81–97
  future trends 92–3
  key components contributing to sensory properties 84–7
  monitoring changes in aroma compounds with time 89–90
  obtaining a complete aroma profile 82–4
  off-notes in a food product 87–9
  sources of information 93–5
  using instrumental data in sensory predictions 90–2
aroma compounds 157
  behaviour in emulsions 158–61
  carbohydrates
    interaction mechanisms with 216–18
    and static partition of 209–10
  dissolving in lipids 158–60
  physical properties and flavour release 290–1
aroma isolates 85
aroma release
  carbohydrates 210–13
  relating to consumer preference 252
  relating to perception 252
aroma retention 252
  carbohydrates 210–13
aroma transport mechanisms 244–6
artificial mouth 241, 252, 267–8, 275, 276–7, 278
artificial sweeteners 17
artificial throat 252, 266–71, 275, 276–7, 278
  aroma release from 269–70
  description 268–9
  effect of sample viscosity on aroma release 270–1
  importance of throat surface 270
artificial tongue 27
Asa foetida 412, 416
aspartame 350
assessors *see* panellists
associative learning 387, 426
astringency 375
Atkins diet 391–2
atmospheric pressure chemical ionisation – mass spectrometry (APCI-MS) 192, 194
atmospheric pressure ionisation (API) 163
auction methods 68
autolysed yeast 388–9
automated aroma analysis 82, 89–90
axons 37

baking 244
balsamic vinegar 392
Balut 389
beer 314, 393
behavioural activation 405–6
benzaldehyde 191, 352, 356–7, 361
β-conglycinin 7S 180
β-cyclodextrin-aroma complexes 218–19
β-glycosides 24
β-lactoglobulin 120, 173–7, 182, 191–2
  binding sites 174–5
  effect of medium on protein–flavour interactions 187–8, 189, 190, 191
  interactions with aroma compounds 274
  stability and denaturation 175–7
  structure 174
beverages *see* liquid food products
bilayers 134–6
binding
  irreversible 182, 192–3
  mechanisms for amylose complexes 221
  reversible 182, 295
binding constants 231
  determinations 183–4
binding sites 193–4
  α-lactalbumin 178
  β-lactoglobulin 174–5
  BSA 178–9, 182
  localisation 184–6
  soy proteins 181
bitter taste 4, 308–9
  experimental design of sensory analysis 70
  gustatory–trigeminal interactions 50–1
  individual differences 310
  PTC/PROP bitterness 311–16
bitter taste receptors 10–12, 24–5, 44, 308, 310–11
  functional analysis 21–4
boiled sweets 299
bolus
  losses of flavour molecules after leaving the bolus 292
  low moisture foods 291–2
bootstrapping techniques 111
bottle-fed infants 407, 408, 421, 427, 429
bottle feeding 384, 385, 414, 427
boundary layer 122, 123
bovine serum albumin (BSA) 178–9, 182
  effect of medium on protein–flavour interactions 188, 189, 190

brain 356–9
development in infancy 382–3
feeding experiences in early life and modification of brain structure 387–8
feelings, emotions and food 395–6
neural processing *see* information processing
and its relationship with flavour 372–83
multisensory perception 372–4
propriosense and flavour learning 379–81
role of senses in flavour perception 374–9
brain scanning techniques 42–3, 54, 394–5
breast-fed infants 407, 408, 421, 427, 429
breast feeding 383–4, 385, 420
breast odour 406–8
breath
flavour profiles after chewing or swallowing 300–1
swallow breath 162, 262, 266, 297–8
volatile concentration at one minute 303
brittle gels 289–90
bulk sweeteners 271–3
business decisions 74–5
butyric acid 419

caffeine 70, 393
calcium imaging 18–26
cAMP 37–8
canonical correlation analysis (CCA) 91, 100, 102–3, 104
capsaicin 48, 49, 50–1, 318
carbohydrates 208–28, 271, 278, 353
aroma interactions with mono- and disaccharides 208–14
dynamic aspects 210–13
matrix physical state 213–14
static partition of aroma compounds 209–10
interaction mechanisms with aroma compounds 216–18, 222
chemical properties of aroma compounds influencing interactions 217–18
physicochemical phenomena 216–17
retention in amylaceous matrices and their derivatives 218–22
β-cyclodextrin–aroma complexes 218–19
starch and derivatives 219–22
structure 214–16
carbon dioxide 49
carbonated fruit drinks 315–16
carboxymethyl cellulose (CMC) 211–13, 329–32
carers 385–6, 387
carry-over effects 70
caseins 179–80, 191
category scales 64
caudal projections 46
Cenovis 389
central information processing 42–3, 46–7, 321
chain length 217
chaos theory 395
Chardonnay wines 64, 66, 67, 74
Cheddar cheese 150, 193
cheeses 193, 352
distribution of fat 152
flavour release 128
lipolysis 150
chemometrics 90
chemoreceptive systems 410–11
inception of nasal chemoreception 411–12
see also olfactory system; trigeminal system
chewing
flavour release 287, 293, 297, 354–5
breath flavour profiles after 300–1
relative importance of chewing and swallowing actions 298–300
texture–aroma interactions 338, 339, 341
chewing efficiency 265
chewing gum 288, 299, 373
children 377
development of brain 382–3
functional value of olfaction in early development 404–10
learning of flavour 383–94
changing preferences after infancy 390–2
early feeding experiences and modification of brain structure 387–8
family structure and diet 390
neonates 383–4
role of carers 385–6
spontaneous food selection 386–7
see also infants
cingulate cortex 41, 43, 52, 53, 359
circular dichroism spectroscopy (CDS) 184–5

citral 191, 263–4, 272, 273
Clausius-Clapeyron equation 234
cluster analysis 91, 108
coffee 223, 314, 393
cognition
  cognitive component of taste 47
  cognitive gustatory pathway 46–7
  and flavour perception 361–2
cold- and menthol-sensitive receptor (CMR1) 48–9
colinearity 99, 101, 103, 104
column centering of data 100
compartments 134
concentration gradients 234–6, 240
conditioning index 99
conferences 95
congruent taste-odour pairs 360, 361
consumer perception 75
consumer preferences
  relating aroma release to 252
  relating sensory analytical data to 72–4
consumer tests 68–9
consumption: resistance to mass transfer during 246–7
continuous phase 156–7
  viscosity 161
controlled release 222, 223, 246
convective diffusion 237–8
co-occurring stimulations 168–9
cooking 370–2
  aroma retention during 243–4
  cooking creates new flavours 371–2
correlation matrix, singular 106, 107
cortical representation 51–3
co-solvents 190
cranberry juice 391
cranial nerves 4–5
Crank equation 125, 249
cream 149
cribriform plate 377
critical concentration 332–4
critical micellar concentration (CMC) 137, 145
critical packing parameter 137–40
critical particle size 249
cross adaptation 24–5
cross-modal matching 64
cross-validation 109
crystallisation 149
cubic phases 143, 144, 145
cubosomes 145
culture 391, 428
cyclodextrins (CD) 215–16, 223, 272
  β-cyclodextrin-aroma complexes 218–19
cylindrical micelles 143, 144, 145
cysteine-aldehyde condensation 191

dairy products
  dairy desserts 262–3
  flavour release 128
  temperature and lipid structures 149
  see also cheeses
decanal 338, 339
denaturation of proteins 175–7
dendritic links 382
descriptive (sensory) analysis (DA) 65–7, 100
detection threshold 63
development of perception 403–36
  early functioning of olfaction 410–20
    discriminative power of early olfaction 414–20
    sensitivity of early olfactory detection 410–14
  functional value of olfaction in early development 404–10
    appetitive movements 408–9
    behavioural activation 405–6
    exploratory and directional behaviours 406–8
    odour sampling responses 408
    physiological effects of odours 409–10
  future trends 428–30
  memory and plasticity of olfactory function in early life 420–8
    early postnatal shaping of odour preferences 420–2
    persistence of early odour acquisitions into later childhood and adulthood 426–8
    preferences established independently of postnatal experience 423–6
dextrose equivalent (DE) 215
diacetyl 212, 340–1
diet, children's 390
difference tests 63
differential scanning calorimetry (DSC) 184
diffusion 121–6
  carbohydrate–flavour interactions 216, 222
  effect of diffusion mechanism on mass transport 236–8
  flavour diffusion in food products 124–6

methods of measurement 124–5
diffusion coefficient 122, 236–7
aroma compounds in different food matrices 125–6
diffusion ordered spectroscopy (RMN Dosy) 125
diffusivity
apparent 125
of aroma compounds 336
water and volatiles during drying 212, 213
digestive processes 409–10
dilution 148
with saliva 162, 166–7
direct isolation techniques 91
direct measurements of diffusion 124–5
directional behaviours 406–8
disaccharides 208, 350–2
aroma interactions with 208–14
discriminative power of early olfaction 414–20
dimensions of olfactory stimulation discriminable to newborns 416–20
discrimination level 414–15
discrimination tests 63–4
disgust 388–90
dispersed phase 156–7
dispersed systems 152
dissolved molecules, aroma binding by 232–3
drinking 379
in vivo aroma release measurements during 262–5
droplet size 160–1
dry/low moisture foods 291–2
drying 212, 213
drying experiments for diffusion measurement 125
duo-trio test 63
Durian fruit 389
dynamic coupled column liquid chromatography (DCCLC) 183
dynamic systems
modelling aroma release 241–3
modelling mass transfer 238–9
dynataste (multichannel flavour delivery system) 350

eating protocol 301
eddy diffusion 237–8
electroencephalographic (EEG) responses 414–15
emotions 395–6
empirical modelling 250–1
emulsifiers 141–2, 147–8, 160
emulsions 156–71
aroma components 157
behaviour of aroma compounds in 158–61
characteristics 156–8
cross-links 169
diffusion coefficient 236
effect on aroma release 276–7
fat mimetics in 193
future trends 163–9
driving principles applied in the elaboration of strategies 165–9
expectations 163–5
improving flavour delivery 162–3
lipid–flavour interactions 141–3, 144, 148–50, 151, 152
phase partitioning in 231–2
taste components 157–8
encapsulation 152, 222
micro and nano 150–1
using liposomes 146–7
entorhinal cortex 41, 43
enzymic degradation 292
epithelial sodium channel (ENaC) 6–7
equilibrium dialysis technique 183
ethanol 352
ethology 404–5
ethyl acetate 120–1, 212, 213
ethyl butyrate 340–1
ethyl octanoate 300, 301
exclusion chromatography 183
experience-dependent plasticity 382
experience-expectant plasticity 382
experimental design 70
exploratory behaviours 406–8
external preference mapping (prefmap) 72–4

facial expressions 417–19
factor analysis (FA) 91
familiarity 416, 417
family paradox 429–30
family structure 390
fat 375
crystallisation and phase partitioning 234
increasing proportion of crystalline fat in emulsions 168
lower fat content 164
PTC/PROP taster groups and fat content 316
replacing by structuring lipids 151–3
see also lipids

fat mimetics 193
feelings 395–6
fibres, trigeminal 47–9, 318
Fick's laws of diffusion 122, 124
fixed vocabulary profiling 100
flavour analysis 251
flavour compounds
  differences between flavour molecules *in vivo* 294–5
  flavour fraction in emulsions 161
  influence of physical properties on flavour release 290–1
  interactions with other constituents 126
flavour integration 49–53
flavour perception *see* perception
flavour preferences *see* preferences
flavour properties 117–21
flavour release 117–32, 287–307
  aroma delivery to the upper airway and nose 295–301
  diffusion and mass transfer 121–6, 127
  factors influencing mobility of flavour compounds in food matrices 126–8
  flavour properties 117–21
  future trends 129, 303–4
  influence of the foodstuff 288–92
  influence of rheological behaviour 329–36, 337
  liquid food products *see* liquid food products
  losses of flavour molecules after leaving the bolus 292
  methods for inducing/monitoring flavour release using lipids 152–3
  odour–taste interactions 353–5
  persistence of 301–3
  replacing fat by structuring lipids 151–3
  transfer of volatiles to the gas phase *in vivo* 293–5
flavour release models 241, 274–5, 302–3, 336
flavour strength 376
flexible lamellar lipids 138
fluorescence 185
fluorescence emission spectrum 185
fluorescence excitation spectrum 185
fluorescence imaging plate reader (FLIPR) 18
fluorescence recovery after fringe pattern photobleaching (FRAPP) 125
foaming 190
focus groups 68
foetus 383
  learning 424
  psychological processes involved in chemosensory acquisitions 425–6
food design 26–7
food matrix *see* matrix
foodstuff: influence on flavour release 288–92
formula feeding 384, 385, 414, 427
Fourier transform infrared (FT–IR) spectroscopy 184, 185
free choice profiling (FCP) 65
fruit drinks, carbonated 315–16
fruits 223
functional groups 218
functional magnetic resonance imaging (fMRI) 42–3, 51, 357, 362–3, 372, 395
fungiform papillae (FP) density 316–18, 319–20, 321

G protein-coupled receptors (GPCR) 12–15, 26, 37, 44, 308
gag reflex 5
garlic 409
gas phase
  pushing emulsions to the gas phase 162
  transfer of volatiles to *in vivo* 293–5
GC 88
GC-MS 88
GC-O 85–6
  assessors 88
gelatine 289–90
gelatinisation of starch 215
gels 169, 248–9
  caging in by gel effect and flavour–carbohydrate interactions 216–17, 223
  and flavour release 289–91
  rheology and flavour release 334–6
  texture and flavour perception 339–41
generalised Procrustes analysis (GPA) 72, 100
generic descriptive analysis 65
genetic influences on taste 308–26
  genetic variations of taste receptors 27–31
  individual differences in taste perceptions 309–11
  PROP taster differences and flavour perceptions 320–1
  PTC/PROP bitterness 311–16
  role of taste anatomy 316–20

supertasters and taste anatomy 319–20
tactile components of tastes 318–19
geranyl acetate 272, 273
glass transition 219–20
global mass transfer coefficient 124
globule size reduction 152
glomeruli 39–40
glucose 208, 209
glucose syrups 215
glutamate 9–10
glycinin 11S 180–1
graphic scales 64
guar gum 211–13, 329–32
gustatory modality 36, 41, 43–7
cortical integration of odours and tastes 51–3
gustatory–trigeminal interactions 50–1
at the cortical level 51
gustducin 12–15

habituation paradigm 415, 416
halo-dumping effect 361–2
HCN family 6
headspace analysis 91
automated methods 90
binding constant determinations 183–4
effect of sucrose on aroma release 272, 273
small sugar solutions 209–10, 211
static *see* static headspace analyses
health 391–2
hearing 375
heat processing *see* cooking
heat-stable flavourings 251
hedonic scales 68
hedonic value 416, 417–19
HEK293 cells 6, 13, 25
Henry's law 272
heterologous expression taste receptors 18–26
current limitations of the system 25–6
hexagonal phases 143, 144, 145
hexanal 338, 339
hexanol 212
high vacuum distillation 87
HLB (hydrophile, lipophile balance) 136–7
homeostasis 395–6
horse meat 389
HSGC 278
human milk 383–4, 405–6, 407, 414, 421
hydration of protein 191
hydrocolloid solutions 236–7
hydrogen bonding 217
hydrolysate infant formula 384, 422
hydrolysis of starch 215
hydrophobic interactions 217, 222
hydrophobic molecules 164
hydrophobic reservoirs 134
hydrophobicity 120
hypothalamus 41, 43

ice-cream 149, 152
in-mouth losses 292
in vitro aroma release 239–43
in vivo aroma release 252, 260–1
from beverages *see* liquid food products
modelling 244–9
physiological factors important for 265–6
inborn unconditional odorants 423
inclusion complexes 213, 215, 217, 220–2, 223
infant formula 384, 385, 414, 427
infants
course of olfactory sensitivity in early development 412–13
discriminative power of early olfaction 414–20
early postnatal shaping of odour preferences 420–2
functional value of olfaction in early development 404–10
appetitive movements 408–9
behavioural activation 405–6
exploratory and directional behaviours 406–8
odour sampling responses 408
physiological effects of odours 409–10
inception of nasal chemoreception 411–12
learning of flavour 383–4
specificity of early olfactory detection 413–14
see also children
information processing 36–61
future trends 53–4
multimodal interactions and flavour integration 49–53, 320–1
cortical representation and flavour integration 51–3
reception of odorants and neural processing 37–43
central integration of information 42–3

peripheral processing 42
reception of taste compounds and neural processing 43–7
central representation of tastes 46–7
peripheral representation of tastes 45–6
trigeminal chemosensitivity 47–9
ingredients: effect on aroma release from beverages 271–7, 278
inositoltriphosphate (IP3) 13–15
instrumental data
matching to sensory data 72, 98–113
case study 105–10
investigating structure of data sets 99–101
relating sensory to instrumental data 101–5
using in sensory predictions 90–2
insula 43, 51–2, 53, 356–7, 359
intense sweeteners 271–3, 350
intensity 416–17
interface
crucial for stability and mass transfer in emulsions 156–7
optimising flavour yield of emulsions 167–8
interface curvatures 136, 137, 138, 139
origins of 139–40
interface phase 141–3
interindividual differences
in aroma release and perception 262–4
and flavour release 296–7, 355
taste perceptions 309–11
internal preference analysis 72
internal standard method 83
internet 93–5
intrapersonal variation 262–4
ionic mechanisms 5–7, 308
irreversible interactions 182, 192–3
irritants 320
PROP testing and perception of 312–14, 315
see also trigeminal system
isoamyl acetate 335, 340
isotope dilution assay 84

jackknife techniques 111

kinetically controlled release 245, 246

labelled line model 15–18, 45
labelled magnitude scale (LMS) 64
lactisole 20, 349–50
lamellar cubic lipids 138, 139–40
lamellar phases 138, 139–40, 144
laminar flow 122–3
large unilamellar vesicles (LUVs) 146
Latin squares designs 70
learning 428
associative 387, 426
foetal 424
see also development of perception
learning of flavour preferences 369–402
flavour perception 372–83
brain and its relationship with flavour 372
development of the brain in infancy 382–3
multisensory perception 372–4
relevance of propriosense 379–81
role of all the senses 374–9
future trends 394–6
how we learn flavour 383–94
changes in food preferences after infancy 390–2
difficult to explain preferences 392–4
early feeding experiences and brain structure modification 387–8
family structure and diet 390
neonates 383–4
neophobia and disgust 388–90
role of carers 385–6
spontaneous food selection 386–7
lecithins 142
food grade 147
leguminous plant proteins 180–2
legumins 180
lemniscal (cognitive) gustatory pathway 46–7
limonene 191, 332, 333
linear discriminant analysis (LDA) 91
linear learning machine (LLM) methods 91
lingual movements 408–9
lipid oxidation 133, 150, 164–5
lipids 119–20, 133–55, 246–7
aroma compounds dissolving in 158–60
aroma–matrices interactions 352–3
classification and properties of 136–43
effect on aroma release from beverages 276–7, 278
key role of molecular organisation 134–6
main types of lipidic structures 143–51
partly water-soluble and phase partitioning in emulsions 231–2
replacing fat by structuring lipids 151–3

lipocalins 173
lipolysis 133
liposomes 141, 143, 144, 146–8, 151
liquid food products 260–84
  development of the artificial throat 266–71
  effect of ingredients on aroma release 271–7
  in vivo aroma release measurements during drinking of beverages 262–5
  modelling aroma release from 247
  physiological factors important for *in vivo* aroma release 265–6
liquid phase analysis 183
London interactions 217
low moisture/dry foods 291–2
lungs, absorption in 295–6

macromolecules
  rheology and flavour release 329–36, 337
  see also polysaccharides; proteins
magneto-encephalography (MEG) 372
magnitude matching 64
maltodextrins 215
mammary pheromone 423
Marmite 389
mass spectrometry 260
mass transfer 121–6, 234–9
  effect of dilution mechanism 236–8
  emulsions
    emulsifiers 160
    interface crucial for 156–7
    viscosity of the continuous phase 161
  factors affecting 234–6
  at interfaces 126, 127, 128
  modelling 238–9
    release from dynamic systems 238–9
    release from static systems 238
  resistance to during consumption 246–7
  under dynamic flow conditions 293–4
mass transfer coefficient 122–4, 126, 127, 229, 235–6
  apparent 240
mastication *see* chewing
mathematical models of flavour release 241, 274–5, 302–3, 336
matrix
  amylaceous matrices 218–22
  aroma–matrices interactions 352–3
  effect of matrix structure on flavour release 289–90
  interactions of flavour compounds with 119–20, 126
  modelling aroma interactions *see* modelling aroma interactions
  physical state of and carbohydrate–flavour interactions 213–14, 219–20
  physical state and structure of 126
McGurk effect 372–3
memory 420–8
menthol 356
menu driven interfaces 111
mesophases 134–6, 143–5
  formation 140–3
metabotropic glutamate receptors (mGluRs) 9, 45
metastable systems 141
  see also emulsions; liposomes
methyl ketones 192
methyl salicylate 388
methylene groups, number of 139
micelles 138, 139, 143–5
microemulsion 143, 144
microencapsulation 150–1
micro-region entrapment 217, 243–4
microvilli 43–4
milk 136
  human 383–4, 405–6, 407, 414, 421
milk formula 384, 385, 414, 427
mint aroma 373, 415
mixed phases, solvent properties of 233–4
'model mouth' systems 241, 252, 267–8, 275, 276–7, 278
model systems 239–43, 252
modelling aroma interactions 229–59
  applications 251
  empirical modelling of aroma release 250–1
  future trends 252
  mass transfer 234–9
  mechanistic modelling of aroma release 239–50
    aroma persistence after swallowing 249–50
    heat processing and storage 243–4
    in vitro aroma release 239–43
    in vivo aroma release 244–9
  phase partitioning 230–4
modified starches 215–16
molecular inclusion 213, 215, 217, 220–2, 223
molecular interactions 217, 222, 223
molecular modelling 186, 193–4
molecular shape 137–40

monitoring changes over time 89–90
monosaccharides 208, 215, 350–2
  aroma interactions with 208–14
monosodium glutamate (MSG) 309–10, 356–7
mouth 293
  flavour delivery from 296–8
  in-mouth losses 292
mouth model systems 241, 252, 267–8, 275, 276–7, 278
mouthpiece: upgrading the management of mobile material in 165–8
mouthspace 162
multicolinearity 101, 103, 104
multidimensional scaling (MDS) 91
multilamellar vesicles (MLVs) 146, 148
multiple emulsion 144
multiple regression analysis (MRA) 91, 98, 101–2
multiple taste qualities 17
multisensory perception 372–4
  multimodal interactions 36, 355–9
    and flavour integration 49–53
  propriosense 379–81
  role of all the senses in flavour perception 374–9
multivariate analysis of variance (MANOVA) 71, 100

naïve newborn model 423
nanoencapsulation 150–1
nanostructured hierarchical fluids 143
nausea 380
negative alliesthesia 410
neohesperidine dihydrochalcone 350
neonates *see* infants
neophobia 388–90
neuronal nicotinic receptor (NnAChR) 48
neurons 37, 39–40, 356, 373, 382
  oscillatory patterns between 395
nicotine 48, 49, 51
non-equilibrium partition model 239, 242–3
nonanal 272, 273
nose, absorption in 295–6
nosespace analysis 162
  texture–aroma interactions 339–41
nuclear magnetic resonance (NMR) 186
nucleus of the solitary tract (NST) 41, 44, 46, 51
nutrition 163–4

octanal 338, 339
odorant receptor cells 37–9
odour binding proteins 173
odour–taste integration 359–61
odour–taste interactions *see* taste–aroma interactions
odour value/odour activity value 86
off-notes 87–9
oil droplets 136
oil-in-water emulsions 152, 165
Okah 389
olfaction
  development of perception *see* development of perception
  using instrumental data in 90–2
olfactory bulb (OB) 37
  odour maps in 39–42
olfactory epithelium 37, 377
  odour and aroma representation in 39
olfactory imprinting 426–8
olfactory system 320–1, 377–9, 410–11
  information processing 36, 37–43
    central integration of information 42–3
    cortical integration of odours and tastes 51–3
    peripheral processing 42
olfactory–trigeminal interactions 49–50
oligosaccharides 208
omission testing 87
optimisation of flavour release 251
oral movements 408–9
orbitofrontal cortex (OFC) 41, 42–3, 46, 51–2, 53, 356–7, 359
orthonasal olfaction 52–3, 288, 411
oxidation 133, 150, 164–5

P2X receptors 48
pair preference tests 68
pair tests 63
palpebral reflex 415
panellists
  evaluating the performance of 71–2
  individual variation 100
  selection 70–1
  training 65, 66–7, 265
papillae 3, 4, 43–4
  fungiform papillae (FP) density 316–18, 319–20, 321
parents 385–6, 387
parsnips 392
partial least squares regression (PLSR) 72, 91, 99, 104–5, 108–10
partition coefficient
  air–product partition coefficient 229, 230–4, 240

air–water partition coefficients 294–5, 301, 303
headspace analysis 183–4
lipids 141
pectin 334
penetration theory 238–9
peptides 352
perception
acknowledged as more than translation of molecular signals 163
consequences of protein–flavour interactions 191–3
development *see* development of perception
intra- and interpersonal variation 262–4
modelling and relating aroma release to 252
odour–taste interactions *see* taste–aroma interactions
PROP taster differences and flavour perceptions 320–1
relevance of sensory analysis to consumer perception 75
of texture and flavour 336–9
perceptual styles 321
peripheral information processing 42, 45–6
peristalsis 271
persistence
of aroma compounds in the breath 264
of early odour acquisitions 426–8
of flavour release 300, 301–3
modelling aroma persistence after swallowing 249–50
pH
β-lactoglobulin stability and denaturation 175–6
effect of medium on protein–flavour interactions 187–8
pharmaceuticals 26
pharynx 302
phase partitioning 230–4
factors affecting 230–1
modelling 231–4
phases: number coexisting in lipid-containing systems 141
phenylalanine 185
phenylthiocarbamide (PTC) 10, 12
genetic variations in taster status 27–31
PTC/PROP bitterness sensitivity 311–16
PTC/PROP tasting and sensitivity to other sensory qualities 312–14
sensory properties within foods 314–16
phospholipase Cβ2 (PLCβ2) 13–15
phospholipids 134–6, 142, 147
physicochemical interactions 350–5
physiological effects of odours 409–10
physiological interactions 351, 355–9
plants 370–1
plasticity of olfactory function 420–8
polar lipids 134–6
polysaccharides 208
rheological behaviour and flavour release 329–36, 337
pontine parabrachalis nucleus (PBN) 41, 46
porus 4
positron emission tomography (PET) 42–3, 372
post-ingestive nausea 380
potato starch 220, 221
power 74–5
power law model for flavour release 303
prandial state 410
preference mapping 69, 72–4
preferences
changes after infancy 390–2
consumer preferences 72–4, 252
early postnatal shaping of odour preferences 420–2
learning of flavour preferences *see* learning of flavour preferences
olfactory preferences established independently of postnatal experience 423–6
popular preferences difficult to explain 392–4
PrefMax approach 73
premature newborns 411–12
preservation of foods 371
pressure
aroma compounds in emulsions 157, 158
β-lactoglobulin stability and denaturation 176–7
primary gustatory cortex (PGC) 41, 46, 53
primary olfactory cortex (POC) 40–3, 53
principal component analysis (PCA) 66, 67, 91, 99, 100, 102, 105–7, 108
principal component correlation 102
principal regression analysis 102
product development 73
product dissolution 248
product extraction 248
propanol 212
propriosense 379–81

propylthiouracil (PROP) 10, 44, 311–21
  PTC/PROP tasting and sensitivity to other sensory qualities 312–14
  sensory properties within foods 314–16
  and taste anatomy 316–20
  taster differences and flavour perceptions 320–1
protein hydrolysate 384, 422
proteins 119–20, 172–207, 232–3, 246–7, 337, 353
  effect on aroma release from beverages 274–5, 278
  effect of medium on protein–flavour interactions 187–91
    impact on protein functionality 190–1
  future trends 193–4
  nature and strength of protein–flavour interactions 182–6
    binding constant determinations 183–4
    flavour binding 182
    structural changes and binding sites localisation 184–6
  perceptive consequences of protein–flavour interactions 191–3
  structure in relation to flavour binding 173–82
    β-lactoglobulin 173–7
    leguminous plant proteins 180–2
    other whey proteins 177–80
psychobiological processes 425–6
psychological interactions 351, 359–62
psychophysical interactions 339–41
purge and trap 82, 83, 87, 88, 91

quality 416
quantification of data 83–4
Quantitative Descriptive Analysis (QDA) 65
quantitative structure activity relationships (QSAR) 186, 194
quantitative structure property relationships (QSPR) 186, 194, 250–1
quinine 17, 70, 310

Raman spectroscopy 184
randomisation tests 111
ratio scaling 64
recognition threshold 63
redundancy analysis (RA) 103–4, 107–8, 109
reference standards 66–7
regression on principal components 102
repertory grid analysis 69
reproducibility 90
respiratory rates 408
retinol binding protein 173
retronasal olfaction 52–3, 411
reverse engineering 73
reversed micelles 138, 139
reversible interactions 182, 295
rewards, foods as 385–6
rheology 329–36, 337
ribonucleotides 9–10
risk, eating and 374
rostral projections 46
rotational diffusivity 125
RV coefficient 72

sac locus 7
saccharin 356–7, 361
salicin 24
saliva 278, 290
  dilution with 162, 166–7
  flavour release 291, 353–4
  interactions with aroma 265–6
salt *see* sodium chloride
salt taste 4, 308–9, 358–9
salt taste receptors 6–7, 44, 308
salting out effect 209, 223, 233, 271, 350–2
sampling, for aroma analysis 82–3, 84–5
saturated vapour pressures 118
scales 64
Schiff base formation 191, 192
Schlich algorithm 107–8, 109
scientific journals 93
secondary gustatory cortex (SGC) 46, 52
secondary olfactory cortex 42, 52
selective diffusion theory 243–4
selectivity of olfactory receptors 38–9
sensitive periods 422
sensory analysis 62–80, 86–7, 359
  current and developing techniques 63–9
    analytical intensity rating tests 64–8
    consumer tests 68–9
    discrimination tests 63–4
  matching sensory data to instrumental data 72, 98–113
    case study 105–10
    investigating structure of data sets 99–101
    relating sensory to instrumental data 101–5

relating analytical sensory data to instrumental or consumer preference data 72–4
sensory testing administration 69–71
experimental design 70
panellist selection 70–1
standard sensory practices 69–70
statistical analysis of data 71–2
evaluating panellist performance 71–2
statistical significance 71
using instrumental data in 90–2
using sensory data for business decisions 74–5
sensory properties: key components contributing to 84–7
Sensory Spectrum 65
shelf-life studies 64
Sherwood equation 249
signal transduction 12–15, 16–17, 26
olfactory processing 37–8
simultaneous distillation extraction (SDE) 82, 83
single cell calcium imaging 18
single nucleotide polymorphisms 27–31
singular correlation matrix 106, 107
slurries 291–2
small unilamellar vesicles (SUVs) 146
snails 389
sniffing 408
soap 133
sodium chloride 17, 163–4, 363
effect of medium on protein–flavour interactions 189
solid phase microextraction (SPME) 82, 88, 91
solid products: modelling of aroma release from 248–9
solubility 120–1
solvent properties of mixed phases 233–4
solutions: flavour release from 291–2
solvent extraction 81, 82, 83
somatosensory information 312–14, 320–1
see also trigeminal system
sour taste 4, 308–9
children and confectionery preferences 391
PROP taster groups and 312, 313, 314, 315
sourness–aroma interaction 347, 349
sour taste receptors 5–6, 44, 308
soy proteins 180–2
effect of medium on protein–flavour interactions 188, 189, 190
soy sauce 358–9
soya-based infant formula 384
specificity of early olfactory detection 413–14
spontaneous food selection 386–7
stabilisers 141–2
stability
β-lactoglobulin 175–7
emulsions 156–7
improving stability 164–5
stagnant-layer theory of interfacial mass transfer 238, 248
standard addition method 83
starch 272
aroma interaction with starch and derivatives 219–22
chemical structure 214
modified starches 215–16
physical structures 214–15
static diffusion 236–8
static headspace analyses 81, 83, 87, 184
effects of ingredients on aroma release from beverages 272, 273, 275, 278
static partition of aroma compounds 209–10, 211
static systems
modelling aroma release 239–40
modelling mass transfer 238
STATIS 100
statistical methods 71–2
matching sensory and instrumental data 98–113
using instrumental data in sensory evaluations 91–2
statistical significance 71, 74–5
steam distillation 87
Stir-bar methods 82, 83, 88, 91
stirred solutions: modelling aroma release from 241–2
Stokes' cell 124
Stokes-Einstein equation 236
strawberry aroma 346, 347, 358–9
sucrose 163–4, 208–9, 211–13, 214, 363
effect on aroma release 272, 273
solubility of ethyl acetate in sucrose solutions 120–1
sucrose:raffinose system 214
super-additive responses 359
supertasters 311, 315, 316
FP densities 317
and taste anatomy 319–20

supervised methods 91–2
supramolecular assemblies 134–6
surface renewal theory 239
surfactants 141–2
swallow breath 162, 262, 266, 297–8
swallowing 266, 271, 287, 297–8
  breath flavour profiles after 300–1
  relative importance of chewing and swallowing actions 298–300
swallowing apnoea 297–8
sweet taste 4, 308–9, 358–9
  PROP taster groups and 312, 313, 314, 315
  sweetness/aroma interaction 346–9
sweet taste receptors 7–9, 24, 44–5, 308, 310
  functional analysis 20
sweeteners 17
  bulk 271–3
  intense 271–3, 350
syrups 334

targeting 148
TAS1R receptors 44–5, 310
  functional analysis 20–1
  signal transduction 13–15
  sweet taste 7–9
  umami taste 9–10
TAS2R receptors 10–12, 18, 19, 25, 26, 44
  bitter adaptation 24–5
  and food design 26–7
  functional analysis 21–4
  genetic variations 27–31
  signal transduction 13–15
taste
  and aroma 373–4
  cortical integration of odours and tastes 51–3
  genetic influences on *see* genetic influences on taste
  linguistic confusion between flavour and taste 369–70
  role in flavour perception 376–7
taste anatomy 316–20
  supertasters and 319–20
  tactile components of taste 318–19
taste–aroma interactions 345–68
  future trends 362–3
  origin of 350–62
    physicochemical interactions 350–5
    physiological interactions 351, 355–9
    psychological interactions 351, 359–62
taste buds 3, 4, 43
  type III cells 17
taste cells 43–5
taste information processing *see* gustatory modality
taste-mGluR4 45
taste–odour integration 359–61
taste quality coding 15–18
taste receptors 3–35, 43–5, 310–11, 376–7
  functional characterisation through calcium imaging 18–26
    assay principle 18–19
    current limitations of heterologous expression system 25–6
    functional expression *in vitro* 24–5
  future trends 26–31
  genetic variations 27–31
  molecular basis of taste perception 5–18
    implication for taste coding 15–18
    signal transduction 12–15
  sense of taste 4–5
  taste receptor assays and food design 26–7
tea 393
temperature
  β-lactoglobulin stability and denaturation 176
  effect of medium on protein–flavour interactions 188
  influence on lipid structures 149
  and phase partitioning 234
temporal dominance of sensations (TDS) technique 68
terms in sensory analysis 65
  comparing consumers' with panellists' terms 69
texture 375–6
texture–aroma interactions 327–44
  rheological behaviour and flavour release 329–36, 337
    mechanisms involved 335–6
    structure and composition of matrix 329–35
  texture measurements and perceived intensity of aroma 336–41
    perception of texture and flavour 336–9
    psychophysical interactions vs texture impact on flavour perception 339–41
thermal denaturation of protein 188–9
thermodynamically controlled release 246
three alternative forced choice test (3AFC) 63

three-dimensional network 335–6
threshold tests 63–4
throat 267
  artificial 252, 266–71, 275, 276–7, 278
  liquid film in 266
time
  monitoring changes over 89–90
  and perception of texture and flavour 338–9
time-intensity (T-I) 67–8, 163, 262–3
  texture and flavour perception 340–1
time-order effects 70
tomatoes 105–10
tongue 4, 302
  artificial 27
touch 375–6
  PROP tasting and 312–14
  tactile components of tastes 318–19
training of panellists 65, 66–7, 265
transducin 13
triangle test 63
trichloroanisole (TCA) 63
trigeminal compounds 69–70
trigeminal fibres 47–9, 318
trigeminal–gustatory interactions 50–1
  at the cortical level 51
trigeminal–olfactory interactions 49–50
trigeminal system 375, 410–11
  information processing 36, 47–9
triglycerides 136
tripe 392
TRPM5 ion channel 13–15
tryptophan 24, 185
turbulent flow 122
two alternative forced choice tests (2AFC) 63
two film theory 289
type III cells 17
tyrosine 185

ultrafiltration technology (UT) 152
umami taste 4, 308–9
umami taste receptors 9–10, 44–5, 308, 310
  functional analysis 21
unstructured line scales 64
unsupervised methods 91
upper airway 293
  flavour molecule transport in 296
urea 190, 310

V-type structures 215
validation of models 110–11
vanilla 392–3, 409, 419, 427
vanilla ice-cream 152
vanillin 191–2
Vegemite 389
velum 262, 263, 293, 296–7
ventral gustatory pathway 46–7
ventro-postero-medial (VPM) nucleus of thalamus 41, 46, 51
vesicles 136, 138
  large unilamellar 146
  multilamellar 146, 148
  small unilamellar 146
vicilins 180
visceral-limbic system 46–7
viscosity
  of continuous phase 161
  and flavour strength 376
  of matrix
    and flavour release 329–36
    and odour 327–8
  optimising flavour yield of emulsions 166–7
  sample viscosity and aroma release from artificial throat 270–1
viscous solutions 339–40
vision 374–5
volatility 217
  see also air–product partition coefficient
VR1 (vanilloid receptor) 48, 49

wash in – wash out principle 296
water
  carbohydrate–flavour interactions 219–20, 221–2
  diffusivity in sugar solutions 212, 213
  lipid affinity for 134, 137
  solubility of ethyl acetate 120–1
water activity 191
water-in-oil emulsions 157–8
weak polar lipids 134, 136
whey proteins 173–80, 191, 192
  effect of medium on protein–flavour interactions 187, 188–9, 190
  interaction with aldehydes 275
wines 64, 66, 67, 74, 288
wintergreen flavour 388

X-ray crystallography 185–6

yeast, autolysed 388–9